Essentials of
OCEANOGRAPHY

Essentials of

OCEANOGRAPHY

TOM GARRISON
Orange Coast College

WADSWORTH PUBLISHING COMPANY

I︎T︎P™ An International Thomson Publishing Company

Belmont • Albany • Bonn • Boston • Cincinnati • Detroit • London • Madrid • Melbourne
Mexico City • New York • Paris • San Francisco • Singapore • Tokyo • Toronto • Washington

Science Publisher: Jack Carey
Development Editor: Mary Arbogast
Editorial Assistant: Kerri Abdinoor
Production Services Coordinator: Gary Mcdonald
Production: Rogue Valley Publications, Mary Douglas
Designer: Kaelin Chappell
Print Buyer: Diana Spence
Art Editor: Myrna Engler-Forkner
Permissions Editor: Jeanne Bosschart
Copy Editor: George Dyke
Photo Researcher: Tom Garrison
Technical Illustrators: Electronic Color conversion and new illustrations by Precision Graphics. Other
    illustrations by Diphrent Strokes; Parrot Graphics; Carlyn Iverson; Tasa Graphic Arts, Inc.;
    Kathryn W. Werhane; Jill Turney; Carole Lawson.
Cover Designer: Jamie Sue Brooks
Cover Photograph: © Tony Stone Images/H. Richard Johnston
Signing Representative: Joann Ludovici
Compositor: Thompson Type
Color Separator: Color Associates
Printer: Banta Company/Menasha

Printed in the United States of America
1 2 3 4 5 6 7 8 9 10—01 00 99 98 97 96 95

For more information, contact Wadsworth Publishing Company:

Wadsworth Publishing Company
10 Davis Drive
Belmont, California 94002, USA

International Thomson Publishing Europe
Berkshire House 168-173
High Holborn
London, WC1V7AA, England

Thomas Nelson Australia
102 Dodds Street
South Melbourne 3205
Victoria, Australia

Nelson Canada
1120 Birchmount Road
Scarborough, Ontario
Canada M1K 5G4

International Thomson Editores
Campos Eliseos 385, Piso 7
Col. Polanco
11560 México D.F. México

International Thomson Publishing GmbH
Königswinterer Strasse 418
53227 Bonn, Germany

International Thomson Publishing Asia
221 Henderson Road
#05-10 Henderson Building
Singapore 0315

International Thomson Publishing Japan
Hirakawacho Kyowa Building, 3F
2-2-1 Hirakawacho
Chiyoda-ku, Tokyo 102, Japan

Library of Congress Cataloging-in-Publication Data

Garrison, Tom, 1942–
    Essentials of oceanography / Tom Garrison.
        p.   cm.
    Includes bibliographical references and index.
    ISBN 0-534-24942-6
    1. Oceanography.   I. Title.
GC11.2.G36   1994
551.46—dc20
                                                    94-20333
                                                    CIP

BRIEF CONTENTS

DETAILED CONTENTS

To my family and my students,
my hope for the future

PREFACE FOR STUDENTS AND INSTRUCTORS

This book was written to provide an *interesting*, clear, and relatively brief overview of the marine sciences. It was designed for college and university students who are curious about the Earth's largest feature, but who may have little or no formal background in science. Oceanography is broadly interdisciplinary; students are invited to see the connections between astronomy, economics, physics, chemistry, history, meteorology, geology, and ecology—areas of study they once considered separate. It's no surprise that oceanography courses have become increasingly popular in the last decade. Together, teachers and students learn to see that the rise of a wave is linked to the twinkle of distant starlight, that a soft summer breeze is shaped by the chemistry of a billion microscopic plants, that the slow drift of a continent molds the evolution of countless organisms.

But nuts-and-bolts science is only part of the story. Studying the ocean awakens in us the sense of wonder we all felt as children when we first encountered the natural world. Students in marine science classes often find themselves immersed in wonderful stories and spectacular pictures, and surrounded by teachers and technicians whose barely contained enthusiasm for their subject overtakes the clock. There is much to tell. The story of the ocean is a story of change and chance; its history is written in the rocks, the water, and the genes of the millions of organisms that have evolved there.

Students bring a natural enthusiasm to their study of this field. Even the most indifferent undergraduate will perk up when presented with stories of encounters with huge waves, photos of divers and giant squids, tales of exploration under the best and worst of circumstances, evidence that chunks of the Earth's surface slowly move, micrographs of glistening diatoms, and figures showing the economic importance of seafood and marine materials. If pure spectacle is required to generate an initial interest in the study of science, oceanography wins hands down!

In the end, however, it is subtlety that triumphs. My goal is to help the students who use this book to gain an oceanic perspective. "Perspective" means being able to view things in terms of their relative importance or relationship to one another. An oceanic perspective lets you see this misnamed planet in a new light and helps you plan for its future. You will see that water, continents, seafloors, sunlight, storms, seaweeds, and society are connected in subtle and beautiful ways.

How This Book Is Organized

A broad view of oceanography is presented in 15 chapters, each free-standing (or nearly so) to allow an instructor to assign chapters in any order he or she finds appropriate. The chapters progress in what seems a reasonable direction: the text begins with a brief history of our understanding of oceanography, then continues through the origin of the Earth and ocean, the structure and character of the ocean floor, a discussion of water and its movements and interactions with the more solid Earth, and an ecologically based discussion of marine life. The text ends with a look at the character of human relationships with the ocean, especially the way marine resources have been managed and mismanaged.

Each chapter begins with an attractive **vignette**—a short observation, eyewitness account, or description of marine scientists at work. The chapters are written in an **engaging style** at a level appropriate to nonscience students. The number of technical terms is kept to a minimum, and, when appropriate to their meanings, the derivations of words are shown. Some of the more complex ideas are initially outlined in broad brush strokes, and the same concepts are discussed again after you have a clear view of the overall situation. **Measurements** are given in both metric and English systems. At the request of a great many students, the units are written out (that is, we write *kilometer* rather than *km*) to avoid ambiguity and for ease of reading.

The **illustration program** is extensive; the photos, charts, graphs, and paintings have been chosen for their utility, clarity, and beauty. **Boxes** in most chapters present commentaries of special interest on unique topics or controversies. Chapters end with a **summary** and a list of important **terms and concepts** that are defined in an extensive **glossary** at the back of the book. **Study questions** are also included in each chapter. The brief **annotated bibliography** that follows will be helpful to anyone who wishes to know more about a particular topic.

Appendixes will help you master measurements and conversions, geological time, latitude and longitude, and chart projections. In case you'd like to join us in our life's work, the last appendix discusses jobs in marine science.

Best of all, the book has been thoroughly **student tested.** You need not feel intimidated by the concepts presented or the words used to describe them. This material has been mastered by students just like you. Read slowly, and go step-by-step through any parts that give you trouble. Your predecessors have found the ideas presented here to be understandable, useful, inspiring, and applicable to their lives. Best of all, they have found the subject to be *interesting*!

Acknowledgments

Jack Carey at Wadsworth, the grand master of college textbook publishing, willed this book into being. His suggestions were combined with those of more than 50 peer reviewers and 150 undergraduate students, all of whom gave of their time and knowledge to help me assemble and organize the material. As always, I am greatly indebted to my long-suffering departmental colleagues Dennis Kelly, Jay Yett, Robert Profeta, William Shea, and Joyce Kai-Mott for putting up with me and doing much of my administrative work through this book's gestation. Thanks also to Stanley Johnson, our dean, and David Grant, our college president, for supporting and encouraging the faculty to write, engage in community service, and conduct research. And again, gold medals should go to my supportive family for their patience with my long hours, late nights, crabby moods, and loud Glenn Gould Bach recordings.

Many of the illustrations for this book came from friends and acquaintances, who maintained their cooperative cheer through a blizzard of increasingly anxious letters and telephone calls. Bruce Hall, diver and friend, again contributed photos, as did colleagues Norman Cole, Ron Romanosky, and Pat Mason. Herbert Kauainui Kane again donated his beautiful pictures of Hawaiian topics. William Cochlan of the Hancock Foundation sent the latest photos of marine viruses, Ted Delaca at the University of Alaska loaned me dramatic photos of the nuclear submarine USS *Pargo* taken during her recent research cruise to the high Arctic, Hank Brandli provided a satellite image of Hurricane Andrew, Ken-ichi Inoue of JAMSTEC brought me up to date on the latest adventures of *Shinkai-6500*, Andreas Rechnitzer shared recollections of his duty aboard *Trieste*, Karen Riedel at DSDP again sent images, and Deborah Day and Cindy Clark at Scripps Institution dug through their archives one more time. The Woods Hole team was also generous: Philip Richardson, David Ross, and Shelly Lauzon all provided photographs and diagrams for inclusion. The United States Coast Guard, United States Navy, Army Corps of Engineers, NASA, and NOAA were patient and understanding of my needs and deadlines.

The Wadsworth team performed the customary miracles. The charge was led by Mary Douglas, production editor and champion of every known means of communication. The text was polished by Mary Arbogast, the best developmental editor on the planet. Myrna Engler-Forkner (production assistant), Kristin Milotich (editorial assistant), and Jeanne Bosschart (permissions editor) kept me and my overheated fax machine on the right path most of the time. And Jack Carey always kept the big picture in view. My thanks to all.

The Ocean's Gift

The ocean's greatest gift to humanity is intellectual—the constant challenge its restless mass presents. Let yourself be swept into this book and the class it accompanies. Ask questions of your instructors, read some of the references, try your hand at the questions at the ends of the chapters. Be optimistic. *Take pleasure in the natural world.* Please write to me when you find errors or if you have comments. Enjoy yourself!

Tom Garrison
Orange Coast College
Costa Mesa, California 92628-5005

REVIEWERS

A STUDENT GUIDE TO LEARNING

Teaching is for me a great joy. I make my living by sharing with students the things I love to think about and talk about, and oceanography is at the top of my list. For example, did you know that the first successes of marine science were closely associated with the history of voyaging? Or that the theory of plate tectonics explains why our ancient planet has very old continents but surprisingly young seafloors, the oldest of which is only about as old as the dinosaurs?

In *Essentials of Oceanography* you'll learn about the history of marine science, about theories of Earth structure and plate tectonics, ocean chemistry and physics, marine biology, marine resources, and environmental concerns. You'll get the latest information on marine environmental issues, whaling treaties, and the destructiveness of California earthquakes and Florida hurricanes.

Sometimes this material can be difficult to understand, so I've included many examples, applications boxes, and illustrations to bring important concepts to life and to enhance studying and learning. I've also tried to tell a few stories (sea tales, if you will) in the hopes that some of my enthusiasm for marine science will rub off on you. The following pages give you a "sneak preview" of what's to come and show you how to get the most out of this exciting book. So get comfortable and read on.

Tom Garrison

"AT LAST, A TEXTBOOK THAT DOES NOT READ LIKE STEREO INSTRUCTIONS"

This was the comment made by a student of mine after reading one of my books. I hope you agree. Studying the ocean awakens in me the sense of wonder I felt as a child when I first encountered the natural world. I've tried to convey this feeling in my writing so you, too, can share my enthusiasm.

Figure 1.12 Captain James Cook, Royal Navy.

Figure 1.13 First contact: Captain James Cook, commanding HMS *Resolution*, off the Hawaiian island of Kaua'i in 1778. "It required but little address to get them to come along side, but we could not prevail upon any one to come on board; they exchanged a few fish they had in the canoes for anything we offered them, but valued nails, or iron above every other thing; the only weapons they had were a few stones in some of the canoes and these they threw overboard when they found they were not wanted."

AFTERWORD

The marine sciences are at the threshold of a new age. The recent revolutions in biology and geology are being assimilated and the road ahead seems clearer. A renaissance in the design of sampling devices, robot submersible vehicles, and data processing has brought new vigor to oceanography. Satellite-borne sensors can provide data in an instant that would have taken years to collect using surface ships. Shipboard technology has become so sophisticated that Wyville Thomson or Fridtjof Nansen would hardly recognize our sensors or sampling devices.

The tools may be different, but the spirits of those who use them remain the same. Today's marine scientists are like all the men and women who have gone before: *We want to know about the ocean.* We haunt our mailboxes for journals bearing the latest research news, search television listings for any new ocean shows, inspect new samples with the enthusiasm of little kids, and share our insights with anyone at the drop of a hat. I am personally delighted that you have traveled with me this far. Those of us who enjoy an oceanographic background (and this now includes you) look at the Earth with greater understanding than we did before we began. The whole concept of an ocean world appeals to us, gives us profound pleasure, and sobers us with a deep sense of responsibility. In no other field of science do so many ideas interweave to form so rich a tapestry.

Our journey together is over, but before we go our separate ways, I have three last ideas to share:

- Change has been a recurrent theme of this book. The Earth's climate has changed with time, as has its atmospheric composition, its ocean chemistry, the size and positions of its continents, and its life forms. The Earth may seem a calm and stable home, but it is really a violent place for inhabitation by such seemingly delicate objects as living things. Even so, life and the ocean have grown old together. The story of the Earth is the story of change and chance; its history is written in the rocks, the water, and the genes of the millions of organisms that have evolved here. We are survivors.

But that survival may now be in question. Change is now progressing at an unnatural rate, and these human-induced changes are imposing stress on natural systems. What we do *with* and *to* the ocean is literally of planetary consequence. In the last century we have developed the physical, chemical, and biological machinery to destroy or rejuvenate the world ocean and all of its life. A painful time of inadvertent global experimentation lies just ahead.

All of us who love the colors and textures of this small wet world need to act to moderate the negative effects of the looming environmental crisis. In Chinese, the written character for the word *crisis* has two components: danger and opportunity. Informed citizens will express their concern, discuss this concern with others, and act whenever possible to minimize the threats and take advantage of new opportunities. Intelligence and beauty must triumph; we have no other rational alternative.

- Appreciation of the ocean doesn't come exclusively from the realm of science. Philosophers, artists, composers and poets have had much to say about the sea. Read Homer's description of the ocean in *The Odyssey* (try books iv, x, and xi). See how Lord Byron's feeling for the ocean colors his poetry (see, for instance, *Childe Harold's Pilgrimage*, stanzas 183 and 184). Read modern poet Robinson Jeffers's powerful *Continent's End*. Share Prospero's marine magic in Shakespeare's *The Tempest*. Find some of the evocative woodcuts of Rockwell Kent and the impressionistic ocean paintings of English artist J. M. W. Turner. Listen to Benjamin Britten's *Four Sea Interludes* from *Peter Grimes*, and Ralph Vaughan-Williams's *Sea Symphony* and *Sinfonia Antartica* (but take care not to blow out your sound system). Read the ocean novels of Herman Melville and Jack London, and try reading the journals and accounts of the famous explorers and scientists you have met in this book. Sit on a quiet beach at night with the stars of the Milky Way shining softly over-

...t that philoso-...ndmass of the ...ists found and ...s Great Barrier ...small islands, ...man habitation ...ndly relations ...emic of dysen-...ile ashore in ...gland around ...on cleanliness ...ncluded cress, ...voided scurvy, ...turies had dec-

...Cook was pro-...772 was given ...*ture*, in which ...n scientific his-...nga and Easter ...e Pacific and ...to circumnav-...e sailed to 70°

south latitude, he never sighted Antarctica. He returned home again in 1775.

Posted to the rank of Captain, Cook set off in 1776 on his third, and last, expedition in *Resolution* and *Discovery*. His commission was to find a northwest passage around Canada and Alaska or a northeast passage above Siberia. He "discovered" the Hawaiian Islands (Hawaiians were there to greet him, of course; see **Figure 1.13**) and charted the west coast of North America. After searching unsuccessfully for a passage across the top of the world, Cook retraced his steps to Hawaii to provision for departure home. On 14 February 1779, after an elaborate farewell dinner with the chief of the Island of Hawaii, Cook and his officers prepared to return to *Resolution*, anchored in Kealakekua Bay. The Englishmen accidentally violated a taboo of property rights relating to their own shoreboat and were beset by the crowd. Cook, among others, was killed in the fracas.

Cook deserves to be considered a scientist as well as an explorer because of his accuracy, thoroughness, and the completeness in his descriptions. He and the researchers aboard took samples of marine life, land plants and animals, the ocean floor, and geological formations, and

CHAPTER-OPENING VIGNETTES

7 ATMOSPHERIC CIRCULATION

Andrew

The costliest natural disaster in the history of the United States began as a cluster of rolling thunderclouds drifting westward across the coast of central Africa. The Earth's rotation had given the clouds a gentle twist, and warm oceanic air rose into their core. Like water spinning down a bathtub drain, humid air was sucked into a growing vortex of clouds. The moisture condensed to form rain, liberating heat energy that caused the storm to grow even larger. On 20 August 1992, near the center of the tropical Atlantic, the storm became Hurricane Andrew.

As the days passed, the hurricane tracked uncertainly toward the west. On two occasions, shearing winds threatened to tear the storm apart; each time it survived and strengthened. Then, on the evening of 23 August, Andrew metamorphosed into a towering black mountain of wind and rain, a rare category-5 storm. Its vast violence sucked the sea surface into a 6-meter (19-foot) dome tens of kilometers across. The wind speed near its center rose to 240 kilometers (150 miles) per hour. In the predawn hours of 24 August, the leading edge of the storm touched the Florida coast near Biscayne Bay. Eyewitnesses described the rising howl of wind—a tearing, clawing force that entered every window and door, that

A neighborhood near Homestead, Florida, bears witness to the shattering force of hurricane winds that reached 350 kilometers (220 miles) per hour.

tore the roofs off their homes and children from their grasp, that hurled cars and mobile homes about like toys, and that lasted more than 90 minutes and reached speeds above 350 kilometers (220 miles) per hour. The division between ocean and land blurred beneath an onrushing wall of windblown seawater. At the height of the storm the hot, wet air glowed yellow-green and was ear-poppingly thin. Devastation was all but absolute.

Before the giant storm died later that week in a series of rattling thunderstorms across the Mississippi Valley, 160,000 people would be homeless, 43 would be dead, and 68,000 businesses would be destroyed; the damage exceeded $30 billion.[1] Scouring currents had rearranged the south Florida coast, sunk ships, shattered harbor installations, uprooted submarine cables, and inundated parts of the Everglades with salt water. Forecasters had not foreseen the ultimate violence of Andrew. Clearly, marine scientists have much to learn about the dynamics of air–ocean interactions.

[1]As this is written, damage estimates from the January 1994 earthquake in southern California are approaching $15 billion.

Hurricane Andrew storms across southern Florida, 24 August 1992.

13 PELAGIC COMMUNITIES

" . . . silver-shining, swift, strong, s[...]"

Among the ranks of marine drifters and swimmers, members of the tuna family are the ocean's fastest and widest-ranging animals. The body of a tuna is dedicated to speed. Its fins retract into slots, its eyes form a smooth surface with the rest of the head, and it may consume as much as 25% of its weight in food each day. Indeed, a tuna uses so much energy that one of its greatest physiological problems is to avoid overheating! The biological equivalent of the legendary Flying Dutchman, these powerful fishes are fated to travel continuously. If they ever stopped they would suffocate, and their massive bodies would fall to the depths.

Tuna and their relatives swim enormous distances and exhibit astonishing bursts of speed. Studies have shown that albacore tuna regularly migrate from the coast of California to Japan and back, a one-way trip of 8,500 kilometers (5,300 miles), moving at an average speed of not less than 26 kilometers (16 miles) per day. Tagged bluefin tuna (a) have traveled at least 7,770 kilometers (4,830 miles) across the North Atlantic in 119

days, that is, o[...] reality, the blu[...] continually de[...] for food. The f[...] 75 kilometers [...] close relative, [...] per hour for a [...]

These magn[...] especially in Ja[...] sashimi, the ra[...] fish market, ol[...] much as $70,0[...] appeared off the California Channel Islands. The largest weighed 458 kilograms (1,008 pounds). Many lucky fishermen paid off home mortgages and boat loans in a week of heroic fishing. But it is the living animal that provides the greatest inspiration: silver-shining, swift, strong and streamlined, these silent nomads slip through the ocean more than a million miles in a lifetime.

The bluefin tuna, the widest-ranging and one of the fastest of the ocean's pelagic animals. Sometimes called a *horse mackerel* because of its enormous size and strength, the record bluefin weighed 458 kilograms (1,008 pounds)!

If you like stories about the sea, you'll like the vignettes that begin each chapter. These short, illustrated tales, observations, or sea stories will whet your appetite for the material to come and familiarize you with basic concepts presented in the chapter. Some vignettes spotlight scientists at work, others describe the experiences of people or animals in the sea.

OUTSTANDING ART IN EACH CHAPTER

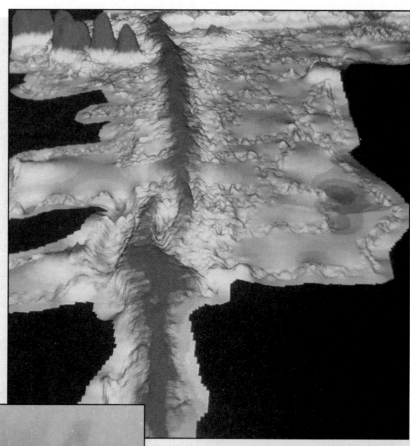

The illustration program is extensive and is used to present information that can be more easily understood in visual form than in words. I selected the photographs, illustrations, charts, graphs, and paintings to help you grasp important concepts and for their beauty and originality.

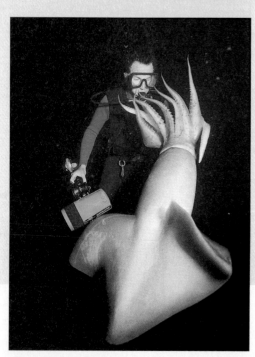

Clockwise from top: A Sea Beam image of an area in the Pacific where new ocean floor is forming. A large squid and a diver investigate each other. The Japanese research submersible *Shinkai 6500* at the start of a deep dive.

Clockwise from top: Arrival of the first
Hawaiians. Residents of Hilo, Hawaii, run for
their lives ahead of a tsunami. Ferdinand
Magellan, whose expedition was first to
circumnavigate the Earth.

APPLICATIONS BOXES

Applications boxes in each chapter present intriguing commentaries, unique topics or controversies in oceanography, cutting-edge field research, or eyewitness accounts of interesting events. I've included them to add to your knowledge base and to give you a taste of the many interesting questions with which oceanographers must grapple.

BOX 9.1   *Surfing*

More than 2 million Americans surf regularly. The thrill of rushing down the face of a growing, breaking wave is exhilarating! People willingly endure cold, boredom, and some danger for a ride lasting only a few seconds. If you're a good swimmer, give it a try. You will need to

paddle your board or swim vigorously to match your speed to that of the advancing wave crest. As the wave rises to break, your forward speed (and sense of timing) will place you on the leading edge of the crest, accelerating downward and forward. The technique takes time to master, but the feeling is worth the effort!

Prejudice compels me to reveal that *real* surfers are body surfers. The buoyant forces on which board surfing depends are diminished in body surfing, but forces within the wave propel the body surfer with greater acceleration. A body surfer's intimate contact with the water and close proximity to the wave surface greatly increase the sensation of movement.

The ultimate trick is to surf the wave completely submerged, as a dolphin does. You push powerfully off the bottom (or swim rapidly toward the surface) as the wave crest approaches, inserting yourself into the freshly breaking wave from below. The rapid flow of water within the toppling wave will ripple your skin, and the sudden acceleration can thrust you from the face of the wave like a wet bar of soap shooting from a fist. Altogether a wonderful oceanic experience!

---

rushes ashore. Along with some wind waves, storm surges and seismic sea waves form surging waves.

Slope alone does not determine the position and nature of the breaking wave. The contour and composition of the bottom can also be important. Bottoms that get shallow gradually can sap waves of their strength because of prolonged friction against the bottom of the low-

jostling chunks of sea ice can also extract energy from a wave. In a few rare cases the shore is configured in such a way that waves don't break at all; the waves have lost virtually all their energy by the time their remnants arrive at the beach.

...ches the shore ... shown in **Fig-**

BOX 5.1   *Deep-Ocean Cores*

The Ocean Drilling Program (ODP), currently the largest and most successful multinational Earth science research project, is the direct successor to the Deep Sea Drilling Project (DSDP), which began in 1968. The two drilling ships commissioned for the projects pioneered oil drilling technology to retrieve "cores," long cylinders of sediment and rock extracted from beneath the seafloor. The cores arrive on the deck of *JOIDES Resolution*—the ship currently in active service—in 9.5-meter (30-foot) sections encased in plastic sheaths. Immediately after retrieval, a core is marked to indicate its original location on the seafloor, coded to distinguish top from bottom, measured, and cut into sections for study and storage. Each section is slit lengthwise; one half is stored for the ODP archives, and the other is taken to the first of many shipboard laboratories for study.

Paleontologists then examine the sediments at the bottom of the core to determine the age of the oldest material. Other researchers measure its density, strength,

molecular composition, radioactivity, and ability to conduct heat. Magnetic specialists read paleomagnetic data from the core to determine the ages of the rock fragments and the latitude at which they probably formed. Sensors are lowered into the hole from which the core was removed to gather additional information on the physical and chemical properties of the site.

ODP scientists have recently recovered fragments of the oldest remaining seafloor. The sample is about 175 million years old, a relic of the Middle Jurassic period, when the continents were massed in one huge cluster. From this and other cores they have learned about cycles of global climate change, information that will be useful in evaluating the present potential for global warming. They have also discovered how fluids move through the lithosphere, found evidence of an ice-free Antarctica, and noted the influence of plate tectonics on worldwide weather and current patterns. From trapped pollen grains, it is even possible to tell what land plants were thriving on the Earth at the time the

core sediments were laid down. As analysis technology evolves, stored cores are restudied and new information obtained.

The size and shape of a typical deep core section may be seen in **a**. This core was taken 250 meters (820 feet) beneath the seafloor off the northwestern coast of Australia by *JOIDES Resolution* on 22 July 1988. The ocean in the area was around 2,000 meters (6,500 feet) deep.

The material in the core progresses from sandstone (near the bottom of the core, at the right side of the photo) to fine claystone (near the top). Foraminiferans and nanofossils are abundant in the sandstone, and small squidlike fossils, easily visible to the unaided eye, are present in the claystone. The sediments date from the early Cretaceous period, about 140 million years ago. The progression of slowly deposited sediments and fossils suggests the seabed in the area has slowly subsided, possibly because of tectonic effects.

Details in a core from DSDP Hole 480, leg 64, are shown in **b**. This core was obtained by the now-retired drilling ship *Glomar Challenger* on 1 January 1979 near

Guaymas in the Gulf of California. The Gulf is an area of active seafloor spreading (see Figure 3.23), and researchers were interested in sampling the bottom in this place where basaltic magma is intruding into soft, wet, young sediments. The sediments here are very deep and have been accumulating at the extremely rapid rate of about 1,200 meters (4,000 feet) every million years.

Glomar Challenger arrived in the area on 29 December 1978 and used a conventional core barrel to penetrate 444 meters (1,456 feet) into diatomaceous ooze and mudstones. Nearly 273 meters (900 feet) of core were recovered from the hole drilled before Hole 480. Because of the disturbance caused by the conventional coring process, researchers were unable to determine the details of fine structure tantalizingly seen in a few lengths of that core.

On 31 December, *Glomar Challenger* moved to a new location 7 kilometers (4.4 miles) northwest. Hole 480, source of the core segment pictured here, was begun later that day. Unlike the previous hole, Hole 480 was drilled using a newly designed hydraulic corer developed by three DSDP engineers. The new device allowed 80% recovery of a core containing essentially undisturbed laminated diatomaceous ooze and muds.

As can be seen in **b**, the core consists of thin alternating brown and gray bands of sediments. These are believed to be annual couplets, formed about 5 million years ago in response to two seasonal events that still occur in the Gulf: the winter rains that introduce terrigenous clays into the region, and seasonal upwelling and northwest winds that produce diatom blooms. Preservation of these fine details depends upon a low level of free oxygen at the seafloor; burrowing animals would normally churn these fine layers into mush, but in a low-oxygen environment these animals are absent; so the thin alternating sheets of clay and diatom tests are preserved.

Note that parts of the core have already been removed for study. Trenches across the width of the core mark places where particular sets of layers have been removed for isotope analysis. A larger sample for paleomagnetic study was taken from the square depression in the sample. Considering the expense and skill necessary to retrieve and analyze them, these small gray, gritty bits of sediment probably cost more than their weight in fine diamonds!

a Researchers examine a deep core taken off the northwestern coast of Australia.

b This core from the Gulf of California shows thin alternating bands of clay and ooze. Voids in the core show where samples were removed for study.

Essentials of
OCEANOGRAPHY

1 HISTORY

". . . small and blue and beautiful . . ."

In the winter of 1968 a lonely crew of astronauts ventured farther from the Earth than anyone has traveled before or since. They rode in a spacecraft named for the Greek god of light and intelligence, Apollo. On the afternoon of Christmas Eve the spacecraft turned to orbit its goal, the calm gray moon. Humans had their first close-up view of Earth's lifeless mate, its airless plaster-of-paris surface dotted by rounded craters, sharp peaks, and dusty plains beneath a perfectly black sky.

Through the 16 hours they spent in the close vicinity of the moon, the apprehensive astronauts' attention was increasingly drawn away from the object of their mission to the shining crescent Earth, their watery home. "The vast loneliness is awe-inspiring," radioed James Lovell. "It makes you realize just what you have back there on Earth." The pictures sent down that day were seen by half a billion people, and the beauty of the distant ocean world was obvious to all. The faraway Earth was "small and blue and beautiful in that eternal silence." Three days later the spacecraft and its relieved occupants shot into Earth's atmosphere at a speed of 40,000 kilometers (25,000 miles) per hour, slowed, and came safely to rest in the welcoming ocean south of Hawaii. For most of the people of Earth, the concept that our planet is a fragile, rare, lovely object dates from that winter. For the first time we saw ourselves as we truly are.

Earth from space. Our water planet shines blue in the darkness, its surface partly hidden by a turbulent layer of clouds.

Ours is not a particularly large planet, it is not unusual in overall composition, and it is not located in some special place. What *is* extraordinary is the liquid water ocean dominating its cloud-shrouded surface. This brilliant blue ocean affects and moderates temperature and dramatically influences weather. Its creatures directly provide at least 2% of humanity's food. From beneath its floor is pumped about one-third of the world's petroleum and natural gas. The ocean borders most of the Earth's largest cities. It is a primary shipping and communication route and a major recreational resource. The dry land on which nearly all of human history has unfolded is hardly visible from space—nearly three-quarters of the planet is covered by water.

Over 97% of the water on or near Earth's surface is contained in the ocean; less than 3% is held in land ice, groundwater, and all the freshwater lakes and rivers (see **Figure 1.1**). The **ocean**[1] may be defined as the vast body of saline water that occupies the depressions of the Earth's surface. Traditionally, we have divided the ocean into artificial compartments called *oceans* and *seas* using the boundaries of continents and imaginary lines such as the equator. In fact there are few dependable natural divisions, only one great mass of water. The Pacific and Atlantic oceans, the Mediterranean and Baltic seas, so named for our convenience, are in reality only temporary features of a single **world ocean**. In this book we refer to the ocean *as a single entity*, with subtly different characteristics at different locations but with very few natural partitions. Marine scientists have descended to its greatest depths, have explored a few of its peaks, and have photographed and sampled some of its floor, but we know more about the topography of the far side of the moon than we know about the 70.78% of the Earth covered by water. The world ocean remains Earth's greatest mystery, a key to its past and present, and the prime link to its future.

On a human scale, the ocean is impressively large—it covers 361 million square kilometers (139 million square miles) of the Earth's surface (see **Figure 1.2**). The average depth of the ocean is about 3,796 meters (12,451 feet), and the volume of seawater is 1.37 billion cubic kilometers (329 million cubic miles) with an average temperature of 3.9°C (39°F). Its mass is a staggering 155 billion billion tons. If the Earth's contours were leveled to a smooth ball, the ocean would cover it to a depth of 2,686 meters (8,810 feet). The volume of the world ocean is presently eleven times the volume of land above sea level. Average land

[1]When an important new term is introduced and defined, it is printed in boldface type. These terms are listed at the end of the chapter and defined in the glossary.

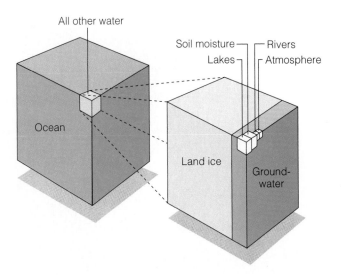

Figure 1.1 The relative amount of water in various locations on or near the Earth's surface. More than 97% of the water lies in the ocean. Ice on land contains about 1.9%, groundwater 0.5%, rivers and lakes 0.02%, and the atmosphere 0.001%.

Table 1.1 Characteristics of the World Ocean		
Area: 361,100,000 square kilometers (139,400,000 square miles)		
Volume: 1,370,000,000 cubic kilometers (329,000,000 cubic miles)		
Average depth: 3,796 meters (12,451 feet)		
Average temperature: 3.9°C (39.0°F)		
Average salinity: 34,482 grams per kilogram (0.56 ounce per pound)		
Most abundant elements (by weight):	Oxygen	(86%)
	Hydrogen	(11%)
	Chlorine	(1.9%)
	Sodium	(1.1%)
	Magnesium	(0.1%)
Age: About 4 billion years		
Future: Uncertain		

elevation is only 840 meters (2,772 feet), but average ocean depth is 4½ times greater!

On a planetary scale, however, the ocean itself is insignificant. Its average depth is a tiny fraction of the Earth's radius—the blue ink representing the ocean on an 8-inch paper globe is proportionally thicker. The ocean accounts for only slightly more than 0.02% of Earth's mass, or 0.13% of its volume. There is much more water trapped within the Earth's hot interior than there is in its ocean and atmosphere. Some characteristics of the world ocean are summarized in **Table 1.1**.

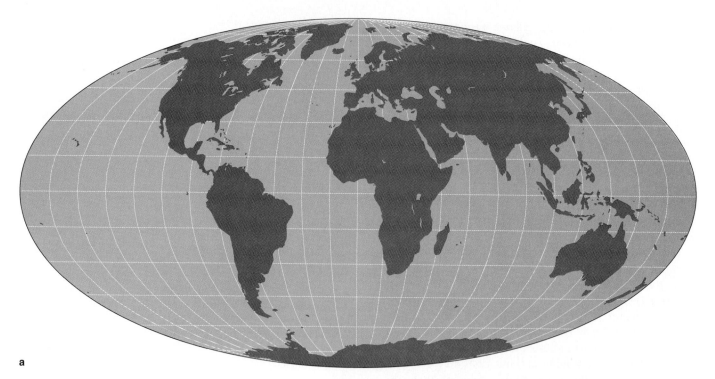

a

Figure 1.2 (a) The proportion of sea versus land, shown on an equal-area projection of the Earth. An equal-area projection is a map drawn to represent areas in their correct relative proportions. (b) A typical scene on Earth. Most of its surface is covered by a liquid water ocean that averages 3,796 meters (12,451 feet) in depth.

b

A BRIEF HISTORY OF MARINE SCIENCE

It has taken a long time for humans to appreciate the nature of the world, but we're a restless and inquisitive lot, and despite the ocean's great size, we have populated nearly every habitable place. This fact was aptly illustrated when the European explorers set out to "discover" the world, only to be met by native peoples at almost every landfall! Clearly, the oceans did not prevent the spread of humanity. The early history of marine science is closely associated with the history of voyaging. Voyaging had a practical aim: to facilitate travel, trade, and warfare.

Voyaging Begins

Ocean transportation offers people the benefits of mobility and greater access to food supplies. Any coastal culture skilled at raft building or small-boat navigation would have economic and nutritional advantages over its less adept competitors. From the earliest period of

Figure 1.3 A Greek ship from about 500 B.C. Such ships were used for trade and to explore the Atlantic outside the Mediterranean.

human history, then, understanding and appreciation of the ocean and its life-forms benefited those patient enough to learn.

The first direct evidence we have for **voyaging**, traveling on the ocean for a specific purpose, comes from records of trade in the Mediterranean Sea. The Egyptians organized shipborne commerce on the Nile River, but the first regular ocean traders were probably the Cretans, or the Phoenicians who inherited maritime supremacy in the Mediterranean after the Cretan civilization was destroyed by earthquakes around 1200 B.C. Skilled sailors, Phoenicians carried their wares through the Straits of Gibraltar to markets as distant as Britain and the west coast of Africa. Considering the simple ships they used, this was quite an achievement.

The Greeks began to explore outside the Mediterranean into the Atlantic Ocean around 900–700 B.C. (**Figure 1.3**). Early Greek seafarers noticed a current running from north to south past Gibraltar. Believing that only rivers had currents, they decided that this great mass of water, too wide to see across, was part of an immense flowing river. The Greek name for this river was *okeanos*. Our word *ocean* is derived from *oceanus*, a Latin variant of that root. Phoenician sailors were also very much at home in this "river," but like the Greeks they rarely ventured out of sight of land.

As they went about their business, early mariners began to record information to make their voyages easier and safer—the location of rocks in a harbor, landmarks and the sailing times between them, the direction of currents. These first **cartographers** (chart makers) were probably Mediterranean traders who made routine journeys from producing areas to markets. Their first charts (drawn about 800 B.C.) were merely notes to jog their memory for obvious features along the route. Today's **charts** are graphic representations that depict primarily water and water-related information. (*Maps* primarily represent land.)

In this early time other cultures also traveled on the ocean. The Chinese began to engineer an extensive system of inland waterways, some of which connected with the Pacific Ocean to make long-distance transport of goods more convenient. The Polynesian peoples had been moving easily between islands off the coasts of Southeast Asia and Indonesia since 3000 B.C. and were beginning to settle the mid-Pacific islands. Though none of these civilizations had contact with the others, each developed methods of charting and navigation. All these early travelers were skilled at telling direction by the stars and by the position of the rising or setting sun.

Curiosity and commerce encouraged adventurous people to undertake ever more ambitious voyages. But these voyages were possible only with the coordination of astronomical direction-finding and knowledge of the shape and size of the Earth, advanced shipbuilding technology, accurate graphic charts (not just written descriptions), and, perhaps most important, a growing understanding of the ocean itself. Marine science, the organized study of the ocean, had its origin in the technical studies of voyagers.

Science for Voyaging

Progress in applied marine science began at the **Library of Alexandria**, in Egypt. Founded in the third century B.C. by Alexander the Great, the library could be considered the first university in the world. This library constituted history's greatest accumulation of ancient writings. Written knowledge of all kinds was warehoused around its leafy courtyards, including the characteristics of nations, trade, natural wonders, artistic achievements, tourist sights, investment opportunities, and other items of interest to seafarers. Traders quickly realized the competitive benefit of this information, and librarians welcomed their interest in return for even more information. Here perhaps was the first instance of cooperation between a university and the commercial community, a partnership that has paid dividends for science and business ever since.

The second librarian at Alexandria (from 235 B.C. until 192 B.C.) was the Greek astronomer, philosopher, and poet **Eratosthenes of Cyrene**. This remarkable man was the first to calculate the circumference of the Earth. The Greek Pythagoreans had realized Earth was spherical by the sixth century B.C., but Eratosthenes was the first to estimate its true size.

Eratosthenes had heard from travelers returning from Syene (now Aswan, site of the great Nile dam) that at noon on the longest day of the year the sun shone directly onto the waters of a deep vertical well. In Alexandria, he noticed that a vertical pole cast a slight shadow on that day. He measured the shadow angle and found it to be a bit more than 7°, about one-fiftieth of a circle. He correctly assumed that the sun is a great distance from the Earth; so the sun's rays would approach Syene and Alexandria in essentially parallel lines. It followed that if the sun were directly overhead at Syene, but not directly overhead at Alexandria, then the surface of the Earth must be curved. But what was the *circumference* of the Earth?

By studying the reports of camel caravan traders, he estimated the distance from Alexandria to Syene at about 785 kilometers (491 miles). Eratosthenes now had the two pieces of information needed to derive the circumference of the Earth by geometry. **Figure 1.4** shows his solution. The size of the units of length (stadia) Eratosthenes used is thought to have been 555 meters (607 yards), and historians estimate his calculation, made in about 230 B.C., was accurate to within about 8% of the true value. Within a few hundred years, most people in the West who had contact with the library or its scholars knew Earth's approximate size.

Cartography flourished. The first workable charts representing a spherical surface on a flat sheet were developed by Alexandrian scholars. Latitude and longitude, systems of imaginary lines dividing the surface of the Earth, were invented by Eratosthenes. **Latitude** lines were drawn parallel to the equator, and **longitude** lines ran from pole to pole. Eratosthenes placed the lines through prominent landmarks and important places, creating a convenient, though irregular, grid (see **Figure 1.5**). Our present regular grid of latitude and longitude was invented by Hipparchus (c.165–c.127 B.C.), a librarian who divided the surface of the Earth into 360 degrees.[2] A later Egyptian-Greek, Claudius Ptolemy (A.D. 90–168), "oriented" charts by placing east to the right and north at the top. Ptolemy's division of degrees into minutes and seconds of arc is still used by navigators.

Ptolemy also attempted to improve on Eratosthenes' surprisingly accurate estimate of the Earth's circumference, but he wrongly calculated a degree as about 80 kilometers (50 miles) instead of the more correct 112 kilometers (70 miles). This error, coupled with his mistake of overestimating the size of Asia, greatly reduced the apparent width of the unknown part of the world between the Orient and Europe. Unfortunately for generations of navigators, Eratosthenes' estimate was forgotten while Ptolemy's persisted.

Though it weathered the dissolution of Alexander's empire, the Alexandrian library did not survive the subsequent period of Roman rule. The last librarian was Hypatia, the first notable woman mathematician, philosopher, and scientist. In Alexandria she was a symbol of science and knowledge, concepts the early Christians identified with pagan practices. The mission of the library, as personified by the last librarian, antagonized the governors and citizens of the city of Alexandria. After years of rising tensions, in A.D. 415 a mob brutally murdered her and burned the library with all its contents. Most of the community of scholars dispersed, and Alexandria ceased to be a center of learning in the ancient world. The academic loss was incalculable, and trade suffered because ship owners no longer had a clearinghouse for updating the nautical charts and information on which they had come to depend. All that remains of the library today is a remnant of an underground storage room. We shall never know the true extent and influence of its collection of over 700,000 irreplaceable scrolls.

Western intellectual development slackened during the so-called Dark Ages that followed the fall of the Roman Empire in A.D. 476. For almost a thousand years, until the European Renaissance, much of the progress in medicine, astronomy, philosophy, mathematics, and

[2]For more information on latitude and longitude, please see Appendix III.

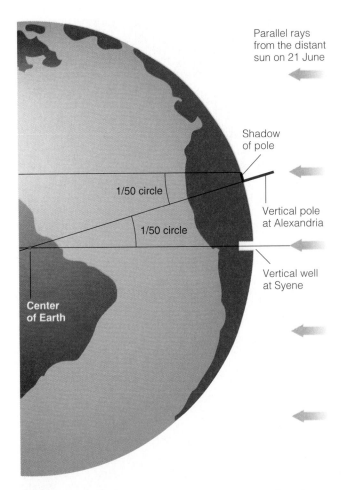

Figure 1.4 A diagram showing Eratosthenes' method for calculating the circumference of the Earth. As described in the text, he used simple geometric reasoning based on the assumptions that the Earth is spherical and that the sun is very far away. (The diagram is not drawn to scale.)

other vital fields of human endeavor was either made by the Arabs or imported by them from Asia. For example, the Arabs used the Chinese-invented compass for navigating caravans over seas of sand. During this time the Vikings raided and explored to the south and west, and the Polynesians continued some of the most extraordinary voyages in history.

Voyages of the Oceanian Peoples

In the history of human migration, no voyaging saga is more inspiring than that of the **Polynesian** colonizations, the peopling of the central and eastern Pacific islands. A profound knowledge of the sea was required for these voyages, and the story of the Polynesians is a high point in our chronology of marine science applied to travel by sea.

The Polynesians are one of four cultures inhabiting some 10,000 islands scattered across nearly 2 million square kilometers (750,000 square miles) of open Pacific ocean (**Figure 1.6**). The Southeast Asian or Indonesian ancestors of the Oceanian peoples, as these cultures are collectively called, spread eastward in the distant past. Although experts vary in their estimates, there is some consensus that by 30,000 years ago New Guinea was populated by these wanderers, and by 20,000 years ago the Philippines were occupied. By around 500 B.C. the so-called cradle of Polynesia—Tonga, Samoa, the Marquesas, and the Society Islands—was settled. Oceanian navigators may already have been using shells attached to a bamboo grid to represent the positions of their islands.

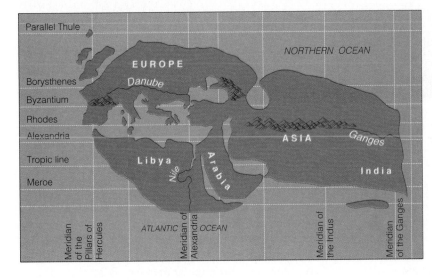

Figure 1.5 The world, according to a chart from the third century B.C. Eratosthenes drew latitude and longitude lines through important places rather than spacing them at regular intervals as we do today. The Alexandrian perception of the world is reflected in the size of the continents and the central position of Alexandria.

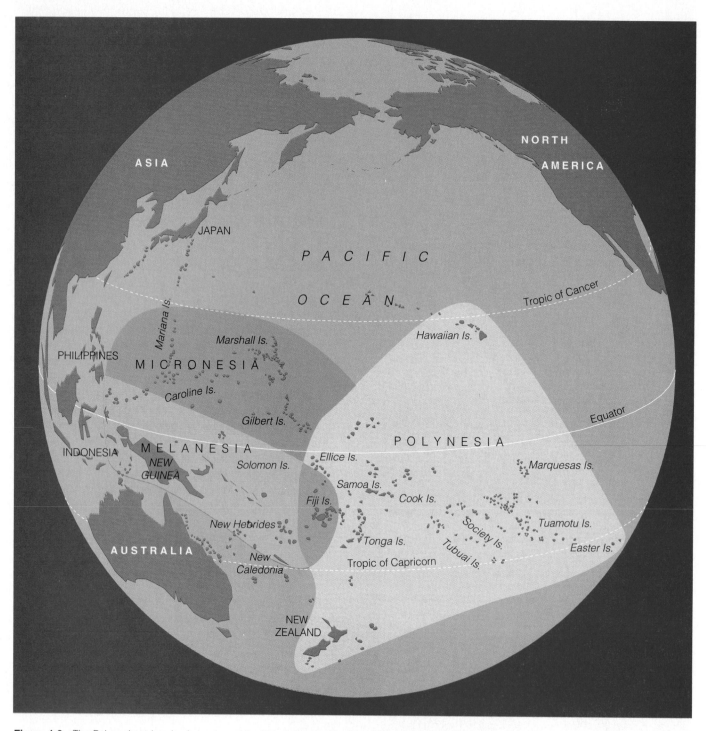

Figure 1.6 The Polynesian triangle. Ancestors of the Polynesians spread from Southeast Asia or Indonesia to New Guinea and the Philippines by about 20,000 years ago. The mid-Pacific islands have been colonized for about 2,500 years, but the explosive dispersion that led to the settlement of Hawaii occurred about A.D. 450–600.

(A Micronesian shell chart from recent times is shown as **Figure 1.7**.)

For a long and evidently prosperous period the Polynesians spread from island to island until the easily accessible islands had been colonized. Eventually, however, overpopulation and depletion of resources became a problem. Politics, intertribal tensions, and religious strife shook society. Groups of people scattered in all directions from some of the "cradle" islands during a period of explosive dispersion. Between A.D. 300 and 600, Polynesians successfully colonized nearly every habitable island within a vast triangular area (see Figure 1.6). Easter

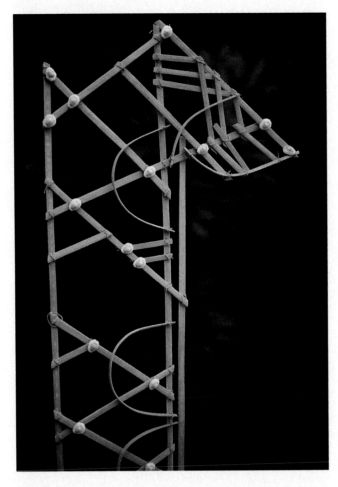

Figure 1.7 A modern Micronesian stick chart. Knots or shells tied at the junctions between bamboo sticks represent islands. Straight strips represent patterns of regular waves; bent strips depict waves curving around the islands.

Island was found against prevailing winds and currents, and the remote islands of Hawaii were discovered and occupied. These were among the last places on Earth to be populated.

How did these risky voyages into unexplored territory come about? Religious warfare may have been the strongest stimulus to colonization. If the losers of a religious war were banished from the home islands under penalty of death, their only hope for survival was to reach a distant and hospitable new land. Seafaring had been a long tradition in the home islands, but these trips called for radical new technology. Great dual-hulled sailing ships, some capable of transporting up to 100 people, were designed and built. New navigation techniques were perfected that depended on the positions of stars barely visible to the north. New ways of storing food, water, and seeds were devised. Whole populations left their home islands in fleets designed especially for long-distance discovery. In some cases, fire was nurtured on

board in case of landfall on an island that lacked volcanic flame. But a new island was only a possibility, a dream. Their gods may have promised the voyagers safe deliverance to new lands, but how many fleets set out from the troubled homelands only to fall victim to storms, thirst, or other dangers?

Yet in that anxious time the Polynesians honed and perfected their seafaring knowledge. To a skilled navigator, a change in the rhythmic set of waves against the hull could indicate an island out of sight over the horizon. The flight tracks of birds at dusk could suggest the direction of land. The positions of the stars told stories, as did the distant clouds over an unseen island. The smell of the water—or its temperature, or salinity, or color—conveyed information, as did the direction of the wind relative to the sun and the type of marine life clustering near the boat. The sunrise colors, sunset colors, the hue of the moon—every nuance had meaning, every detail had been passed in ritual from father to son. The greatest Polynesian minds were navigators, and reaching Hawaii was their greatest achievement.

Of all islands colonized by the Polynesians, Hawaii is farthest away, across an ocean whose guide stars were completely unknown to the southern navigators. The Hawaiian Islands are isolated in the northern hemisphere. There are no islands of any significance for more than two thousand miles to the south. Moreover, Hawaii lies beyond the equatorial doldrums, a hot and often windless stretch across which these pioneers must somehow have paddled. And yet some fortunate and knowledgeable people colonized Hawaii sometime between A.D. 450 and 600. Try to imagine their feelings of relief and justification upon reaching a promised paradise under a new night sky (**Figure 1.8**). Think of that first approach to the high islands of Hawaii, the first unlimited drink of fresh water, the first solid Earth after months of uncertainty.

Within 100 years of their first arrival, Hawaiian navigators were routinely piloting vessels on regular return trips to the Marquesas and the Society Islands (Tahiti and others). Some of the trips were undertaken to import needed food species to the newly found islands, but others were made to recruit new citizens and leaders to "green-clad Hawaii."

At a time when seafarers of other civilizations sailed beside the comforting bulk of a charted coast, Polynesians looked to the open sea for sustenance, deliverance, and hope. Their great knowledge of the ocean protected them.

The Age of European Discovery

Half a world from their Polynesian counterparts Renaissance Europeans—having been jolted by internal awakening and external reality—set out to explore the world

Figure 1.8 Polynesian voyagers reach the island of Hawaii. "Looking anew at the clouds we saw a sight difficult to comprehend. What had appeared as an unusual cloud formation was now revealed as the peak of a gigantic mountain, a mountain of unbelievable size, a white mountain—a pillar that seemed to support the sky! We watched in wonder until nightfall. Then to the south of that mountain a dull red glow lighted the underside of the lifting clouds, revealing the shape of another mountain. It brightened as the night darkened. That mountain seemed to be burning! No one slept that night. Our two ships thrashed along in the night wind, and the dreadful red beacon lighted our way."

Figure 1.9 Prince Henry of Portugal, the Navigator. In the mid-1400s, Henry established a center at Sagres for the study of marine science and navigation ". . . through all the watery roads."

by sea. They did not undertake exploration for its own sake, however; any voyage had to have a material goal. Trade between east and west had long been dependent on arduous and insecure desert caravan routes through the Arabian desert. This commerce was cut off in 1453 when the Turks captured Constantinople, and an alternate ocean route was needed.

A European visionary who thought that ocean exploration held the key to great wealth and successful trade was **Prince Henry the Navigator**, third son of the royal family of Portugal (**Figure 1.9**). Prince Henry established a center at Sagres for the study of marine science and navigation ". . . through all the watery roads." Although he personally was not well traveled (he went to sea only twice in his life), captains under his patronage explored from 1451 to 1470, compiling detailed charts wherever they went. Henry's explorers pushed south into the unknown and opened the west coast of Africa to commerce. He sent out small, maneuverable ships designed for voyages of discovery and manned by well-trained crews, and his mariners used the **compass**, an instrument that points to magnetic north, for navigation. Although Arab traders had brought the compass from China in the twelfth century, navigators still considered it a magical tool. They concealed the compass in a special box (predecessor to today's binnacle) and consulted it out of plain view. Henry's students knew the Earth was round but, thanks to the errors of Claudius Ptolemy, they were wrong in their understanding of its true size.

A master mariner (and skilled salesman), **Christopher Columbus** "discovered" the New World quite by accident. Native Americans had been living on the continent for about 11,000 years, and the Norwegian Vikings had made about two dozen visits to a functioning colony on the continent 500 years before his noisy arrival, yet Columbus gets the credit. Why? Because his interesting souvenirs, exaggerated stories, inaccurate charts, and promises of vast wealth excited the imagination of royal courts. Columbus made North America a media event without ever sighting it!

Columbus wasn't trying to discover new lands. His intention was to pioneer a sea route to the rich and fabled lands of the east discovered more than 200 years earlier in the overland travels of Marco Polo. As "Admiral of the Ocean Sea," Columbus was to have a financial interest in the trade routes he blazed. He was familiar with Prince Henry's work and, like all other competent contemporary navigators, knew the Earth was spherical. By sailing west he believed he could come close to his eastern destination, the latitude of which he thought he knew. Because he depended on Ptolemy's data, however, Columbus made the *smallest* estimate of the size of the Earth by any navigator in modern history; he assumed the Earth to be only about 14% ocean.

Not surprisingly, Columbus mistook the New World for his goal of India or Japan. He explained the notable absence of wealthy cities by claiming he struck the coast too far north or south of his desired latitude. He made three more trips to the New World but died still believing that he had found islands off the coast of Asia. He never saw the mainland of North America and never realized the size and configuration of the continents whose future he had so profoundly changed.

Other explorers quickly followed, and Columbus's error was soon rectified. Charts drawn as early as 1507 included the New World (**Figure 1.10**). Such charts perhaps

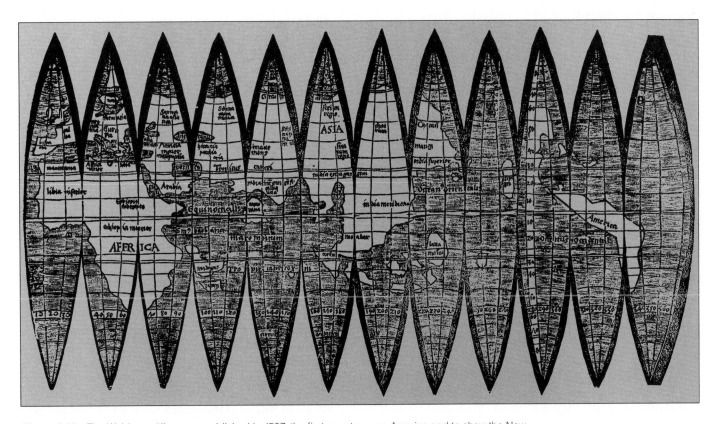

Figure 1.10 The Waldseemüller map, published in 1507, the first map to name America and to show the New World as separate from Asia. The deep gores are designed to form a globe about 4 inches in diameter.

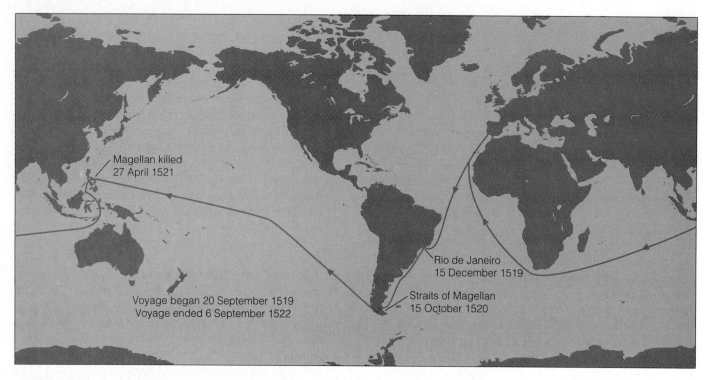

Figure 1.11 Track of the Magellan expedition, the first voyage around the world. Magellan himself did not survive the voyage; only 18 out of 270 sailors managed to return after three years of voyaging.

inspired **Ferdinand Magellan**, a Portuguese navigator in the service of Spain, to believe that he could open a westerly trade route to the Orient. Unfortunately, the chartmakers estimated that the Americas and the Pacific Ocean were much smaller than they actually are. (Compare Figure 1.10 with Magellan's route, shown in **Figure 1.11**.) Of 270 sailors who set out with Magellan from Seville on five ships, only 18 managed to return after three years, on a worm-eaten and barely floating vessel laden with valuable spices. Magellan was not among the voyagers—he had been killed in a battle with the natives on Mactan Island in the Philippines. His crew, however, had proved it was possible to circumnavigate the globe.

The Magellan expedition's return to Spain in 1522 marks the end of the Age of European Discovery. An unpleasant era of exploitation of the human and natural resources of the Americas followed, in which native empires were destroyed and objects of priceless archaeological value were melted into coin to fund European warfare and greed.

VOYAGING FOR SCIENCE

British seapower rose after the Age of Discovery to compete with the colonial aspirations of France and Spain.

Sailing ships require dependable supply and repair stations, especially in remote areas. The great powers sent out expeditions to claim appropriate locations, preferably inhabited by friendly natives eager to help provision ships half a globe from home. The French sent Admiral Louis de Bougainville into the South Pacific in the mid-1760s. His discovery of what is now called French Polynesia opened the area to the powerful European nations. The British followed almost immediately.

James Cook

Scientific oceanography begins with the departure from Plymouth Harbor in 1768 of HMS *Endeavour* under command of **James Cook** of the British Royal Navy (**Figure 1.12**). An intelligent and patient leader, Cook was also a skillful navigator, cartographer, writer, artist, diplomat, sailor, scientist, and dietician. The primary reason for the voyage was to assert the British presence in the South Seas, but the expedition had numerous scientific goals as well. First Cook conveyed several members of the Royal Society (a scientific research group) to Tahiti to observe the transit of Venus across the disk of the sun. These measurements verified calculations of planetary orbits made earlier by Edmund Halley (later of comet fame) and others. Then Cook turned south into unknown territory to

Figure 1.12 Captain James Cook, Royal Navy.

Figure 1.13 First contact: Captain James Cook, commanding HMS *Resolution*, off the Hawaiian island of Kaua´i in 1778. "It required but little address to get them to come along side, but we could not prevail upon any one to come on board; they exchanged a few fish they had in the canoes for anything we offered them, but valued nails, or iron above every other thing; the only weapons they had were a few stones in some of the canoes and these they threw overboard when they found they were not wanted."

search for a hypothetical southern continent that philosophers believed had to exist to balance the landmass of the northern hemisphere. Cook and the scientists found and charted New Zealand, mapped Australia's Great Barrier Reef, marked the positions of tens of small islands, made notes on the natural history and human habitation of these distant places, and initiated friendly relations with many chiefs. Cook survived an epidemic of dysentery contracted by ship's company while ashore in Batavia (Djakarta) and sailed home to England around the world in 1771. Because of his insistence on cleanliness and ventilation, and because his provisions included cress, sauerkraut, and citrus extracts, his sailors avoided scurvy, a Vitamin-C deficiency disease that for centuries had decimated crews on long voyages.

The Admiralty was deeply impressed. Cook was promoted to the rank of Commander and in 1772 was given command of the ships *Resolution* and *Adventure*, in which he embarked on one of the great voyages in scientific history. On this second voyage he charted Tonga and Easter islands, discovered New Caledonia in the Pacific and South Georgia in the Atlantic. He was first to circumnavigate the world at high latitudes. Though he sailed to 70°

south latitude, he never sighted Antarctica. He returned home again in 1775.

Posted to the rank of Captain, Cook set off in 1776 on his third, and last, expedition in *Resolution* and *Discovery*. His commission was to find a northwest passage around Canada and Alaska or a northeast passage above Siberia. He "discovered" the Hawaiian Islands (Hawaiians were there to greet him, of course; see **Figure 1.13**) and charted the west coast of North America. After searching unsuccessfully for a passage across the top of the world, Cook retraced his steps to Hawaii to provision for departure home. On 14 February 1779, after an elaborate farewell dinner with the chief of the Island of Hawaii, Cook and his officers prepared to return to *Resolution*, anchored in Kealakekua Bay. The Englishmen accidentally violated a taboo of property rights relating to their own shoreboat and were beset by the crowd. Cook, among others, was killed in the fracas.

Cook deserves to be considered a scientist as well as an explorer because of his accuracy, thoroughness, and the completeness in his descriptions. He and the researchers aboard took samples of marine life, land plants and animals, the ocean floor, and geological formations, and they reported their characteristics in their logbooks and journals. His navigation, aided by the newly invented chronometer, was outstanding; his charts of the Pacific were accurate enough to be used by the Allies in invasions of the Pacific islands during World War II. He drew accurate conclusions, did not exaggerate his findings in his reports, and opened friendly diplomatic relations with many native populations. Cook recorded and successfully interpreted events in natural history, anthropology, and oceanography. Unlike most captains of his day, he cared for his men. He was a thoughtful and clear writer. This first marine scientist peacefully changed the map of the world more than any explorer or scientist in history.

The Sampling Problem

Marine science advances by the analysis of samples. Sampling of floor sediments or bottom water is not an easy task in the deep ocean. The line used to suspend the sampling device snakes back and forth as currents strike it, and the weight of the line makes it difficult to tell when the sampler has hit bottom. Deploying and recovering the line is laborious and time-consuming, and sometimes the sampling device does not work properly. Early bottom-sampling devices (such as those used by Cook) were simple wax-covered lead weights lowered to shallow bottoms to pick up sediments and test the suitability of anchorages. Later devices took deep-water samples, extracted cores from the sediments, grabbed samples of the bottom, or scooped biological specimens from the ocean floor.

The first researchers to attack the deep-water sampling problem successfully were British explorers Sir John Ross and his nephew Sir James Clark Ross. During an expedition to scout the Northwest Passage in 1818, Sir John Ross obtained a bottom sample from 1,919 meters (3,296 feet) near Greenland by using a clamping sampler to trap the specimen. Sir James Clark Ross, discoverer of the Ross Sea and the area of Antarctica known as Victoria Land, obtained depth **soundings** (depth measurements) of 4,433 meters and 4,893 meters (14,545 feet and 16,054 feet) in the South Atlantic.

Sampling techniques improved through the century. Using a sounding method perfected in the late 1840s by a U.S. Navy midshipman, American Commodore Matthew Maury used a long lightweight line and lead weight to discover the Mid-Atlantic Ridge, an important hidden range of mountains. Fridtjof Nansen perfected the deep-water sampling bottle bearing his name near the end of the century. Even today, in spite of modern advances, deep sampling remains difficult. Remotely operated vehicles can work at great depths and return samples and pictures to the surface, but their electronic complexity makes them delicate and expensive to operate.

SCIENTIFIC EXPEDITIONS

Great as his contributions undoubtedly were, Cook's three voyages (and those of the Rosses) were not purely scientific expeditions. These men were British naval officers engaged in Crown business, concerned with charting, "foreign relations," and natural phenomena as they applied to Royal naval matters. The first genuine only-for-science expedition may well have been the *Challenger* expedition of 1872–76, but the United States got into the act first with a hybrid expedition in 1838.

The United States Exploring Expedition

After a ten-year argument over its potential merits, the **United States Exploring Expedition** was launched in 1838. This was primarily a naval expedition, but its captain was somewhat more free in maneuvering orders than Cook had been. The work of the scientists aboard the flagship *Vincennes* and the expedition's five other vessels helped to establish the natural sciences as reputable professions in America. Had it not been for the combative and disagreeable personality of its leader, Lieutenant Charles Wilkes (**Figure 1.14**), this expedition might have become as famous as those of Cook or the upcoming *Challenger* voyage.

The expedition departed on a four-year circumnavigation. Its goals included showing the flag, whale scouting, mineral gathering, charting, observing, and pure exploration. One unusual goal was to disprove a peculiar theory that the Earth was hollow and could be entered through huge holes at either pole.

Wilkes's team explored and charted a large sector of the east Antarctic coast and it made observations that confirmed the landmass as a continent. A map of the Oregon Territory produced in 1841, one of 241 maps and charts drawn by members of the expedition, proved especially valuable when connected to the map of the Rocky Mountains prepared the following year by Captain John C. Fremont. Hawaii was thoroughly explored, and Wilkes led an ascent of Mauna Loa, one of the two peaks of Hawaii's tallest volcano. The expedition returned with many scientific specimens and artifacts, which formed the nucleus of the collection of the newly established Smithsonian Institution in Washington, D.C. No evidence of polar holes was found.

Upon their return, Wilkes and his "scientifics" prepared a final report totaling 19 volumes of maps, text, and illustrations. The report is a landmark in the history of American scientific achievement.

Matthew Maury

At about the time the Wilkes expedition returned, **Matthew Maury** (**Figure 1.15**), a Virginian and U.S. naval

Figure 1.14 Lieutenant Charles Wilkes soon after his return from the United States Exploring Expedition.

Figure 1.15 Matthew Fontaine Maury (1806–73). This photograph was probably taken in 1853.

officer, was appointed director of the navy's Bureau of Charts and Instruments. There he studied a huge and neglected treasure trove of ships' logs with their many regular readings of temperature and wind direction. By 1847 Maury had assembled much of this information into coherent wind and current charts. Maury began to issue these charts free to mariners in exchange for logs of their own new voyages.

Slowly a picture of planetary winds and currents began to emerge. Maury himself was a compiler, not a scientist, and he was vitally interested in the promotion of maritime commerce. His understanding of currents was built on the work of Benjamin Franklin, who nearly 100 years earlier had noticed the peculiar fact that the fastest ships were not always the fastest ships—that is, the ship's speed through the water did not always correlate with out-and-return time on the European run. Franklin's cousin, a Nantucket merchant named Tim Folger, noted Franklin's puzzlement and provided him with a rough chart of the "Gulph Stream" that he (Folger) had worked out. By staying within the stream on the outbound leg and adding its speed to their own—and by avoiding it on their return—captains could traverse the Atlantic much more quickly. It was Franklin who published, in 1769, the first chart of any current (**Figure 1.16**).

But it was Maury who was the first person to sense the worldwide pattern of surface winds and currents. His work became famous in 1849 when the California gold rush made his sailing directions invaluable for fast trips around Cape Horn. His crowning achievement, *The Physical Geography of the Seas*, a book explaining his discoveries, was published in 1855. Maury, considered by many to be the father of physical oceanography, was perhaps the first man to undertake the systematic study of the ocean as a full-time occupation.

The *Challenger* Expedition

The first sailing expedition devoted completely to marine science was conceived by a professor of natural history at Scotland's University of Edinburgh, Charles Wyville Thomson, and his Canadian-born student of natural history, John Murray. Stimulated by their own curiosity and by the inspiration of Charles Darwin's 1831–36 voyage in HMS *Beagle*, they convinced the Royal Society and the British government to provide a Royal Navy ship and trained crew for a "prolonged and arduous voyage of exploration across the oceans of the world." Thomson and Murray even coined a word for their enterprise: *oceanography*. Though the term literally

Figure 1.16
Benjamin Franklin's 1769 chart of the Gulf Stream system. His cousin, Timothy Folger, discovered that Yankee whalers had learned to use the Gulf Stream to their advantage. Others, especially English shipowners, were slower to learn. Folger, himself a sea captain, wrote that Nantucket whalers ". . . in crossing it have sometimes met and spoke with those packets who were in the middle of and stemming it. We have informed them that they were stemming a current that was against them to the value of three miles an hour and advised them to cross it, but they were too wise to be counseled by simple American fishermen."

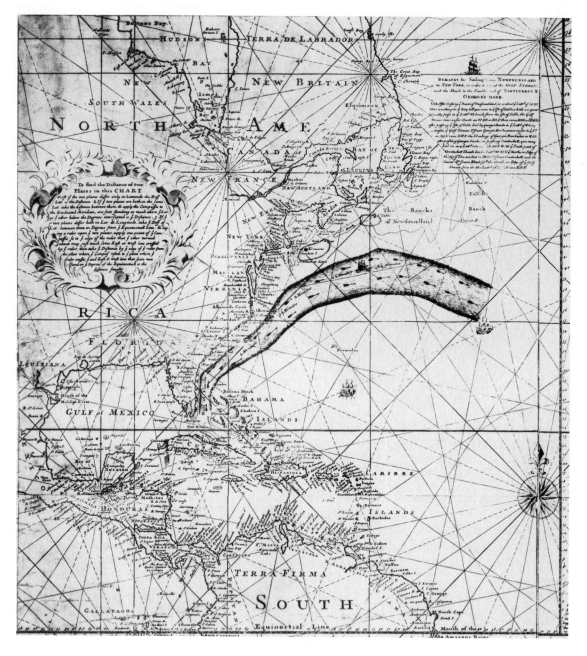

implies only marking or charting, it has come to mean the science of the ocean. The government and the Royal Society agreed to the endeavor provided that a proportion of any financial gain from discoveries was handed over to the Crown. This arranged, the scientists made their plans.

HMS *Challenger*, a 2,306-ton steam corvette (**Figure 1.17**), set sail on 7 December 1872 on a four-year voyage that took them around the world and covered 127,600 kilometers (79,300 miles). Although the captain was a Royal naval officer, the six-man scientific staff directed the course of the voyage. *Challenger*'s track is shown in **Figure 1.18**.

One important mission of the *Challenger* **expedition** was to investigate Edinburgh professor Edward Forbes's contention that life below 550 meters (1,800 feet) was impossible because of high pressure and lack of light. The steam winch on board made deep sampling practical, and samples from depths as great as 8,185 meters (26,850 feet) were collected off the Philippines. Through the course of 492 deep soundings with mechanical grabs and nets at 362 stations (including 133 dredgings) Forbes was proven resoundingly wrong. With each hoist, animals new to science were strewn on the deck; in all, staff biologists discovered 4,717 new species! **Figure 1.19** shows one of the large trawl nets used for some of these discoveries.

Figure 1.17 Lt. Pelham Aldrich, first lieutenant of HMS *Challenger,* kept a detailed journal of the *Challenger* expedition. With accuracy and humor he kept this record in good weather and bad, and had the patience and skill to include watercolors of the most exciting events. This is part of the first page of his journal.

Figure 1.18 HMS *Challenger*'s track from December 1872 to May 1876.

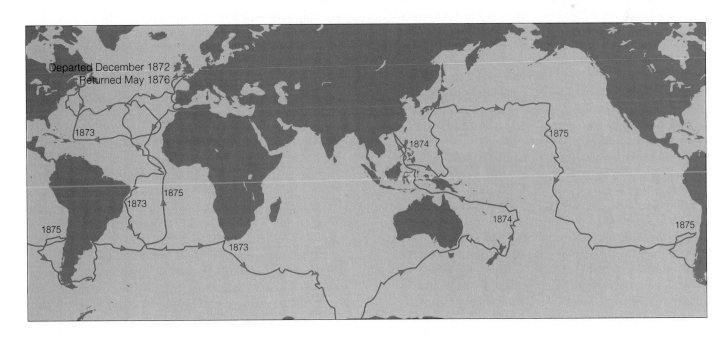

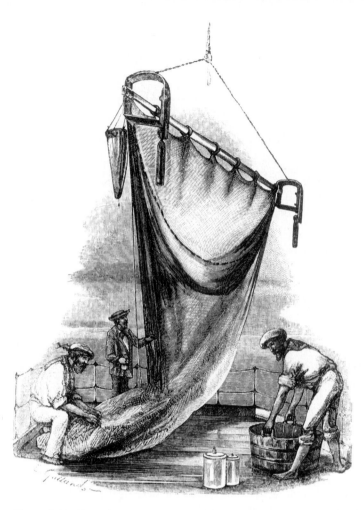

Figure 1.19 Emptying a trawl net, an engraving from the *Challenger Report.*

The scientists also took salinity, temperature, and water density measurements during these soundings. Each reading contributed to a growing picture of the physical structure of the deep ocean. They completed at least 151 open-water trawls and stored 77 samples of seawater for detailed analysis ashore. The expedition collected new information on ocean currents, meteorology, and the distribution of sediments; the locations and profiles of coral reefs were also charted. Thousands of pounds of specimens were brought to British museums for study. Manganese nodules—brown lumps of mineral-rich sediments—were discovered on the seabed, sparking interest in deep-sea mining.

This first pure oceanographic investigation was an unqualified success. The discovery of life in the depths of the oceans stimulated the new science of marine biology. The scope, accuracy, thoroughness, and attractive presentation of the researchers' written reports made this expedition a high point in scientific publication. The *Challenger Report*, the record of the expedition, was published between 1880 and 1895 by Sir John Murray in a well-written and magnificently illustrated 50-volume set; it is

still used today. Indeed, it was the 50-volume *Report*, rather than the cruise, that provided the foundation for the new science of oceanography. The expedition's many financial spin-offs indicated that pure research was a good investment, and the British government realized quick profits from the exploitation of newly discovered mineral deposits on islands. The *Challenger* expedition remains history's longest continuous scientific oceanographic expedition.

With successes like these the pace of exploration accelerated. American naturalist Alexander Agassiz, sailing in 1877 on the U.S. Coast and Geodetic Survey ship *Blake*, collected data corroborating the *Challenger* material at 355 deep-sea stations. The distribution of manganese nodules was found to be widespread. Further work by Agassiz and his students around the turn of this century on the survey ship *Albatross* helped train a generation of influential American marine biologists. In 1886 the Russians entered the field of marine exploration with the three-year cruise of *Vitiaz* under the leadership of S. O. Makarov; their main contribution was a careful analysis of salinity and temperature of North Pacific water.

TWENTIETH-CENTURY VOYAGING FOR SCIENCE

In this century oceanographic voyages became more technically ambitious and expensive. Scientist-explorers sought out and investigated places that once had been too difficult to reach. New electronic and optical devices aided navigation and sampling. In the last half of the century high-speed shipboard computers have made it possible for marine scientists to analyze data while still at sea.

Polar oceanography made dramatic advances early in the century. Newly designed ships and new methods of food storage had made polar exploration possible in the last years of the nineteenth century. In 1893 Fridtjof Nansen began studying the north polar ocean in *Fram*, a ship designed specifically to withstand the crushing pressure of sea ice (**Figure 1.20**). In the next 20 years Nansen and others probed the polar ocean depths. Researchers confirmed the feeding relationships between whales and plankton and collected much data about the whale population of the southern ocean—not out of scientific curiosity, but as a source of oil and baleen (whalebone).

In 1925 the German *Meteor* **expedition**, which crisscrossed the South Atlantic for two years, introduced modern optical and electronic equipment to oceanographic investigation. Its most important innovation was to use an **echo sounder**, a device that bounces sound waves off the ocean bottom, to study the depth and contour of the seafloor (see **Figure 1.21**). The echo sounder revealed to *Meteor* scientists a varied and often extremely

a

b

Figure 1.20 (a) Fridtjof Nansen, pioneering Norwegian oceanographer and polar explorer. (b) His 123-foot schooner *Fram* ("forward"). With 13 men, *Fram* sailed on 22 June 1893 to the high Arctic with the specific purpose of being frozen into the ice. *Fram* was designed to slip up and out of the frozen ocean, and it drifted with the pack ice to within about 4° of the North Pole. The whole harrowing adventure took nearly four years. The ship's 1,650-kilometer (1,025-mile) drift proved that no Arctic continent existed beneath the ice. In 1908 Nansen became the first professor of oceanography, a post created for him at Christiania University.

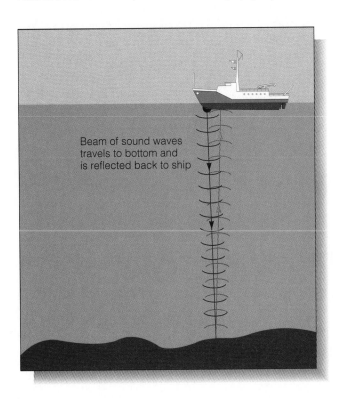

Beam of sound waves
travels to bottom and
is reflected back to ship

Figure 1.21 Echo sounders sense the contour of the seafloor by beaming sound waves to the bottom and measuring the time required for the sound waves to bounce back to the ship. If the round-trip travel time and wave velocity are known, distance to the bottom can be calculated. This technique was first used on a large scale by the German research vessel *Meteor* in the 1920s.

rugged bottom profile, rather than the flat floor they had anticipated.

Atlantis, launched in 1931, was the first U.S. research ship built specifically for ocean studies. Investigations by her scientists confirmed Matthew Maury's findings of a mid-Atlantic ridge and helped to discover its extent. The 32-meter (104-foot) schooner *E. W. Scripps*, under the direction of Harald Sverdrup, began a wide-ranging program of chemical, biological, and geophysical exploration off the coast of southern California in 1937. These voyages led to publication in 1942 of *The Oceans*, the first modern reference work on all phases of marine science.

In October 1951 a new HMS *Challenger* began a two-year voyage that would make precise depth measurements in the Atlantic, Pacific, and Indian oceans and in the Mediterranean Sea. With echo sounders, measurements that would have taken the crew of the first *Challenger* nearly 4 hours to complete could be made in seconds. *Challenger II*'s scientists discovered the deepest part of the ocean's deepest trench, naming it Challenger Deep in honor of their famous predecessor. In 1960 U.S. Navy Lieutenant Don Walsh and Jacques Piccard descended into the Challenger Deep in *Trieste*, a Swiss-designed, blimplike bathyscaphe (see **Figure 1.22**).

But in many ways the last voyage to be discussed here is the most portentous of all. In 1968 the drilling ship *Glomar Challenger* (see **Box 1.1**) set out to test a controversial hypothesis about the history of the ocean floor. It was capable of drilling into the ocean bottom beneath more than 6,000 meters (20,000 feet) of water and recovering samples of seafloor sediments. These long and revealing plugs of seabed provided confirming evidence for seafloor spreading and plate tectonics. (The details will be found in Chapter 3.) In 1985, deep-sea drilling duties were taken over by the much larger and more technologically advanced ship *JOIDES Resolution*.[3] She contains equipment capable of drilling in water 8,100 meters (27,000 feet) deep and houses the most completely equipped geological laboratories ever put to sea.

THE RISE OF OCEANOGRAPHIC INSTITUTIONS

The demands of scientific oceanography have become greater than the capability of any single voyage. Oceanographic institutions, agencies, and consortia evolved in part to ensure continuity of effort. The first of these coordinating bodies was founded by Prince Albert I of Monaco, who endowed his country's oceanographic laboratory and museum in 1906. The most famous alumnus of Albert's Musée Océánographique is Jacques Cousteau, coinventor in 1943 of the scuba underwater breathing system. Monaco also became the site of the International Hydrographic Bureau, founded in 1921 as an association of maritime nations. This bureau published one of the first general charts of the ocean showing bottom contours.

In the United States, the three preeminent oceanographic institutions are the Lamont-Doherty Earth Observatory of Columbia University (founded in 1949), the Scripps Institution of Oceanography (founded in La Jolla, California, and affiliated with the University of California in 1912), and the Woods Hole Oceanographic Institution on Cape Cod (founded in 1930), which is associated

---

[3]JOIDES stands for Joint Oceanographic Institutions for Deep Earth Sampling.

Figure 1.22 The bathyscaphe *Trieste* seen on the surface. *Trieste* reached the ocean's deepest spot in 1960.

a

b

Figure 1.23 (a) The Woods Hole Oceanographic Institution, Woods Hole, Massachusetts. Marine science has been an important part of this small Cape Cod fishing community since Spencer Fullerton Baird, then assistant secretary of the Smithsonian Institution, established the U.S. Commission of Fish and Fisheries there in 1871. The Marine Biological Laboratory was founded in 1888; the Oceanographic Institution in 1930. (b) The Scripps Institution of Oceanography, La Jolla, California. Begun in 1892 as a portable laboratory-in-a-tent, Scripps was founded by William Ritter, a biologist at the University of California. Its first permanent buildings were erected in 1905 on a site purchased with funds donated by philanthropic newspaper owner E. W. Scripps and his daughter, Ellen.

with the neighboring Marine Biological Laboratory (founded in 1888) (**Figure 1.23**).

The U.S. government has been active in oceanographic research. Within the Department of the Navy are the Office of Naval Research, the Office of the Oceanographer of the Navy, the Naval Oceanic and Atmospheric Research Laboratory, and the Naval Ocean Systems Command. These agencies are responsible for oceanographic research related to national defense. The National Oceanic and Atmospheric Administration (**NOAA**), founded within the Department of Commerce in 1970, includes the National Ocean Service, the National Weather Service, the National Marine Fisheries Service, and the Office of Sea Grant. NOAA seeks to facilitate commercial uses of the ocean. The National Aeronautics and Space Administration (**NASA**), organized in 1958, has recently become an important institutional contributor to marine science. For four months in 1978, NASA's *Seasat*, the first oceanographic satellite, beamed oceanographic data to Earth. Other ocean-observing satellites are scheduled for launch from the space shuttle in the near future. Satellite oceanography is an important frontier, and many discoveries made by satellites are discussed in later chapters.

CHAPTER SUMMARY

Earth is a water planet, possibly one of few in the galaxy. An ocean covering 71% of its surface has greatly influenced its rocky crust and atmosphere. The ocean dominates the Earth, and its average depth is about 4½ times the average height of the continents above sea level.

BOX 1.1 ● *Glomar Challenger*

The history of marine science took a long leap forward in 1968 when the drilling ship *Glomar Challenger* returned the first complete cores of deep-sea sediments. A few of these cores—from several sites in the South Atlantic—yielded samples of sediments down to the solid oceanic crust. The deepest sediments, and thus the oceanic crust immediately below, were shown to be surprisingly young—only about 180 million years old. Yet, the oldest continental crust had been dated at more than 3.8 billion years. What could explain this dramatic discrepancy?

Glomar Challenger had been conceived and built in the early 1960s by an international consortium of oceanographic institutions and the U.S. National Science Foundation to test the then-radical hypothesis that continents are moved across the Earth's surface by seafloor spreading. The relative youth of ocean beds was a

central prediction of that theory. Researchers using *Glomar Challenger*'s sophisticated deep-water drilling technology labored for the next 15 years collecting core samples, some of them more than 1,700 meters (5,600 feet) long. The age of these samples confirmed the fact of seafloor spreading, a central principle of the theory of plate tectonics.

In 1983, *Glomar Challenger* was retired, her drilling program successfully completed. She had traveled more than 600,000 kilometers (375,000 miles), had drilled 1,092 holes at 624 sites, and had recovered a total of 96 kilometers (57.6 miles) of deep-sea cores for study. Thanks in large part to her efforts, by the time of her decommissioning, the theory of plate tectonics was firmly established in the mainstream of geological understanding.

Glomar Challenger, operated by the Deep Sea Drilling Project from 1968 to 1983. The 122-meter (400-foot) ship used computers to maintain her position with the precision needed to complete the drilling of cores up to a mile long.

The early history of marine science is closely associated with the history of voyaging. The first marine studies had a practical aim: to facilitate travel, trade, and warfare. Later the search for new knowledge became a goal in itself. The first part of this chapter focused on *marine science for voyaging*; it looked at some of the voyagers and their voyages, the inventions that made their adventures possible, and some of the discoveries they made. The second part discussed *voyaging for marine science*, including the British *Challenger* expedition, the first wholly-for-science oceanographic research voyage. The contributions of a few of the founders of modern marine science were summarized, and the rise of oceanographic institutions was outlined.

cartographers
Challenger expedition
charts
Columbus, Christopher
compass
Cook, James
echo sounder
Eratosthenes of Cyrene
latitude

Library of Alexandria
longitude
Magellan, Ferdinand
Maury, Matthew
Meteor expedition
NOAA (National Oceanic and Atmospheric Administration)
ocean

oceanography
oceanus
Polynesia
Prince Henry the Navigator
Seasat
soundings
United States Exploring Expedition
voyaging
world ocean

Study Questions

1. Which hemisphere contains the greatest percentage of ocean? Is most of Earth's water in the ocean?

2. Which is greater: the average depth of the ocean floor, or the average height of the continents above sea level?

3. How did the library at Alexandria contribute to the development of marine science? What happened to most of the information accumulated there? Why do you suppose the residents of Alexandria became hostile to the librarians and the many achievements of the library?

4. What were the stimuli to Polynesian colonization? How were the long voyages accomplished?

5. What were the main stimuli to European voyages of exploration during the Age of Discovery? Why did it end?

6. What were the contributions of Captain James Cook? Does he deserve to be remembered more as an explorer or as a marine scientist?

7. What was the first purely scientific oceanographic expedition, and what were some of its accomplishments?

8. Who was probably the first person to undertake the systematic study of the ocean as a full-time occupation? Are his contributions considered important today?

9. Sketch briefly the major developments in marine science since 1900. Do individuals, separate voyages, or institutions figure most prominently in this history?

10. In your opinion, where does the future of marine science lie?

For Further Study

The World Ocean

Borgese, E. M. 1974. *The Drama of the Oceans*. New York: Abrams. Powerful prose, beautiful photographs.

Carson, R. 1951. *The Sea Around Us*. New York: Houghton Mifflin. The book that introduced many of us to the wonders of the ocean when we were young.

Kennish, M. J., ed. 1989. *Practical Handbook of Marine Science*. Boca Raton, FL: CRC Press. An excellent summary of physical and biological oceanographic data, primarily in the form of graphs and tables. A thorough bibliography follows the introduction.

Mangone, G. J. 1986. *Concise Marine Almanac*. Van Nostrand-Reinhold. A wealth of tabular information.

The History of Marine Science

Beaglehole, J. C. 1974. *The Life of Captain James Cook*. Stanford: Stanford University Press. The definitive biography of Captain James Cook.

Bellwood, P. S. 1991. "The Austronesian Dispersal and the Origin of Languages." *Scientific American*, July, 88–93. The Polynesian voyages are reflected in their languages.

Boorstin, D. 1983. *The Discoverers*. New York: Random House. Elegantly written, very informative sections on exploration and on the instruments invented to make the discoveries possible.

Kane, H. 1991. *Voyagers*. Bellevue, WA: Whalesong. Deftly written and stirringly illustrated book on the Hawaiian colonization and other Polynesian topics. Perhaps hard to find, but not to be missed!

Linklater, E. 1972. *Voyage of the* Challenger. Garden City, NY: Doubleday. Nicely written account, beautifully illustrated with contemporary photographs and excerpts from the logs.

Matkin, J. 1992. *At Sea with the Scientifics: The* Challenger *Letters of Joseph Matkin*. P. F. Rehbock, ed. Honolulu: University of Hawaii Press. A detailed account of the *Challenger* expedition from a member of the below-decks crew.

Schlee, S. 1973. *The Edge of an Unfamiliar World—A History of Oceanography*. New York: Dutton. Complete and well written.

Viola, H. J., and C. Margolis, eds. 1987. *Magnificent Voyagers: The United States Exploring Expedition 1838–1842*. Washington, DC: Smithsonian Institution Press.

Wilford, J. N. 1981. *The Mapmakers*. New York: Knopf. An entertaining history of cartography from Eratosthenes to satellite mapping.

2 ORIGINS

Solitude

Joshua Slocum, shown here rounding the tip of Cape Horn in his 37-foot sloop *Spray*, was the first person to sail around the world alone. After a difficult voyage of more than three years and 46,000 miles, he returned to Newport, Rhode Island, and tied *Spray* to the same cedar spike driven in the bank that held her when she was first launched. "I could bring her no nearer home."

The ocean has always challenged the human spirit. Meeting that challenge *alone* is a supreme triumph of humanity over the uncontrollable and unpredictable forces of nature. Whatever the reasons for their voyages, all who travel by sea alone have experienced profound solitude, loneliness, helplessness, and—if things went well—the exaltation of success.

The first man to sail alone around the world was Joshua Slocum, a Massachusetts sailor who went to sea when he was 14. He was 51 when he began his solitary voyage. He started from Boston on 24 April 1895 in the *Spray*, an 11-meter (37-foot) sloop he had rebuilt from a derelict hulk. More than three years and 74,000 kilometers (46,000 miles) later, on 3 July 1898, he tied the boat to the cedar spike driven in the bank that held her when she was first launched. "I could bring her no nearer home," he said.

In his book, *Sailing Alone Around the World*, Slocum wrote about the peace and heightened awareness that intimate contact with the ocean can bring.

> The fog lifted just before night, and I was afforded a look at the sun just as it was touching the sea. I watched it go down and out of sight. Then I turned my face eastward, and there, apparently at the very end of the bowsprit, was the smiling full moon rising out of the sea. Neptune himself coming over the bows could not have startled me more. "Good evening, sir," I cried; "I'm glad to see you." Many a long talk since then I have had with the man in the moon; he had my confidence on the voyage.
>
> About midnight the fog shut down again denser than ever before. One could almost "stand on it." It continued so for a number of days, the wind increasing to a gale. The waves rose high, but I had a good ship. Still, in the dismal fog I felt myself drifting into loneliness, an insect on a straw in the midst of the elements. I lashed the helm, and my vessel held her course, and while she sailed I slept.

> During these days a feeling of awe crept over me. My memory worked with startling power. The ominous, the insignificant, the great, the small, the wonderful, the commonplace—all appeared before my mental vision in magical succession. Pages of my history were recalled which had been so long forgotten that they seemed to belong to a previous existence. I heard all the voices of the past laughing, crying, telling what I had heard them tell in many corners of the Earth. The loneliness of my state wore off when the gale was high and I found much work to do. When fine weather returned, then came the sense of solitude, which I could not shake off. . . .

The ocean is as vast, as quiet, as furious, as inspiring as it was in Slocum's day, but our understanding of it has grown immensely. I think he would have enjoyed the insights that nearly a century of progress in marine science has added to the things he saw, heard, and felt.

Source: Slocum, J. 1899. *Sailing Alone Around the World*. Reprint. New York: Sheridan House, 1972.

MARINE SCIENCE

Marine science (also called oceanography) is the process of discovering unifying principles in data obtained from the ocean, its associated life-forms, and the bordering lands. It draws on several disciplines, synthesizing the fields of geology, physics, biology, chemistry, and engineering as they apply to the ocean and its surroundings. *Marine geologists* focus on questions such as the composition of the inner Earth, the mobility of the crust, and the characteristics of seafloor sediments. Some of their work touches on areas of intense scientific and public concern, including earthquake prediction and the distribution of valuable resources. *Physical oceanographers* try to understand wave dynamics, currents, and ocean-atmosphere interaction. Their predictions of long-term climate trends are becoming increasingly important as pollutants change Earth's atmosphere. *Marine biologists* work with the nature and distribution of marine organisms, the impact of oceanic and atmospheric pollutants on the organisms, the isolation of disease-fighting drugs from marine species, and the yields of fisheries. *Chemical oceanographers* study the ocean's dissolved solids and gases and the relationships of these components to the geology and biology of the ocean as a whole. *Marine engineers* design and build oil platforms, ships, harbors, and other structures that enable us to use ocean resources wisely. Other specialists study weather forecasting, ways to increase the safety of navigation, methods to generate electricity, and much more. Virtually all marine scientists specialize in one area of research, but they also must be familiar with related specialties and appreciate the linkages among them. **Figure 2.1** shows marine scientists in action.

a

b

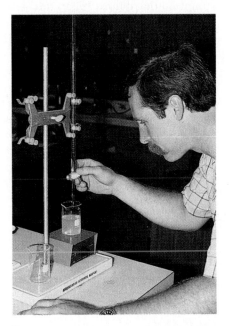

c

Figure 2.1 Marine scientists at work. (a) A marine biology student looks for organisms in bottom grab sample. The sampler is to his right. (b) An observer enters the hatch of *Alvin*, a deep submersible, to visit thermal vents 2,261 meters (7,416 feet) below the surface off the coast of Oregon. (c) A marine chemist tests a water sample to find the concentrations of dissolved elements.

Marine scientists today are asking critical questions about the origin of the ocean, the age of its basins, and the nature of the life-forms it has nurtured. We are fortunate to live at a time when scientific study may be able to answer some of those questions. **Science** is a systematic *process* of asking questions about the observable world and testing the answers to those questions. Scientists gather and study data, but data themselves aren't science. Science is the process of interpreting data by constructing a general explanation with which they are compatible.

Scientists begin with an informed guess called a working **hypothesis**, a speculation about the natural world that can be verified or disproven by observations and experiments. Hypotheses consistently supported by observation or experiment are advanced to the status of **theory**, a statement of relationship accepted by most scientists. The largest constructs, known as **laws**, are principles explaining events in nature that have been observed to occur with unvarying uniformity under the same conditions. Theories and laws in science do not arise fully formed, nor all at once. Scientific thought progresses as a continuing chain of questioning, testing, and matching theories to observations. A theory is strengthened if new facts support it. If not, the theory is modified or a new explanation is sought. The power of science lies in the ability of the process to operate *in reverse*; that is, in the use of a theory or law to make predictions and anticipate new facts to be observed.

The **scientific method** is the orderly process by which theories are verified or rejected. It is based on the assumption that nature "plays fair"—that the answers to our questions about nature are *ultimately knowable* as our powers of questioning and observing improve.

Nothing is ever proved absolutely true by the scientific method. Theories may change as our knowledge and powers of observation change; thus, all scientific understanding is tentative. The conclusions about the natural world that we reach by the process of science may not always be popular or immediately embraced, but if those conclusions consistently match observations, they may be considered true.

This book shows some of the results of the scientific process as it has been applied to the world ocean. It presents facts, interpretations of facts, examples, stories, and some of the crucial discoveries that have led to our present understanding of the ocean and the world on which it formed. As the results of science change, so will the ideas and interpretations presented in books like this one.

ORIGINS

We have always wondered about our origins, how the Earth was formed, and how the ocean arose. In the last 50 years, researchers using the scientific method have determined a tentative age for the ocean, Earth, and universe. They have developed hypotheses about how matter is assembled, how stars and planets are formed, and even how life may have arisen. Many of the details are still sketchy, of course, but these hypotheses have predicted some important recent discoveries. Perhaps the most dramatic discoveries in natural science this century have been those dealing with the origin and history of the universe.

The universe apparently had a beginning. The **big bang**, as that event is modestly named, probably occurred about 15 billion years ago. All of the mass and energy of the universe is thought to have been concentrated at a geometric point at the beginning of time, the moment when the expansion of the universe began. We don't know what initiated the expansion, but it continues today and will almost certainly continue for billions of years.

The very early universe was unimaginably hot, but as it expanded it cooled. About a million years after the big bang, temperatures fell enough to permit the formation of atoms from the energy and particles that had predominated up to that time. Most of these atoms were hydrogen, then as now the most abundant form of matter in the universe. About a billion years after the big bang, this matter began to congeal into the first galaxies and stars.

Galaxies and Stars

A **galaxy** is a huge rotating aggregation of stars, dust, gas, and other debris held together by gravity. There are perhaps 50 billion galaxies in the universe and 50 billion stars in each galaxy. Our galaxy is named the **Milky Way galaxy** (*galaktos* = milk). A galaxy very similar to our own is shown in **Figure 2.2**.

The **stars** that make up a galaxy are massive spheres of incandescent gases. They are usually intermingled with diffuse clouds of gas and debris. In spiral galaxies like the Milky Way, the stars are arrayed in spiral arms radiating from the galactic center. Our part of the Milky Way is populated with many stars, but distances within a galaxy are so huge that the star nearest the sun is about 42 trillion kilometers (26 trillion miles) away. Astronomers tell us there are as many galaxies in the universe as there are stars in our own galaxy, and more stars in the Milky Way than grains of sand on a beach!

Our sun is a typical star. The sun and its family of planets, called the **solar system**, are located about three-fourths of the way out from the galaxy's center in a spiral arm called the Orion arm. We orbit the galaxy's brilliant core, taking about 230 million years to make one orbit even though we are moving at about 280 kilometers per second (half a million miles an hour). The Earth has made about twenty circuits of the galaxy since the ocean formed.

Our planet and our sun probably had a common origin. The members of the solar system are thought to have condensed from a thin cloud—the **solar nebula**—that had been enriched with heavy elements created and released by the explosive deaths of nearby stars. By about 5 billion years ago, the solar nebula was a rotating disk-shaped mass of about 75% hydrogen, 23% helium, and 2% other material including heavier elements, gases, dust, and ice (see **Figure 2.3**). Like a spinning skater bringing in her arms, the nebula spun faster as it condensed. Material concentrated near its center became the **protosun**. Much of the outer material eventually became **planets**, the smaller bodies that orbit a star and do not shine by their own light.

The new planets formed in the disk of dust and debris surrounding the young sun through a process known as **accretion**—the clumping of small particles into large masses. Bigger clumps with stronger gravity pulled in most of the condensing matter. Near the protosun, where temperatures were highest, the first materials to solidify were substances with high boiling points, mainly metals and certain rocky minerals. The planet Mercury, closest to the sun, is mostly iron because iron is a solid at high temperatures. Somewhat farther out, in the cooler regions, magnesium, silicon, water, and oxygen condensed. Methane and ammonia accumulated in the frigid outer zones. The Earth's array of water, silicon–oxygen compounds, and metals results from its position within that accreting cloud. The planets of the outer solar system—Jupiter, Saturn, Uranus, and Neptune—are composed mostly of methane and ammonia ices because those gases can congeal only at cold temperatures.

The period of accretion lasted perhaps 50 to 70 million years. The protosun became a star when its internal temperature became high enough to fuse atoms of hydrogen into helium. The violence of these nuclear reactions sent radiation sweeping past the inner planets, clearing the area of excess particles and ending the period of rapid accretion. Gases like those we now see on the giant outer planets may once have surrounded the inner planets, but this rush of solar energy and particles stripped them away.

Earth and Ocean

The young Earth, formed by the accretion of cold particles, was probably homogeneous throughout. Then, late

Figure 2.2 The great galaxy in Andromeda, which is very similar in size and structure to our own Milky Way galaxy. Both are members of a loose aggregation of about 20 galaxies wonderfully called the "local group."

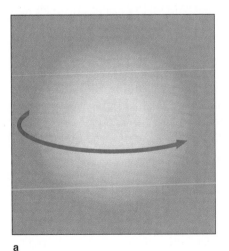
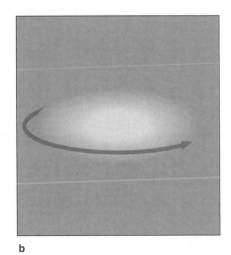
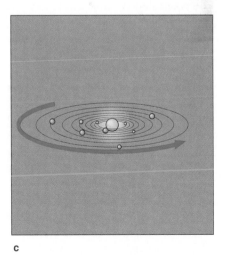

a   b   c

Figure 2.3 The formation of the solar system from the solar nebula. Because the nebula was rotating (a), it contracted into a disk (b). The protosun condensed at the center (c), and when the planets accreted, their orbits all lay in nearly the same plane.

Figure 2.4 Outgassing. Volcanic gases emitted by fissures add water vapor, carbon dioxide, nitrogen, and other gases to the atmosphere. Volcanism was a major factor in altering the Earth's original atmosphere; the action of photosynthetic plants was another.

in the accretion phase the Earth's surface was heated by the impact of falling meteors and debris. This heat, combined with heat from the decay of radioactive elements accumulating within the newly assembled planet, caused the whole planet to melt. Gravity pulled most of the heavy iron inward to form the planet's core. The sinking iron released huge amounts of gravitational energy, which, through friction, heated the Earth even more. At the same time, a slush of lighter minerals—silicon, magnesium, aluminum, and oxygen-bonded compounds—migrated toward the surface, forming the Earth's crust. This important process, called **density stratification**, lasted perhaps 100 million years.

Then the Earth began to cool. Its first hard surface is thought to have formed about 4.6 billion years ago.[1]

Radiation from the energetic young sun stripped away the planet's outermost layer of gases, its first atmosphere, but soon gases that had been trapped inside the forming planet burped to the surface to begin making the present atmosphere. This volcanic venting of volatile substances—including water vapor—is called **outgassing** (**Figure 2.4**). As the hot vapors rose, they condensed into clouds in the cool upper atmosphere. Earth's gravity kept this new atmosphere near the surface of the young planet.

The Earth's surface was so hot that no water could settle there, and no sunlight could penetrate the thick clouds. A visitor approaching from space 4.5 billion years ago would have seen a vapor-shrouded sphere blanketed by lightning-stroked clouds. After millions of years the upper clouds cooled enough for some of the outgassed water to form droplets. Hot rains fell toward the Earth, only to boil back into the clouds again. As the surface became cooler, water collected in basins and dissolved minerals from the rocks. Some of the water evaporated, cooled, and fell again. The world ocean was gradually accumulating.

These heavy rains may have lasted for about 10 million years. Large amounts of water vapor and other gases continued to escape through volcanic vents during that time and for millions of years thereafter. The ocean grew deeper. Evidence suggests that the Earth's crust grew thicker as well, perhaps in part from chemical reaction with oceanic compounds.

The physical expanse and distribution of the early ocean is a matter of some controversy. Most researchers hold that masses of rock have always protruded through the ocean surface to form continents. However, some recent studies suggest that water may have covered the Earth's entire surface for some 200 million years before the continents emerged. Although most of the ocean was in place about 4 billion years ago, ocean formation continues very slowly even today—about $\frac{1}{10}$ cubic kilometer ($\frac{1}{40}$ cubic mile) of new water is added to the ocean each year, mostly as steam from volcanic vents.

The composition of that early atmosphere—consisting of methane, ammonia, carbon dioxide, water vapor, and other gases—was much different from today's. Beginning about 3.5 billion years ago this mixture was gradually al-

---

[1]The age estimates presented in this chapter are derived from interlocking data obtained by many researchers using different sources. One source is meteorites, chunks of rock and metal formed at about the same time as the sun and planets and out of the same cloud. Many have fallen to Earth in recent times. Signs of radiation within these objects, combined with the rate of radioactive decay of unstable atoms in meteorites, in moon rocks, and in the oldest rocks on Earth, allow astronomers to make reasonably accurate estimates of how long ago these objects formed.

tered to the present composition, mostly nitrogen and oxygen, by chemical change and by the oxygen-producing action of photosynthesizing organisms, ancestors of today's green plants.

Eventually, sunlight pierced the clouds for the first time. The face of the Earth was exposed in daytime to the radiance of the young sun, and at night to the twinkle of ancient stars like those that formed its elements.

The Origin of Life

Life, at least as we know it, would be inconceivable without large quantities of water. Water has the ability to retain heat, moderate temperature, dissolve many chemicals, and suspend nutrients and wastes. These characteristics make it a mobile stage for the intricate biochemical reactions that allowed life to begin and prosper on Earth.

As early as 1929, biologist J. B. S. Haldane proposed that exposing a primitive atmosphere to ultraviolet radiation or lightning might produce some of the same carbon compounds found in living things. Building on this idea in a classic 1953 experiment, Stanley Miller mixed together water vapor, ammonia, methane, and hydrogen—gases that were thought to be present in the early atmosphere of Earth—and passed an electric spark through them for a week. In that short time, his apparatus produced a number of different amino acids and other organic compounds (**Figure 2.5**). The electric discharge had provided energy for the formation of these simple molecules. Since then other mixtures thought to reflect more accurately the early atmosphere of Earth have been tested, with similar results. When exposed to ultraviolet light, heat, and electrical spark, these mixtures produce simple sugars and most of the biologically important amino acids. They even produce small proteins and nucleotides (components of the molecules that transmit genetic information between generations).

Did *life* form? No, these compounds are only the building blocks of life, but these experiments do tell us something about the commonality and unity of life on Earth. The fact that these crucial compounds can so easily be synthesized *and are present in virtually all living forms* is probably not coincidental. Those compounds are "permitted" by physical laws and by the chemical composition of this planet. The experiments also underscore the special role of water in life processes. The fact that all life, from a jellyfish to a dusty desert weed, depends on saline water within its cells to dissolve and transport chemicals is certainly significant. It strongly suggests that simple self-replicating—living—molecules arose somewhere in the early ocean.

The early steps in the evolution of living organisms from simple organic building blocks—a process known as **biosynthesis**—are still speculative. One idea, popular-

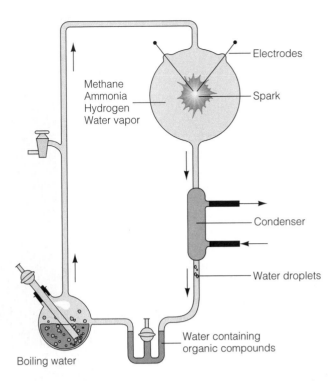

Figure 2.5 Stanley Miller's apparatus for producing organic molecules from a mixture of gases believed to be similar to the Earth's primitive atmosphere.

ized in the 1950s, suggests that life may have originated in shallow tidal pools at the ocean's edge (**Figure 2.6**). Evaporation of water from pools would have concentrated the amino acid and nucleotide building blocks into a rich organic "soup." Grains of sand or tiny bubbles would have provided handy surfaces on which larger chemical combinations could be assembled. Sunlight would have supplied the energy for these reactions. Accumulating in protected pools, reacting aggregates could have become progressively more complex, eventually evolving into biochemical systems capable of reproducing. A more recent analysis of ancient climate suggests the surface of the ocean could have been frozen, and that biosynthesis may have occurred at seafloor vents where warm seawater brought a rich broth of minerals from beneath the crust.

Wherever it happened, a similar biosynthesis cannot occur today. Living things have changed the conditions in the ocean and atmosphere, and those changes are not consistent with any new origin of life. For one thing, green plants have filled the atmosphere with oxygen, a compound that can disrupt any unprotected large molecule. For another, some of this oxygen (as ozone) now blocks much of the ultraviolet radiation from reaching the surface of the ocean. And finally, the many tiny organisms present today would gladly scavenge any large organic molecules as food.

Figure 2.6 Environments for biosynthesis? Scientists believe that life may have originated either in tidal pools (as shown here) or near hydrothermal vents.

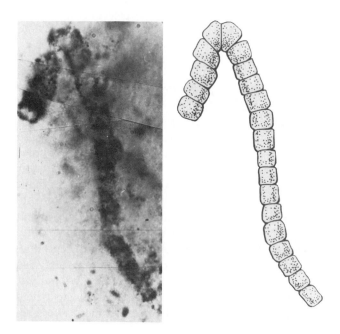

Figure 2.7 Fossil of a bacteria-like organism (with artist's reconstruction) that photosynthesized and released oxygen into the atmosphere. Among the oldest fossils ever discovered, this microscopic filament from northwestern Australia is about 3.5 billion years old.

How long ago might life have begun? The oldest fossils yet found, from northwestern Australia, are between 3.4 and 3.5 billion years old (**Figure 2.7**). They are remnants of fairly complex bacteria-like organisms, indicating that life must have originated even earlier, probably only a few hundred million years after a stable ocean formed. Life and the Earth have grown old together; each has greatly influenced the other.

AN OCEAN WORLD

Water planets are probably uncommon in the universe, but water itself is not scarce. In the solar system, for example, Jupiter has hundreds of times more water than Earth does. Astronomers have even located water molecules drifting in free space. But *liquid* water is unexpected.

Consider the conditions necessary for a large permanent ocean of liquid water to form on a planet. An ocean world must move in a nearly circular orbit around a stable star. The distance of the planet from the star must be just right to provide a temperature environment in which water is liquid. Unlike most stars, a water planet's sun must not be a double or multiple star, or the orbital year would have irregular periods of intense heat and cold. The materials that accreted to form the planet must have included both water and substances capable of forming a solid crust. The planet must be large enough that its gravity will keep the atmosphere and ocean from drifting off into space.

Special conditions were also necessary for the formation of life here. Earth's gravity is strong enough to retain an ocean, but not strong enough to crush the life forms that came from it. The planet has a magnetic field provided by an iron core to deflect radiation that would otherwise harm the genetic instructions of living things.

A single moon provides gentle tides to encourage life forms to leave the ocean and reside on land. The atmosphere is relatively clear—so that sunlight penetrates to the surface—but moist enough to form rains and winds that drive air and ocean currents. Furthermore, the upper air contains ozone, which protects against the most harmful ultraviolet rays. This combination is probably *exceedingly rare* in the galaxy. We should enjoy and protect the water planet we call home.

CHAPTER SUMMARY

We have learned much about our planet using the scientific method, a systematic way of asking and answering questions about the natural world. Marine science applies the scientific method to the ocean, the Earth of which it is a part, and the living organisms dependent on it.

Most of the atoms that make up the Earth and its inhabitants were formed within stars. Stars form in the dusty spiral arms of galaxies and spend their lives changing hydrogen and helium to heavier elements. As they die, some stars eject these elements into space by cataclysmic explosions. The sun and the planets, including Earth, probably condensed from a cloud of dust and gas enriched by the recycled remnants of exploded stars.

The ocean is not a remnant of that cloud, however. Most of the ocean formed later, as water vapor trapped in the Earth's outer layers escaped to the surface through volcanic activity during the planet's youth. Life originated in the ocean soon after its formation; life and the Earth have grown old together.

Terms and Concepts to Remember

accretion	laws	scientific method
big bang	marine science	solar nebula
biosynthesis	Milky Way galaxy	solar system
density	outgassing	stars
stratification	planets	theory
galaxy	protosun	
hypothesis	science	

Study Questions

1. Can the scientific method be applied to speculations about the natural world that are not subject to test or observation?

2. What are the major specialties within marine science?

3. What is biosynthesis? Where and when do researchers think it might have occurred on our planet? Could it happen again this afternoon?

4. Would you expect ocean worlds to be relatively abundant in the galaxy? Why or why not?

5. How old is Earth? On what is that estimate based?

6. What is density stratification? Does the Earth show evidence of density stratification? If so, when did the stratification occur?

7. Is the ocean a relatively new feature of the Earth, or has it been around for most of Earth's history? Where did its water come from?

For Further Study

The Nature of Science

McCain, G., and E. M. Segal. 1988. *The Game of Science*. 5th ed. Pacific Grove, CA: Brooks/Cole. A lighthearted summary that conveys the underlying seriousness of its content.

Morrison, P., and P. Morrison. 1987. *The Ring of Truth—An Inquiry into How We Know What We Know*. New York: Random House. An accomplished physicist discusses how scientists search for answers.

The Origin of the Earth and the Ocean

Cloud, Preston. 1988. *Oasis in Space—Earth's History from the Beginning*. New York: Norton.

Sagan, C. 1980. *Cosmos*. New York: Random House. The most popular scientific book ever published, and with good reason. A magnificently written and illustrated description of the cosmos. A must for anyone wishing to understand the Earth's place in the universe.

Trefil, J. 1985. *Space, Time, Infinity*. Washington, DC: Smithsonian Books. Great illustrations, clear and well-written text.

The Origin of Life

Bjerklie, D., B. Hillenbrand, and J. O. Jackson. 1993. "How Did Life Begin?" *Time*, 11 October, 68–74. Excellent summary article on the most recent speculations.

Horgan, J. 1991. "In the Beginning . . ." *Scientific American*, February, 116–25. Thoughts on the origin of life by a number of researchers.

3 : EARTH STRUCTURE AND PLATE TECTONICS

Breakthroughs

Scientific investigators sometimes speak of a *breakthrough*, the moment when the answer to a complex problem presents itself. In October 1987 the scientists and crew of the Scripps Institution research vessel *Melville* were present for a breakthrough of a different sort—the eruption of an undersea volcano directly beneath their ship! *Melville* was on an expedition to collect rock and water samples from the MacDonald Seamount, a submerged volcano in French Polynesia, located 1,100 kilometers (700 miles) west of Pitcairn Island. When the research team arrived on station, they noticed large patches of greenish-brown water containing fine particles of volcanic ash, which suggested recent volcanic activity. The crew was lowering their gear to the top of the seamount, which rises to within 40 meters (130 feet) of the surface, when huge bubbles of gas and steam suddenly engulfed the ship, making "horrendous clangs and clamors" as they burst against *Melville*'s hull. Chocolate-colored water containing steaming lava balls too hot to hold in bare hands streamed to the surface. The ship's depth recorder and hull-mounted water temperature sensor were damaged, but no one was injured. The eruption lasted about 5 minutes.

The MacDonald Seamount is the last—and youngest—in a chain of submerged and emergent volcanoes stretching 2,000 kilometers (1,200 miles) across the floor of the South Pacific. Here magma, molten rock, rises from deep inside the Earth. The

composition of gases within the young rock would tell *Melville* researchers about the forces and conditions that cause island chains to form, and something about the history of the Pacific floor itself. Samples they obtained that morning helped to confirm recent theories about oceanic *hot spots*, themselves a verification of the theory of plate tectonics. Seldom do researchers have the good fortune to be in exactly the right place at exactly the right time; scientific breakthroughs are generally less threatening to life and property!

Birth of an island.

Powerful forces inside our planet are continually form-ing the major features of its surface. These forces deter-mine the outlines and locations of the continents and ocean floors. They build and destroy mountains, raise islands, power volcanoes, form deep trenches, and (through earthquakes) influence the lives of millions of people. Few discoveries in marine science have been as exciting to geologists as the recent breakthroughs in our understanding of how these internal forces work. They have pieced together a view of Earth's interior—based on measurements of heat leaking from its depths, on the chemical composition of volcanic gases, on the study of shocks from distant earthquakes, on local variations in the pull of gravity, and even on the analysis of meteorites thought to have coalesced from the same material as the Earth. They have discovered that the Earth is *layered*; it looks a bit like the inside of an onion (see **Figure 3.1**).

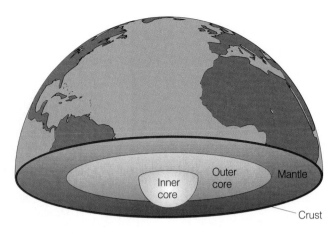

Figure 3.1 Cross section of the Earth, showing the major internal layers.

THE LAYERS OF THE EARTH

Each layer has different chemical and physical character-istics. The uppermost layer is the lightweight, brittle, aptly named **crust**. The crust beneath the ocean differs in thickness, composition, and age from the crust of the continents. The thin **oceanic crust** is primarily **basalt**, a heavy, dark-colored rock composed mostly of oxygen, silicon, magnesium, and iron. By contrast, the most com-mon material in the thicker **continental crust** is **granite**, a familiar speckled rock composed mainly of oxygen, sil-icon, and aluminum. The deepest crust extends only 35 kilometers (22 miles) below the Earth's surface. The **mantle**, the layer beneath the crust, is thought to consist mainly of oxygen, magnesium, and silicon. Most of the Earth is mantle; it accounts for 68% of the Earth's mass and 83% of its volume. The outer and inner **cores**, which consist mainly of iron, lie beneath the mantle at the Earth's center.

No samples have ever been taken from the layers be-neath the crust.[1] Even so, researchers have much indirect evidence about the chemical composition, temperature, and thickness of each layer. They have gained an under-standing of the processes that form the features of Earth's surface by studying the characteristics of the various lay-ers. Different conditions of temperature and pressure prevail at different depths, and these conditions influence the physical properties of the materials subjected to them. The behavior of a rock is determined by three factors: temperature, pressure, and the rate at which a deforming force (stress) is applied. Depending on these three fac-tors, a rock may behave in a brittle manner (by breaking), it may deform plastically (flow without fracturing), or it may deform elastically (bend or shrink but return to its original shape once the stress is released).[2]

Geologists have devised another classification of the Earth's interior based on physical rather than chemical properties:

The **lithosphere** (*lithos* = rock)—the Earth's cool, rigid outer layer—may be up to 100 kilometers (60 miles) thick. *It is comprised of the brittle continental and oceanic crusts and the uppermost cool and rigid portion of the mantle.*

The **asthenosphere** (*asthenos* = soft) is *the thin, hot, slowly flowing layer of upper mantle below the lithosphere.* Ex-tending to a depth of about 700 kilometers (430 miles), the asthenosphere is characterized by its ability to de-form plastically under stress. Its thick fluidity has been compared to cold taffy.

The **mesosphere** (*mesos* = middle) is *the rigid middle and lower mantle extending to the core.* Though it is hotter than the asthenosphere, the greater pressure at this depth probably prevents it from flowing. The mesosphere and asthenosphere have similar chemical compositions but very different physical properties.

The core is divided into two parts: The outer core is a viscous liquid with a density about four times that of the crust; the inner core is a solid with a maximum density about six times that of crustal material. Both parts are extremely hot, with an average temperature of about 5,500°C (9,900°F). Recent evidence indicates that the core may be as hot as 6,600°C (12,000°F) at its center, hotter than the surface of the sun!

---

[1]The deepest hole drilled so far is being bored by researchers in Siberia. They have reached a depth of 12.06 kilometers (7.54 miles), where temperatures are 245°C (475°F) and pressure often causes the hole casing to collapse.

[2]This may help: A candy cane is brittle, Silly Putty is plastic, and a rubber ball is elastic.

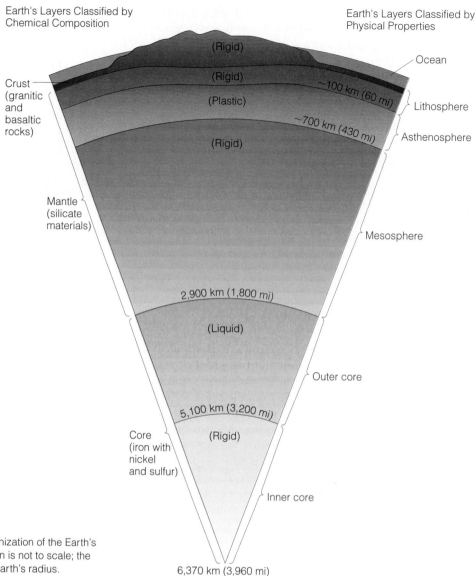

Earth's Layers Classified by Chemical Composition

Earth's Layers Classified by Physical Properties

(Rigid)

(Rigid)

(Plastic)

~100 km (60 mi)

~700 km (430 mi)

(Rigid)

Ocean

Lithosphere

Asthenosphere

Crust (granitic and basaltic rocks)

Mantle (silicate materials)

Mesosphere

2,900 km (1,800 mi)

(Liquid)

Outer core

5,100 km (3,200 mi)

Core (iron with nickel and sulfur)

(Rigid)

Inner core

6,370 km (3,960 mi)

Figure 3.2 The chemical and physical organization of the Earth's layers compared. Note that this representation is not to scale; the lithosphere accounts for only about 1.5% of Earth's radius.

Figure 3.2 compares the chemical and physical classifications of the layers, and **Figure 3.3** shows the lithosphere and asthenosphere in more detail. Note in Figure 3.3 that the rigid sandwich of crust and upper mantle—the lithosphere—floats on (and is supported by) the denser plastic asthenosphere. As we shall see in a moment, recent research has shown that slabs of the Earth's relatively cool and solid surface—its lithosphere—float and move independently of one another over the hotter, partially molten asthenosphere layer directly below.

Internal Heat

The interior of the Earth is hot. The main source of that heat is **radioactive decay**, a process that generates heat when unstable forms of elements are transformed into new elements. As noted in Chapter 2, radioactive decay within the newly formed Earth released heat that contributed to the melting of the original mass. Most of the melted iron sank toward the core, releasing huge amounts of energy. By now almost all of the heat generated by the formation of the core has dissipated, but radioactive elements within the Earth still continue to decay and produce new heat. Today most of the radioactive heating takes place in the crust and upper mantle, rather than in the deeper layers. Some of this heat journeys toward the surface by **conduction**, a process analogous to the slow migration of heat along a skillet's handle. Some heat also rises by **convection** in the asthenosphere. Convection occurs when a fluid is heated, expands and becomes less dense, and rises. (Convection causes air to rise over a warm radiator.)

Thus, even after 4.6 billion years, heat continues to flow out from within the Earth. This heat powers the

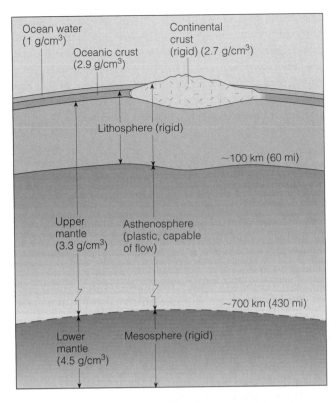

Figure 3.3 The layering of the lithosphere and asthenosphere in more detail. Densities are shown in grams per cubic centimeter (g/cm³).

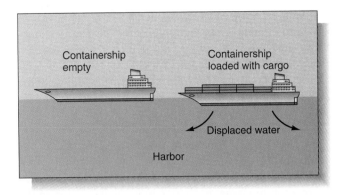

Figure 3.4 The principle of buoyancy. A ship sinks until it displaces a volume of water equal in weight to the weight of the ship and its cargo.

construction of mountains and volcanoes, causes earthquakes, moves continents, and shapes ocean basins.

Isostatic Equilibrium

As you can see in Figure 3.3, the structure of oceanic lithosphere differs from that of continental lithosphere. Below the ocean, the crust averages only about 7 kilometers (4 miles) thick, but beneath the continents the crust thickness averages 35 kilometers (22 miles) and increases to 70 kilometers (44 miles) at the highest mountain ranges. The granitic continental crust is light enough to project above sea level, but the heavy basaltic oceanic crust is almost always submerged.

Why do large regions of continental crust stand high above sea level? If the asthenosphere is nonrigid and deformable, why don't mountains sink because of their weight and disappear? Another look at Figure 3.3 will help to explain the situation. The mountainous parts of continents have "roots" extending into the asthenosphere. The continental crust and the rest of the lithosphere "float" on the denser asthenosphere. The situation is analogous to buoyancy, the principle that explains why ships float.

Buoyancy is the ability of an object to float in a fluid by displacing a volume of that fluid equal in mass to the floating object's own mass. *A steel ship floats because its shape displaces a volume of water equal in weight to its own weight plus the weight of its cargo.* Thus, an empty containership displaces a smaller volume of water than the same ship when fully loaded (**Figure 3.4**). The water supporting the ship is not "strong" in the mechanical sense; water does not support a ship in the same way a strong steel bridge supports the weight of your car. Buoyancy, rather than mechanical strength, supports the ship and her cargo.

Any region of a continent that projects above sea level is supported in the same way. As an extreme example, consider the continent containing Mount Everest, highest of Earth's mountains at 8.84 kilometers (29,007 feet) above sea level. Mount Everest and its neighboring peaks are not supported by the *mechanical* strength of the materials within the Earth; nothing on (or in) our world is that strong. *The mountainous upper surface of the continent floats high above sea level because the lithosphere of which it is a part sinks into the plastic asthenosphere until it has displaced a volume of asthenosphere equal in mass to its own mass.* The continent's mountains rest at great height, in balance with their subterranean underpinnings but susceptible to rising or falling as erosion or crustal stresses dictate. Lower regions are supported by shallower roots. Like a ship floating in water, the entire continent stands in **isostatic equilibrium** (*isos* = equal + *stasis* = standing).

Antarctica provides an excellent example of how continents maintain isostatic equilibrium in the face of changing conditions. Most of the southern continent is covered by a layer of ice as much as 4 kilometers (13,000 feet) thick. The great weight of the ice has pushed much of the continental crust below sea level (**Figure 3.5**). If this ice were to melt, Antarctica would slowly rise from the water in much the same way a ship rises while being unloaded.

Unlike the asthenosphere on which lithosphere floats, crustal rock does not flow at normal surface temperatures. A ship reacts to any small change in weight with a correspondingly small change in vertical position in the water, but an area of continent or ocean floor cannot react to every small weight change because its edges are me-

chanically bound to adjacent crustal masses. When the force of uplift or downbending exceeds the mechanical strength of the adjacent rock, the rock will fracture along a plane of weakness—a **fault**. The adjacent crustal fragments will move vertically in relation to each other. This sudden adjustment of the crust to isostatic forces by fracturing, or faulting, is one cause of earthquakes.

EARTHQUAKES AND EARTHQUAKE WAVES

On the last Friday of March, 1964, at 1736 (5:36 P.M.), one of the largest earthquakes ever recorded struck 144 kilometers (90 miles) east of Anchorage, Alaska (**Figure 3.6**). The release of energy ruptured the surface of the Earth for 800 kilometers (500 miles) between the small port of Cordova in the east and Kodiak Island in the west. In some places the vertical movement of the crust was 3.7 meters (12 feet); one small island was lifted 38 feet. Horizontal movement caused the greatest damage: 65,000 square kilometers (25,000 square miles) of land abruptly moved west. In 4½ minutes of violent shaking, Anchorage had moved sideways 2 meters (6.6 feet), and the town of Seward had moved 14 meters (46 feet)! A seismic (earthquake-generated) sea wave destroyed two harbors. Over 75% of the state's commerce was disrupted, and thousands were made homeless. Damage exceeded $750 million, and, considering the violence of the earthquake, it's a wonder that only 115 lives were lost.

Seismic waves—huge low-frequency pulses of energy generated by the forces that caused the earthquake—

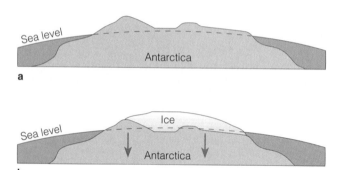

Figure 3.5 The principle of isostatic equilibrium applied to Antarctica. Without an ice cover (a), the continent would rise higher above sea level in a situation analogous to the empty containership in Figure 3.4. (b) The great weight of ice blanketing the continent forces its roots deeper into the asthenosphere, so that more of its mass is below sea level. The ice is analogous to the loaded cargo in Figure 3.4.

Figure 3.6 Alaska earthquake damage, 27 March 1964. The collapse of Fourth Avenue in Anchorage. The earthquake caused the ground to liquefy and slide, undermining buildings and causing the pavement to sink more than 3 meters (10 feet) in places.

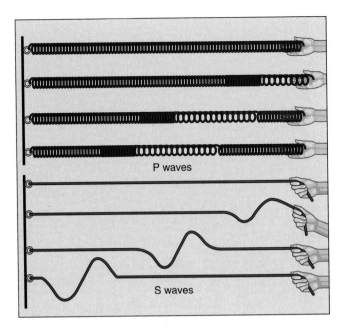

Figure 3.7 P waves (primary waves) are compressional waves like those seen in a Slinky that is alternately stretched and compressed. S waves (secondary waves) are side-to-side waves like those seen in a shaken rope. Both kinds of waves are associated with earthquakes.

spread rapidly in all directions. Geological research stations over much of the world saw the extraordinarily large waves arrive, and many of the 800 seismographs on-line worldwide were physically damaged. The shaken citizens of Anchorage could not have known at the time, but "their" earthquake helped to solidify our present model of the interior of the Earth.

Earthquake waves can reveal information about the chemical and physical nature of Earth's interior. We use the same kind of analysis to select a ripe watermelon. If we tap the outside and hear a "tick," we suspect the melon isn't ripe. A "thunk," on the other hand, indicates a winner.

Two kinds of seismic waves are particularly useful for this kind of analysis of the Earth's interior structure. One kind of wave, the **P wave** (or primary wave), is a compressional wave similar in behavior to a sound wave. Rapidly pushing and pulling a very flexible spring (like a Slinky) generates P waves. The **S wave** (or secondary wave) is a transverse wave like that seen in a rope shaken side to side. Both kinds of seismic waves are shown in **Figure 3.7**. P waves and S waves are generated simultaneously at the source of an earthquake. P waves travel through the Earth nearly twice as fast as S waves; so they arrive first at the **seismograph** (*seismos* = earthquake + *graph* = to write), an instrument that senses and records earthquakes. Liquids are unable to transmit the side-to-side force of S waves but do propagate compressional P waves. Solid rock

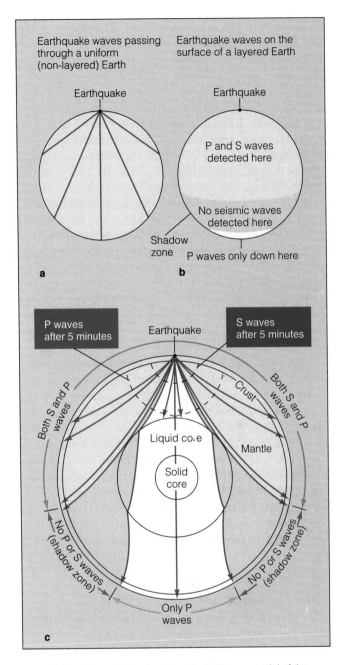

Figure 3.8 How earthquakes contributed to our model of the layered Earth. (a) If the Earth were homogeneous throughout, seismic waves would travel in straight-line paths at constant speed. (b) Actually, the Earth has a dense core, producing a shadow zone in which no seismic waves are detected. (c) The patterns of reflected and refracted waves helped geophysicists deduce the structure of the Earth's layers.

transmits both kinds of waves. Analysis of the characteristics of seismic waves returning to the Earth's surface after passage through the interior suggests which parts of the interior are solid, liquid, or plastic (**Figure 3.8**).

You might think that seismic waves would travel at constant speeds from an earthquake and that their paths through the interior would be straight lines (as in **Figure 3.8a**). The discovery in 1905 that earthquake waves were bent by their passage through the Earth, and that their speeds were different from those anticipated, gave the first firm indication of the varying chemical and physical properties of the Earth's interior layers. In 1906, English geologist Richard Oldham found that P waves arrived at a seismograph farthest away from an earthquake (that is, on the opposite side of the globe) much more slowly than expected. He also found that no S waves survived deep passage through the Earth. Oldham deduced that a dense fluid structure, or core, must exist within the Earth to absorb the S waves and slow down the P waves. He further predicted that a **shadow zone (Figure 3.8b)**, a wide band from which seismic waves were nearly absent, would encircle the side of the Earth opposite the location of the earthquake. The shadow zone would be formed by refraction (bending) of the P waves by the outer liquid core, as shown in **Figure 3.8c**. The existence of the shadow zones and liquid core were verified by seismographic analysis in 1914.

More sensitive seismographs were developed in the 1930s. In 1935 Dutch seismologist Inge Lehmann suggested that the very faint, very low-frequency P waves discovered opposite the earthquake site had speeded up as they passed through an inner core, indicating that it was a solid (see again Figure 3.8c). Measurements of subtle differences in the pull of gravity, plus a more accurate estimate of the Earth's mass (derived from precise timings of the orbits of artificial satellites) gave further clues to the layered structure of the outer Earth. By the early 1960s another new generation of sensitive seismographs stood ready to provide geologists with an even better understanding of the Earth's inner workings.

TOWARD A NEW UNDERSTANDING OF THE EARTH

These insights were certainly welcome! In the early 1960s geologists were deeply divided by their theories of interior characteristics. The biggest problem was explaining the jigsaw-puzzle fit of some of the continents. People had been trying to explain this coincidence for a very long time.

A Puzzling Fit

As Leonardo da Vinci noticed on early charts, in some regions the continents looked as if they would fit together like jigsaw-puzzle pieces if the intervening ocean were removed. In 1620 Francis Bacon also wrote of a "certain correspondence" between shorelines on either side of the South Atlantic. In 1885 Edward Suess, a respected German scientist, suggested that the Southern Hemisphere's continents might once have been a single large land mass. He based his belief in part on the similarities of fossils found on these continents, especially fossils of the fern *Glossopteris*. Suess was not taken seriously by his colleagues because he could not explain how the continents had moved. Still, the correspondence of South America and Africa had a certain graphic appeal (**Figure 3.9**).

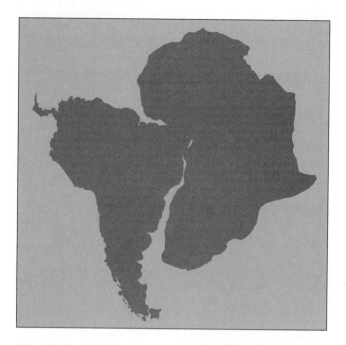

Figure 3.9 Soon after charts began including the New World, some scientists noticed that the coastlines of South America and Africa seemed to fit together in jigsaw-puzzle fashion.

As they probed the submerged edges of the continents, geologists found that the ocean bottom nearly always sloped gradually out to sea for some distance and then dropped steeply to the deep-ocean floor. They realized that these shelflike continental edges were extensions of the continents themselves. In the few locations where they had measurements, researchers found that the fit between South America and Africa, impressive at the shoreline, was even better along the submerged edges of the continents.

Though so accurate a fit almost certainly could not have occurred by chance, no one had yet proposed a mechanism that could separate whole continents into moving pieces. If such a mechanism did exist, its gradual operation would surely require a great deal of time.

Continental Drift

Into the fray stepped **Alfred Wegener**, a busy German meteorologist, polar explorer, astronomer, and geologist.

In a lecture in 1912 he proposed a startling and original theory, **continental drift**. Wegener suggested that all the Earth's land had once been joined into a single supercontinent surrounded by an ocean. He called the land mass **Pangaea** (*Pan* = all + *gaea* = Earth) and the surrounding ocean **Panthalassa** (*pan* = all + *thalassa* = ocean). Wegener thought Pangaea had broken into pieces about 200 million years ago. Since then, the pieces had moved to their present positions and were still moving.

Wegener's evidence included the apparent shoreline fit of continents across the North and South Atlantic and new information on offshore contours obtained by contemporary oceanographic expeditions. He pointed to Suess's *Glossopteris* fossils; to areas of erosion apparently caused by the same glacier in tropical areas now widely separated (South Africa, India, and Australia); and Ernest Shackleton's 1908 discovery of coal, the fossilized remains of tropical plants, in frigid Antarctica. Wegener even suggested that volcanic activity was powered by the friction of continental movement.

Unlike anyone before him, Wegener also proposed a mechanism to account for the hypothetical drift. He believed that the heavy continents were slung toward the equator on the spinning Earth by a centrifugal effect. This force, coupled with the tidal drag on the continents from the combined effects of sun and moon, would account for the phenomenon of drifting continents, he thought.

Wegener was dismissed as a crank. His detractors claimed, with some justification, that he had carefully selected only those data supporting his hypothesis, ignoring contrary evidence. Where, for instance, were the "wakes" or "tracks" through old seabed that the migrating continents would leave? But a few geologists sided with Wegener. These "drifters" were hesitant to embrace the centrifugal force theory, yet they were unable to propose an alternate power source that could move the massive granitic continents.

The greatest block to the acceptance of continental drift lay in geology's view of the Earth's mantle. The available evidence seemed to suggest that a deep, solid mantle supported the crust and mountains mechanically (not isostatically) from below. Drift would be impossible with this kind of subterranean construction. But then a few perceptive seismic researchers noticed that the upper mantle reacted to earthquake waves as if it were a plastic mass, not a rigid solid. Perhaps such a layer would resemble a slug of iron heated in a blacksmith's forge; it would deform with pressure and even flow slowly. But established geologists dismissed this interpretation, saying that the mountains would simply fall over or sink without rigid underpinnings. By 1926 the "drifters" were in full retreat. When Wegener died on an expedition across Greenland in 1930, his theory was already in eclipse.

The Idea Transformed

The concept of continental drift refused to die, however; those neatly fitted continents provided a haunting reminder of Wegener to anyone looking at an Atlantic chart. In 1935 a Japanese scientist, Kiyoo Wadati, speculated that earthquakes and volcanoes near Japan might be associated with continental drift. In 1940, seismologist Hugo Benioff plotted the locations of deep earthquakes at the edges of the Pacific. His charts revealed the true extent of the **Pacific Ring of Fire**, a circle of violent geologic activity surrounding much of the Pacific Ocean. Seismographs were now beginning to reveal a worldwide pattern of earthquakes and volcanoes. Deep earthquakes did not occur randomly over the Earth's surface but were concentrated in zones that extended in lines along the Earth's crust. Benioff, Wadati, and others wondered what could cause such an orderly pattern of deep earthquakes. Many of the lines corresponded with a worldwide system of oceanic ridges, the first of which was plotted in 1928 by *Meteor* oceanographers working in the middle of the North Atlantic. **Figure 3.10** is a plot of about 30,000 earthquakes. Notice the odd pattern they form—almost as if the Earth's lithosphere is divided into sections! Benioff's sensitive seismographs also began to gather strong evidence for a partially molten, nonrigid layer in the upper mantle. Could the continents somehow be sliding on that?

Other seemingly unrelated bits of information were accumulating. **Radiometric dating** of sediments and rocks was perfected after World War II. This technique is based on the discovery that unstable, naturally radioactive elements lose particles from their nuclei and change into new, stable elements. The radioactive decay occurs at a constant rate, and measuring the ratio of radioactive to stable atoms in a sample provides its age. To the surprise of many geologists, the maximum age of the ocean floor and its overlying sediments was radiometrically dated to less than 200 million years, only about 4% of the age of Earth.[3] The centers of the continents are much older: Some parts of the continental crust are more than 3.9 billion years old, about 85% of the age of Earth. *Why was oceanic crust so young?*

Attention quickly turned to the deep-ocean floors, the complex profiles of which were now being revealed by **echo sounders**, devices that measure depth by bouncing high-frequency sound waves off the bottom (see Figure 1.21). In particular, scientists aboard the Lamont Geological Observatory deep-sea research vessel *Vema* (a converted three-masted schooner) invented deep survey techniques as they went. They probed the bottom with

[3]For a discussion of geological time, please see Appendix II.

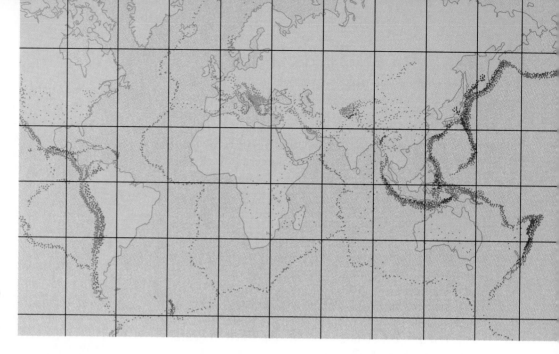

Figure 3.10 The global distribution of seismic events from January 1977 through December 1986. The locations of earthquakes are colored red, green, and blue to represent event depths of 0 to 70 kilometers, 70 to 300 kilometers, and below 300 kilometers, respectively.

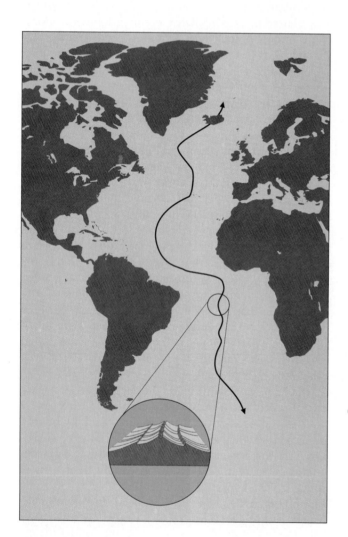

Figure 3.11 The Mid-Atlantic Ridge, showing its conformance to the coastlines of the continents.

powerful echo sounders and looked beneath sediments with reflected pressure waves generated by surplus Navy depth charges dropped gingerly overboard. The overall shape of the Mid-Atlantic Ridge was slowly revealed. The ridge's conformance to shorelines on either side of the Atlantic (**Figure 3.11**) raised many eyebrows. Ocean floor sediments were shown to be thickest at the edge of the Atlantic and thinnest near this mid-ocean ridge.

Vema and other research ships also compiled more complete, accurate charts of the submerged edges of continents. At Cambridge University, Sir Edward Bullard used a computer to process these data to achieve the best possible fit of the continental jigsaw-puzzle pieces around the Atlantic. The fit was astonishingly good (**Figure 3.12**).

Mantle studies were keeping pace. The first links in the Worldwide Standardized Seismograph Network, begun during the International Geophysical Year in 1957, were beginning to report data from seismic waves reflected and refracted through the planet's inner layers. This information verified the existence of a layer in the upper mantle that caused a decrease in the velocity of seismic waves. This finding strongly suggested that the layer was not solid. Perhaps the lithosphere was isostatically balanced in this plastic layer, and perhaps continents could move around in it *if* a suitable power source existed.

The Breakthrough: From Seafloor Spreading to Plate Tectonics

In 1960 Professor Harry Hess of Princeton University proposed a radical idea to explain the features of the ocean floor and the "fit" of the continents. He suggested

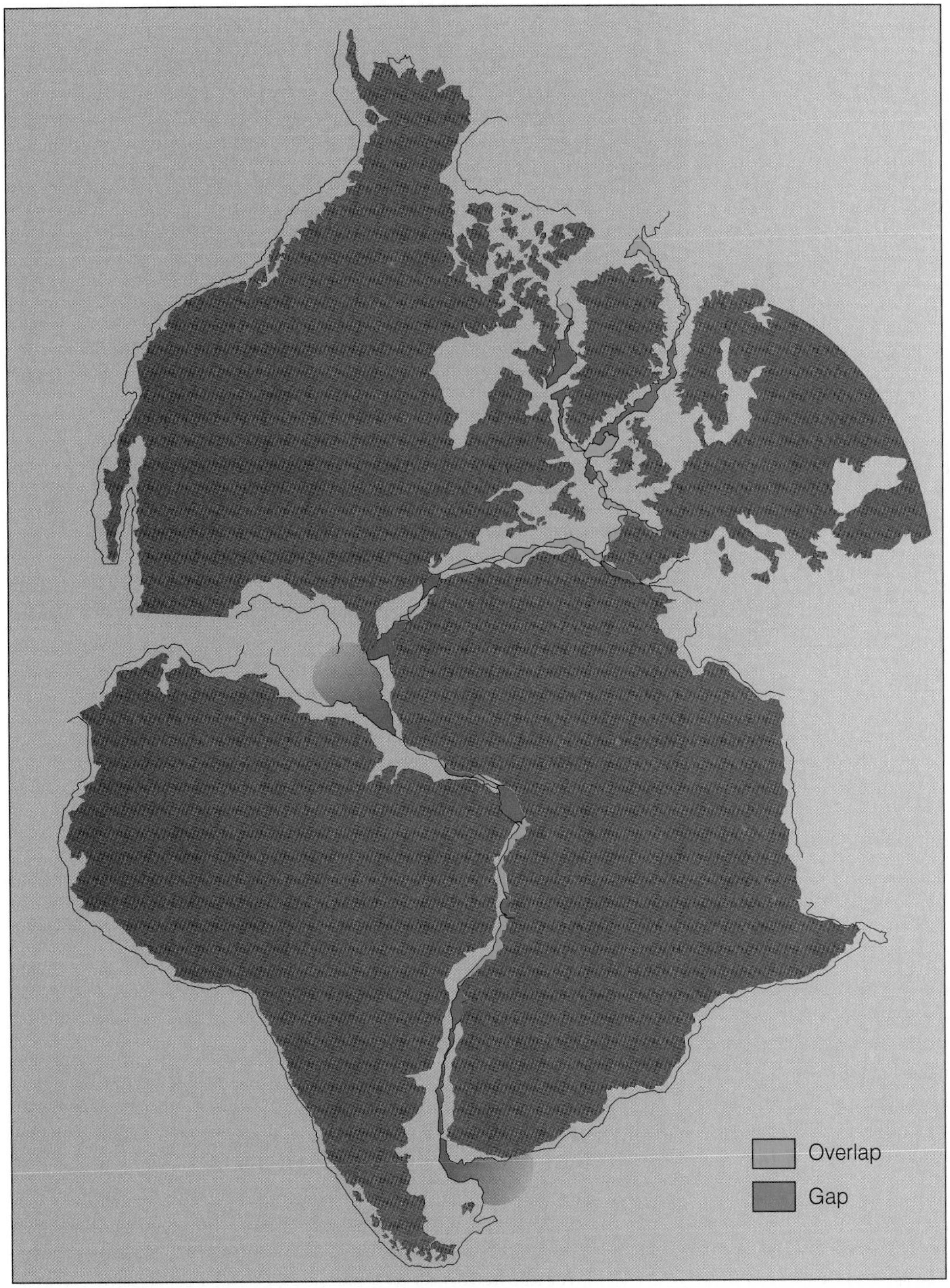

Figure 3.12 The fit of all the continents around the Atlantic, as calculated by Sir Edward Bullard.

Overlap

Gap

that new seafloor forms at the Mid-Atlantic Ridge (and the other newly discovered ocean ridges) and spreads outward from this line of origin. Continents would be pushed aside by the same forces that cause the ocean to grow. This motion could be powered by **convection currents**, slow-flowing circuits of material within the asthenosphere.

Seafloor spreading, as the new theory was called, pulled many loose ends together. If the mid-ocean ridges were **spreading centers** and sources of new ocean floor rising from the asthenosphere, they should be hot. They were. If the new oceanic crust cooled as it moved from the spreading center, it should shrink in volume and become more dense, and the ocean should be deeper farther from the spreading center. It was. Sediments at the edges of the ocean basin should be thicker than those

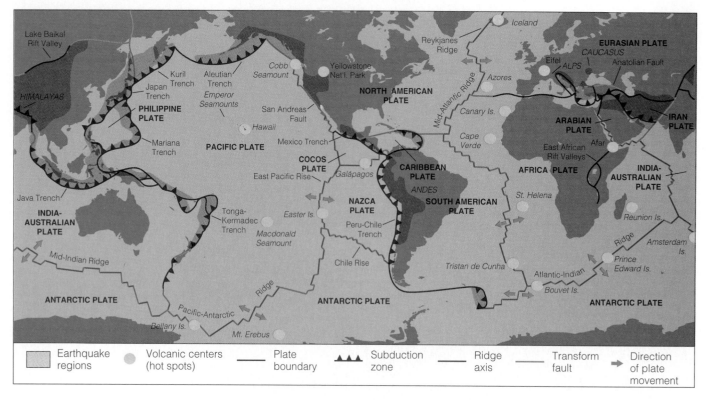

Figure 3.13 The major lithospheric plates, showing their directions of relative movement and the location of the principal hot spots. Note the correspondence of plate boundaries and earthquake locations.

near the spreading centers. They were, and they were also older.

But did this mean that the Earth was continuously expanding? Since there was no evidence for a growing Earth, the creation of new crust at spreading centers would have to be balanced by the destruction of crust somewhere else. Then researchers discovered that the crust plunges down into the mantle along the periphery of the Pacific. The process is known as **subduction**, and these areas are called **subduction zones** (or Wadati–Benioff zones in honor of their discoverers). The zones of concentrated earthquakes (see again Figure 3.10) were found in regions of crustal formation (spreading centers) and crustal destruction (subduction zones).

In 1965 the ideas of continental drift and seafloor spreading were integrated into the overriding concept of **plate tectonics** (*tekton* = builder), primarily by the work of **J. Tuzo Wilson**, a geophysicist at the University of Toronto. In this theory Earth's outer layer consists of about a dozen separate lithospheric **plates**, each about 70 to 100 kilometers (43 to 65 miles) thick, floating on the asthenosphere. When heated from below, the fluid asthenosphere expands, becomes less dense, and rises. It turns aside when it reaches the lithosphere, however, and it drags the plates laterally until it turns under again to complete the circuit. The large plates include both continental and oceanic crust. The plates, which jostle about

like huge flats of ice on a warming lake, are shown and named in **Figure 3.13**. Plate movement is slow in human terms, averaging about 5 centimeters (2 inches) a year. The plates interact at converging, diverging, or slipping junctions, sometimes forcing one another below the surface or wrinkling into mountains. Most of the million or so earthquakes and volcanic events each year occur along plate boundaries.

Through the great expanse of geologic time, this slow movement remakes the surface of the Earth, expands and splits continents, and forms and destroys ocean basins. The light, ancient granitic continents ride high in the lithospheric plates, rafting on the moving asthenosphere below. Plate movement may be caused by friction against the plate from convection currents flowing in the asthenosphere. In subduction, heavy basaltic ocean floor (and its overlying layer of sediment) plunges into the mantle at a subduction zone to be partially remelted. The subducting plate may be very slightly more dense than the upper asthenosphere on which it rides, and so it is pulled downward into the mantle by gravity. Melted material from the subducted plate may separate into lighter components and rise, reemerging through a volcano. Because the ocean floor itself acts as a vast conveyor belt, transporting accumulated sediment to subduction zones where the seafloor sinks into the asthenosphere, no marine sediments are of great age. This process has progressed since

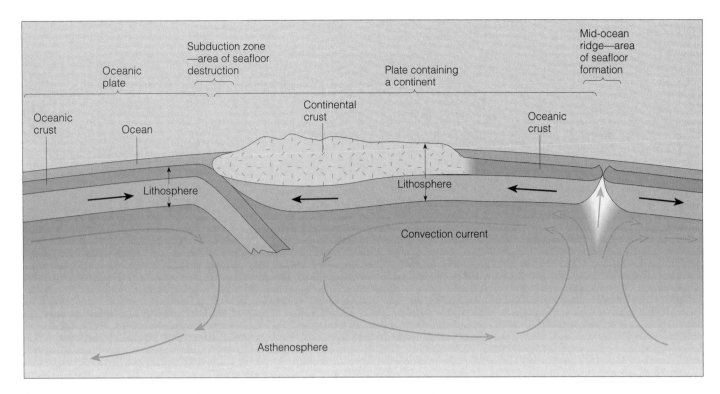

Figure 3.14 An overview of the plate tectonic process.

the Earth's crust first solidified. Literally and figuratively it all fits; **Figure 3.14** presents an overview of the process. This new understanding of the ever-changing nature of the Earth has given fresh meaning to historian Will Durant's warning: "Civilization exists by geological consent, subject to change without notice."

After a series of raucous scientific meetings in 1966 and 1967, the revolution in geology entered a period of rapid consolidation. In 1968 *Glomar Challenger* drilled her first deep-ocean crustal cores and provided the confirmation of plate tectonics. Researchers found supporting data from many sources that tended to confirm Wilson's surprising synthesis. Every scientist had to reexamine his or her specialty in light of this new information. Zoologists found new explanations for the unusual animals of Australia. Biologists discovered a new source of the isolation required for the formation of new species by natural selection. Paleontologists found an explanation for similar fossils on different continents. Resource specialists could at last explain why coal deposits were buried in Antarctica. Some geologists were pleased, some were skeptical. All were anxious to explore further to prove or disprove the tenets of this new theory.

This historical overview is useful in transmitting the sense of discovery and excitement surrounding the twentieth-century revolution in geology. Let's now investigate the workings of plate tectonics in more detail.

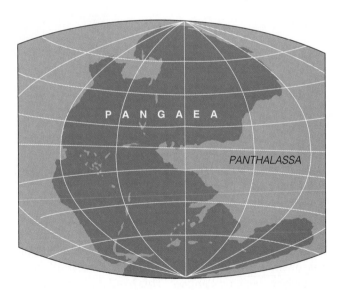

Figure 3.15 Reconstruction of Pangaea as it is thought to have appeared 200 million years ago.

PLATE TECTONICS: A CLOSER LOOK

Plate tectonics shares the concept of Pangaea with the older theory of continental drift. **Figure 3.15** depicts Pangaea as it may have looked about 200 million years ago,

Figure 3.16 The formation of a new plate boundary: the breakup of Pangaea. (a) As the lithosphere began to crack, a rift formed beneath the continent, and molten basalt from the asthenosphere began to rise. (b) As the rift continued to open, the two new continents were separated by a growing ocean basin. Volcanoes and earthquakes occur along the active rift area, which is the mid-ocean ridge. (c) A new ocean forms.

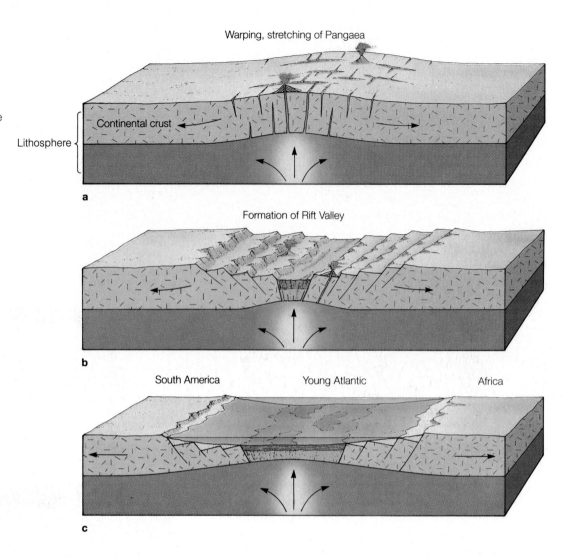

when it was beginning to break up. The present continents of North America and Eurasia, and Africa and South America are shown as part of the old supercontinent. Plate tectonics can be illustrated by what geologists think happened next.

Because it is much thicker, continental crust is only about half as efficient as oceanic crust in conducting heat from Earth's interior to the surface. Heat from the lower crust and mantle accumulates beneath the continents. For reasons not yet fully understood, excess heat collected along lines beneath Pangaea. This heat, probably generated from the decay of radioactive elements, caused the asthenosphere near the line to expand and rise, thus lifting and fracturing the lighter, solid lithosphere above. The plate and its embedded supercontinent split in two, and a new plate boundary was formed between the pieces (**Figure 3.16** shows the separation of South America and Africa).

The convection currents in the asthenosphere turned aside when they reached the brittle lower lithosphere.

The continental fragments (North America and Eurasia, and South America and Africa) moved apart, and a new ocean basin began to form between the diverging plates. As the broken plate separated at this new spreading center, molten rock called **magma** rose into the crustal fractures. (Magma is called *lava* when found above ground.) Some of the magma solidified in the fractures; some erupted from volcanoes on the seafloor. Together these processes produced new oceanic crust. As may be seen in **Figure 3.17**, the same process continues today. About 20 cubic kilometers (4.8 cubic miles) of new ocean crust forms each year.

As it moved away from the active spreading center, the oceanic lithosphere cooled and shrank slightly. This shrinkage caused the ocean to become deeper farther from the growing Mid-Atlantic Ridge.

The Atlantic's rate of spreading was—and continues to be—a jerky 5 centimeters (2 inches) a year. This young ocean, which began less than 200 million years ago, therefore was about 25 meters (85 feet) narrower when

Columbus sailed than it is today. A continent moves about as fast as your fingernails grow.

Divergent Plate Boundaries

The spreading center at the Mid-Atlantic Ridge is a **divergent plate boundary**, a line along which two plates are moving apart. Oceanic crust forms along divergent plate boundaries. For example, in the South Atlantic a large new ocean basin has formed between the diverging plates (as shown in **Figure 3.18**). A long mid-ocean ridge divided by a central rift valley traverses the ocean floor roughly equidistant from the shorelines in both the North and South Atlantic, terminating beneath the ice cap north of Iceland. The lithospheric plates west of the ridge extend all the way to the eastern edge of the Pacific Ocean, and they include the North and South American continents. The two plates east of the ridge include Africa and all of Eurasia. The Eurasian Plate wraps around the Earth to contact the western edge of the Pacific Plate. (Look again at Figure 3.13.)

Plate divergence is not confined to the Atlantic, nor has it been limited to the last 200 million years. As may also be seen in Figure 3.13, the Mid-Atlantic Ridge has counterparts in the Pacific and Indian oceans. The Pacific floor, for example, diverges along the East Pacific Rise and the Pacific Antarctic Ridge, spreading centers that form the eastern and southern boundaries of the great Pacific

Figure 3.17 The process of seafloor spreading was investigated firsthand in 1979 in an exploration of the Galápagos rift. The heat that the researchers measured, the lack of sediment covering at the ridge, and the characteristic pillow-shaped lava formations found on the ridge by the deep submersible *Alvin* were consistent with the theory of plate tectonics.

Figure 3.18 The South Atlantic in cross section, showing the mid-ocean ridge.

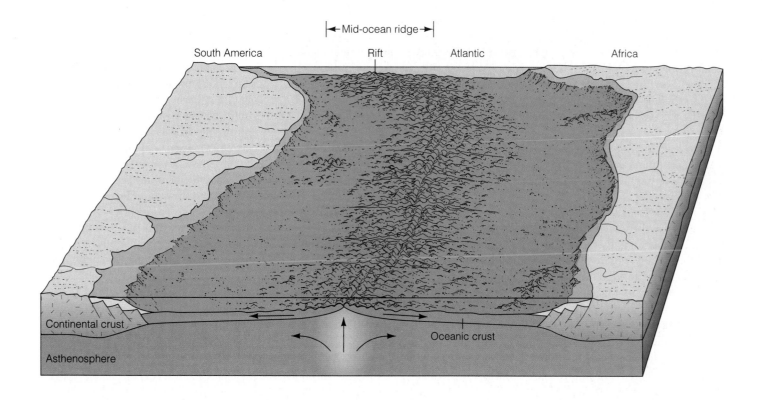

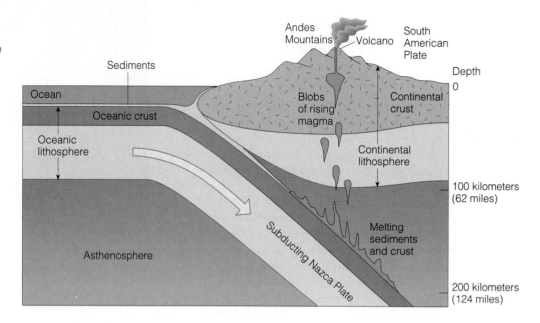

Figure 3.19 A cross section through the west coast of South America, showing the convergence of a continental plate and an oceanic plate. The oceanic plate begins to melt as it subducts. Blobs of magma rising from the melting zone power Andean volcanoes.

Plate. In East Africa rift valleys have formed relatively recently as plate divergence begins to separate another continent. As happened in the Red Sea, the ocean will invade when the rift becomes deep enough.

Convergent Plate Boundaries

Crust is destroyed at **convergent plate boundaries**, regions of violent geologic activity where plates are pushing together. South America, embedded in the westward-moving South American Plate, encounters the Pacific's Nazca Plate as it moves eastward. The relatively thick and light continental lithosphere of South America rides up and over the heavy oceanic lithosphere of the Nazca Plate, which is subducted into the deep trench that parallels the west coast of South America. **Figure 3.19** is a cross section through these plates.

The subducting plate's periodic downward lurches cause earthquakes. Some of the oceanic crust and its sediments will melt as the plate plunges downward, forming a magma rich in water and carbon dioxide. In places this magma rises through overlying layers to the surface and causes volcanic eruptions. The active volcanoes of Central America and South America's Andes Mountains are a product of this activity, as are the area's numerous earthquakes. The North American Cascade volcanoes, including Mount St. Helens, result from similar processes. Most of the subducted crust mixes with the mantle, some of it eventually reaching into the mantle to depths of at least 400 kilometers (250 miles)!

In the previous example continental crust met oceanic crust. What happens when two *oceanic* plates converge? One of the colliding plates will usually be older, and therefore cooler and denser, than the other. This heavier plate will slip below the lighter one into the asthenosphere. The ocean bottom is distorted in these areas to form deep trenches, the ocean's greatest depths. Water and carbon dioxide trapped with the melting rock of the subducting plate help to form a relatively light magma and vigorous volcanoes, but the volcanoes emerge from the seafloor rather than from a continent. These volcanoes appear in patterns of curves on the overriding oceanic crust; when they emerge above sea level, they form curving chains of islands (see **Figure 3.20**). The arc shape results from the geometric constraints of forces applied on the surface of a sphere. Some geophysicists believe all of Earth's continental crust may have originated from granitic rock produced in this way. The island arcs may have coalesced to form larger and larger continental masses.

The many island arcs and peripheral trenches of the western and northern Pacific shown in **Figure 3.21** result from the convergence of two oceanic plates. Note the correlation of these areas with Figure 3.10's plot of earthquakes. Subduction at converging oceanic plates was responsible for the great Alaska earthquake of 1964. Plate convergence (and divergence) is faster in the Pacific than in the Atlantic, in a few places reaching a rate of 18 centimeters (7 inches) a year. You can now clearly see the source of the Pacific Ring of Fire.

Two plates bearing continental crust can also converge. The most spectacular example of such a collision,

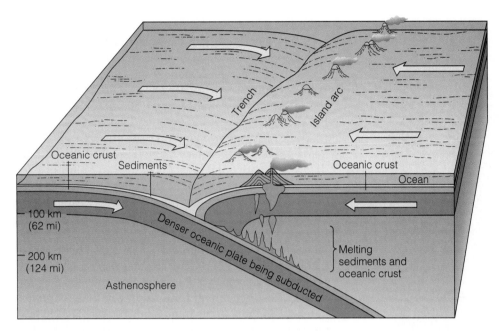

Figure 3.20 The formation of an island arc along a trench as two oceanic plates converge. The arc shape results from the geometric constraints of forces applied to the surface of a sphere. The volcanic islands form as blobs of magma reach the seafloor.

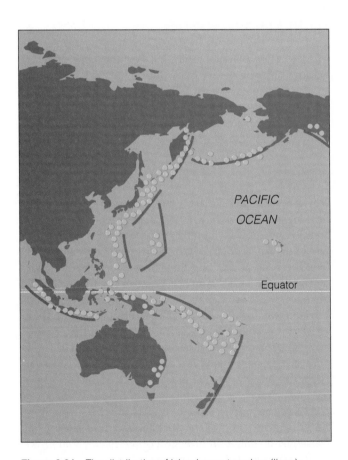

Figure 3.21 The distribution of island arcs, trenches (lines), and volcanoes (dots) in the western Pacific, where two plates converge.

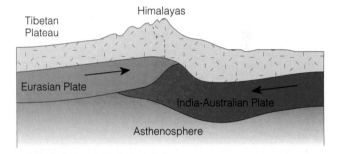

Figure 3.22 A cross section through southern China, showing the convergence of two continental plates. Neither plate was dense enough to subduct; instead, their compression and folding uplifted the plate edges to form the Himalayas.

between the India–Australian and Eurasian plates some 45 million years ago, formed the Himalayas. Neither plate edge is being subducted; instead, both are compressed, folded, and uplifted, as **Figure 3.22** shows. Notice the massive supporting "root" beneath the emergent mountain needed for isostatic equilibrium. Here is an explanation for the mountaintop marine fossils that catastrophists mistook to be evidence of the biblical Flood: The mountains and fossil shells formed on shallow submerged seabeds that were uplifted by plate convergence. The lofty top of Mount Everest is made of rock formed from sediments deposited long ago in a shallow sea!

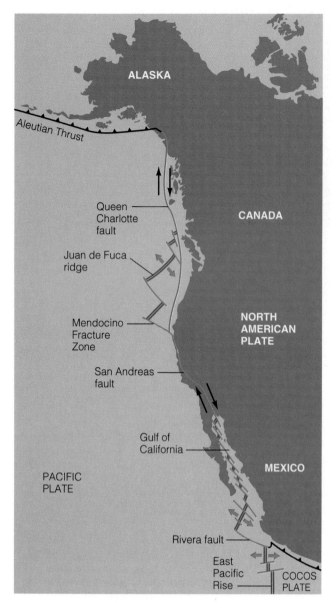

Figure 3.23 A long transform plate boundary, which includes California's San Andreas Fault.

Transform Plate Boundaries

In some places, crustal plates shear laterally past one another. These areas are called **transform plate boundaries**. Crust is neither produced nor destroyed at this type of junction, but the potential for earthquakes can be great as the plate edges slip past one another. The eastern boundary of the Pacific Plate is a long transform fault system. California's San Andreas Fault (**Figure 3.23**) is merely the most famous of the many faults marking the junction between the Pacific and North American plates. The Pacific Plate moves steadily, but its movement is stored elastically at the North American Plate boundary until friction is overcome. Then the Pacific Plate lurches in abrupt

jerks to the northwest along much of its shared border with the North American Plate, an area that includes the major population centers of California. These jerks cause California's famous earthquakes. Because of this movement, western California is gradually sliding north along the rest of North America; some 50 million years from now, it will encounter the Aleutian Trench.

There are, then, two kinds of plate divergences: Divergent oceanic crust (such as that in the mid-Atlantic) and divergent continental crust (as in the Rift Valley of East Africa). And there are three kinds of plate convergences: Oceanic crust toward continental crust (west coast of South America), oceanic crust toward oceanic crust (northern Pacific), and continental crust toward continental crust (Himalaya Mountains). Transform boundaries mark locations at which crustal plates move past one another (San Andreas Fault). Each of these movements produces a distinct topography, and each zone contains potential dangers for its human inhabitants. The characteristics of plate boundaries are summarized in **Table 3.1**.

THE CONFIRMATION OF PLATE TECTONICS

The theory of plate tectonics has had the same effect on geology that the theory of evolution has had on biology. In each case a catalog of seemingly unrelated facts was unified by a powerful central idea. Many discoveries contributed to our understanding of plate tectonics. Some compelling evidence for plate tectonics is outlined below.

Hot Spots

Hot spots are stationary sources of heat in the upper mantle. Hot spots are not always located at plate boundaries, and no one knows why their source of heat is localized or what anchors them in place. As lithospheric plates slide over these fixed locations, they are weakened from below by rising heat and magma. A volcano can form over the hot spot, but because the plate is moving, the volcano is carried away from its source of magma after a few million years and becomes inactive. It is replaced at the hot spot by a new volcano a short distance away. A chain of volcanoes and volcanic islands results (**Figure 3.24**).

Figure 3.25 shows the most famous of these "assembly line" chains, which extends from the old eroded volcanoes of the Emperor Seamounts to the still-growing island of Hawaii. In fact, the abrupt bend in the chain was caused by a change in direction of the Pacific Plate from a largely northward to a more westward movement about 40 million years ago. The next Hawaiian island that will come into being—already named Loihi—is building on the ocean floor to the southeast. Now about 1,000 meters

Table 3.1 Characteristics of Plate Boundaries

Plate Boundary		Plate Movement	Seafloor	Events Observed	Example Locations
Divergent plate boundaries	Ocean-ocean	Apart	Forms by seafloor spreading	Ridge forms at spreading center. Ocean basin expands, plate area increases. Many small volcanoes and/or shallow earthquakes.	Mid-Atlantic Ridge, East Pacific Rise
	Continent-continent		New ocean basin may form as continent splits	Continent spreads, central rift collapses, ocean will intrude	East African Rift
Convergent plate boundaries	Ocean-continent	Together	Destroyed at subduction zones	Dense oceanic lithosphere plunges beneath less dense continental. Earthquakes trace path of downmoving plate as it descends into asthenosphere. A trench forms. Subducted plate partially melts. Magma rises to form continental volcanoes.	Western South America
	Ocean-ocean			Older, cooler, more dense crust slips beneath less dense crust. Strong quakes. Deep trench forms in arc shape. Subducted plate heats in upper mantle, magma rises to form curving chains of volcanic islands.	Aleutians, Marianas
	Continent-continent		(N/A)	Collision between masses of granitic continental lithosphere. Neither mass is subducted. Plate edges are compressed, folded, uplifted; one may move beneath the other.	Himalayas, Alps
Transform plate boundaries		Past each other	Neither created nor destroyed	A line (fault) along which lithospheric plates move past each other. Strong earthquakes along fault.	San Andreas Fault

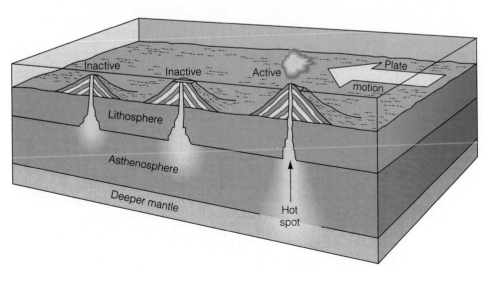

Figure 3.24 Formation of a volcanic island chain by the movement of an oceanic plate over a stationary hot spot. The age of the islands increases toward the left. New islands will continue to form over the hot spot.

Figure 3.25 The Hawaiian chain, islands formed one by one as the Pacific Plate slid over a hot spot. The oldest known member of the chain, the Meiji Seamount, formed about 70 million years ago (70 mya), and the "bend" in the chain shows that the plate changed direction about 40 million years ago. The island of Hawaii still has active volcanism, but the next island that will come into being in the chain, Loihi, has begun building on the ocean floor.

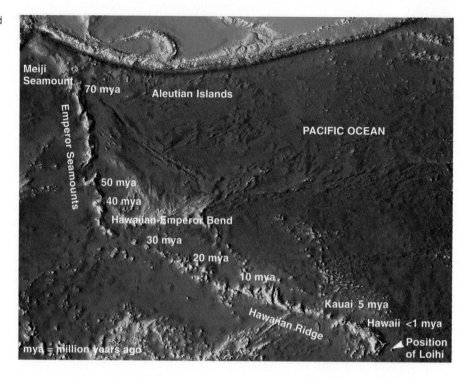

Figure 3.26 A group of atolls in the Tuamotu Archipelago in the South Pacific, looking southeast as seen from *Apollo 7* from an altitude of 153 kilometers (96 miles).

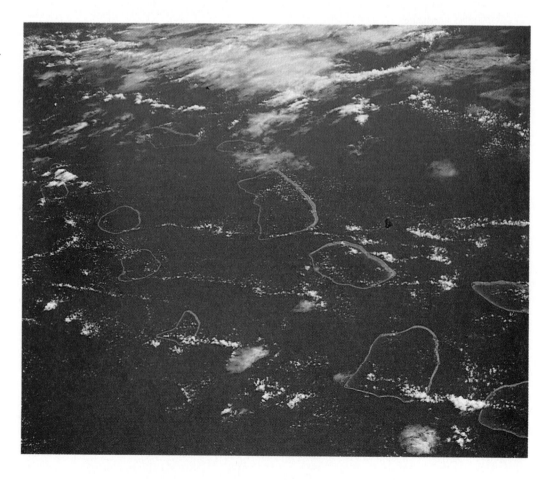

(3,200 feet) beneath the surface, Loihi will break the surface about 30,000 years from now.

There are other hot spots in the Pacific. The island chains formed by their activity also "jog" in the Hawaiian pattern, indicating that they are positioned on the same lithospheric plate. Chains of undersea volcanoes in the Atlantic, centered on the Mid-Atlantic Ridge, suggest a similar process is at work there. Hot spots can exist beneath continental crust as well; Yellowstone National Park is believed to be over a hot spot beneath the westward-moving North American Plate.

The configuration and length of all these chains of volcanoes and geothermal sites are consistent with the theory of plate tectonics.

Atolls and Guyots

Atolls are ring-shaped islands of coral reefs and reef-derived sediment centered over submerged inactive volcanoes (see **Figure 3.26**). The coral animals that build atolls can live only in the upper sunlit layer of seawater. How, then, did their skeletal remains end up at a depth of 1,280 meters (4,222 feet) within Eniwetok Atoll? This surprising discovery was made in 1954, when bore holes were being drilled in preparation for the first hydrogen bomb tests. Plate tectonics suggests an answer. Coral animals can build atop the skeletons of their dead predecessors at a rate of about 1 centimeter (½ inch) each year. Coral animals living on a volcano's flanks can grow upward as

the crust beneath the volcano slowly cools and contracts (sinks) during its movement away from the warm spreading center where it formed (see Figure 14.16). A deep column of coral skeletons can accumulate if the rate of sinking is less than about 1 centimeter per year. Thus, the coral record traces plate subsidence for millions of years into the past, supporting the plate tectonics theory.

Guyots were discovered by Harry Hess during his service as commander of a U.S. Navy transport in the Pacific during the Second World War. With the ship's echo sounder he found chains of odd, flat-topped, submerged volcanic mountains, and he named them after Princeton's first professor of geology, Arnold Guyot. More than 500 guyots (pronounced ghee-OH) have been located, and plate tectonics neatly explains the formation of most of them. Like atolls, they were once volcanic peaks standing above sea level. As the plate on which they are riding cooled, contracted, and was carried away from the spreading center, they became inactive and stopped growing. They were "shaved" flat by wave action as they sank beneath the ocean surface. Evidence of ancient beaches along their rims suggests that this hypothesis is correct. **Figure 3.27** shows a sinking progression of guyots. They never formed atolls, either, because they began in water too cold for coral or because they moved and sank too rapidly for coral growth to keep up. Indeed, an atoll can become a guyot if its rate of subsidence increases, coral growth slows, or plate motion takes it into colder water.

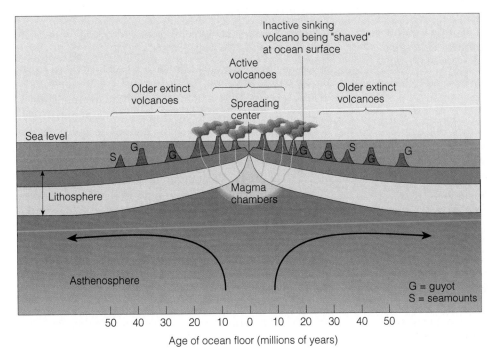

Figure 3.27 The process by which guyots (G) and seamounts (S) form. Guyots have flat tops because they have grown tall enough to be "shaved" by waves at the ocean's surface. Seamounts have a similar origin, but they retain their more pointed volcano shape because they never reach the surface.

Age and Distribution of Sediments

If the ocean basins are genuinely ancient, and if the processes that produce sediments have been operating for most or all of that time, both the thickness and age of sediments on the ocean floor should be great. They are not. The young spreading ridges are almost free of sediment, and the oldest edges of the basins support layers of sediment 15 to 20 times thinner than the age of the ocean itself would suggest.

Powerful, low-frequency echo sounders have probed the depth and structure of these sediments in many locations. Core sampling has shown that sediments resting directly on the basalt floor are almost always youngest near the spreading centers and oldest near subduction zones. The oldest sediments of the ocean basins are rarely more than 160 million years old. These data are consistent with the plate tectonics idea that ocean basins are continuously recycled.

Oceanic Ridges

The location and configuration of the oceanic ridges are clear evidence of past events. The volcanic nature of ridge islands like Iceland, the shape of the longitudinal rifts splitting the ridge tops, and the sinking of the seabed as new oceanic crust cools and travels outward are all consistent with the theory of plate tectonics. The distribution of transform faults and fracture zones along the oceanic ridges (features described in Chapter 4) also supports plate tectonics theory.

Heat Flow

Heat flows rapidly from the ocean bottom at mid-ocean ridges and slowly in the vicinity of trenches. This suggests the presence of hot, new crust at the ridge and cool, older crust at the trenches, just as plate tectonics predicts.

Terranes

Buoyant continental and oceanic plateaus (fragments of granitic rock and sediments) can be rafted along with a plate and scraped off onto a continent when the plate is subducted. This process is similar to what happens when a sharp knife is scraped across a table top to remove pieces of cool candle wax. The wax accumulates and wrinkles on the knife blade in the same way land masses and ocean sediments accumulate against the face of a continent as the lithosphere in which they are embedded reaches a plate boundary. Plateaus, isolated segments of seafloor, ocean ridges, ancient island arcs, and parts of continental crust that collect on the face of a continent are called **terranes**. The thickness and low density of terranes

prevent their subduction. A simplified account of terrane accumulation is diagrammed in **Figure 3.28**.

Terranes are surprisingly common. New England, much of North America west of the Rocky Mountains, and all of Alaska appear to be composed of this sort of crazy-quilt assemblage of material, some of which has evidently arrived from thousands of miles away in the Southern Hemisphere!

Fossils

Suess's *Glossopteris* fossils were part of a complex collection of plants that flourished in low, swampy areas near the margins of glaciers. The *Glossopteris* flora, as the fossil plants are called, is common to southern South America, southern Africa, India, Australia, and Antarctica. It is extremely unlikely that identical intricate, interdependent communities of plants could have arisen simultaneously in all these places. A more likely explanation is a site of common origin in Pangaea. The breakup of Pangaea would account for the present distribution of the *Glossopteris* flora.

Animal fossils also support the idea of an ancient supercontinent. Fossils of *Mesosaurus*, a half-meter- (2-foot-) long aquatic reptile, are found only in eastern South America and southwestern Africa. Again, it is extremely unlikely that this animal could have evolved simultaneously in two widely separated locations. It is equally unlikely that this small shallow-water reptile could have swum across 5,500 kilometers (2,500 miles) of open ocean to establish itself on both sides of the Atlantic.

Paleomagnetism

The young ocean basins hold the most convincing evidence for plate tectonics. This evidence was obtained through the magnetic analysis of rocks formed over the last 200 million years.

A compass needle points toward the magnetic north pole because it aligns with the Earth's magnetic field. Tiny particles of iron-bearing magnetic minerals are found in basaltic magma. These minerals act like miniature compass needles; as they cool, their magnetic fields align with the Earth's magnetic field. Thus the orientation of the Earth's magnetic field at the time becomes frozen in the rock as it solidifies. Any later change in the strength or direction of the Earth's magnetic field will not significantly change the characteristics of the field trapped within the solid rocks. The "fossil," or remanent, magnetic field of a rock is known as **paleomagnetism** (*paleos* = ancient).

A **magnetometer** measures the amount and direction of residual magnetism in a rock sample. In the late 1950s geophysicists towed sensitive magnetometers just above

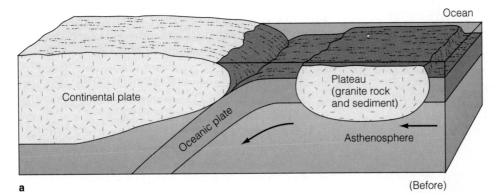

Continental plate

Oceanic plate

Plateau (granite rock and sediment)

Ocean

Asthenosphere

a   (Before)

Figure 3.28 Terrane formation. Oceanic plateaus are not subducted into the trench with the oceanic plate. Instead, they are "scraped off," causing uplifting and mountain building as they strike a continent.

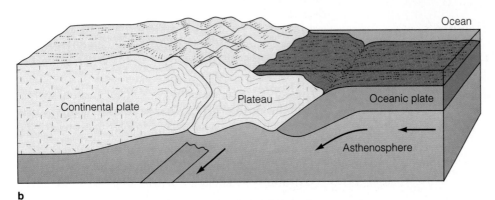

Continental plate

Plateau

Oceanic plate

Ocean

Asthenosphere

b

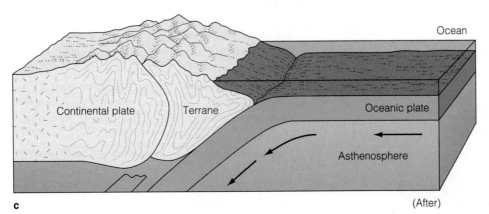

Continental plate

Terrane

Oceanic plate

Ocean

Asthenosphere

c   (After)

the ocean floor to detect the weak magnetism frozen in the rocks. When plotted on charts, the data revealed an odd pattern of symmetrical magnetic "stripes" or bands on both sides of a spreading center (**Figure 3.29a**). The tiny compass needles contained in the rocks of some bands join with the Earth's present magnetic orientation to enhance the strength of the local magnetic field, while the needles in rocks in adjacent bands weaken it. What could cause such a pattern?

In 1963 geologists Drummond Matthews and Frederick Vine proposed a clever interpretation. They knew that the Earth's magnetic field reverses at irregular intervals. In a time of reversal a compass needle would point south instead of north, and any particles of magnetic material in fresh seafloor basalt at a spreading center would be imprinted with the reversed field. The alternating magnetic stripes represent rocks with alternating magnetic polarity—one band having normal polarity (magnetized in the same direction as today's magnetic field direction), the next band having reversed polarity (opposite from today's direction). These researchers realized that the pattern of alternating weak and strong magnetic fields was symmetrical because freshly magnetized rocks born at the ridge are spread apart and carried away from the ridge by plate movement (**Figure 3.29b**).

The Earth's magnetic field reverses about once every 300,000 to 500,000 years. At least ten "flips" have been documented for the last 4 million years. No one yet knows what causes these reversals, or what the effects are on the Earth during the change. Similar magnetic patterns have

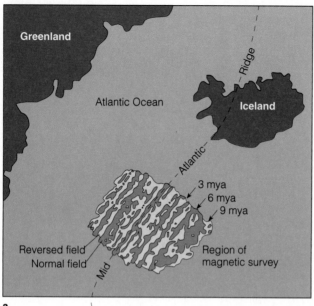

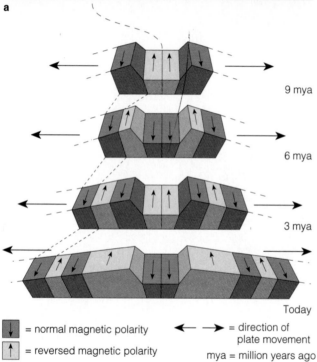

= normal magnetic polarity   ←—→ = direction of plate movement

= reversed magnetic polarity   mya = million years ago

b

Figure 3.29 Patterns of paleomagnetism and their explanation by plate tectonics theory. (a) When scientists conducted a magnetic survey of a spreading center, the Mid-Atlantic Ridge, they found bands of reversed magnetic fields frozen in the rocks. (b) The molten rocks forming at the spreading center take on the polarity of the planet at the time they are cooling, and they then move slowly in both directions from the center. When the Earth's magnetic field reverses, the polarity of newly formed rocks changes, creating symmetrical bands of opposite polarity.

been found on land and independently dated by other means. By 1974, scientists had compiled charts showing the paleomagnetic orientation—and the age—of the seafloors of the eastern Pacific and the Atlantic (see **Figure 3.30**). Plate tectonics beautifully explains these patterns, and the patterns themselves are among the most compelling of all arguments for the theory.

Paleomagnetic data have recently been used to measure spreading rates, to calibrate the geologic time scale, to reconstruct continents, and to understand the movement of terranes. Paleomagnetism has been among the most productive specialties in geology for the past two decades.

PROBLEMS AND IMPLICATIONS

The theory of plate tectonics reveals much about the nature of the Earth's surface. In case you feel geophysicists have all the answers, however, consider just a few of the theory's unsolved problems:

- Why should long *lines* of asthenosphere be any warmer than adjacent areas?

- If plate movement depends on "drag" within asthenosphere convection cells, why should the plastic material flow parallel to the plate bottoms for long distances, instead of cooling and sinking near the spreading center?

- Are plate movements due entirely to motion of the asthenosphere? Is deeper mantle convection also involved? Could the plates themselves play a role in their own movement?

- Has seafloor spreading always been a feature of the Earth's surface? Has a previously thin crust become thicker with time, permitting plates to function in the ways described here?

- There is much evidence of tectonic movement prior to the breakup of Pangaea. Will the process continue indefinitely, or are there cycles within cycles?

Though there is clearly much to learn, plate tectonics is already an especially powerful predictive theory. Discoveries and insights made by the researchers mentioned in this chapter, and hundreds of others, have borne out the intuition of Alfred Wegener. Our understanding of the process will evolve as more data become available, but there seems very little chance that geologists will ever return to the dominant pre-1960 view of a stable and motionless crust. Plate tectonics theory shows us the picture of an actively cycling Earth and an ever-changing surface with *a single world ocean* changing shape and shifting positions as the plates slowly move.

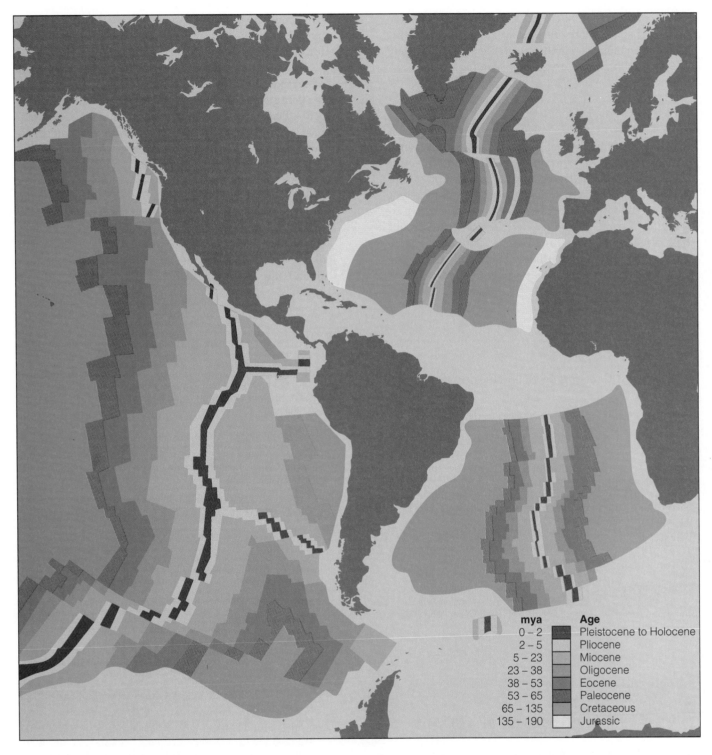

	mya	Age
	0 – 2	Pleistocene to Holocene
	2 – 5	Pliocene
	5 – 23	Miocene
	23 – 38	Oligocene
	38 – 53	Eocene
	53 – 65	Paleocene
	65 – 135	Cretaceous
	135 – 190	Jurassic

Figure 3.30 The age of the eastern Pacific and Atlantic ocean floors. The magnetic patterns are an expression of seafloor spreading over the last 200 million years. (mya = millions of years ago).

The configuration of the ocean basins—discussed in the next chapter—is the result of plate tectonic activity. Armed with an understanding of the theory, the variety of features will make more sense to you.

CHAPTER SUMMARY

The Earth is composed of concentric spherical layers, with the least dense layer on the outside, and the most dense at the core. The layers may be classified by chemical composition into crust, mantle, and core; or by physical properties into lithosphere, asthenosphere, mesosphere, and core. The lithosphere, the outermost solid shell, consists of granitic and basaltic crust bonded to a denser solid region immediately below. It floats on the hot, plastic (deformable) asthenosphere. Geologists have confirmed the existence and basic properties of the layers by analysis of seismic waves, which are generated by the forces that cause large earthquakes.

The theory of plate tectonics explains the distribution of earthquake location, the curious jigsaw-puzzle fit of the continents, and the patterns of magnetism in surface rocks. Plate tectonics theory suggests that the Earth's surface is not a static arrangement of continents and ocean, but a dynamic mosaic of jostling lithospheric plates. The plates have converged, diverged, and slipped past one another since the Earth's crust first solidified, driven by slow, heat-generated currents flowing in the asthenosphere. Most major continental and seafloor features are shaped by plate movement. Plate tectonics explains why our ancient planet has surprisingly young seafloors, the oldest of which is only as old as the dinosaurs, that is, about ⅓ the age of the Earth.

Terms and Concepts to Remember

asthenosphere	granite	plates
atoll	guyot	radioactive decay
basalt	hot spot	radiometric dating
buoyancy	isostatic	S wave
conduction	equilibrium	seafloor spreading
continental crust	lithosphere	seismic waves
continental drift	magma	seismograph
convection	magnetometer	shadow zone
convection currents	mantle	spreading center
convergent plate	mesosphere	subduction
boundary	oceanic crust	subduction zone
core	P wave	terrane
crust	Pacific Ring of Fire	transform plate
divergent plate	paleomagnetism	boundary
boundary	Pangaea	Wegener, Alfred
echo sounders	Panthalassa	Wilson, John Tuzo
fault	plate tectonics	

Study Questions

1. How are Earth's internal layers classified? Which method is more useful in explaining plate tectonics?

2. How is crust different from lithosphere?

3. Where are the youngest rocks in the seabed? The oldest? Why?

4. On what points was Wegener correct? Wrong?

5. Would the most violent earthquakes be associated with spreading centers or with subduction zones? Why?

6. Describe the mechanism that powers the movement of the lithospheric plates.

7. Why is paleomagnetic evidence thought to be the lynchpin in the plate tectonics argument? Can you think of any objections to the Drummond/Vine interpretation of the paleomagnetic data?

8. What biological evidence supports plate tectonics theory?

9. What evidence can you cite to support the theory of plate tectonics? What questions remain unanswered? Which side would you take in a debate?

10. Why are the continents about 20 times older than the oldest ocean basins?

For Further Study

Bonatti, E. 1994. "The Earth's Mantle Beneath the Ocean." *Scientific American*, March, 44–51. How the mantle's convection forces shape the Earth's surface and perhaps affect its rotation.

Bott, M. H. P. 1982. *The Interior of the Earth: Its Structure, Constitution, and Evolution.* New York: Elsevier. A prime technical reference.

Burchfiel, B. 1983. "The Continental Crust." *Scientific American*, September, 130–42.

Courtillot, V., and J. Besee. 1987. "Magnetic Field Reversals, Polar Wander, and Core–Mantle Coupling." *Science*, 237 (no. 4819): 1140–47. The jerky movement of the magnetic poles may be correlated to episodes of faster or slower continental drift. Includes new information on the transfer of heat from the core to the mantle and on the formation of convection cells.

Dietz, R. S. 1961. "Continent and Ocean Basin Evolution by Spreading of the Sea Floor." *Nature*, 190:854–57. A historic paper by one of the originators of the theory of seafloor spreading.

Francheteau, J. 1983. "The Oceanic Crust." *Scientific American*, September, 114–29. Companion piece to the Burchfiel article, above.

Frohlich, C. 1989. "Deep Earthquakes." *Scientific American*, January, 48–55. How can rock fail at temperatures and pressures that prevail hundreds of kilometers down? Summarizes Wadati's contributions.

Glen, W. 1982. *The Road to Jaramillo: Critical Years in the Revolution in Earth Science*. Stanford, CA: Stanford University Press. An entertaining history of the theory of plate tectonics.

Heirtzler, J. R. 1968. "Sea Floor Spreading." *Scientific American*, December, 60–70. The first major review article on the subject. Suggests that changes in the Earth's rotational motion may be responsible in part for seafloor spreading and even for the occasional reversal of poles.

Kerr, R. 1991. "Do Plumes Stir Earth's Entire Mantle?" *Science*, 252 (no. 5010):1068–69. Rising plumes and sinking plates may span the entire mantle, not just the 670-kilometer-thick upper mantle.

MacDonald, G. A., A. T. Abbott, and F. L. Peterson. 1983. *Volcanoes in the Sea—The Geology of Hawaii*. 2d ed. Honolulu: University of Hawaii Press. New information on Loihi, and a chapter on plate tectonics as it relates to the origin of the islands.

Menard, H. W. 1986. *The Ocean of Truth: A Personal History of Global Tectonics*. Princeton, NJ: Princeton University Press. History of the geological revolution by one of its major scientists. Menard's last book; thoughtful and philosophically written.

Powell, C. S. 1991. "Peering Inward." *Scientific American*, June, 100–111. Update on mantle studies.

Schwarzbach, M. 1986. *Alfred Wegener, the Father of Continental Drift*. Madison, WI: Science Tech. A biography (translated from German) that offers many insights into Wegener's life.

Stanley, S. M. 1986. *Earth and Life Through Time*. New York: Freeman. Discusses the biological as well as geological ramifications of plate tectonics.

4 OCEAN BASINS

Ticks and Tones

Although a few people have visited the deep seabed, their experiences have necessarily been limited in time and area; the rigors of the environment don't encourage leisurely field trips. Thus, almost all of what we know about the ocean basins comes from remote sensing: robots returning samples, tethered video cameras recording pictures, echo sounders coating the ocean floor with ticks and tones.

The first images of the ocean floor were fuzzy photos taken with leaky cameras attached to very long ropes. The pictures were of poor quality, covered a very small area, and were difficult to place in context. Today the most successful methods for visualizing the seabed use reflected sound instead of light. By the 1950s precision acoustical sounders were able to measure water depths to an accuracy of about 1.8 meters (6 feet). The images they produced were little more than wavy lines representing the contour of the bottom directly under the track of the research ship carrying the device. Despite these limitations, echo sounders gave oceanographers their first clear overview of basin topography.

In the 1980s sophisticated acoustical equipment, high-speed computers, and satellite navigation systems with pinpoint accuracy have made it possible to examine the seafloor in unprecedented detail. The Sea Beam multibeam echo sounder system, which generated this relief map of the fast-spreading East Pacific Rise, can cover tens of thousands of square kilometers of the seafloor in a single 30-day cruise. Land-based computers later process the data into the kind of smooth, three-dimensional contour map seen here. In an instant, geologists can see patterns that once would have required hours or days of concentrated study to discern. As the precision of our equipment improves, the basins begin to yield their deeply hidden secrets.

A Sea Beam image of the East Pacific Rise near 9°N, an area where the ocean basin is spreading at the rate of about 8 centimeters (3 inches) a year. Note the unusually offset ridge axis at the center of the image and the volcanoes (here in red false color) at the top.

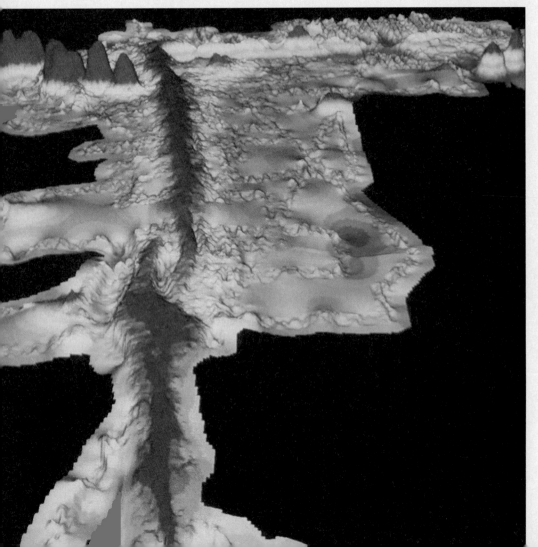

Most people think that an ocean basin is shaped like a giant bathtub. They imagine that the continent drops off steeply just beyond the surf, and that the ocean is deepest somewhere out in the middle. Until the mid-1800s natural scientists generally shared this view, but soundings made in the late 1840s revealed shallow areas near the center of the Atlantic, casting doubt on the deepest-at-the-center hypothesis. The *Challenger* expedition of the late 1870s, which took 492 soundings around the world, found submerged mountains in the mid-Atlantic that were connected into an extensive ridge. In the 1920s, *Meteor* researchers using echo sounders traced this imposing mid-ocean ridge for much of its length.

Echo sounder studies made before and during World War II discovered other unexpected features. For example, *Challenger* scientists had discovered that shallow extensions of the continents intruded into the Atlantic; echo soundings showed that this was a worldwide phenomenon. They also found that the deepest parts of the ocean lie near basin *edges* rather than at their centers.

Echo sounders powerful enough to probe beneath the layers of sediment on the ocean floor revealed hidden hills, mountains, and steep valleys. These surveys have shown that the seabeds are complex structures indeed (**Figure 4.1**). Near shore, the features of the ocean bottom are similar to the geologic features of continents, but features of the deep-ocean floor seen from submersibles usually differ from those on land. This should come as no surprise; after all, continental crust and oceanic crust have dramatically different histories. As you would expect from reading the last chapter, the processes of plate tectonics figure prominently in these histories, and thus in the shape of the ocean floor.

We now know that the undersea edges of continents are made of relatively light granitic rock buried beneath layers of sediment, and that the deep-ocean floor is heavier basalt (also covered with sediment). The theory of seafloor spreading explained why granite and basalt are distributed in that way, and a more detailed picture of the ocean floor began to emerge as plate tectonics theory

Figure 4.1 Typical cross sections of the Atlantic and Pacific ocean basins. The vertical axis is greatly exaggerated.

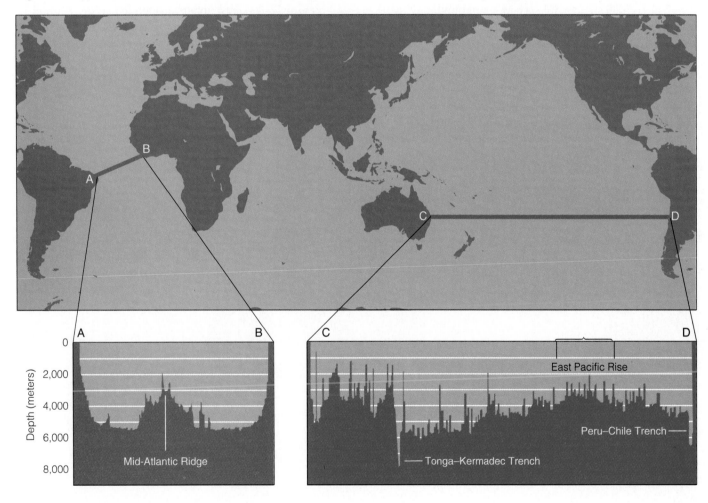

BOX 4.1 ● *Deep Tools*

Terrestrial geologists lead a relatively easy life. They may suffer desert heat, tropical rain, or Antarctic cold, but at least they have *direct* access to their goal. Marine geologists, on the other hand, almost never have the opportunity to study their targets in context. They rarely get to handle the rocks around the specimens they're after; they rarely even see their specimens in place. Usually they must be content to study rocks collected from the ocean bottom by dredges or robots. A few hearty geologists may use scuba equipment to survey a nearshore area and obtain their own study specimens. With the proper mix of gases and adequate training, a scuba diver can descend to about 60 meters (200 feet). With special mixtures of breathing gases and dry suits inflated with argon to avoid heat loss, divers have reached a depth of 300 meters (984 feet). Below that depth a research submersible is better equipped to withstand the cold and pressure. The frail human body is no match for the depths where the most important formations lie.

And so marine geologists venture forth in pressure-proof hulls—hulls that are insulated, temperature-controlled, and dry. Once on the bottom, they look through a thick plastic window toward their unreachable goal less than a meter away. Their touch is extended by the arms of remote manipulators, their sight by finely detailed television pictures. But the urge to turn over the next stone, to probe under the surface layer, is often impossible to satisfy. So near, yet so far away.

Transporting these marine geologists to work presents daunting obstacles. The most difficult problem is to reach extreme depths, but, amazingly, scientists have visited the bottom of the deepest ocean basin. On 23 January 1960 U.S. Navy Lieutenant Don Walsh and Dr. Jacques Piccard descended to a depth of 10.92 kilometers (6.78 miles) in the Challenger Deep, an area of the Mariana Trench discovered in 1951 by the British oceanographic research vessel *Challenger II*. The vehicle used in the descent was the **bathyscaphe** *Trieste* (Figure 1.22, and **Figure a**), a deep-diving submersible designed like a blimp with a very strong and thick (and cramped) steel crew sphere suspended below. A blimp uses helium gas for bouyancy, but a gas would be compressed by water pressure; so gasoline, which is relatively incompressible, was used to provide lift.

a *Trieste* is 15.7 meters (59.2 feet) in length, and the pressure hull protecting her two-person hull is 13 to 18 centimeters (5 to 7 inches) thick. She made 130 dives and earned three world records. Her last voyage was in 1963.

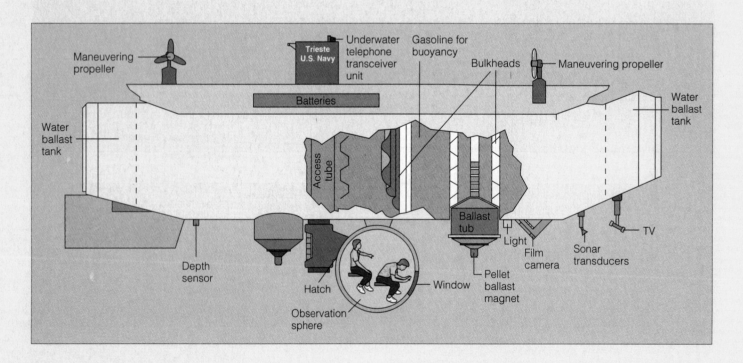

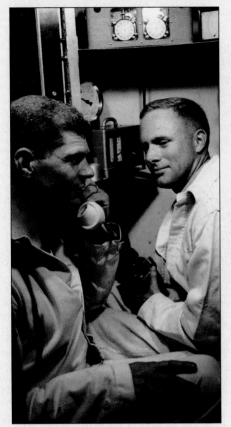

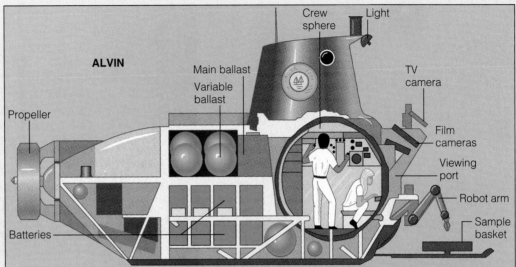

ALVIN

Propeller
Batteries
Main ballast
Variable ballast
Crew sphere
Light
TV camera
Film cameras
Viewing port
Robot arm
Sample basket

b Andreas Rechnitzer and Don Walsh prepare *Trieste* for a deep dive.

The state of the art in deep submersibles.

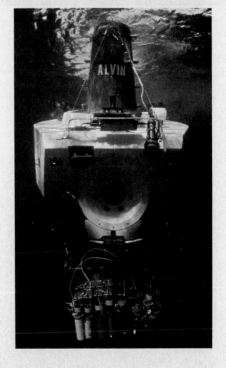

c *Alvin*, the best known and oldest of the six deep-diving manned research submarines now in operation, has made more than 2,500 dives in 28 years of operation. *Alvin* carries three people and is capable of diving to 4,000 meters (13,120 feet).

After a series of increasingly deep practice dives (**Figure b**), the fateful day arrived. The descent out of the ocean's sunlit upper zone was slow at first. The bathyscaphe actually stopped when it encountered the thermocline, a zone of denser water that caused the vessel to become more buoyant. When a little gasoline was valved from the maneuvering tank, the explorers began falling—at a rate of about 4 feet per second—into a dark, calm, cold realm never before seen.

Six hours after setting out, the vessel came to rest on a carpet of fine, cream-colored sediment where pressure was over 1,100 kilograms per square centimeter (8 tons per square inch). No rocks were visible. The voyagers took temperature readings and spotted a pair of bottom fish and some crustaceans before ascending. Both scientists later wrote about their feeling of awe in the presence of the totally unknown. With the possible exception of the moon, no place visited by humans has been more hazardous to explore.

Trieste was not very maneuverable and lacked the ability to gather the large amount of data that could justify her costly operation. She made 130 dives, the last in 1963, and is currently on display in the Navy Yard Museum in Washington, D.C. A new generation of research vehicles has followed. The most famous of these is *Alvin* (**Figure c**), a vehicle operated jointly by the Woods Hole Oceanographic Institution, the National Science Foundation, the Office of Naval Research, and the National Oceanic and Atmospheric Administration. *Alvin* has explored the Mid-Atlantic Ridge near the Azores at 2,700 meters (9,000 feet), measuring rock temperatures, collecting water samples for chemical analysis, and photographing the rounded basaltic extrusions known as pillow lava. Tethered, remotely operated robots, such as *Jason, Jr.* (used to photograph the RMS *Titantic* in 1985), are becoming more popular with

researchers; they're cheaper and safer than people-carrying vehicles.

The newest deep-diving vehicle is Mitsubishi Heavy Industries' *Shinkai 6500* (**Figure d**). Now the deepest-diving submersible in operation, *Shinkai 6500* can safely descend to a depth of 6,500 meters (21,300 feet) and thus can reach all but the deepest trench floors, or about 98% of the world ocean bottom. The U.S. Navy has made one of its modern nuclear submarines available to marine scientists (**Figure e**).

d *Shinkai 6500*, built in 1981 by Mitsubishi Heavy Industries, is currently the world's newest and deepest-diving submersible. Three people can be accommodated in the vehicle's titanium pressure hull. *Shinkai 6500* reached its design depth of 6,500 meters (21,320 feet) on 11 August 1989.

e Fast, strong, and silent, a nuclear submarine makes an ideal platform for oceanographic research. The U.S. Navy made one of its nuclear submarines available to marine scientists during the summer of 1993. The research cruise aboard USS *Pargo*, a *Sturgeon*-class submarine strengthened for operation in ice, began in Groton, Connecticut, and ended in Bergen, Norway. More than 9,060 kilometers (4,900 nautical miles) of underway data were collected in the deep Arctic Ocean beneath the pack ice. Water samples were taken for chemical and biological analysis (as seen here), current probes were released and tracked, and buoys were placed to monitor weather and surface water conditions. The data from the cruise will be placed in the public domain and results published in unclassified literature. More cruises are planned, one as early as the summer of 1995.

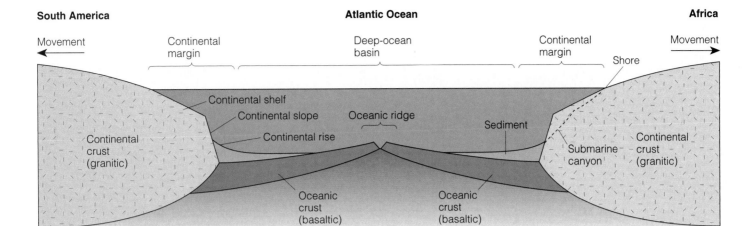

South America   **Atlantic Ocean**   **Africa**

Figure 4.2 Cross section of a typical ocean basin flanked by passive continental margins. (The vertical scale has been exaggerated to emphasize the basin contours.)

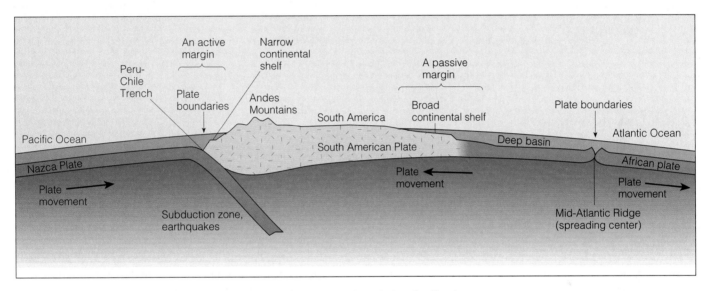

Figure 4.3 Typical continental margins bordering the leading (tectonically active) and trailing (tectonically passive) edges of a moving continent (exaggerated vertical scale).

developed in the early 1960s. Deep soundings, echo sounders, and on-site observations have all contributed to our present understanding of ocean floor shape and structure. An example is shown in **Figure 4.2**. Notice the abrupt transition between the thick granitic rock of the continents and the relatively thin basalt of the deep-sea floor. Nearshore ocean floors are similar to the adjacent continents because they share the same granitic basement. The transition to basalt marks the true edge of the continent and divides ocean floors into two major provinces. The submerged outer edge of a continent is called the **continental margin**. The deep-sea floor beyond the continental margin is properly called the **ocean basin**.

CONTINENTAL MARGINS

You learned in Chapter 3 that lithospheric plates converge, diverge, or slip past each other. As you might expect, continental margins are greatly influenced by this tectonic activity (see **Figure 4.3**). Continental margins facing the edges of *diverging* plates are called **passive margins** because relatively little earthquake or volcanic activity is associated with them. They surround the Atlantic, so passive margins are sometimes referred to as *Atlantic-type* margins. Continental margins near the edges of *converging* plates (or near places where plates are slipping past each other) are called **active margins** because of their earthquake and volcanic activity. Because of their

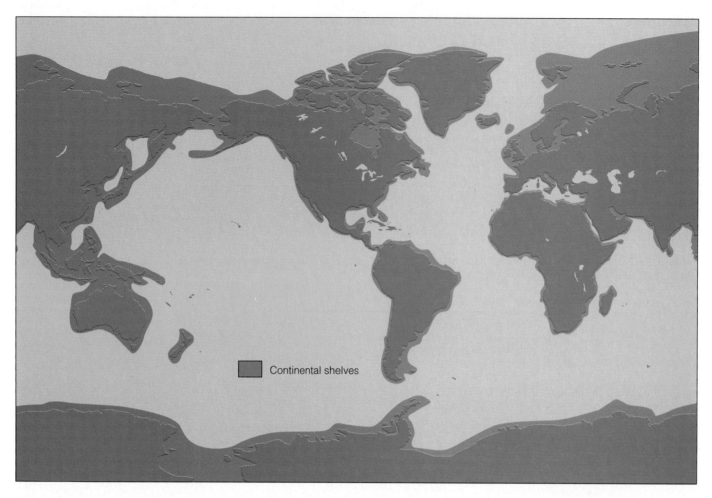

Figure 4.4   The worldwide distribution of continental shelves. Great Britain is part of the continent of Europe, and Alaska connects to Siberia.

prevalence in the Pacific, active margins are sometimes referred to as *Pacific-type* margins. Active and passive margins west and east of South America are shown in Figure 4.3. Note that active margins coincide with plate boundaries, but passive margins do not. Passive margins are also found outside the Atlantic, but active margins are confined mostly to the Pacific.

Continental margins have two main divisions: a shallow, nearly flat continental shelf close to shore and a more steeply sloped continental slope to seaward.

Continental Shelves

The shallow submerged extension of a continent is called the **continental shelf**. Continental shelves, like the adjacent continents, are underlain by granitic continental crust. They are much more like the continent than the deep-ocean floor, and they may have hills, depressions, and mineral and oil deposits similar to those on the dry land nearby. Earth's continental shelves are shown in **Fig-**

ure 4.4. Taken together, the area of the continental shelves is equivalent to 18% of Earth's dry land area.

Figure 4.5 shows a passive-margin continental shelf characteristic of Atlantic Ocean edges. The shelf extends far from shore in a gentle incline, typically 1.7 meters per kilometer (about 9 feet per mile), much less than the slope of a well-drained parking lot. Shelves along the margin of the Atlantic Ocean can reach about 350 kilometers (220 miles) in width and end at a depth of about 150 meters (500 feet), where a steeper dropoff begins.

The passive-margin shelves of the Atlantic Ocean formed as the fragments of Pangaea were carried away from each other by seafloor spreading. The continental lithosphere, thinned during initial rifting, cooled and contracted as it moved away from the spreading center, submerging the trailing edges of the continents and forming the shelves.

Most of the material comprising a shelf comes from erosion of the adjacent continental mass. Rivers assist in passive shelf-building by transporting huge loads of sed-

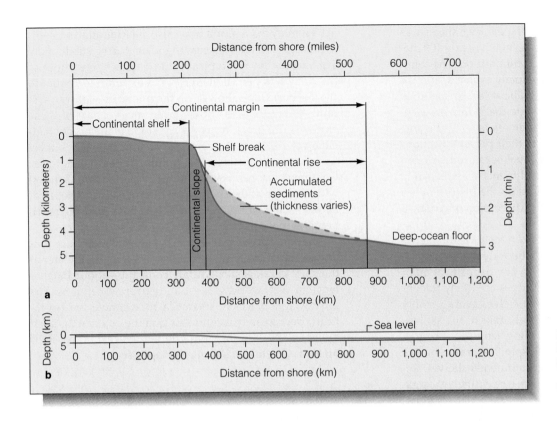

Distance from shore (miles)

Distance from shore (km)

Figure 4.5 The features of a broad passive continental margin. (a) Vertical exaggeration 150:1. (b) No vertical exaggeration.

iments to the shore from far inland. In some places the sediments accumulate behind natural dams formed by ridges of granitic crust. The weight of the sediment isostatically depresses the continental edges and allows the sediment load to grow even thicker. Sediment at the outer edge of a shelf can be up to 15 kilometers (9 miles) thick and 150 million years old.

The width of a shelf is usually determined by proximity to a plate boundary. You can see in Figure 4.3 that the shelf at the *passive* margin (east of South America) is broad, while the shelf at the *active* margin (west of South America) is very narrow. The widest shelf, 1,280 kilometers (800 miles) across, lies north of Siberia in the tectonically quiet Arctic Sea. Shelf width depends not only on tectonics, but also on marine processes: Fast-moving ocean currents can sometimes prevent sediments from accumulating. For example, the east coast of Florida has a very narrow shelf because there is no offshore dam and because the swift current of the nearby Gulf Stream scours eroded sediment away.

The shelves of the active Pacific margins are generally not as broad and flat as Atlantic shelves. An example is the abbreviated shelf off the west coast of South America, where the steep western slope of the Andes Mountains continues nearly uninterrupted beneath the sea into the depths of the Peru–Chile Trench (see again Figure 4.3). Active-margin shelves have more varied topography than passive-margin shelves; the character of con-

tinental shelves at an active margin is determined more by faulting, volcanism, and tectonic deformation than by sedimentation. Off the coast of southern California, for example, the shelves are broken into a series of deep depressions and high ridges that were formed by faulting and folding. The crests of some of the ridges emerge off the southern California coast, forming islands such as Catalina, San Clemente, Anacapa, and others.

A few Pacific shelves are broad, however. As in the Atlantic, natural offshore dams trap sediments and form shelves; but in the Pacific the sediment-trapping dams are more commonly offshore chains of volcanoes or lines of coral reefs. Volcanic activity east of China and Southeast Asia has formed a broad basin that is now filling with sediments and is one of the largest shelves in the Pacific (see Figure 4.4).

Because of their gentle slope, continental shelves are greatly influenced by changes in sea level. About 120 million years ago, high rates of seafloor spreading were associated with the expansion in volume of the mid-ocean ridges. The enlarged ridges displaced seawater and caused sea level to rise. During that time, sea level may have been about 300 meters (1,000 feet) higher than it is at present, flooding about 35% of the continents and resulting in proportionally greater shelf areas. In contrast, around 18,000 years ago, at the height of the last **ice age**, or major period of glaciation, the ridges had become less active, and massive ice caps covered huge regions of the continent. The water that formed the thick ice sheets was derived from the ocean, and sea level fell nearly 100 meters (about 300 feet) below its present position. The continental shelves were almost completely exposed, and the surface area of the continents was about 18% greater than it is today. Rivers and waves cut into the sediments accumulated during periods of higher sea level and transported some coarse sediments to their present locations at the shelves' outer edges. Sea level began to rise again when the ice caps melted, and sediments again began to accumulate on the shelves. (More on the history of sea level change will be found in the discussion of coasts in Chapter 11.)

The continental shelves have been the focus of intense exploration for natural resources. Because shelves are the submerged margins of continents, any deposits of oil or minerals along the continental edges are likely to continue offshore. Water depth over shelves averages only about 75 meters (250 feet); so large areas of the shelves are accessible to mining and drilling activities. Many of the techniques used to find and exploit natural resources on land can also be used on the continental shelves. Resource development requires intense scientific investigations, and our understanding of the geology of the shelves has benefited greatly from the search for offshore oil and natural gas.

Continental Slopes

The **continental slope** is the transition between the gently tilting continental shelf and the deep-ocean floor. Continental slopes are formed by sediments that reach the built-out edge of the shelf and fall over the side. At active margins a slope may also include marine sediments scraped off a descending plate during subduction. The inclination of a typical continental slope is about 4° (or 70 meters per kilometer, 370 feet per mile), equal to the steepest slope allowed on the interstate highway system. As Figure 4.5b shows, even the steepest of these slopes is not precipitous—a 25° slope is the greatest incline yet discovered. In general, continental slopes at active margins are steeper than those at passive margins. Continental slopes average about 20 kilometers (13 miles) wide and end at the continental rise, usually at a depth of about 3,700 meters (12,000 feet). The bottom of the continental slope is the true edge of a continent.

The **shelf break** marks the abrupt transition from continental shelf to continental slope. The depth of water at the shelf break is surprisingly constant—about 150 meters (500 feet) worldwide—but there are exceptions. The great weight of ice on Antarctica, for example, has isostatically depressed that continent, and the depth at the shelf break is 300–400 meters (1,000–1,300 feet). The shelf break in Greenland is similarly depressed.

Submarine Canyons

Submarine canyons are V-shaped indentations incised into the continental shelf and slope, often terminating on the deep-sea floor in a fan-shaped wedge of sediment (**Figure 4.6**). More than a hundred submarine canyons nick the edge of nearly all of Earth's continental shelves. The canyons generally trend at right angles to the shoreline (and shelf edge), sometimes beginning very close to shore. Congo Canyon actually extends into the African continent as a deep estuary at the mouth of the Congo

Figure 4.6 A submarine canyon.

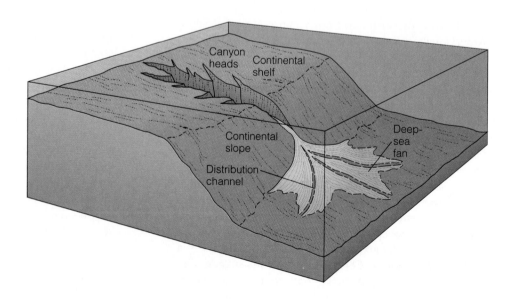

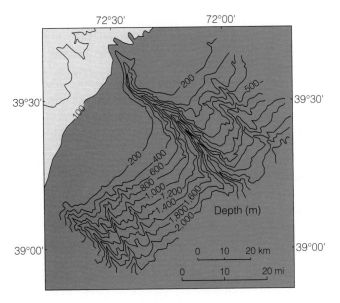

Figure 4.7 A topographic map of Hudson Canyon.

Figure 4.8 A turbidity current flowing down a submerged slope off the island of Jamaica. The propeller of a submarine caused the turbidity current by disturbing sediment along the slope.

River. These enigmatic features are large: Submarine canyons similar in size and profile to Arizona's Grand Canyon have been discovered!

Hudson Canyon, a typical large canyon on a passive margin, is shown in **Figure 4.7**. Like many submarine canyons, Hudson Canyon is located just offshore of the mouth of a river or stream, in this case, New York's Hudson River. Because of their similarity to canyons on land, submarine canyons appear to have been created by erosion; so marine geologists initially thought the canyons were carved into the shelves by stream erosion at times of lower sea level. But most researchers agree that sea level has never fallen more than 200 meters (660 feet) below its present level in the last 600 million years. Stream erosion could account for the shape of the uppermost parts of the canyons; however, as submarine canyons can be traced to depths in excess of 3,000 meters (10,000 feet), stream erosion could not have played a direct role in cutting their lower depths. What, then, caused the submarine canyons to form? As is so often the case in the history of marine science, the answer to this riddle was associated with a peculiar event.

A sharp earthquake struck the Grand Banks south of Newfoundland at 1531 (3:31 P.M.) Greenwich time on 18 November 1929. The quake's **epicenter**, the point on the surface of the Earth directly above the focus of an earthquake, was located on the continental slope at a depth of about 2,200 meters (7,200 feet). Ordinarily such an earthquake would not attract much attention, but a number of important transatlantic communication cables lay on the ocean floor south of the earthquake area. One by one these cables failed. A segment of the Western Union cable nearest the epicenter broke almost immediately, and a French cable about 160 kilometers (100 miles) south of the Western Union cable parted at 1835 (6:35 P.M.). Another Western Union cable to the south snapped early the next morning, and cables even farther south were damaged later that day. Officials had expected an occasional random failure, but a *sequence* of cable failures, from north to south along 480 kilometers (300 miles) of seafloor in the mid-Atlantic, was baffling.

Local landslides or sediment liquefaction triggered by the earthquake probably caused the first cable breaks, and although no one thought of it right away, an abrasive underwater "avalanche" almost certainly caused the more distant breaks. These avalanche-like sediment movements, called **turbidity currents**, are caused when turbulence mixes sediments into water above a sloping bottom. The sediment-filled water is denser than the surrounding water; so the thick muddy fluid runs down the slope. In a 1952 paper Bruce Heezen and Maurice Ewing estimated that the turbidity current that caused the 1929 breaks could have been traveling at speeds up to 97 kilometers per hour (60 miles per hour). These estimates have since been challenged, and recent analysis suggests a more sedate (though still impressive) 27 kilometers per hour (17 mph). **Figure 4.8** is a rare photograph of a turbidity current.

What is the connection between turbidity currents and submarine canyons? Some geologists have suggested that the canyons have been formed by abrasive turbidity currents plunging down the canyons. Small amounts of debris

Figure 4.9 A continuous cascade of sediment at the head of San Lucas submarine canyon (off the coast of Baja California, Mexico), which may be eroding the narrow gorge in conjunction with occasional turbidity currents. The sandfall pictured here is about 2 meters (7 feet) across.

may cascade continuously down the canyons (**Figure 4.9**), but earthquakes can shake loose huge masses of sediment, which rush down the edge of the shelf, scouring the canyon deeper as they go. In this way the canyons can be cut to depths far below the reach of streams, even during the low sea level of the ice ages.

Continental Rises

Along passive margins, the oceanic crust at the base of the continental slope is covered by an apron of accumulated sediment called the **continental rise** (see again Figure 4.5a). Sediments from the shelf slowly descend to the ocean floor along the whole continental slope, but most of the sediments that form the continental rise are transported to the area by turbidity currents. The width of the rise varies from about 100 to 1,000 kilometers (63 to 630 miles), and its slope is gradual —about one-eighth that of the continental slope. One of the widest and thickest continental rises has formed in the Bay of Bengal at the mouths of the Ganges–Brahmaputra River, the most sediment-laden of the world's great rivers.

DEEP-OCEAN BASINS

Away from the margins of continents, the structure of the ocean floor is quite different. Here the seafloor is a blan-

ket of sediments overlying basaltic rocks up to 5 kilometers (3 miles) thick. Deep-ocean basins comprise more than half of Earth's surface.

The deep-ocean floor consists mainly of oceanic ridge systems and the adjacent sediment-covered plains. Deep basins may be rimmed by trenches or by masses of sediment. Flat expanses are interrupted by islands, hills, active and extinct volcanoes, and active zones of seafloor spreading. The sediments on the deep floor reflect the history of the surrounding continents, the biological productivity of the overlying water, and the ages of the basins themselves.

Oceanic Ridges

If the ocean evaporated, the oceanic ridges would be Earth's most remarkable and obvious feature. An **oceanic ridge** is a mountainous chain of young basaltic rock at the active spreading center of an ocean. Stretching 65,000 kilometers (41,000 miles), more than 1½ times the Earth's circumference, oceanic ridges girdle the globe like seams surround a softball (**Figure 4.10**). The rugged ridges, which often are devoid of sediment, rise about 2 kilometers (1.25 miles) above the seafloor. In places they project above the surface to form islands such as Iceland, the Azores and Easter Island. Oceanic ridges and their associated structures account for 23% of the world's solid surface area (all the land above sea level accounts for 29%). Although these features are often called mid-ocean ridges, less than 60% of their length actually exists at mid-ocean.

As we saw in our discussion of plate tectonics, the rift zones associated with oceanic ridges are sources of new ocean floor where lithospheric plates diverge. The oceanic ridges are widest where they are most active. The youngest rock is located at the active ridge center, and rock becomes older with distance from the center. As crust cools, it shrinks and subsides. Slowly spreading ridges have a steeper profile than rapidly spreading ones because slowly diverging seafloor cools and shrinks closer to the spreading center. **Figure 4.11**, a topographic map of the North Atlantic, clearly shows the great extent of the Mid-Atlantic Ridge, a typical oceanic ridge. Features on the older basalt floor farthest from spreading center tend to be obscured by a layer of sediment that thickens away from the center.

As can be seen in Figure 4.11, the Mid-Atlantic Ridge is fractured at more or less regular intervals by transform faults. A fault is a break in a rock mass along which movement has occurred, and **transform faults** are planes along which rock masses slide horizontally past one another (**Figure 4.12**). When segments of a ridge system are offset, the fault connecting the axis of the ridge is a transform fault. Shallow earthquakes are common along transform

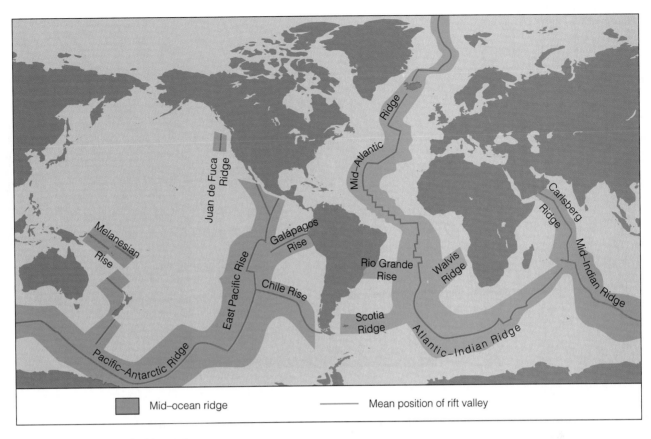

Figure 4.10 The oceanic ridge system.

Legend:
☐ Mid–ocean ridge   —— Mean position of rift valley

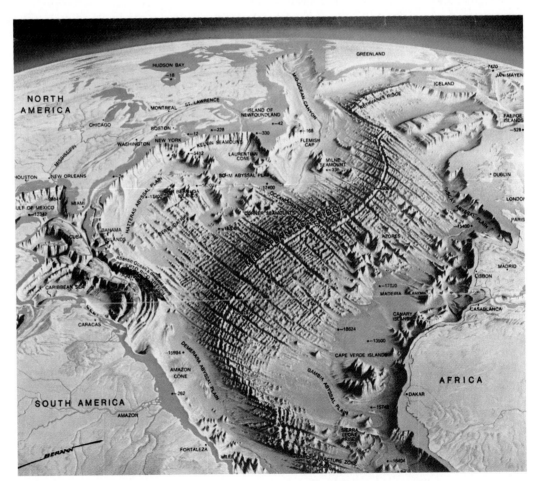

Figure 4.11 Map of a portion of the Atlantic Ocean floor showing some major oceanic features: mid-ocean ridge, transform faults, fracture zones, submarine canyons, seamounts, continental rises, trenches, and abyssal plains.

Figure 4.12 Transform faults and fracture zones along an oceanic ridge.

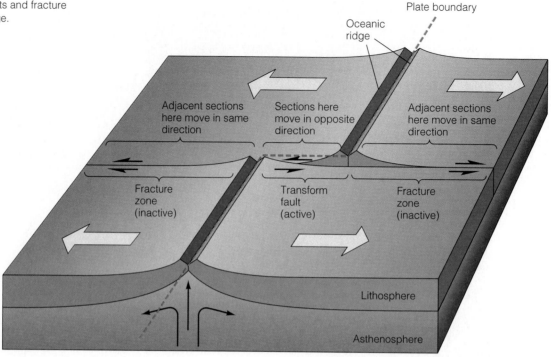

Plate boundary

Oceanic ridge

Adjacent sections here move in same direction

Sections here move in opposite direction

Adjacent sections here move in same direction

Fracture zone (inactive)

Transform fault (active)

Fracture zone (inactive)

Lithosphere

Asthenosphere

faults. Since the ocean floor cannot expand evenly on the surface of a sphere, plate divergence on a sphere can only be irregular and asymmetrical, and transform faults and fracture zones result. **Fracture zones** are seismically inactive areas that show evidence of past transform fault activity. While segments of a lithospheric plate on either side of a transform fault move in *opposite* directions from each other, the plate segments adjacent to a fracture zone move in the *same* direction, as Figure 4.12 shows. Earthquakes rarely occur along fracture zones.

Hydrothermal Vents

Perhaps the strangest features of the ocean basins are the **hydrothermal vents**. Hot springs were discovered on oceanic ridges in 1977 by Robert Ballard and J. F. Grassle of the Woods Hole Oceanographic Institution. Diving in *Alvin* at 3 kilometers (1.8 miles) near the Galápagos Islands along the East Pacific Rise—an oceanic ridge—they came across rocky chimneys up to 20 meters (66 feet) high, from which dark, mineral-laden water was blasting at 350°C (660°F) (**Figure 4.13**). Only the great pressure at this depth prevented the escaping water from flashing to steam. These "black smokers," as they were quickly nicknamed, fascinate marine geologists. They believe that water descends through fissures and cracks in the ridge floor until it comes into contact with very hot rocks associated with active seafloor spreading.

There the superheated, chemically active water dissolves minerals and gases and escapes upward through the vents by convection.

Since that first discovery, vents have been found on the Mid-Atlantic Ridge east of Miami, in the Sea of Cortez east and south of Baja California, and off the coast of Washington State and southern Oregon. Scientists now believe that hydrothermal vents may be very common on oceanic ridges, especially in zones of rapid seafloor spreading. In July 1990 vents were discovered in fresh water, at the bottom of Lake Baikal in southern Siberia. This discovery suggests that the world's oldest and deepest lake may someday become an ocean as Asia slowly breaks apart.

Not all vents form chimneys of mineral deposits; some are simply cracks in the seabed, or porous mounds, or broad segments of ocean-ridge floor through which warm mineral-laden water percolates upward. Cooler vents result when hot, rising water mixes with cold bottom water before reaching the surface. Water temperature in the vicinity of most hydrothermal vents averages 8–16°C (46–61°F), much warmer than usual for ocean-floor water, which has an average temperature of 3–4°C (37–39°F). We will study the unusual communities of animals that populate the vents in Chapter 14.

A volume of water equal to the volume of the world ocean is thought to circulate through the hot oceanic crust at spreading centers every 1 to 10 million years! The

Figure 4.13 A "black smoker" discovered at a depth of about 2,800 meters (9,200 feet) along the East Pacific Rise.

continental margins and the oceanic ridges about 3,700 to 5,500 meters (12,000 to 18,000 feet) below the surface (see again Figure 4.11). The Canary Abyssal Plain, a huge plain west of the Canary Islands in the North Atlantic, has an area of about 900,000 square kilometers (350,000 square miles).

Abyssal plains are extraordinarily flat. A 1947 survey by the Woods Hole Oceanographic Institution ship *Atlantis* found that a large Atlantic abyssal plain varies no more than a few meters in depth over its entire area. Such flatness is caused by the smoothing effect of the layers of sediment, which often exceed 1 kilometer (3,300 feet) in thickness. Most of the sediment that forms the abyssal plains appears to be of terrestrial or shallow-water origin, not derived from biological activity in the ocean above. Some of it may have been transported to the plains by turbidity currents. These deep sediment layers mask irregularities in the underlying ocean crust, but echo sounders can "see" through this sediment to reveal the complex topography of the basaltic basin floor below. The broad shoulders of the Mid-Atlantic Ridge extend beneath this cloak of sediment almost as far as the bordering continental slopes.

Abyssal plain sediments may not be thick enough to cover the underlying basaltic floor near the edges bordering the oceanic ridges. Here the plains are punctuated by **abyssal hills**, small sediment-covered extinct volcanoes or intrusions of once-molten rock less than 200 meters (650 feet) high. These abundant features are thought to be associated with seafloor spreading. Abyssal hills occur alone or in groups, and some occur in lines parallel to the flanks of the nearby oceanic ridge. Abyssal plains and abyssal hills account for nearly all of the area of deep-ocean floor that is not part of the oceanic ridge system.

Trenches and Island Arcs

A **trench** is an arc-shaped depression in the deep-ocean floor. These flat-bottomed creases in the seafloor occur where a converging oceanic plate is subducted. The water temperature at the bottom of a trench is slightly cooler than the near-freezing temperatures of the adjacent flat ocean floor, an observation that suggests that trenches are underlain by cold ocean crust sinking into the upper mantle. Trenches (and their associated island arcs topped by erupting volcanoes) are the most active geological features on Earth. Great earthquakes and tsunami (huge waves we will discuss in Chapter 9) are born in them. **Figure 4.14** shows the distribution of the ocean's major trenches. Not surprisingly, most are around the edges of the active Pacific.

Trenches are the deepest places in the Earth's crust, punching 3 to 6 kilometers (1.8–3.7 miles) deeper than

heat and chemicals issuing from these structures may play important roles in the chemical composition of seawater and the atmosphere, and in the formation of mineral deposits.

Abyssal Plains and Abyssal Hills

A quarter of the Earth's surface consists of abyssal plains and abyssal hills. *Abyssal* is an adjective derived from a Greek word meaning "without bottom." While this is obviously not literally true, you can appreciate how the term came into use following the *Challenger* Expedition's laborious soundings of these extremely deep areas.

Abyssal plains are flat, cold, featureless expanses of sediment-covered ocean floor found on the periphery of all oceans. They are most common in the Atlantic, less so in the Indian Ocean, and relatively rare in the active Pacific (where peripheral trenches trap most of the sediments flowing off the continents). They lie between the

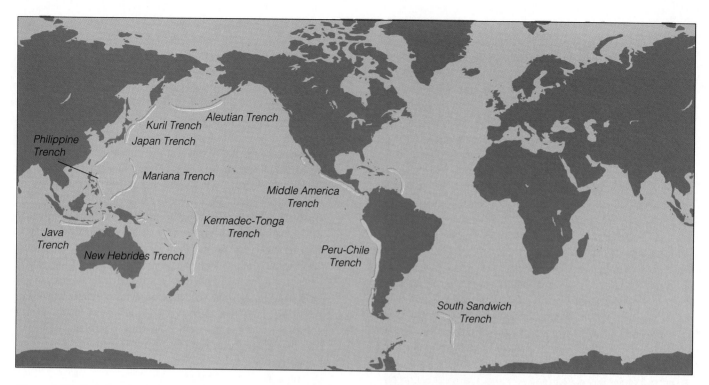

Figure 4.14 Trench regions of the world.

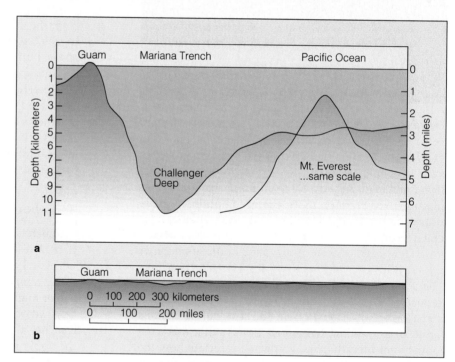

Figure 4.15 The Mariana Trench. (a) Comparing the Challenger Deep and Mount Everest at the same scale shows that the deepest part of the Mariana Trench is about 20% deeper than the mountain is high. (b) The Mariana Trench shown without vertical exaggeration.

the adjacent basin floor! The ocean's greatest depth is in the Mariana Trench of the western Pacific, where *Trieste* set the world's deep-diving record (see Box 4.1). Near the place where *Trieste* came to rest, the ocean bottom is 11,022 meters (36,163 feet) below sea level, 20% deeper than Mount Everest is high (see **Figure 4.15**). The Mari-

ana Trench is about 70 kilometers (44 miles) wide and 2,550 kilometers (1,600 miles) long, typical dimensions for these structures.

From above, trenches are curving chains of individual V-shaped indentations. The trenches are curved because of the geometry of plate interactions on a sphere. The convex sides of these curves face the open ocean (see again Figure 4.14). The trench walls on the island side of the depressions are steeper than those on the seaward side, indicating the direction of plate subduction. The sides of trenches become steeper with depth, normally reaching angles of about 10°–16° before flattening to a floor underlain by thick sediment. (Parts of the concave wall of the Kermadec–Tonga Trench are the world's steepest, at 45°.) No continental rise occurs along coasts with trenches because the sediment that would form the rise ends up at the bottom of the trench.

Island arcs, curving chains of volcanic islands and seamounts, are almost always found paralleling the concave edges of trenches. As you may remember from Chapter 3, trenches and island arcs are formed by tectonic activity associated with subduction. The descending lithospheric plate contains some materials that melt (as the plate sinks into the mantle) and then rise to the surface as magmas and lavas, which form the chain of islands behind the trench. The Aleutian Islands, the Lesser Antilles, and the Marianas Islands are island arcs. (See Figure 3.20 for a review of the processes involved in their construction.)

Seamounts and Guyots

The ocean floor is dotted with thousands of "islands" that do not rise above the surface of the sea. These projections are called **seamounts**. Seamounts are circular or elliptical, more than a kilometer (0.6 mile) in height, with relatively steep slopes of 20° to 25°. (Abyssal hills, in contrast, are much more abundant, almost always smaller, and not as steep.) Seamounts may be found alone or in groups of from 10 to 100. Though many form at hot spots, most are thought to be submerged inactive volcanoes that formed at spreading centers (see again Figure 3.27). Movement of the lithosphere away from spreading centers has carried them outward and downward to their present positions. As many as 10,000 seamounts are thought to occur in the Pacific, about half the world total.

Guyots are flat-topped seamounts that once were tall enough to approach or penetrate the sea surface. Generally they are confined to the west-central Pacific. The flat top suggests that they were eroded by wave action when they were near sea level. Their plateau-like tops eventually sank too deep for wave erosion to continue wearing them down. Like the more abundant seamounts, most guyots were formed near spreading centers and transported outward and downward by seafloor spreading (see again Figure 3.27). As noted in Chapter 3, chains of seamounts and guyots are compelling evidence for the theory of plate tectonics.

AN OVERVIEW OF OCEAN BOTTOM TOPOGRAPHY

Figure 4.16 is a **hypsographic curve** (*hypsos* = height + *graphos* = to write), a plot of the area of Earth's surface above any given elevation above or depth below sea level. Note that more than half of the Earth's solid surface is at least 3,000 meters (10,000 feet) below sea level. *The average depth of the ocean is much greater than the average elevation of the continents.* The average depth of the world ocean is 3,790 meters (12,430 feet), but the average height of the continents is only 840 meters (2,770 feet). This distribution is largely a result of the difference in density between granitic rock and basalt; the lighter granitic continents "float" in isostatic equilibrium above the heavy rock of the ocean basins. The shape of the hypsographic curve results from the smaller volume of continental crust relative to oceanic crust. **Figure 4.17** shows the relative amounts of Earth's surface composed

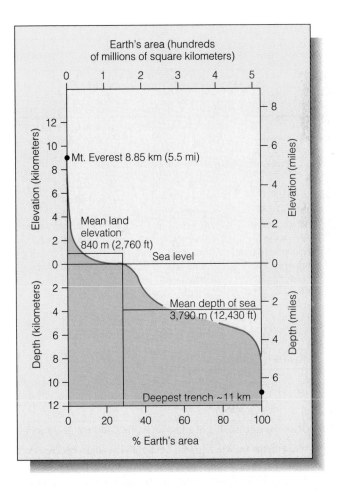

Figure 4.16 The hypsographic curve, a graph showing the distribution of elevations and depths on the Earth.

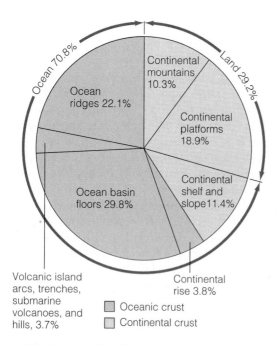

Ocean 70.8%

Land 29.2%

Continental mountains 10.3%

Ocean ridges 22.1%

Continental platforms 18.9%

Ocean basin floors 29.8%

Continental shelf and slope 11.4%

Volcanic island arcs, trenches, submarine volcanoes, and hills, 3.7%

Continental rise 3.8%

☐ Oceanic crust
☐ Continental crust

Figure 4.17 Features of the Earth's solid surface, shown as percentages of the planet's total surface.

of continental and oceanic structures and the important subdivisions of each.

A comprehensive view of the ocean floor is shown in Appendix IV. Continental margins, oceanic ridges, transform faults, abyssal plains, and most other features discussed in this chapter are also shown there.

CHAPTER SUMMARY

The ocean floor can be divided into two regions: continental margins and deep-ocean basins. The continental margin, the relatively shallow ocean floor nearest the shore, consists of the continental shelf and the continental slope. The continental margin shares the structure of the adjacent continents, but the deep-ocean floor away from land has a much different origin and history. Prominent features of the deep-ocean basins include rugged oceanic ridges, flat abyssal plains, occasional deep trenches, and curving chains of volcanic islands. Most of the ocean floor is blanketed with sediment. The processes of plate tectonics, erosion, and sediment deposition have shaped the continental margins and ocean basins.

Terms and Concepts to Remember

abyssal hill	fracture zone	seamount
abyssal plain	hydrothermal vent	shelf break
active margin	hypsographic	submarine canyon
bathyscaphe	curve	transform fault
continental margin	ice age	trench
continental rise	island arc	turbidity current
continental shelf	ocean basin	
continental slope	oceanic ridge	
epicenter	passive margin	

Study Questions

1. Why did people think an ocean was deepest at its center? What changed their minds?

2. Draw a rough outline of an ocean basin. Label the major parts.

3. What do the facts that granite underlies the edges of continents and that basalt underlies the deep-ocean basins, suggest?

4. The terms *leading* and *trailing* are also used to describe continental margins. How do you suppose these words relate to *active* and *passive*, or *Atlantic-type* and *Pacific-type* used in the text?

5. Why are abyssal plains relatively rare in the Pacific?

6. Answer this question if you have already read Chapter 3: Your time machine has been programmed to deliver you to Frankfurt, Germany, on a chilly evening in January 1912, to hear Wegener's lectures on continental drift. What two illustrations from this chapter would you take with you to cheer him up after the lecture? Why did you select those particular illustrations?

For Further Study

Edmond, J., and K. Von Damm. 1983. "Hot Springs on the Ocean Floor." *Scientific American*, April, 78–93. Information on the "black smokers" and the associated organisms.

Emery, K. O. 1969. "The Continental Shelves." *Scientific American*, September, 107–22. Paper by one of the notable researchers on the topic.

Kennett, J. P. 1982. *Marine Geology*. Englewood Cliffs, NJ: Prentice-Hall. A standard text.

Macdonald, K. C., and P. J. Fox. 1990. "The Mid-Ocean Ridge." *Scientific American*, June, 72–79. The article discusses complex ridge offset patterns and the reason most ridges have an undulating upper surface.

Oceanus 34 (no. 4, Winter 1991–92) is dedicated to the topic of mid-ocean ridges.

Shepard, F. 1969. *The Earth Beneath the Sea*. Rev. ed. New York: Atheneum. A firsthand account from the grandfather of modern marine geology.

Sieboldt, E., and W. H. Berger. 1994. *The Sea Floor*. 2d ed. New York: Springer-Verlag. A broad overview of recent research in plate tectonics, marine sediments, climatology, and the uses to which the seafloor is put.

Stanley, S. M. 1986. *Earth and Life Through Time*. New York: Freeman. Information on dating of sediments.

Willie, P. J. 1976. *The Way the Earth Works*. New York: Wiley. Concise summary of floor anatomy and plate tectonics.

5 SEDIMENTS

The Memory of the Ocean

In the early 1960s, researchers aboard the American research ship *Chain* discovered a layer of sediments of unknown composition beneath the floor of the Mediterranean Sea. For the next ten years scientists used powerful echo sounders to trace the mysterious layer, which appeared everywhere they probed this inland sea. The M-reflector, as it came to be called, covered the basement of the Mediterranean like a thick blanket of snow in a mountain valley.

On 24 August 1970 geologists aboard the deep-sea drilling ship *Glomar Challenger* inspected the first samples from the layer. They had expected to see finely compacted particles of silts and clays; instead they found pea-sized gypsum (calcium sulfate) gravel, which forms when seawater evaporates. Gypsum occurs in muddy sediments on arid coasts and on land at the sites of dry lakes and ponds. It had never been reported on the deep-sea floor and does not occur in gravel deposits on the surrounding continents. Other cores showed evidence of algal mats, which form only in water less than 10 meters (33 feet) deep. All these puzzling samples were taken about 200 meters (700 feet) beneath the seabed in 2,000 meters (6,600 feet) of water. What could account for them?

From the evidence contained in these sediment samples, two scientists aboard *Glomar Challenger*, William B. F. Ryan of the Lamont-Doherty Earth Observatory and Kenneth J. Hsü of the Swiss Federal Institute of Technology, pieced together a controversial new history of the Mediterranean Sea. About 6 million years ago, they said, the shallow Strait of Gibraltar—the narrow passage between the Mediterranean and the Atlantic—was blocked by an uplift of the land—a dam, in effect. Water could no longer flow into the Mediterranean; so the sea east of Gibraltar gradually evaporated, leaving behind a desert of gypsum-rich salts—the future M-reflector. Half a million years later, however, either the strait subsided or the level of the Atlantic rose, and the ocean flooded back into the basin. In little more than a century, waters equal to 1,000 Niagaras falling across the Gibraltar precipice refilled the Mediterranean Sea!

At least that's how Ryan and Hsü reconstructed the tale of the drill cores. But some scientists who have studied the same evidence doubt that the Mediterranean was ever completely dry. They suggest that most of the floor was studded by alkaline lakes, or that perhaps the entire bottom was covered by very shallow, very salty water. The M-reflector could still have formed in this scenario.

Sediments have many stories to tell, some of them still open to interpretation. Sediments are the memory of the ocean because they provide a record—if we can read it—of the recent history of the ocean, and therefore of the planet itself.

Did huge waterfalls along the Gibraltar Precipice refill the Mediterranean Sea?

Most of the ocean floor is being slowly dusted by a continuing rain of sediments. **Sediment** is particles of organic or inorganic matter that accumulate in a loose, unconsolidated form. The particles originate from the weathering and erosion of rocks, from the activity of living organisms, from volcanic eruptions, from chemical processes within the water itself, and even from space. Accumulation rates vary from a few centimeters per year to the thickness of a dime every thousand years.

Marine sediments occur in a broad range of sizes and types. Beach sand is sediment; so is the mud of a quiet bay and the mix of silt and tiny shells found on the continental margins. Less familiar sediments are the fine clays of the deep-ocean floor, the biologically derived oozes of abyssal plains, and the nodules and coatings that form around hard objects on the seafloor. The origin of these materials, and the distribution and sizes of the particles, depends on a combination of physical and biological processes.

What do sediments look like? That depends on where you look. **Figure 5.1** shows the Mid-Atlantic Ridge, 48°N, at a depth of 2,629 meters (8,626 feet). Sponges and some relatives of coral grow from young rocky outcrops only lightly powdered with sediment. Contrast that rough ridge with the smooth seafloor shown in **Figure 5.2**. The sediment here is about 200 meters (660 feet) thick.

The colors of marine sediments are often quite striking. Sediments of biological origin are white or cream-colored, with deposits high in silica tending toward gray. Deep-sea clays are often a chocolate-brown color, from the rusting (oxidation) of iron within the sediments to form iron oxide. Other clays are shades of green or tan. Nodular sediments are a dark sooty brown or black. Some nearshore sediments contain decomposing organic material and smell of hydrogen sulfide, but most are odorless.

CLASSIFYING SEDIMENTS

Particle size is frequently used to classify sediments. The scheme shown in **Table 5.1** was first devised in 1898 and with refinements has been used by geologists, soil scientists, and oceanographers ever since. In this classification the coarsest particles are boulders, which are more than 256 millimeters (about 10 inches) in diameter. Although boulders, cobbles, and pebbles occur in the ocean, most marine sediments are made of finer particles: **sand**, **silt**, and **clay**. The smaller the particle, the more easily it can be transported by streams, waves, and currents. As sediment is transported, it tends to be sorted by size; coarser grains, which are only moved by turbulent flow, tend to remain behind finer grains, which are more readily

Figure 5.1 Sediment near the crest of the Mid-Atlantic Ridge. The rock outcrops are dusted only lightly with sediment; sponges and gorgonians (a relative of coral) attach to the rocks.

Figure 5.2 Sediment on the continental rise off the Sahara. In this geologically quiet area, the sediment surface is smooth, and a layer almost 200 meters thick has collected. The small dark object is a sea cucumber.

moved. The clays, particles less than 0.004 millimeter in diameter, can remain suspended for very long periods and may be transported great distances by ocean currents before they are deposited.

A layer of sediment can contain particles of similar size, or it can be a mixture of different-sized particles. Sediments composed of particles of one size, say fine sand, are said to be **well-sorted sediments**. Sediments with a mixture of sizes are **poorly sorted sediments**. Sorting is

Table 5.1 Particle Sizes and Settling Rate in Sediment

Type of Particle	Diameter	Settling Velocity	Time to Settle 4 km (2.5 mi)
Boulder	>256 mm (10 in.)	—	—
Cobble	64–256 mm (>2½ in.)	—	—
Pebble	4–64 mm (⅛–2½ in.)	—	—
Granule	2–4 mm (½2–⅛ in.)	—	—
Sand	0.062–2 mm	2.5 cm/sec (1 in./sec)	1.8 days
Silt	0.004–0.062 mm	0.025 cm/sec (⅟₁₀₀ in./sec)	6 months
Clay	<0.004 mm	0.00025 cm/sec	50 years[a]

[a]Though the theoretical settling time for individual clay particles is very long, clay particles in the ocean can interact chemically with seawater, clump together, and fall at a faster rate.

Table 5.2 Classification of Marine Sediments by Source of Particles

Sediment Type	Source	Examples	Distribution	Percent of All Ocean Floor Area Covered
Terrigenous	Erosion of land, volcanic eruptions, blown dust	Quartz sand, clays, estuarine mud	Dominant on continental margins, abyssal plains, polar ocean floors	~45%
Biogenous	Organic; accumulation of hard parts of some marine organisms	Calcareous and siliceous oozes	Dominant on deep-ocean floor (siliceous ooze below about 5 km)	~55%
Hydrogenous (authigenic)	Precipitation of dissolved minerals from water	Manganese nodules, phosphorite deposits	Present with other more dominant sediments	<1%
Cosmogenous	Dust from space, meteorite debris	Tektite spheres, glassy nodules	Mixed in very small proportion with more dominant sediments	0%

Source: Kennett, 1982; Weihaupt, 1979; Sverdrup, Johnson, and Fleming, 1942.

a function of the energy of the environment—the exposure of an area to the action of waves, tides, and currents. Well-sorted sediments occur in an environment where energy fluctuates within narrow limits. Sediments of the calm deep-ocean floor are typically well sorted (as in Figure 5.2). Poorly sorted sediments form in environments where energy fluctuates over a wide spectrum. The mix of rubble at the base of a rapidly eroding shore cliff is a good example of poorly sorted sediment. Sediments that have been transported by turbidity currents (which can transport a wide range of grain sizes) tend to be poorly sorted.

Another way to classify marine sediments is by their origin. Such a scheme was first proposed in 1891 by Sir John Murray and A. F. Renard after a thorough study of sediments collected during the *Challenger* expedition. A modern modification of their organization is shown in **Table 5.2**. This scheme separates sediments into four cat-

egories by source: terrigenous, biogenous, hydrogenous (or authigenic), and cosmogenous.

Terrigenous Sediments

Terrigenous (*terra* = Earth + *generare* = to produce) sediments are the most abundant. As the name implies, terrigenous sediment originates on the continents or islands near them. Quartz and clay are two common components of such marine sediments. Quartz, an important mineral in granite, is hard, relatively insoluble, very durable, and can withstand lengthy weathering and transport. Quartz sands washed from the adjacent continents are important constituents of the sediments along continental margins. Feldspar, another important mineral in granite, ultimately decomposes to clay. These tiny particles, which are the chief component of soils, are carried to the ocean by wind, rivers, or streams. Because of their small size,

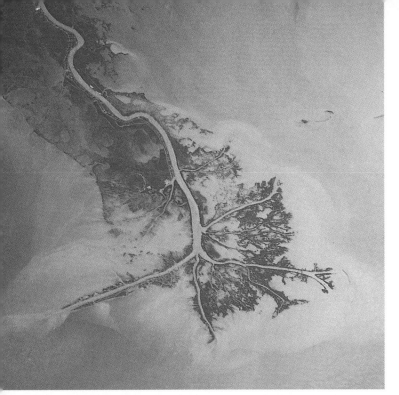

a

b

Figure 5.3 Sources of terrigenous sediments. (a) Rivers are the main source of terrigenous sediments. This photo from space shows sediment carried by the Mississippi River into the Gulf of Mexico after a flood. (b) The wind may transport ash from a volcanic eruption for hundreds of miles and deposit it in the ocean. This ash cloud was caused by the summer 1991 eruption of Mt. Pinatubo in the Philippines.

they are easily transported across the continental shelf to settle slowly to the deep-ocean floor. Although estimates vary, it appears that about 15 billion metric tons (16.5 billion tons) of terrigenous sediments are transported in rivers to the sea each year, with an additional 100 million metric tons transported annually from land to ocean as fine airborne dust and volcanic ash (see **Figure 5.3**).

Biogenous Sediments

Biogenous (*bio* = life + *generare* = to produce) sediments are the next most abundant marine sediment. The siliceous (silicon-containing) and calcareous (calcium carbonate-containing) compounds that make up these sediments of biological origin were originally dissolved in the ocean at mid-ocean ridges or brought to the ocean in solution by rivers. The siliceous and calcareous materials were then extracted during the normal activity of tiny plants and animals in developing protective shells and skeletons. Some of this sediment derives from larger mollusk shells or from stationary colonial animals such as corals, but most of the organisms that produce biogenous sediments drift free in the water as plankton. After the death of their owners, the hard structures fall slowly to the bottom and accumulate in layers. Biogenous sediments are most abundant where ample nutrients encourage high biological productivity, usually near continental margins. Organic molecules within these sediments can form oil and natural gas.

Hydrogenous Sediments

Hydrogenous (*hydro* = water + *generare* = to produce) sediments are minerals that have precipitated directly from seawater. The sources of the dissolved minerals include submerged rock and sediment, leaching of the fresh crust at oceanic ridges, material issuing from hydrothermal vents, or substances flowing to the ocean in river runoff. As we shall see, the most prominent hydrogenous sediments are manganese nodules, which litter abyssal plains, and phosphorite nodules, seen along some continental margins. Hydrogenous sediments are also called **authigenic** (*authis* = in place, "on the spot") because they were formed in the place they now occupy. Though they usually accumulate very slowly, rapid deposition of hydrogenous sediments is possible—in a rapidly drying lake, for example.

Cosmogenous Sediments

Cosmogenous (*cosmos* = universe + *generare* = to produce) sediments, which are of extraterrestrial origin, are the least abundant. These particles enter the Earth's high

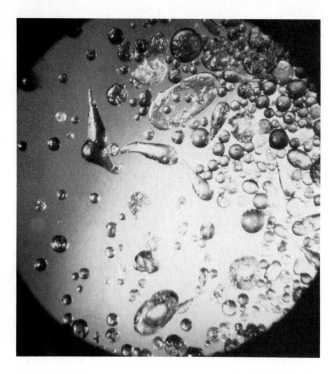

Figure 5.4 Tektites, particles of extraterrestrial origin. These specimens of sculptured glass were separated from sediments off Africa's Ivory Coast. They range from 0.2 to 0.8 millimeter in length. Glassy dust much finer in size, as well as nut-sized chunks, also fell on the Earth.

atmosphere as blazing meteors or as quiet motes of dust. Their rate of accumulation is so slow that they never accumulate as distinct layers—they occur as isolated grains in other sediments, rarely constituting more than 1% of any layer. Estimates vary widely, but between 1,000 and 50,000 metric tons (2 million to 110 million pounds) of this material are thought to fall into the ocean from the high atmosphere every day. Most of this material consists of extremely fine particles of iron and iron-rich minerals that dissolve before reaching the seabed, but occasionally this sediment includes translucent oblong particles of glass known as **tektites** (**Figure 5.4**). Tektites are thought to have been formed by the violent impact of large meteors or asteroids on the crust of Earth or the moon. The impact melts some of the crustal material and splashes it into space; the material melts again as it rushes through the atmosphere, producing the various raindrop shapes shown in the photo. Tektites do not dissolve easily and usually reach the ocean floor. Most are less than 1.5 millimeters (⅟₁₆ inch) long.

Sediments on the ocean floor only rarely come from a single source; most sediment deposits are a mixture of particles. The patterns and composition of sediment layers on the seabed are of great interest to researchers studying conditions in the overlying ocean. Different marine environments have characteristic sediments, and these sediments preserve a record of past and present conditions within those environments.

THE DISTRIBUTION OF MARINE SEDIMENTS

The sediments on the continental margin are generally different in quantity, character, and composition from those on the deeper basin floors. Continental shelf sediments—called **neritic** sediments (*neritos* = of the coast)—consist primarily of terrigenous material. Deep-ocean floors are covered by finer sediments than those of the continental margins, and a greater proportion of deep-sea sediment is of biogenous origin. Sediments of the slope, rise, and deep-ocean floor that originate in the ocean are called **pelagic** sediments (*pelagios* = of the sea). The distribution and average thickness of the marine sediments in each oceanic region are shown in **Table 5.3**. Note that 72% of all marine sediment is associated with continental slopes and rises.

The Sediments of Continental Margins

The bulk of terrigenous sediment is eroded and carried to streams, where it is transported to the ocean. Currents distribute sand and larger particles along the coast, while wave action carries the silts and clays to deeper water. When wave action is too weak to disturb the bottom, the sediment comes to rest. Ideally, these processes produce an orderly sorting of particles by size, from relatively large grains near the coast to relatively small grains near the shelf break. Shelf deposits are subject to further modification and erosion as sea level fluctuates; larger particles may be moved toward the shelf edge when sea level is low, as it was in the ice ages. Poorly sorted sediments are also found as glacial deposits. In polar regions, glaciers and ice shelves give rise to icebergs, which "raft" particles of all sizes and then break, melt, and distribute their mixtures of rocks, gravel, sand, and silt onto high-latitude continental margins and deep-ocean floor. Turbidity currents also disrupt the orderly sorting of sediments on the continental margin by transporting coarse-grained particles away from coastal areas and onto the deep-ocean floor.

The rate of sediment deposition on continental shelves is variable, but it is almost always greater than the rate of sediment deposition in the deep ocean. Near the mouths of large rivers, 1 meter (about 3 feet) of sediment may accumulate every 1,000 years. Along the east coast of the United States, however, many large rivers terminate in

Table 5.3 The Distribution and Average Thickness of Marine Sediments

Region	Percent of Ocean Area	Percent of Total Volume of Marine Sediments	Average Thickness
Continental shelves	9%	15%	2.5 km (1.6 mi)
Continental slopes	6%	41%	9 km (5.6 mi)
Continental rises	6%	31%	8 km (5 mi)
Deep-ocean floor	78%	13%	0.6 km (0.4 mi)

Sources: Emery in Kennett, 1982 (Table 11-1); Weihaupt, 1979; Sverdrup, Johnson, and Fleming, 1942.

estuaries, which trap most of the sediment brought to them. The continental shelf of eastern North America is therefore covered mainly by sediments laid down during the last ice age, when sea level was lower.

In addition to terrigenous material, the continental margins almost always contain biogenous sediments. Organic productivity in coastal waters is often quite high, and the skeletal remains of creatures living on the bottom or in the water above mix with the terrigenous sediments.

Sediments can build to impressive thickness on continental shelves. In some cases, shelf sediments undergo **lithification**—conversion into sedimentary rock by pressure or by cementing—and are thrust above sea level by tectonic forces to form mountains or plateaus. The top of Mount Everest is a shallow-water biogenic marine limestone (a calcareous rock). Much of the Colorado Plateau, with its many stacked layers, was formed by sedimentary deposition and lithification beneath a shallow continental sea, beginning about 570 million years ago. The Colorado River has cut and exposed the uplifted beds to form the Grand Canyon. Hikers walking from the canyon rim down to the river pass through spectacular examples of continental shelf sedimentary deposits. Their journey takes them deep into an old ocean floor!

The Sediments of Deep-Ocean Basins

The Atlantic Ocean bottom is covered by sediments to an average thickness of about 1 kilometer (3,300 feet), while the Pacific floor has an average sediment thickness of less than 0.5 kilometer (1,650 feet). The difference occurs because the Atlantic Ocean is smaller in area and is fed by more large, sediment-laden rivers, and because in the Pacific Ocean many oceanic trenches trap sediments moving toward basin centers. The composition and thickness of pelagic sediments also vary with location, being thickest on the abyssal plains and thinnest (or absent) on the oceanic ridges.

Turbidites   Mudslides occur in many places along the margin of the continental shelf. Loose sediment periodically rushes down the continental slope in turbidity currents, the erosive force of which is thought to help cut submarine canyons. These underwater mudflows can reach the continental rise and often continue moving onto an adjacent abyssal plain before eventually coming to rest. The resulting deposits are called **turbidites**—layers of coarse-grained terrigenous sediment interleaved with the finer sediments typical of the deep-sea floor.

Clays   About 38% of deep-sea sediments are clays and other fine terrigenous particles. As we have seen, the finest terrigenous sediments are easily transported by wind and water currents. Microscopic waterborne particles and tiny bits of wind-borne dust and volcanic ash settle slowly to the deep-ocean floor, forming fine brown, olive-colored, or reddish clays. As Table 5.1 shows, the velocity of particle settling is directly proportional to particle size, and clay particles fall very slowly indeed. Terrigenous sediment accumulation on the deep-ocean floor may be less than a millimeter every thousand years.

Oozes   Seafloor samples taken farther from land usually show a greater proportion of biogenous sediments than those obtained near the continental margins. This is not because biological productivity is higher *farther* from land (the opposite is usually true), but because there is less terrigenous material far from shore and thus a greater *proportion* of biogenous material reaches the deep-sea bed.

Deep-ocean sediment containing at least 30% biogenous material is called an **ooze** (surely one of the most descriptive terms in the marine sciences). Oozes are named after the dominant remnant organism of which they are composed. The organisms contributing their remains to deep-sea oozes are small, single-celled, drifting plantlike organisms and the single-celled animals that feed on them. The hard shells and skeletal remains of these creatures

a

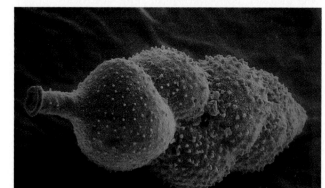

b

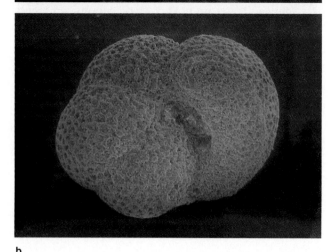

c

Figure 5.5 Organisms that contribute to calcareous ooze.
(a) A living foraminiferan, an amoeba-like organism. The shell
of this beautiful foram, genus *Hastigerina*, is surrounded by a
bubble-like capsule. It is one of the largest of the planktonic
species with spines reaching nearly 5 centimeters (2 inches) in
length. (b) The shells of two smaller foraminiferans are visible in
these scanning electron micrographs: the bottle-shaped, bottom-
dwelling *Uvigerina* and the snail-like, planktonic *Globigerina*.
(c) Coccoliths, individual plates of coccolithophores, a form of
planktonic algae. Because of their tendency to dissolve, calcareous oozes very rarely occur at bottom depths below 4,500 meters
(14,800 feet).

are of relatively dense glasslike silica or calcium carbonate (limey) substances. When these organisms die, their
shells settle slowly toward the bottom, mingle with fine-
grained terrigenous silts and clays, and accumulate as
ooze. The silica-rich residues give rise to **siliceous ooze**,
the calcium-containing material to **calcareous ooze**.

Oozes accumulate slowly—at a rate of about 1 to 6 centimeters (½ to 2½ inches) per 1,000 years—but they collect
almost ten times more quickly than deep-ocean terrigenous clays. The production of any ooze therefore depends
on a delicate balance between the abundance of organisms at the surface, the number that settle to the bottom,
and the rate of accumulation of terrigenous sediment.

Calcareous ooze forms mainly from shells of the
amoeba-like **foraminiferans** (**Figure 5.5a** and **b**), small drifting mollusks called **pteropods**, and tiny plantlike organisms known as **coccolithophores** (**Figure 5.5c**). Though
these creatures live in nearly all surface ocean water, calcareous ooze does not accumulate everywhere on the
ocean floor, because in some areas the shells are dissolved into the seawater. At great depths, seawater contains more CO_2 and becomes slightly acid. This acidity,
combined with the increased solubility of calcium carbonate in cold water under pressure, dissolves the shells
as they fall. At a certain depth, the **calcium carbonate
compensation depth**, the rate at which calcareous sediments *accumulate* equals the rate at which those sediments *dissolve*. Below this depth the tiny skeletons of calcium carbonate dissolve as they fall (or soon after they

a

b

Figure 5.6 Scanning electron micrographs of siliceous oozes, which are most common at great depths. (a) Shells of radiolarians, an amoeba-like organism. Radiolarian oozes are found primarily in equatorial regions. (b) Diatoms, single-celled algae. Diatom oozes are most common at high latitudes.

reach the bottom); so no calcareous oozes form. Calcareous sediment dominates at bottom depths less than about 4,500 meters (14,800 feet), the usual calcium carbonate compensation depth. Sometimes a "snow line" can be seen on undersea peaks: Above the line the white sprinkling of calcareous ooze is visible; below it, the "snow" is absent. About 48% of all deep-ocean sediments are calcareous oozes.

Siliceous (silicon-containing) ooze predominates at greater depths and in colder polar regions. Siliceous ooze is formed from the hard parts of another amoeba-like animal, the beautiful glassy **radiolarian** (**Figure 5.6a**), and single-celled algae, the **diatoms** (**Figure 5.6b**). The silica within these organisms can dissolve in deep water (which is usually deficient in silicon), but silica dissolves much more slowly than calcium carbonate does. Slow dissolution, combined with very high diatom productivity in surface waters, leads to the buildup of siliceous ooze. Diatom ooze is most common in the Antarctic, where the strong ocean currents and seasonal upwelling support large populations of diatoms. Radiolarian oozes occur in equatorial regions, notably in the zone of equatorial upwelling west of South America (see Figure 5.8). About 14% of all deep-ocean sediments are siliceous oozes.

Some deep-sea oozes have been uplifted by geologic processes and are now visible on land. The calcareous white chalk cliffs of Dover in eastern England are partially lithified deposits, composed largely of foraminiferans and coccolithophores. Fine-grained siliceous deposits called *diatomaceous earth* are mined from other deposits. This fossil material is a valued component in paints, filters, and mildly abrasive car and tooth polishes.

Hydrogenous Materials Hydrogenous sediments also accumulate on deep-sea floors. They are found with terrigenous or biogenous sediments; they very rarely form sediments by themselves. Most hydrogenous sediments are thought to originate from chemical reactions that occur on particles of the dominant sediment.

The most famous hydrogenous sediments are manganese **nodules**, which were discovered by the hardworking crew of HMS *Challenger*. The nodules consist primarily of manganese and iron oxides but also contain cobalt, nickel, chromium, copper, molybdenum, and zinc. They form in ways not fully understood by marine chemists, growing at an average rate of only 1 to 10 millimeters (0.04 to 0.4 inch) per *million* years, one of the slowest chemical reactions in nature. Though most are irregular lumps the size of a lemon or a potato, some nodules exceed 1 meter (3.3 feet) in diameter. Manganese nodules often form around nuclei such as sharks' teeth, bits of bone, microscopic alga and animal skeletons, or tiny crystals. The cross section of a manganese nodule is shown in **Figure 5.7a**. Bacterial activity may also play a role in the development of a nodule. Between 20% and 50% of the Pacific Ocean floor may be strewn with nodules (**Figure 5.7b**). Why don't these heavy lumps disappear

a   b

Figure 5.7 Manganese nodules. (a) A cross section cut through a manganese nodule, showing the concentric layers of manganese and iron oxides. This nodule is about 11 centimeters (4 inches) long, a typical size. (b) Cannonball-sized manganese nodules littering the abyssal Pacific.

beneath the constant rain of accumulating sediment? Possibly the continuous churning of the underlying sediment by creatures living there keeps the dense lumps on the surface, or perhaps slow currents in areas of nodule accumulation waft particulate sediments away.

Evaporites are an important group of hydrogenous deposits that include many salts important to humanity. These salts precipitate as water evaporates from isolated arms of the ocean or from landlocked seas or lakes. For thousands of years people have collected sea salts from evaporating pools or deposited beds. Evaporites are forming today in the Gulf of California, the Red Sea, and the Persian Gulf. The first evaporites to precipitate as water's salinity increases are the carbonates such as calcium carbonate, from which limestone is formed. Calcium sulfate, which gives rise to gypsum, is next. Crystals of sodium chloride (table salt) will form if evaporation continues.

Figure 5.8 The general pattern of sediments on the ocean floor.

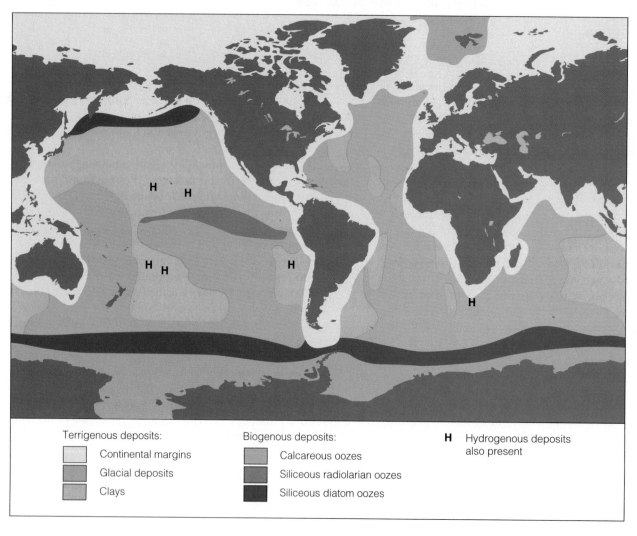

Terrigenous deposits:
- Continental margins
- Glacial deposits
- Clays

Biogenous deposits:
- Calcareous oozes
- Siliceous radiolarian oozes
- Siliceous diatom oozes

H Hydrogenous deposits also present

a

The Distribution of Sediments

Figure 5.8 is a simplified look at the distribution of marine sediments. Notice especially the lack of radiolarian deposits in much of the deep North Pacific, the strand of siliceous oozes extending west from equatorial South America, and the broad expanses of the Atlantic, South Pacific, and Indian ocean floors covered by calcareous oozes. As you might expect, the poorly sorted glacial deposits are found only at high latitudes.

This map summarizes more than a century of effort by marine scientists. Studies of sediments will continue because of their importance to natural resource development and because of the details of Earth's history that remain locked beneath their muddy surfaces.

STUDYING SEDIMENTS

Methods of Study

Deep-water cameras have allowed researchers to photograph bottom sediments. The first of these cameras was simply lowered on a cable and triggered by a trip wire. Other more elaborate cameras have been taken to the seafloor on towed sleds or deep submersibles.

Actual samples usually provide more information than photographs do. HMS *Challenger* scientists used weighted wax-tipped poles and other tools attached to long lines to obtain samples, but today's oceanographers have more sophisticated equipment. Shallow samples may be taken using a **clamshell sampler** (named because of its method of operation, not its target; see **Figure 5.9**). Deeper samples are taken by a **piston corer** (**Figure 5.10**),

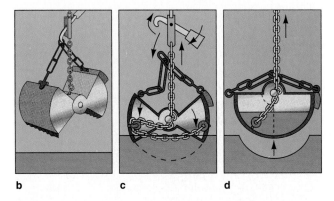

b c d

Figure 5.9 (a) A clamshell sampler (b) before sampling, (c) during sampling, (d) and after the sample has been taken. Note that the sample is relatively undisturbed.

a device capable of punching through as much as 25 meters of sediment and returning an intact plug of material. Using a rotary drilling technique similar to that used to drill for oil, the drilling ship *Glomar Challenger* has returned much longer core segments, some more than 1,100 meters (3,608 feet) in length! As we saw in Chapter 1, *Glomar Challenger* drilled more than 1,000 holes for the Deep Sea Drilling Project, a program begun in 1968 with funds from the National Science Foundation. These cores are stored in a core library, a very valuable scientific resource (**Figure 5.11**). Analysis of sediments and fossils from the Deep Sea Drilling Project cores helped verify the theory of plate tectonics, has shed light on the evolution of life-forms, and has helped researchers to decipher the history of changes in Earth's climate over the last 100,000 years. A newer and larger ship, *JOIDES*

a

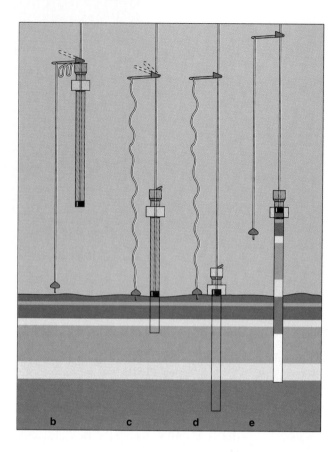

b   c   d   e

Figure 5.10 (a) A piston corer. (b) The corer is allowed to fall to the bottom. (c) The corer reaches the bottom and continues, forcing a sample partway into the cylinder. (d) Tension on the cable draws a small piston within the corer toward the top of the cylinder, and the pressure of the surrounding water forces the piston deeper into the sediment. (e) The corer and sample being hauled in.

Figure 5.11 Sediment cores in storage. Cores are sectioned longitudinally, placed in trays, and stored in hermetically sealed cold rooms. The Gulf Coast Repository of the Ocean Drilling Program, located at Texas A&M University (pictured here), stores about 75,000 sections taken from more than 50 kilometers (30 miles) of cores recovered from the Pacific and Indian oceans. Smaller core libraries are maintained at the Scripps Institution in California (Pacific and Indian oceans) and at the Lamont-Doherty Earth Observatory in New York State (Atlantic Ocean).

Figure 5.12 *JOIDES Resolution*, the deep-sea drilling ship operated by the Joint Oceanographic Institutions for Deep Earth Sampling. The vessel is 124 meters (470 feet) long, with a displacement of over 16,000 tons. The rig can drill to a depth of 9,150 meters (30,000 feet) below sea level.

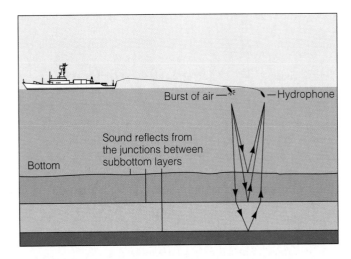

Figure 5.13 A typical method of continuous seismic profiling. A moving ship trails a sound transmitter and receiver at a distance sufficient to minimize the ship's noise. Collapsing bubbles from a burst of compressed air act as a sound source. The sound is reflected from sediment layers beneath the surface and is detected by a sensitive hydrophone for analysis.

Resolution (**Figure 5.12**), is carrying this work forward as part of the Ocean Drilling Program, an international research consortium.

Powerful new continuous seismic profilers have also been used to determine the thickness and structure of layers of sediment on the continental shelf and slope and to assist in the search for oil and natural gas (see **Figure 5.13**). Recent improvements in computerized image processing of the echoes returning from the seabed now permit detailed analysis of these deeper layers.

Sediments as Historical Records

In 1899 the British geologist W. J. Sollas theorized that deep-sea deposits could reveal much of the planet's his-

tory. In the era before plate tectonics theory, this certainly seemed reasonable; the deep-ocean bottom was thought to be a calm, changeless place where an unbroken accumulation of sediment could be probed to discover the entire history of the ocean. Unfortunately for this promising idea, difficulties began to crop up almost immediately. For one thing, the sediments should have been much *thicker* than early probes indicated. If Earth's ocean is truly older than a few hundred thousand years, and if life has existed within it for most of that time, the sediment layer should be thicker than had been observed. Another difficulty lay in the uneven distribution of sediments. Sollas thought the center of an ocean basin should contain the thickest layers of sediment, yet the raised mid-Atlantic bottom was nearly naked. There didn't seem to be any difference in the nature of the overlying seawater that could account for the variations in thickness and composition of the sediments across the bottom of the Atlantic. Oozes were especially puzzling; the organisms that form ooze grow well at the surface of the middle Atlantic, yet the mid-Atlantic floor seemed to bear little ooze.

Today we know the tectonic reasons for these discrepancies, but turn-of-the-century geologists were understandably confused. The sedimentologists who followed Sollas learned to read the history of the ocean and atmosphere in the composition, thickness, and other characteristics of each sediment layer. Analysis of the layers by photographs, samples, and sonic profiling is called **stratigraphy** (*stratum* = layer + *graph* = a drawing). The thick, relatively undisturbed sediments of the deep

BOX 5.1 • *Deep-Ocean Cores*

The Ocean Drilling Program (ODP), currently the largest and most successful multinational Earth science research project, is the direct successor to the Deep Sea Drilling Project (DSDP), which began in 1968. The two drilling ships commissioned for the projects pioneered oil drilling technology to retrieve "cores," long cylinders of sediment and rock extracted from beneath the seafloor. The cores arrive on the deck of *JOIDES Resolution*—the ship currently in active service—in 9.5-meter (30-foot) sections encased in plastic sheaths. Immediately after retrieval, a core is marked to indicate its original location on the seafloor, coded to distinguish top from bottom, measured, and cut into sections for study and storage. Each section is slit lengthwise; one half is stored for the ODP archives, and the other is taken to the first of many shipboard laboratories for study.

Paleontologists then examine the sediments at the bottom of the core to determine the age of the oldest material. Other researchers measure its density, strength, molecular composition, radioactivity, and ability to conduct heat. Magnetic specialists read paleomagnetic data from the core to determine the ages of the rock fragments and the latitude at which they probably formed. Sensors are lowered into the hole from which the core was removed to gather additional information on the physical and chemical properties of the site.

ODP scientists have recently recovered fragments of the oldest remaining seafloor. The sample is about 175 million years old, a relic of the Middle Jurassic period, when the continents were massed in one huge cluster. From this and other cores they have learned about cycles of global climate change, information that will be useful in evaluating the present potential for global warming. They have also discovered how fluids move through the lithosphere, found evidence of an ice-free Antarctica, and noted the influence of plate tectonics on worldwide weather and current patterns. From trapped pollen grains, it is even possible to tell what land plants were thriving on the Earth at the time the

a Researchers examine a deep core taken off the northwestern coast of Australia.

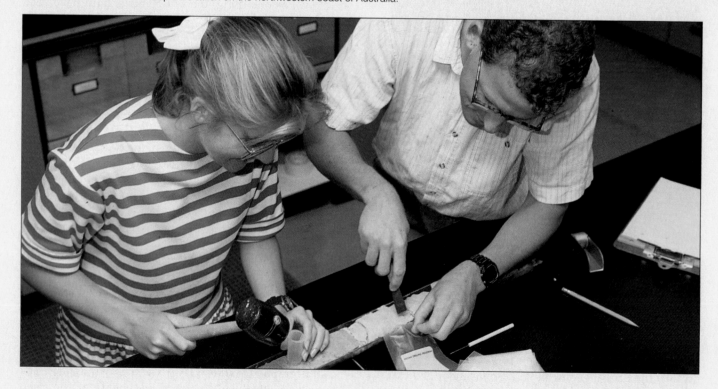

core sediments were laid down. As analysis technology evolves, stored cores are restudied and new information obtained.

The size and shape of a typical deep core section may be seen in **a**. This core was taken 250 meters (820 feet) beneath the seafloor off the northwestern coast of Australia by *JOIDES Resolution* on 22 July 1988. The ocean in the area was around 2,000 meters (6,500 feet) deep.

The material in the core progresses from sandstone (near the bottom of the core, at the right side of the photo) to fine claystone (near the top). Foraminiferans and nanofossils are abundant in the sandstone, and small squidlike fossils, easily visible to the unaided eye, are present in the claystone. The sediments date from the early Cretaceous period, about 140 million years ago. The progression of slowly deposited sediments and fossils suggests the seabed in the area has slowly subsided, possibly because of tectonic effects.

Details in a core from DSDP Hole 480, leg 64, are shown in **b**. This core was obtained by the now-retired drilling ship *Glomar Challenger* on 1 January 1979 near

Guaymas in the Gulf of California. The Gulf is an area of active seafloor spreading (see Figure 3.23), and researchers were interested in sampling the bottom in this place where basaltic magma is intruding into soft, wet, young sediments. The sediments here are very deep and have been accumulating at the extremely rapid rate of about 1,200 meters (4,000 feet) every million years.

Glomar Challenger arrived in the area on 29 December 1978 and used a conventional core barrel to penetrate 444 meters (1,456 feet) into diatomaceous ooze and mudstones. Nearly 273 meters (900 feet) of core were recovered from the hole drilled before Hole 480. Because of the disturbance caused by the conventional coring process, researchers were unable to determine the details of fine structure tantalizingly seen in a few lengths of that core.

On 31 December, *Glomar Challenger* moved to a new location 7 kilometers (4.4 miles) northwest. Hole 480, source of the core segment pictured here, was begun later that day. Unlike the previous hole, Hole 480 was drilled using a newly designed hydraulic corer developed by three DSDP engineers. The new device allowed 80% recovery of a core containing essentially undisturbed laminated diatomaceous ooze and muds.

As can be seen in **b**, the core consists of thin alternating brown and gray bands of sediments. These are believed to be annual couplets, formed about 5 million years ago in response to two seasonal events that still occur in the Gulf: the winter rains that introduce terrigenous clays into the region, and seasonal upwelling and northwest winds that produce diatom blooms. Preservation of these fine details depends upon a low level of free oxygen at the seafloor; burrowing animals would normally churn these fine layers into mush, but in a low-oxygen environment these animals are absent; so the thin alternating sheets of clay and diatom tests are preserved.

Note that parts of the core have already been removed for study. Trenches across the width of the core mark places where particular sets of layers have been removed for isotope analysis. A larger sample for paleomagnetic study was taken from the square depression in the sample. Considering the expense and skill necessary to retrieve and analyze them, these small gray, gritty bits of sediment probably cost more than their weight in fine diamonds!

b This core from the Gulf of California shows thin alternating bands of clay and ooze. Voids in the core show where samples were removed for study.

Figure 5.14

Interpreting changes in climate from analysis of a deep-sea core. Scientists know that certain species of foraminiferans are abundant only when the water temperature is warm, others only when it is cold. By plotting the relative numbers of these species at various depths in a core sample, scientists can depict major changes in climate, which have been dated at about 11,000 and 65,000 years ago. The dashed line represents the end of the last ice age; numbers at right are radiocarbon ages in years.

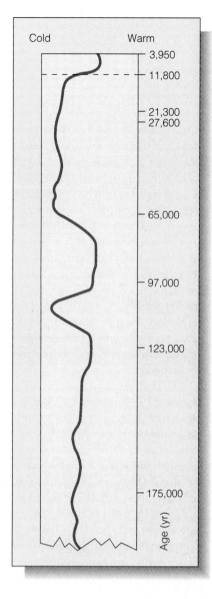

restrial dust or volcanic ash. At these times sunlight could easily reach the ocean surface, and plants and animals grew rapidly. Sediments that contain once-living materials can be dated by radiocarbon analysis. **Figure 5.14** shows a record of recent ice ages reflected in the proportion of foraminiferan species and their ages. Bottom sediments may also contain magnetic particles symmetrically banded along a spreading ridge, potent paleomagnetic evidence of plate tectonics. **Figure 5.15** shows the age of the Pacific Ocean floor using data obtained largely from analyses of the overlying sediment. Note that sediments get older with increasing distance from the East Pacific Rise spreading center. So the reasonable theory of Sollas has fallen short because the ocean basins are continually being recycled by tectonic forces. Therefore, the "memory" of the sediments is not ancient and in fact is continually being erased by ocean floor subduction.

Still, marine sediments in the modern basins can shed light on unexpected details of the last 180 million years of Earth's history. One of the oddest details is the unexplained extinction of up to 52% of known marine animal species (and the dinosaurs) at the end of the Cretaceous period, 65 million years ago. Researchers have proposed hypotheses—such as a sudden and violent increase in worldwide volcanism or the impact of one or more very large meteors or comets—to explain this catastrophe. The clouds of dust and ash thrown into the atmosphere by any of these events would have drastically reduced incident sunlight and greatly affected the lives of organisms and the photosynthetic base of ecosystems. Oceanographers are presently searching for evidence of the cause of the Cretaceous extinctions in layers of deep sediments.

CHAPTER SUMMARY

The ocean floor is covered in most places by layers of sediment. The sediment is composed of particles from land, from biological activity in the ocean, from chemical processes within water, and even from space. The blanket of marine sediment is thickest at the continental margins and thinnest over the active oceanic ridges.

Sediments may be classified by particle size, source, location, or color. Terrigenous sediments, the most abundant, originate on continents or islands near them. Biogenous sediments are composed of the remains of once-living organisms. Hydrogenous sediments are precipitated directly from seawater. Cosmogenous sediments, the ocean's rarest, come to the seabed from space.

The position and nature of sediments provide important clues to Earth's recent history, and valuable resources can sometimes be recovered from them.

basins, away from zones of seafloor spreading, are the most useful to modern stratigraphers. Light-colored bands of sediment many centimeters thick represent times when the atmosphere was heavy with airborne ash from volcanic eruptions. Darker layers are rich in clay minerals and may contain hydrocarbons. These and other distinct bands often extend intact for great horizontal distances and can be identified in cores taken at widely separated locations. These bands are very useful in the correlation of units from core to core.

The distribution, depth, and composition of sediment layers tell of conditions in the past. Oozes might have accumulated rapidly when the atmosphere was free of ter-

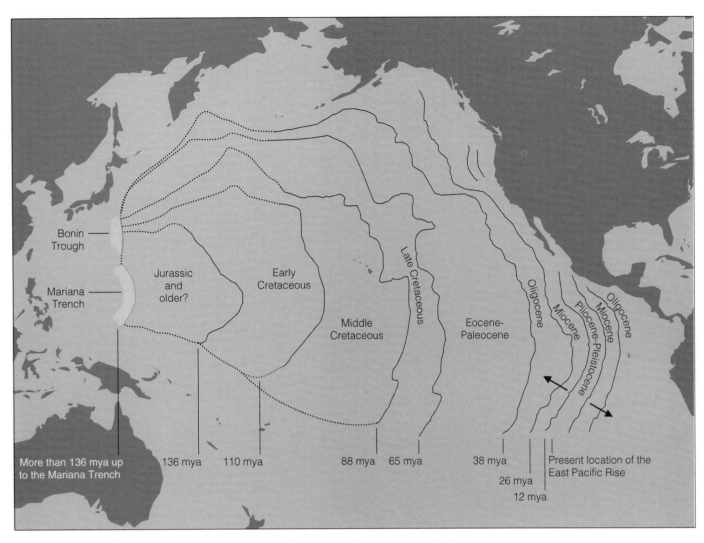

Figure 5.15 The age of portions of the Pacific Ocean floor, based on core samples of sediments just above the basalt seabed, in millions of years ago (mya). The youngest sediments are found near the East Pacific Rise, and the oldest close to the eastern side of the troughs and trenches. Contrast this figure with Figure 3.30.

Terms and Concepts to Remember

authigenic
  sediment
biogenous
  sediment
calcareous ooze
calcium carbonate
  compensation
  depth
clamshell sampler
clay
coccolithophore
cosmogenous
  sediment

diatom
evaporites
foraminiferan
hydrogenous
  sediment
lithification
neritic sediment
nodules
ooze
pelagic sediment
piston corer
poorly sorted
  sediment

pteropods
radiolarian
sand
  sediment
siliceous ooze
silt
stratigraphy
tektite
terrigenous
  sediment
turbidites
well-sorted
  sediment

Study Questions

1. List the four types of marine sediments. Explain the origin of each.

2. How are neritic sediments generally different from pelagic ones?

3. What sediments accumulate most rapidly? Least rapidly? Why?

4. How are sediments collected for study?

5. Can marine sediments tell us about the history of the ocean from the time of its origin? Why, or why not?

6. Where are sediments thickest? Are there any areas of the ocean floor free of sediments?

For Further Study

Dietz, R. S. 1978. "IFO's (Identified Flying Objects)." *Sea Frontiers* 24 (no. 6): 341–46. Discusses the origin of tektites.

Heezen, B., and C. Hollister. 1971. *The Face of the Deep*. New York: Oxford University Press. Magnificent compendium of pictures of the deep-sea bed. Wonderfully written by two of the pioneers in deep-sea research.

Kennett, J. P. 1982. *Marine Geology*. Englewood Cliffs, NJ: Prentice-Hall. A standard text.

Nafe, J. E., and C. L. Drake. 1963. "Physical Properties of Marine Sediments." In *The Sea*, vol. 3, 794–815, ed. M. N. Hill. New York: Interscience. Technical.

Prospero, J. 1985. "Records of Past Continental Climates in Deep-Sea Sediments." *Nature* 315 (23 May): 279–80. Subtitle could be: "History in the Mucking."

6

WATER

Icebergs

Water occurs in three states: liquid, solid, and gas. Oceanographers are most familiar with water's liquid form, but about 6% of the world ocean is covered by ice. A minute fraction of this ice is contained in the fantastic shapes of icebergs. Southern ocean icebergs originate as huge ice sheets attached to the Antarctic continent; lengths of up to 8 kilometers (5 miles) are not unusual, with flat tops rising 45 meters (150 feet) above sea level. Arctic icebergs are typically smaller in area but can be higher, pinnacled (or *castellated*, after their castle-turret shapes), and extraordinarily beautiful:

> On the afternoon of the day we knew the storm had passed, I stood on the starboard side of the bridge at a window, with the heavy protective glass lowered. I rested my forearms on the sill, feeling the warmth of the bridge heaters around my legs and a slipstream of cool air past my face. The first icebergs we had seen, just north of the Strait of Belle Isle, listing and guttered by the ocean, seemed immensely sad, exhausted by some unknown calamity. We sailed past them.
>
> I occasionally drew back from the starboard window to make a sketch, or to bring the binoculars up to my eyes. I marveled as much at the behavior of light around the icebergs as I did at their austere, implacable progress through the water. They took their color from the sun, and from the clouds and the water. But they also took their dimensions from the light: the stronger and more direct it was, the greater the contrast upon the surface of the ice, of the ice itself with the sea. And the more finely etched were the dull surfaces of their walls. The bluer the sky, the brighter their outline against it. . . .
>
> Where the walls entered the water, the surf pounded them, creating caverns, grottoes, and ice bridges, strengthening an impression of sea cliffs. At the waterline the ice gleamed aquamarine against its own gray-white walls above. Where meltwater had filled cracks or made ponds, the pools and veins were milk-blue, or shaded to brighter marine blues, depending on the thickness of the ice. If the iceberg had recently fractured, its new face glistened greenish blue—the greens in the older, weathered faces were grayer. In twilight the ice took on the colors of the sun: rose, reddish yellows, watered purples, soft pinks. The ice both reflected the light and trapped it within its crystalline corners and edges, where it intensified. . . . How utterly still, unorthodox, and wonderous they seem.

Source: Barry Lopez, *Arctic Dreams* (New York: Scribner's, 1986).

An iceberg drifts in the northern polar ocean.

Because it is familiar and abundant, we don't always appreciate water's unusual characteristics. If liquid methane, ammonia, or any other flowing substance dominated the surface of Earth, physical conditions here would be strikingly different, and life as we know it would be impossible.

As you may recall from Chapter 2, Earth's surface waters are thought to have escaped from the crust and mantle through the process of outgassing. Outgassing of substances other than water, and the chemical dissolution by water of crustal material, have added salts and other solids and gases to the ocean. In this chapter we will investigate the structure of pure water, look at the chemistry of seawater with its many dissolved substances, and discuss some of water's physical properties and their implications for the ocean and Earth as a whole.

THE WATER MOLECULE

A **molecule** is a group of atoms held together by chemical bonds. **Chemical bonds**, the energy relationships between atoms that hold them together, are formed when **electrons**—tiny, negatively charged particles found toward the outside of an atom—are shared between atoms or transferred. Water's familiar chemical formula, H_2O, shows that two atoms of hydrogen (H) are present for each atom of oxygen (O). Because of the way a water molecule's oxygen electrons are distributed, the overall geometry of the molecule is a bent or angular shape. The angle of the chemical bonds formed by the two hydrogen atoms and the central oxygen atom is about 105°. The formation of a water molecule is depicted in **Figure 6.1**.

The angular shape of the water molecule makes it electrically unbalanced, or **polar**. Each water molecule can be thought of as having a positive (+) end and a negative (−) end because positively charged particles at the center of the hydrogen atoms—called **protons**—are left partially exposed when the negatively charged electrons bond more closely to oxygen. The polar water molecule acts something like a magnet; its positive end attracts particles having a negative charge, and its negative end attracts particles having a positive charge. When water comes in contact with compounds whose elements are held together by the attraction of opposite electrical charges (most salts, for example), the polar water molecule will separate that compound's component elements from each other. This explains why water can dissolve so many other compounds so easily.

The polar nature of water also permits it to attract other water molecules. When a hydrogen atom in one water molecule is attracted to the oxygen atom of an adjacent water molecule, a **hydrogen bond** forms. The resulting loosely held webwork of water molecules is shown in **Figure 6.2**. Hydrogen bonds greatly influence the properties of water by allowing individual water mole-

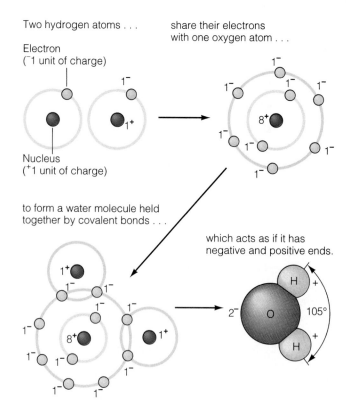

Figure 6.1 The formation of a water molecule.

cules to stick to each other, a property called **cohesion**. Cohesion gives water an unusually high surface tension, which results in a surface "skin" capable of supporting needles, razor blades, and even walking insects. It also causes the capillary action that makes water spread through a towel when one corner is dipped in water. **Adhesion**, the tendency of water to stick to other materials, allows water to adhere to solids—that is, to make them wet. The absorption of red light by hydrogen bonds is also what gives pure water—and thick ice—its pale bluish hue.

THE DISSOLVING POWER OF WATER

Water is a powerful solvent; it will eventually dissolve nearly any substance. No wonder, then, that seawater and most other liquids in nature are water solutions. Water's dissolving power is due to the polar nature of the water molecule. Consider how water dissolves sodium chloride (or NaCl), the most common salt.[1] In solid crystals of this salt, the sodium atoms in NaCl have lost electrons, and the chlorine atoms have gained them. The resulting charged

---

[1] Na, the chemical symbol for sodium, is derived from *natrium*, the Latin name for sodium.

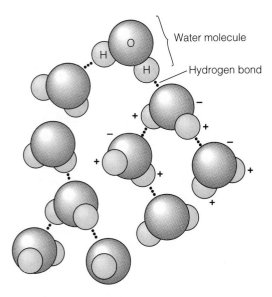

Figure 6.2 Hydrogen bonds in liquid water. The attractions between adjacent dipolar water molecules form a webwork of hydrogen bonds. These bonds are responsible for cohesion and adhesion, the properties of water that cause surface tension and wetting. Hydrogen bonds between water molecules also make it difficult for individual molecules to escape from the surface.

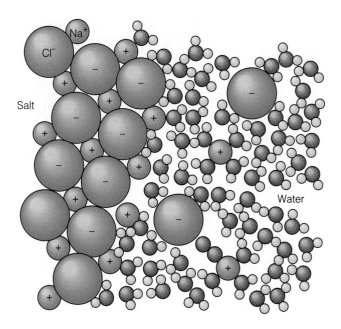

Figure 6.3 Salt in solution. When a salt such as NaCl is put in water, the positively charged hydrogen end of the dipolar water molecule is attracted to the negatively charged chloride ion, and the negatively charged oxygen end is attracted to the positively charged sodium ion. The ions are surrounded by water molecules that are attracted to them, and they become solute ions in the solvent.

atoms are called ions. When liquid water is present, the polarity of water causes the sodium ion (Na^+) to separate from the chloride ion (Cl^-) (**Figure 6.3**). The ions move away from the salt crystal, permitting water to attack the next layer of NaCl. Note that NaCl does not exist as "salt" in seawater; its components are separated when salt crystals dissolve and are joined when crystals re-form.

By contrast, oil doesn't dissolve in water, even if the two are very thoroughly shaken together. When oil is dispersed in water it forms a mixture because molecules of oil are nonpolar in character. This means that oil has no positive or negative charges to attract the polar water molecule. In a way this is fortunate: Living tissues would readily dissolve in water if the oils within their membranes didn't blunt water's powerful attack.

SEAWATER

The total quantity (or concentration) of dissolved inorganic solids in water is its **salinity**. The ocean's salinity varies from about 3.3% to 3.7% by weight—depending on such factors as evaporation, precipitation, and freshwater runoff from the continents—but the average salinity is usually given as 3.5%. The world ocean contains some 5,000 trillion kilograms (5.5 trillion tons) of salt. If the ocean's water evaporated completely, leaving its salts behind, the dried residue could cover the entire planet with an even layer 45 meters (150 feet) thick! Most of the

dissolved solids in seawater are salts that have been separated into ions. Sodium and chloride are the most abundant of these.

The Components of Salinity

Because about 3.5% of seawater consists of dissolved substances, boiling away 100 kilograms of seawater theoretically produces a residue weighing 3.5 kilograms. Because variations of 0.1% are significant, oceanographers prefer to use parts-per-thousand notation (‰) rather than percent (%, parts-per-hundred) in discussing these substances.[2] The seven ions listed below oxygen and hydrogen in **Table 6.1** make up more than 99% of this residual material. When seawater evaporates, its ionic components combine in many different ways to form table salt (NaCl), epsom salts ($MgSO_4$), and other mineral salts.

Seawater also contains minor constituents. The ocean is sort of an "Earth tea"; nearly every element present in the crust and atmosphere is also present in the oceans, though sometimes in extremely small amounts. Only 14 elements have concentrations in seawater larger than one part per million. Elements present in amounts less than

[2]Note that 3.5% = 35‰. If you began with 1,000 kilograms of seawater, you would expect 35 kilograms of residue.

Table 6.1 Major Constituents of Seawater at 34.32‰ Salinity		
Constituent	Concentration in Parts-per-Thousand (‰) or Grams-per-Kilogram (g/kg)	Percent by Weight
Water Itself		
Oxygen	857.8	85.8%
Hydrogen	107.2	10.7%
The Most Abundant Ions		
Chloride (Cl^-)	18.980	1.9%
Sodium (Na^+)	10.556	1.1%
Sulfate (SO_4^{2-})	2.649	0.3%
Magnesium (Mg^{2+})	1.272	0.1%
Calcium (Ca^{2+})	0.400	0.04%
Potassium (K^+)	0.380	0.04%
Bicarbonate (HCO_3^-)	0.140	0.01%
	999.377 g/kg	99.9%

Source: Adapted from Walton-Smith, 1974.

0.001‰ (one part per million, or ppm) are known as **trace elements**.

The Source of the Ocean's Salts

Remembering the effectiveness of water as a solvent, you might think that the ocean's saltiness has resulted from the ability of rain, groundwater, or crashing surf to dissolve crustal rock. Much of the sea's dissolved material (solutes) originated in that way, but is crustal rock the source of all the ocean's solutes? An easy way to find out would be to investigate the composition of salts in river water and compare these figures to those of the ocean as a whole. If crustal rock is the only source, the salts in the ocean should be like those of concentrated river water. But they are not. River water is usually a dilute solution of calcium and bicarbonate ions, while the principal ions in seawater are chloride and sodium. The magnesium content of seawater would also be higher if seawater were simply concentrated river water. The proportions of salts in isolated salty inland lakes, such as Utah's Great Salt Lake or the Dead Sea, are indeed much different from the proportions of salts in the ocean. Thus weathering and erosion of crustal rocks cannot be the sole source of sea salts.

The components of ocean water whose proportions are *not* accounted for by the weathering of surface rocks are called **excess volatiles**. To find the source of these excess volatiles we must look to the Earth's deeper layers. The upper mantle appears to contain more of the substances found in seawater (including the water itself) than are found in surface rocks, and their proportions are about the same as those found in the ocean. As you read in Chapter 3, convection currents slowly churn Earth's mantle, causing the movement of tectonic plates. Because of this activity, some deeply trapped volatile substances escape to the exterior, outgassing through volcanoes and rift vents. These excess volatiles include carbon dioxide, chlorine, sulfur, hydrogen, fluorine, nitrogen, and, of course, water vapor. This material, along with residue from surface weathering, accounts for the chemical constituents of today's ocean.

Some of the ocean's dissolved materials are hybrids of the two processes of weathering and outgassing. Table salt, sodium chloride, is an example. The sodium ions come from the weathering of crustal rocks, while the chlorine ions come from the mantle by way of volcanic vents and outgassing from mid-ocean rifts. As for the lower-than-expected quantity of magnesium ions in the ocean, recent research at a spreading center east of the Galápagos Islands suggests that mid-ocean rifts may play a role in reducing the magnesium content and increasing the calcium content of seawater. The water that circulates through new ocean floor at these sites is apparently stripped of magnesium and a few other elements. The magnesium seems to be incorporated into mineral deposits, but calcium is added as hot water dissolves adjacent rocks.

The Principle of Constant Proportions

In 1865 the chemist Georg Forchhammer noted that although the total *amount* of dissolved solids (salinity) might vary between samples, the *ratio* of major salts in samples of seawater from many locations was constant. In other words, the percentage of various salts in seawater is the same in samples from many places regardless of how salty the water is. This constant ratio is known as **Forchhammer's Principle** or the **principle of constant proportions**. Forchhammer was also the first to observe that seawater contains fewer silica and calcium ions than concentrated river water—and the first to realize that removal of these compounds by marine animals and plants to form shells and other hard parts might account for part of the difference. The English chemist William Dittmar, working with HMS *Challenger* samples ten years after they had been collected, confirmed Forchhammer's principle of constant proportions. Building on Forchhammer's and Dittmar's work, and taking advantage of improved analytical techniques, researchers have established a reliable way to determine salinity.

Determining Salinity

Water's salinity by weight would seem an easy property to measure. Why not simply evaporate a known weight of seawater and weigh the residue? This simple method yields imprecise results because some salts will not release all the molecules of water associated with them. If these salts are heated to drive off the water, other salts (carbonates, for example) will decompose to form gases and solid compounds not originally present in the water sample.

Modern analysis depends instead on determining the sample's chlorinity. **Chlorinity** is a measure of the total weight of chlorine, bromine, and iodine ions in seawater. Because chlorinity is comparatively easy to measure, and because the proportion of chlorinity to salinity is constant, marine chemists have devised the following formula to determine salinity:

Salinity in ‰ = 1.80655 × chlorinity in ‰

Chlorinity is about 19.4‰, so salinity is around 35‰.

Seawater samples can be obtained by methods ranging from tossing a clean bucket over the side of the ship to sophisticated tube-and-pump systems. Typically, water samples are collected using a string of sampling bottles similar to those perfected early in this century by the Norwegian scientist and explorer Fridtjof Nansen. A **Nansen bottle (Figure 6.4)** is open at both ends. A series of bottles is lowered on a wire to known depths. Each bottle is then triggered to close by a brass weight (appropriately called a *messenger*) sent sliding down the line. The bottles are hauled to the surface and their contents analyzed.

Until recently, marine chemists used a delicate chemical procedure involving a silver nitrate solution to measure the chlorinity of seawater. Conversion to salinity was made by a set of mathematical tables. The procedure was calibrated against a standard sample of seawater of precisely known chlorinity. Today's marine scientists use an

Figure 6.4 The Nansen bottle. (a) A Nansen bottle string being deployed. (b) When the messenger slides down the line, it triggers the bottle to flip upside down, closing the end valve at the same time. Nansen bottles are being replaced by sampling bottles that can hold more water and that are more resistant to accidental triggering.

a

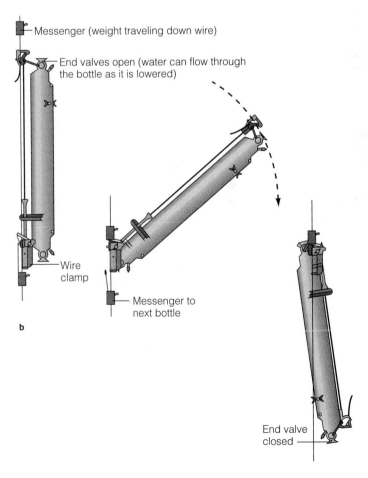

Messenger (weight traveling down wire)

End valves open (water can flow through the bottle as it is lowered)

Wire clamp

Messenger to next bottle

b

End valve closed

Figure 6.5 A portable induction salinometer. The instrument determines salinity by comparing the electrical conductivity of a seawater sample with the conductivity of laboratory-standard seawater. Water is drawn into the sampling cell through tubing visible at the top right side of the salinometer.

Table 6.2 Approximate Residence Times for Constituents of Seawater

Constituent	Residence Time (years)
Chloride (Cl^-)	100,000,000
Sodium (Na^+)	68,000,000
Magnesium (Mg^{2+})	13,000,000
Potassium (K^+)	12,000,000
Sulfate (SO_4^{2-})	11,000,000
Calcium (Ca^{2+})	1,000,000
Carbonate (CO_3^{2-})	110,000
Silicon (Si)	20,000
Water (H_2O)	3,500
Manganese (Mn)	1,300
Aluminum (Al)	600
Iron (Fe)	200

Source: Data from Broecker and Peng, 1982; Bruland, 1983; Riley and Skirrow, 1975.

electronic device called a **salinometer**, which measures the electrical conductivity of seawater (see **Figure 6.5**). Conductivity varies with the concentration and mobility of ions present and with water temperature. Circuits in the salinometer adjust for water temperature, convert conductivity to salinity, and then display salinity. Salinometers are also calibrated against a sample of known conductivity and salinity. The best salinometers can determine salinity to an accuracy of 0.005%. Some salinometers are designed for remote sensing: The electronics stay aboard ship while the sensor coil is lowered over the side.

Chemical Equilibrium and Residence Times

If outgassing and the chemical weathering of rock are continuing processes, shouldn't the ocean become progressively saltier with age? Landlocked seas and some lakes usually become saltier as they grow older, but the ocean does not. The ocean appears to be in **chemical equilibrium**; that is, the proportion *and amounts* of dissolved salts per unit volume of ocean are nearly constant. Evidently, whatever goes in must come out somewhere else.

Geologists in the 1950s developed the concept of a *steady state ocean*. The idea suggests that ions are added to the ocean at the same rate as they are being removed. This theory helps explain why the ocean is not growing saltier. The idea was quantified by T. F. W. Barth, who in 1952 devised the concept of **residence time**, the average length of time an element spends in the ocean. Residence time for a particular element may be calculated by this equation:

$$\text{Residence time} = \frac{\text{Amount of element in the ocean}}{\substack{\text{The rate at which the element} \\ \text{is added to (or removed} \\ \text{from) the ocean}}}$$

Additions of salts from the mantle or from the weathering of rock are balanced by subtractions of minerals being bound into sediments. Dissolved salts precipitate out of the water, and the silicon- and calcium carbonate-containing hard parts of living organisms drift slowly down to the seabed. Some of these sediments are removed from the ocean and drawn into the mantle at subduction zones by the cycling of crustal plates. Input from runoff and outgassing equals outfall (binding into sediments) for each dissolved component.

The residence time of an element depends on its chemical activity. Atoms (or ions) of some elements, such as aluminum and iron, remain in seawater for a relatively short time before becoming incorporated into sediments; others, such as chloride, sodium, and magnesium, remain in water for millions of years. The approximate residence times for the major constituents of seawater are shown in **Table 6.2**. Even seawater itself has a residence time (see **Box 6.1**).

Mixing Time

If constituent minerals are added to ocean water at rates that are less than the ocean's mixing time, they will become evenly distributed through the ocean. Because of

Ocean water has a residence time. Scientists estimate that about 334,000 cubic kilometers (80,000 cubic miles) of pure water evaporates from the ocean each year. Assuming a total ocean volume of 1,370 million cubic kilometers (329 million cubic miles), we can calculate that about 0.024% of the water in the ocean vaporizes each year. Thus, if it were not replenished by precipitation and runoff from land, the ocean would evaporate completely in about 3,500 years.

For a complete picture, though, more than just oceanic water should be considered. The Earth's total surface water volume—including the ocean, lakes, rivers, groundwater, and ice caps—is about 1,399 million cubic kilometers (336 million cubic miles). The total yearly evaporation/precipitation estimate is 396,000 cubic kilometers (95,000 cubic miles) of pure water. From these figures we again calculate a recycling time of about 3,500 years. With the exception of water subducted into the mantle along with sediments at subduction zones, then, water molecules have a residence time in the ocean of about 3,500 years.

The Earth's ocean is more than 4 billion years old. Thus, on average, individual water molecules have evaporated from and returned to the ocean more than a million times since the world ocean formed in its basins!

Table 6.3 Major Gases in the Atmosphere and Ocean

Gas	Percent of Gas in Atmosphere, by Volume	Percent of Dissolved Gas in Seawater, by Volume	Concentration in Seawater in Parts-per-Million, by Weight
Nitrogen (N_2)	78.08%	48%	10–18 ppm
Oxygen (O_2)	20.95%	36%	0–13 ppm
Carbon dioxide (CO_2)[a]	0.035%	15%	64–107 ppm

Source: Data from Weihaupt, 1979; Hill, 1963.

[a]Also present in seawater as carbonic acid, carbonate ions, and bicarbonate ions.

the vigorous activity of currents, the **mixing time** of the ocean is thought to be on the order of 1,000 years. Thus the ocean has been mixed hundreds of thousands of times during its long history. The relatively long residence times of seawater's major constituents thus assure thorough mixing, which is the basis of Forchhammer's principle of constant proportions.

DISSOLVED GASES

Gases in the air readily dissolve in seawater at the ocean's surface. Plants and animals living in the ocean require these dissolved gases to survive; no marine animal has the ability to break down water molecules to obtain oxygen directly, and no marine plant can manufacture enough carbon dioxide to support its own metabolism. In order of their relative abundance, the major gases found in seawater are nitrogen, oxygen, and carbon dioxide (see **Table 6.3**). The proportions of dissolved gases in the ocean are very different from the proportions of the same gases in the atmosphere.

Unlike solids, gases dissolve most readily in *cold* water. A cubic meter of chilly polar water usually contains a greater volume of dissolved gases than a cubic meter of warm tropical water.

Nitrogen

About 48% of the dissolved gas in seawater is nitrogen. (In contrast, the atmosphere is slightly more than 78% nitrogen by volume.) The upper layers of ocean water are usually saturated with nitrogen—that is, additional nitrogen will not dissolve. Living organisms require nitrogen to build proteins and other important biochemicals, but they cannot use the free nitrogen in the atmosphere and ocean directly. It must first be *fixed* into usable chemical forms by specialized organisms. Though some species of bottom-dwelling bacteria can manufacture usable nitrates from the nitrogen dissolved in seawater, most of the nitrogen compounds needed by living organisms must be recycled among the organisms themselves.

Oxygen

About 36% of the gas dissolved in the ocean is oxygen, but there is about 100 times more gaseous oxygen in the Earth's atmosphere than is dissolved in the whole ocean. An average of 6 milligrams of oxygen is dissolved in each liter of seawater (that is, 6 parts per million parts of oxygen per liter of seawater, by weight). Yet this small amount of oxygen is a vital resource for animals that extract oxygen with gills. The primary source of the ocean's dissolved oxygen is its photosynthetic plants. Since photosynthesis requires sunlight, most of the available oxygen lies near the ocean's surface (**Figure 6.6**). Because marine

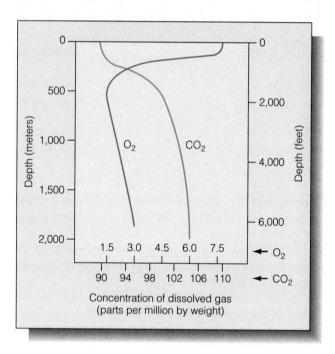

Figure 6.6 How concentrations of oxygen and carbon dioxide vary with depth. Oxygen is abundant near the surface due to the photosynthetic activity of marine plants. Oxygen amounts decrease below the sunlit layer because of the respiration of marine animals and bacteria. In contrast, plants use carbon dioxide during photosynthesis; so surface levels of CO_2 are low. Photosynthesis cannot take place in the dark; so CO_2 given off by animals and bacteria tends to build up at depths below the sunlit layer.

plants are so abundant and photosynthetically active, much more free oxygen diffuses from ocean to atmosphere than moves from atmosphere to ocean.

Carbon Dioxide (CO_2)

The amount of carbon dioxide in the atmosphere is very small (0.03%) because CO_2 is in great demand by photosynthetic plants as a source of carbon for growth. Carbon dioxide is very soluble in water, though; the proportion of dissolved CO_2 in water is about 15% of all dissolved gases. Because CO_2 combines chemically with water to form a weak acid (H_2CO_3, carbonic acid), water can hold perhaps 1,000 times more carbon dioxide than either nitrogen or oxygen at saturation. Carbon dioxide is quickly used by marine plants; so dissolved quantities of CO_2 are almost always much less than this theoretical maximum. Even so, at the present time there is about 60 times as much CO_2 dissolved in the ocean as in the atmosphere. Much more CO_2 moves from atmosphere to ocean than from ocean to atmosphere, in part because some dissolved CO_2 forms carbonate ions, which are locked into sediments, minerals, and the shells and skeletons of living organisms.

Figure 6.6 illustrates how carbon dioxide and oxygen concentrations vary with depth. Carbon dioxide concentrations increase with increasing depth, but oxygen concentrations usually decrease through the mid-depths and then rise again toward the bottom. High concentrations of oxygen at the surface are usually caused by the photosynthesis of plants in the ocean's brightly lit upper layer. Since plants require carbon dioxide for metabolism, surface CO_2 concentrations tend to be low. A decrease in oxygen below the sunlit upper layer is usually due to the respiration of bacteria and marine animals, activity that tips the balance in favor of carbon dioxide. Oxygen levels are slightly higher in deeper water because fewer animals are present to take up oxygen reaching these depths and because oxygen-rich polar water is the greatest source of deep water.

ACID–BASE BALANCE

Water can separate to form hydrogen ions (H^+) and hydroxide ions (OH^-). In pure water, these two ions are present in equal concentrations. An imbalance in the proportion of ions produces an acidic (or basic) solution. An **acid** is a substance that *releases* a hydrogen ion in solution; a **base** is a substance that *combines with* a hydrogen ion in solution. Basic solutions are also called alkaline solutions.

The acidity or alkalinity of a solution is measured in terms of the **pH scale**, which measures the concentration of hydrogen ions in a solution. An excess of hydrogen ions (H^+) in a solution makes that solution acidic. An excess of hydroxide ions (OH^-) makes a solution basic. **Figure 6.7** shows a pH scale and the pH of a few familiar solutions. The scale is logarithmic, which means that a change of one pH unit represents a tenfold change in hydrogen ion concentration. Thus, a modern nonphosphate detergent is 1,000 times more alkaline than seawater, and black coffee is 100 times more acidic than pure water. Pure water, which is neutral (neither acidic nor basic) has a pH of 7; lower numbers indicate greater acidity (more H^+ ions), and higher numbers indicate greater alkalinity (fewer H^+ ions).

Seawater is slightly alkaline; its average pH is about 7.8. This seems odd because of the large amount of CO_2 dissolved in the ocean. If dissolved CO_2 combines with water to form carbonic acid, why is the ocean mildly alkaline and not slightly acidic? When dissolved in water, CO_2 is actually present in several different forms. Carbonic acid (H_2CO_3) is only one of these. In water solutions, some carbonic acid breaks down to produce the hydrogen ion (H^+), the bicarbonate ion (HCO_3^-), and the carbonate ion (CO_3^{2-}). This behavior acts to **buffer** the water, preventing broad swings of pH when acids or bases are introduced.

Though seawater remains slightly alkaline, it is subject to some variation. In areas of rapid plant growth, for example, pH will rise because CO_2 is used by the plants for photosynthesis. Because temperatures are generally warmer at the surface, less CO_2 can dissolve in the first place. Thus, surface pH in warm productive water is usually around 8.5.

At middle depths and in deep water, more CO_2 may be present. Its source is the respiration of animals and bacteria. With cold temperatures, high pressure, and no photosynthetic plants to remove it, this CO_2 will lower the pH of water, making it more acidic with depth. Thus, deep, cold seawater below 4,500 meters (15,000 feet) has a pH of around 7.5. This lower pH can dissolve calcium-containing marine sediments—you may recall from Chapter 5 that sediments containing calcium carbonate are rarely found in deep water. A drop to pH 7 can occur at the deep-ocean floor when bottom bacteria consume oxygen and produce hydrogen sulfide.

WATER AND HEAT

Heat is energy produced by the random vibration of atoms or molecules. On the average, water molecules in hot water vibrate more rapidly than water molecules in cold water. Heat and temperature are not the same thing. Heat tells us *how many* and *how rapidly* molecules are vibrating. Temperature records only *how rapidly* the molecules

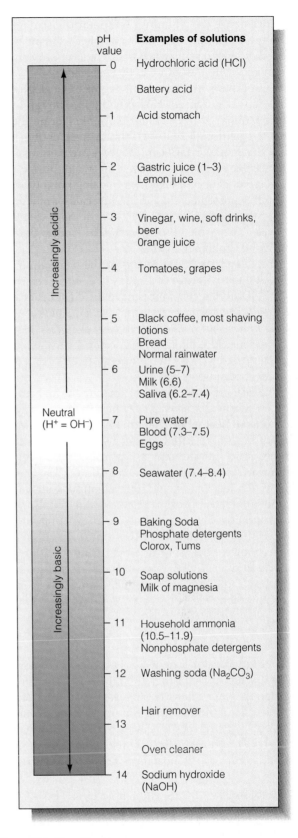

Figure 6.7 The pH scale.

of a substance are vibrating. **Temperature** is an object's response to an input (or removal) of heat. The amount of heat required to bring a substance to a certain temperature varies with the nature of that substance.[3]

Temperature is measured in **degrees**. One degree Celsius (°C) = 1.8 degrees Fahrenheit (°F). Though we are more familiar with the older Fahrenheit scale (named after the eighteenth-century German inventor of the mercury thermometer), Celsius degrees are more useful in science because they are based on two of pure water's most significant properties: its freezing point (0°C) and its boiling point (100°C). The Celsius scale—until recently called the centigrade scale because of the 100° interval between base points—was invented by Anders Celsius, a Swedish astronomer, in 1742.

Heat capacity is a measure of the heat required to raise the temperature of 1 gram of a substance by 1°C (1.8°F). Different substances have different heat capacities: *Not all substances respond to identical inputs of heat by rising in temperature the same number of degrees*. Heat capacity is measured in calories. A **calorie** is the amount of heat required to raise the temperature of 1 gram (0.035 ounce) of pure water by 1°C.[4]

The heat capacity of water is among the highest of all known substances; so water can absorb (or release) large amounts of heat while changing relatively little in temperature. Anyone who waits by a stove for water to boil knows much about water's heat capacity! By contrast with water, ethyl alcohol has a much lower heat capacity. If both liquids absorb heat from identical stove burners at the same rate, pure ethyl alcohol—the active ingredient in alcoholic beverages—will rise in temperature about twice as fast as an equal mass of water. Beach sand has an even lower heat capacity. Sand requires as little as 0.2 calorie to rise 1°C (1.8°F); so beaches can get too hot to stand on with bare feet even while the water remains pleasantly cool.

Temperature and Density

The uniqueness of pure water becomes even more apparent when we consider the effect of a temperature change on water's **density** (its mass per unit of volume). The density of pure water is 1 gram per cubic centimeter (1 g/cm^3); granite rock is heavier, with a density of about 2.7 g/cm^3, and air is lighter, with a density of about 0.0012 g/cm^3. Most substances become denser (weigh

more per unit of volume) as they get colder. Cold air sinks and warm air rises because the rapidly vibrating molecules of warm air are farther apart and therefore occupy more space than the same number of cool air molecules. The cold air falls because it is denser. Like air, pure water generally becomes denser as heat is removed and temperature falls, but water's density curve shows a curious quirk as the temperature approaches the freezing point.

A **density curve** shows the relationship between the temperature of a substance and its density. Most substances become progressively denser as they cool; their temperature–density relationships are linear (that is, appear as a straight line on graphs). But **Figure 6.8** shows the unusual temperature–density relationship of pure water. Imagine heat being removed from some water in a freezer. Initially, the water is at room temperature, point A on the graph. As expected, the density of water increases as its temperature drops along the line from point A toward point B. Approaching point B, the density increase slows, reaching a maximum at point B of 1 g/cm^3 at 3.98°C (39.2°F). Oddly, water then becomes slightly *less* dense as cooling continues, until point C (0°C, 32°F) is reached. At point C the water begins to freeze—to change state by crystallizing to ice.

State is an expression of the internal form of a substance. Water exists in three states: liquid, gas (water vapor), and solid (ice). If the freezer continues to remove heat from the water at point C, the water will change from liquid to solid state. Through this transition from water to ice—from point C to point D—the density of the water *decreases* abruptly. Ice is therefore lighter than an equal volume of water. Ice increases in density as it gets colder than 0°C, but no matter how cold it gets, ice never reaches the density of liquid water. Being less dense than water, ice "freezes over" as a floating layer instead of "freezing under" like the solid forms of virtually all other liquids.

As we'll see in a moment, the implications of water's high heat capacity, and the ability of ice to float, are vital in maintaining Earth's moderate surface temperature. But first let's take a closer look at the transition from point C to point D in Figure 6.8.

Freezing Water

During the transition from liquid to solid state at the freezing point, the bond angle between the oxygen and hydrogen atoms in water expands from about 105° to slightly more than 109°. This change allows ice to form a crystal lattice (see **Figure 6.9**). The space taken by 24 water molecules in the solid lattice could be occupied by 27 water molecules in liquid state, so water expands about 9% as the crystal forms. Because the molecules are packed less efficiently, ice is less dense than liquid water,

---

[3]This example will help: Which has a higher temperature—a candle flame or a bathtub of hot water? The flame. Which contains more heat? The tub. The molecules in the flame vibrate very rapidly, but there are relatively few of them. The molecules of water in the tub vibrate more slowly, but there are a great many of them; so the total amount of heat energy in the tub is greater.

[4]A nutritional Calorie, the unit we see on cereal boxes, equals 1,000 of these calories. A gram is about ten drops of seawater.

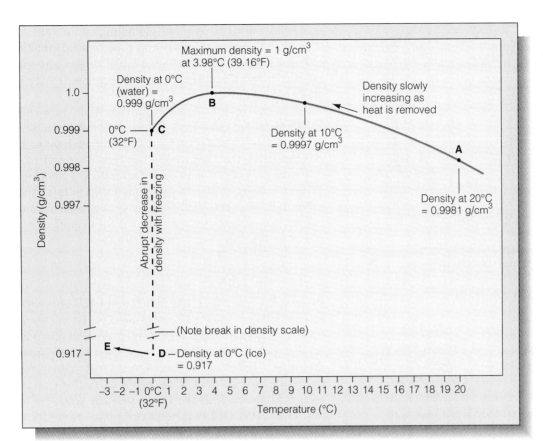

Figure 6.8 The relationship of density to temperature for pure water. Note that points C and D both represent 0°C (32°F) but different densities—and thus different states of water. Because the density of ice is less than the density of liquid water, ice floats.

Maximum density = 1 g/cm³ at 3.98°C (39.16°F)

Density at 0°C (water) = 0.999 g/cm³

0°C (32°F)

Density slowly increasing as heat is removed

Density at 10°C = 0.9997 g/cm³

Density at 20°C = 0.9981 g/cm³

Abrupt decrease in density with freezing

(Note break in density scale)

D — Density at 0°C (ice) = 0.917

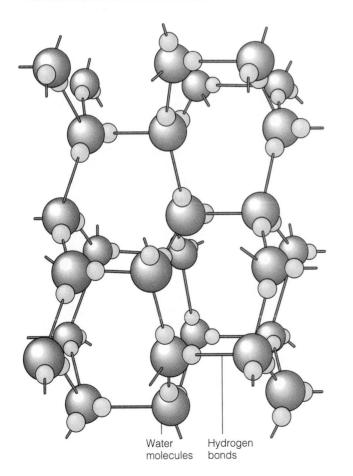

Water molecules   Hydrogen bonds

Figure 6.9 The lattice structure of an ice crystal, showing its hexagonal arrangement at the molecular level. The hexagonal pattern of a snowflake is a visible reflection of this same structure.

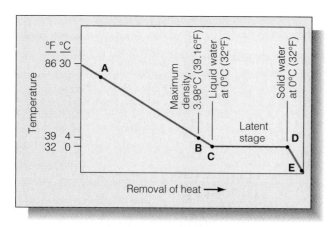

Figure 6.10 A graph of temperature versus heat removal as water freezes (or ice melts). The horizontal line between points C and D represents the latent heat of fusion, when heat is being removed but temperature is not changing. (Note that points A–E on this graph are the same as those in Figure 6.8.)

and it floats. Ice at 0°C (32°F) weighs only 0.917 g/cm³, whereas liquid water at 0°C weighs 0.999 g/cm³.

The ice lattice is not a flat sheet but a three-dimensional network. The water molecules in the ice lattice are still linked by hydrogen bonds, but they now form regular hexagons. This hexagonal pattern explains the lovely six-sided symmetry of snowflakes and other ice crystals.

The transition from liquid water to ice crystal (point C to point D in Figure 6.8) requires continued removal of heat energy; the change in state does not occur instantly throughout the mass when the cooling water reaches 0°C (32°F). Again, consider water in a freezer. **Figure 6.10**, a plot of heat removal *vs.* temperature, details the water's progress to ice. As in Figure 6.8, point A represents warm water just placed into the freezer. The removal of heat does not stop when the water reaches point C, but the decline in temperature does. Even though heat continues to be removed, the water will not get colder until all of it has changed state from liquid (water) to solid (ice). Heat may therefore be removed from water without the water dropping in temperature when water is changing state—that is, when it is freezing. Indeed, the continued removal of heat is what makes the state change possible. Heat is released as bonds form to make ice, and that heat must be removed to allow more ice to form.

The removal of heat from point A to point C in Figures 6.8 and 6.10 produces a *measurable* lowering of temperature detectable by a thermometer. Removing just 1 calorie of heat from a gram of liquid water causes its temperature to drop 1°C. This detectable decrease in heat is called **sensible heat** loss (because it can be sensed by a thermometer). But the loss of heat as water freezes between points C and D is not measurable by a thermometer. Removing a calorie of heat from freezing water at 0°C (32°F) won't change its temperature at all; 80 calories of

heat energy must be removed per gram of pure water at 0°C (32°F) to form ice. This heat is called the **latent heat of fusion** (from *latere* = to be hidden). The straight line between points C and D in Figure 6.10 represents water's latent heat of fusion.

No more ice crystals can form when all the water in the freezer has turned to ice. If the removal of heat continues, the ice will get colder and will soon reach the temperature inside the freezer, point E in Figures 6.8 and 6.10.

Latent heat of fusion is also a factor during thawing. When ice melts, it *absorbs* large quantities of heat (the same 80 calories per gram) and does not change in temperature until the entire mass has turned to liquid. This is why ice is so effective in cooling drinks.

Evaporating Water

When water evaporates, individual water molecules diffuse into the air. Since each water molecule is hydrogen-bonded to adjacent molecules, heat energy is required to break those bonds and allow the molecule to fly away from the surface. Evaporation cools a moist surface because departing molecules of water vapor carry this energy away with them. (This is why we perspire when we're hot. The heat energy required to evaporate water from our skin is taken away from our bodies, thus cooling us.)

Hydrogen bonds are quite strong, and the amount of energy required to break them—known as the **latent heat of evaporation**—is high. At 585 calories per gram at 20°C (68°F), water has the highest latent heat of evaporation of any known substance. As before, the term *latent* applies to heat input that does not cause a temperature change but does produce a change of state—in this case from liquid to gas.

About a meter of water evaporates each year from the surface of the ocean, a volume of water equivalent to 350,000 cubic kilometers (84,000 cubic miles). The great quantities of solar energy that cause this evaporation are carried from the ocean by the escaping water vapor. When a gram of water vapor condenses back into liquid water, the same 585 calories is again available to do work. As we shall see in the next chapter, winds, storms, ocean currents, and wind waves are all powered by that heat.

Why the big difference between water's latent heat of fusion (80 calories per gram) and its latent heat of evaporation (585 calories per gram)? Only a small percentage of hydrogen bonds are broken when ice melts, but *all* must be broken during evaporation. Breaking these bonds requires additional energy in proportion to their number.

Seawater and Pure Water

The solids dissolved in seawater change its thermal characteristics, lowering its heat capacity by about 4%. Only

0.96 calorie of heat energy is needed to raise the temperature of seawater by 1°C.

The dissolved solids also interfere with the formation of the ice lattice, acting as "antifreeze" to lower the freezing point. The saltier the water, the lower the freezing point. The temperature of maximum density moves toward the freezing point as salinity increases (see **Figure 6.11**), finally coinciding with the freezing point at a salinity of 24.7‰ (−1.33°C). Seawater at 35‰, typical ocean salinity, freezes at −1.91°C (28.6°F). Seawater's density simply increases smoothly with decreasing temperature until it freezes. The crystals that form are pure water ice, with the seawater salts excluded. The leftover cold, salty water is very dense.

Seawater evaporates more slowly than fresh water under identical circumstances because the dissolved salts tend to attract and hold water molecules. The latent heat of evaporation, however, is essentially the same for both fresh water and seawater. Salts are left behind as seawater evaporates. The remaining cool, salty water is also very dense.

Global Thermostatic Effects

The **thermostatic properties** (*therme* = heat + *stasis* = standing still) of water are those properties that act to moderate changes in temperature. Water temperature rises as sunlight is absorbed and changed to heat, but, as we've seen, water has a very high heat capacity; so its temperature will not rise very much even if a large quantity of heat is added. This tendency of a substance to resist change in temperature with the gain or loss of heat energy is called **thermal inertia**. To investigate the impact of water's thermostatic properties on conditions at the Earth's surface we need to look at the planet's overall heat balance.

Only about 1 part in 2,200 million of the sun's radiant energy is intercepted by Earth, but that amount averages 7 million calories per square meter per day at the top of the atmosphere, or, for the Earth as a whole, an impressive 17 trillion kilowatts (23 trillion horsepower)! About half of this light reaches the surface. As may be seen in **Figure 6.12**, on a global basis about 26% of the incoming light is reflected back into space from clouds or scattered by ice, water droplets, or other particles in the atmosphere. Another 19% is absorbed by water vapor, dust, and CO_2 in the atmosphere. About 4% bounces off the shining sea surface, ice and snow, or rocks and soil. Only 51% is absorbed by the Earth's land and water surface. How much light penetrates the ocean depends greatly on the angle at which it approaches, the sea state (surface turbulence), the presence of an ice covering or light-colored foam, and other factors.

The 51% of solar energy striking land and sea is converted to heat and then transferred into the atmosphere

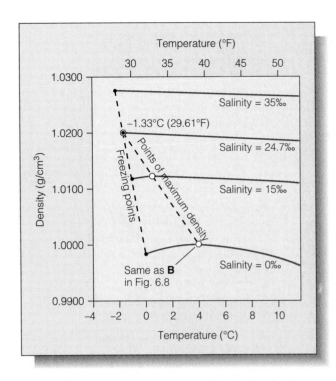

Figure 6.11 The dependence of freezing temperature and temperature of maximum density upon salinity. As we saw at point B in Figure 6.8, pure water is densest at 3.98°C (39.16°F), and its freezing point is 0°C (32°F). Seawater with 15‰ salinity is densest at 0.73°C (33.31°F), and its freezing point is −0.80°C (30.56°F). The temperature of maximum density and freezing point coincide at −1.33°C (29.61°F) in seawater with a salinity of 24.7‰. At salinities greater than 24.7‰, the density of water always decreases as temperature increases.

by conduction, radiation, and evaporation. The atmosphere, like the land and ocean, eventually radiates this heat back into space in the form of long-wave (infrared) radiation. The heat input and outflow "account" for the Earth can be thought of as a **heat budget**. As in your personal financial budget, income must eventually equal outgo. Over long periods of time the total *incoming* heat (plus that from earthly sources) equals the total *outgoing* heat; so the Earth is in **thermal equilibrium**. It is growing neither significantly warmer nor colder.[5] Heat input comes mainly from the sun; heat outflow can occur only as heat radiates into the cold of space.

Liquid water's thermal characteristics prevent broad swings of temperature during day and night, and, through a longer span, during winter and summer. Heat is stored in the ocean during the day and released at night. A much greater amount of heat is stored through

[5]Changes in heat balance do occur over short periods of time. Increasing amounts of carbon dioxide and methane in the Earth's atmosphere may be contributing to an increase in surface temperature called the *greenhouse effect*. More on this subject may be found in Chapter 15.

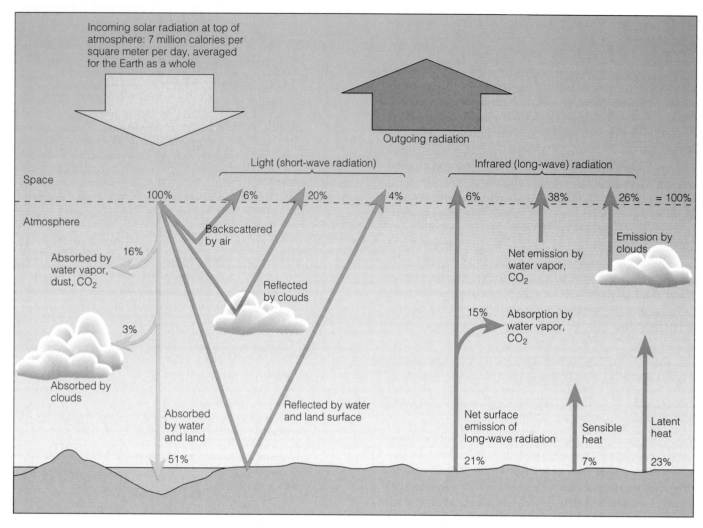

Figure 6.12 The heat budget for Earth. On an average day, about half of the solar energy arriving at the upper atmosphere is absorbed at the Earth's surface. Light (short-wave) energy absorbed at the surface is converted into heat. Heat leaves the Earth as infrared (long-wave) radiation. Since input equals output over long periods of time, the heat budget is balanced.

the summer and given off during the winter. If our ocean were made of alcohol—or almost any other liquid—summer temperatures would be much hotter, and winters bitterly cold. Sea ice in the polar regions also contributes to thermal inertia. Because water expands and floats when it freezes, ice can absorb the morning warmth of the sun, melt, then refreeze at night, giving back to the atmosphere the heat it stored through the daylight hours. The *heat content* of the water changes through the day; its *temperature* does not. The same principle applies to the seasonal formation and melting of polar ice. More than 18,000 cubic kilometers (4,300 cubic miles) of polar ice thaws and refreezes each year. Seasonal extremes are moderated by the immense amounts of heat energy that are alternately absorbed and released without a change in temperature. Without these properties of ice, tempera-

tures on the Earth's surface would change dramatically with minor changes in atmospheric transparency or solar output.

THE DENSITY STRUCTURE OF THE OCEAN

The density of water is mainly a function of its salinity and temperature. Cold, salty water is denser than warm, less salty water. The density of seawater varies between 1.020 and 1.030 g/cm^3, indicating that a liter of seawater weighs between 2% and 3% more than a liter of pure water (1.00 g/cm^3) at the same temperature. Seawater's density increases with increasing salinity, increasing pressure, and decreasing temperature. **Figure 6.13** shows the relationship between temperature, salinity, and density. Notice that two samples of water can have the *same* density at different combinations of temperature and salinity.

Ocean water tends to form into stable layers, with the heaviest water at the bottom—a form of density stratification. Curiously, even the deepest of these layers originates at the ocean's surface. Very cold and salty water produced during the formation of sea ice at the polar ocean surface is denser than the surrounding water and sinks until it reaches a layer of water of equal density, or the seabed. In a few marginal basins, the most notable being the Mediterranean Sea, evaporation can also produce salty, dense water that sinks toward the bottom. In contrast, warm fresh water entering the ocean at a river mouth is much less dense and floats for miles above the cooler salty layers below. Early explorers of South America were amazed to find a surface layer of fresh water far to sea, and they discovered the Amazon River by following this fresh surface layer to its source.

Much of the ocean is divided into three density zones. The **surface zone**, or **mixed layer**, is the upper layer of ocean in which temperature and salinity are relatively constant with depth because of the action of waves and currents. The surface zone consists of water in contact with the atmosphere and exposed to sunlight; it contains the ocean's least dense water, and it accounts for about 2% of total ocean volume. Depending on local conditions, the surface zone may reach a depth of 1,000 meters or be absent entirely. Beneath it is the **pycnocline** (*pyknos* = strong + *clinare* = slope, to lean), a zone in which density increases with increasing depth. This zone isolates surface water from the more dense layer below. The pycnocline contains about 18% of all ocean water. The **deep zone** lies below the pycnocline, at depths below about 1,000 meters (3,000 feet) in mid-latitudes (40°S to 40°N). There is little additional change in water density with increasing depth through this zone. This deep zone contains about 80% of all ocean water.

Temperature is the most important factor affecting density at mid-latitudes because the change in temperature with depth is much more pronounced than the change in salinity with depth. **Figure 6.14** shows the general relationship of temperature with depth in the open sea. The surface zone is well mixed, with little decrease in temperature with depth; in the next layer, temperature drops rapidly with depth; beneath it lies the deep zone of cold, stable water. The middle layer, the zone in which temperature changes rapidly with depth, is called the **thermocline** (*therme* = heat). This thermocline is the major cause of the pycnocline mentioned above.

Thermoclines are not identical in form for all areas or latitudes. Temperate and tropical ocean areas cradle a warm surface layer whose bottom boundary is the top

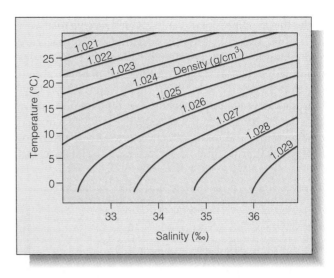

Figure 6.13   The complex relationship between the temperature, salinity, and density of seawater. Note that two samples of water can have the same density at different combinations of temperature and salinity.

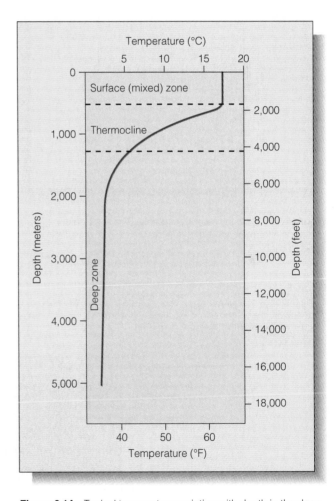

Figure 6.14   Typical temperature variation with depth in the deep ocean at mid-latitudes.

of the thermocline. Polar waters, which receive relatively little solar warmth, are not stratified by temperature and generally lack a thermocline. **Figure 6.15** contrasts polar, tropical, and temperate thermal profiles, showing that the thermocline is primarily a mid- and low-latitude phenomenon. Thermocline depth and intensity also vary with season, local conditions (storms and so on), currents, and many other factors. Divers often notice minor thermoclines that form near the surface, but

most temperate and tropical oceans also contain a deep main thermocline.

Below the thermocline, water is very cold, ranging in temperature from –1°C to 3°C (30.5°F to 37.5°F). Because this deep and cold layer contains the bulk of ocean water, the average temperature of the world ocean is a chilly 3.5°C (38°F).

Dissolved salts contribute to the ocean's density structure, especially in cool regions where precipitation is rapid or along coasts exposed to freshwater runoff. The **halocline** (*halos* = salt) is a zone of rapid salinity increase with depth. The halocline often coincides with the thermocline, and the combination produces a pronounced pycnocline.

Figure 6.16 shows how the thermocline and halocline combine to form the pycnocline. **Figure 6.17** records changes in the position of the pycnocline with latitude.

REFRACTION, LIGHT, AND SOUND

The ring of light sometimes seen around the moon and the safe concealment of a submarine may not seem related, but both events depend on **refraction**, the bending of waves. Light and sound are wave phenomena. When a light wave or sound wave leaves a medium of one density—such as air—and enters a medium of a different density—such as water—at an angle other than 90°, it is bent from its original path. The reason for this bending is that light or sound waves travel at different speeds in the different media.

The situation is analogous to a line of marchers walking along a desert highway with arms intertwined. The marchers can walk faster if they stay on the pavement than if they walk in the sand next to the highway. Their speed on the pavement, then, is greater than their speed in the sand. As long as they stay on the pavement, they won't change direction. But if their marching angle grad-

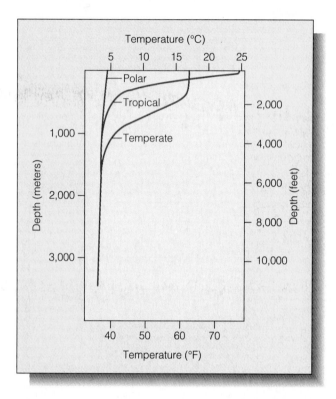

Figure 6.15 Typical temperature profiles at polar, tropical, and middle (temperate) latitudes. Note that polar waters lack a thermocline.

Figure 6.16 Formation of the pycnocline, which depends on the characteristics of the thermocline and halocline.

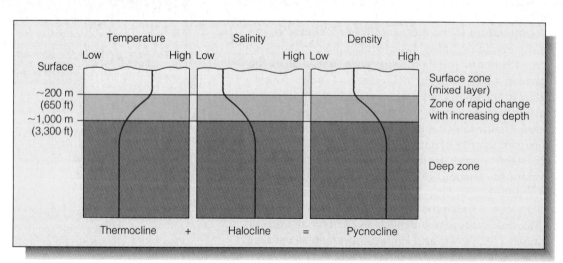

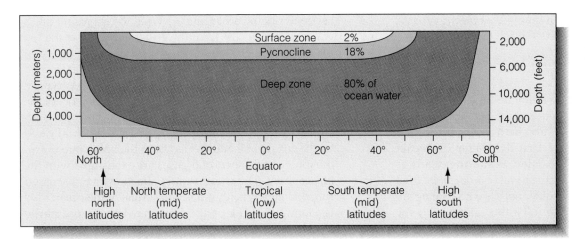

Figure 6.17 Variation in the average depth of vertical density zones with latitude.

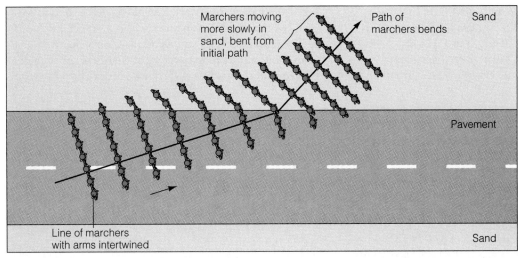

a

Figure 6.18 An analogy for refraction. The lines of marchers represent light or sound waves; the pavement and sand represent different media. (a) If the marchers head off the pavement at an angle, their path will bend (refract) as they hit the sand because some in each line will be walking more slowly than others. (b) If they march straight off the pavement, the lines will slow down and become closer together—but not bend—as they hit the sand.

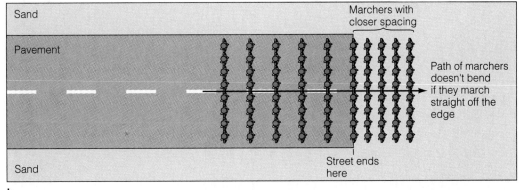

b

ually takes them off the edge (into a medium in which their speed is *lower*), the people who reach the sand first will suddenly slow down, and the line will pivot quickly off the highway. They have been refracted. Their progress is depicted in **Figure 6.18a**. Note that the transition from one medium to another must occur at an angle other than 90° for refraction to occur; our marchers will not change direction if they march straight off the asphalt into the sand. (They will still slow down, however—see **Figure 6.18b**.)

Examples of the refraction of light by water are all around you. A pencil sticking out of a glass of water looks bent because of refraction; the submerged steps of a swimming pool ladder look closer than they are because of refraction; and refraction magnifies objects, causing divers to exaggerate the size of the fish that got away.

Light in the Ocean

As we've seen, sunlight has a difficult time reaching and penetrating the ocean. Clouds and the sea surface reflect light, and atmospheric gases and particles scatter and absorb it. Once past the sea surface, light is rapidly weakened by scattering and absorption. **Scattering** occurs as light is bounced between air or water molecules, dust particles, water droplets, or other objects before being absorbed. The greater density of water (along with the greater number of suspended and dissolved particles) makes scattering more prevalent in water than in air. The **absorption** of light is governed by the structure of the water molecules it happens to strike. When light is absorbed, molecules vibrate and the light's energy is converted to heat.

Even perfectly clear seawater is not perfectly transparent. If it were, the sun's rays would illuminate the great-est depths of the ocean, and seaweed forests would fill its warmed basins. The thin film of lighted water at the top of the surface zone is called the **photic zone** (*photos* = light). In clear tropical waters the photic zone may extend to a depth of 200 meters (660 feet), but a more typical value for the open ocean is 100 meters (330 feet). All the production of food by photosynthetic marine plants occurs in this thin, warm surface zone. Here water is heated by the sun, heat is transferred from the ocean into the atmosphere and space, and gases are exchanged with the atmosphere. The thermostatic effects we've discussed function largely within this zone. Most of the ocean's life is found here. The photic zone may be extraordinarily thin, but it is also extraordinarily important.

The ocean below the photic zone lies in blackness. Except for light generated by living organisms, the region is perpetually dark. This dark water beneath the photic zone is called the **aphotic zone** (*a* = without + *photos* = light).

The light energy of some colors is converted into heat nearer to the surface than the light energy of other colors. **Figure 6.19** shows this differential absorption by color.

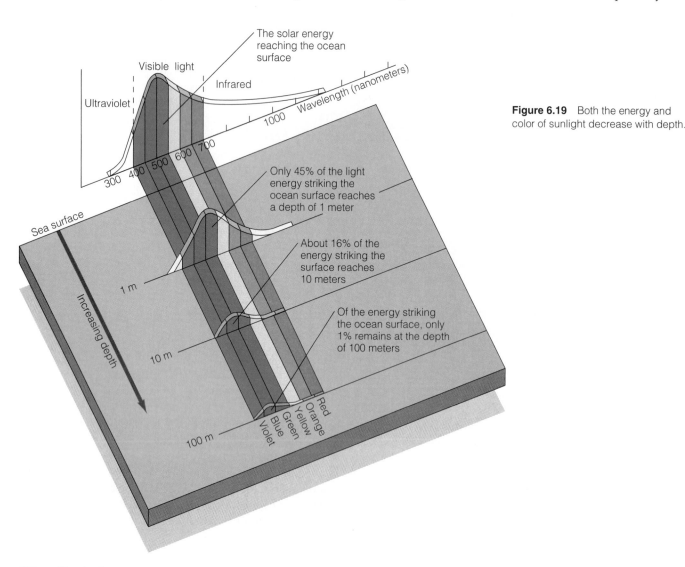

Figure 6.19 Both the energy and color of sunlight decrease with depth.

The solar energy reaching the ocean surface

Visible light

Infrared

Ultraviolet

Wavelength (nanometers)

1000

300 400 500 600 700

Sea surface

Only 45% of the light energy striking the ocean surface reaches a depth of 1 meter

1 m

About 16% of the energy striking the surface reaches 10 meters

Increasing depth

10 m

Of the energy striking the ocean surface, only 1% remains at the depth of 100 meters

100 m

Red
Orange
Yellow
Green
Blue
Violet

Notice that after 1 meter (3.3 feet) of travel, only 45% of the light energy remains, most of it in the green and blue wavelengths. After 10 meters (33 feet) 85% of the light has been absorbed, and after 100 meters (330 feet) just 1% remains. The dimming light becomes bluer with depth because the red, yellow, and orange wavelengths have already been absorbed. Even in the clearest conditions, sunlight rarely penetrates below 250 meters (820 feet).

From above, clear ocean water looks blue because blue light can travel through water far enough to be scattered back through the surface to our eyes. Divers near the surface see an even brighter blue color. Because nearly all red light is converted to heat in the first few meters of ocean water, red objects a short distance beneath the surface look gray. A diver working at a depth of 10 meters (33 feet) who cuts his hand will see gray blood rather than red because there is not enough red light at that depth to reflect from blood's red pigment and stimulate his eye. The underwater pictures of red organisms you've seen are possible only because the diver has brought along a source of white light (which contains all colors).

Sound in the Ocean

Sound is a form of energy transmitted by rapid pressure changes in an elastic medium. Sound energy decreases as it travels through seawater because of spreading, scattering, and absorption. Energy loss due to spreading is proportional to the square of the distance from the source. Scattering occurs as sound bounces off bubbles, suspended particles, organisms, the surface, the bottom, or other objects. Eventually sound is absorbed and converted by molecules into a very small amount of heat. Absorption of sound is proportional to the square of the frequency of the sound—higher frequencies are absorbed sooner. Sound waves can travel for much greater distances through water than light waves can before being absorbed. Because sound travels through water so efficiently, many marine animals use sound rather than light to "see" in the ocean.

The speed of sound in seawater of 35‰ salinity is about 1.5 kilometers per second (3,345 miles per hour), almost five times the speed of sound in air. The speed of sound in seawater increases as temperature and pressure increase. Sound travels faster at the warm ocean surface than it does in deeper, cooler water. Its speed decreases with depth, eventually reaching a minimum at about 1,000 meters (3,300 feet). Below that depth, the effect of increasing pressure offsets the effect of decreasing temperature; so speed increases again. Near the bottom of an ocean basin, the speed of sound may be higher than at the surface. Though important in the behavior of oceanic sound, these variations amount to only 2% or 3% of the average speed of sound in seawater. The relationship between depth and sound speed is shown in **Figure 6.20**.

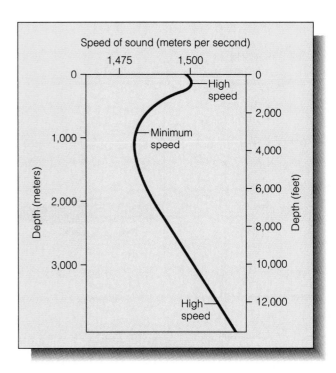

Figure 6.20 The relationship between water depth and the speed of sound.

Sofar Layers and Shadow Zones

The depth at which the speed of sound reaches its *minimum* varies with conditions, but is usually located near 1,200 meters (3,900 feet) in the North Atlantic or about 600 meters (2,000 feet) in the North Pacific. Transmission of sound in this minimum-velocity layer is very efficient because refraction tends to cause sound energy to remain within the layer. The outer edges of sound waves escaping from this layer will enter water in which the speed of sound is higher, speed up, and cause the wave to pivot back into the minimum velocity layer, as diagrammed in **Figure 6.21**. Upward-traveling sound waves that are generated within the minimum-velocity layer will tend to be refracted downward, and downward-traveling sound waves will tend to be refracted upward. In short, sound waves bend *toward* layers of lower sound velocity and so tend to stay within the zone. Therefore, loud noises made at this depth can be heard for thousands of kilometers.

Navy depth charges detonated in the minimum-velocity layer in the Pacific have been heard 3,680 kilometers (2,280 miles) from the explosion. In a recent test, sound generated by a U.S. Navy ship in the Indian Ocean was heard at the Oregon coast (see **Box 6.2**)! In the early 1960s the U.S. Navy experimented with the use of sound transmission in the minimum-velocity layer as a life-saving tool. The idea was that survivors in a life raft would drop a small charge into the water, which was set to explode at the proper depth. A number of widely spaced listening stations ashore would compare the differences in the

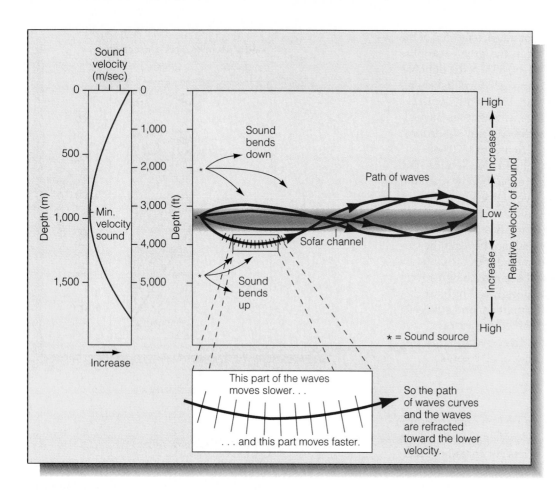

Figure 6.21 The sofar channel, in which sound waves travel at minimum speed. Sound transmission is particularly efficient—sounds can be heard for great distances—because refraction tends to keep sound waves within the channel.

Figure 6.22 The shadow zone. The thin high sound-velocity layer that forms at a depth of about 80 meters (260 feet) deflects sound. The shadow zone creates a good place for submarines to hide from sonar.

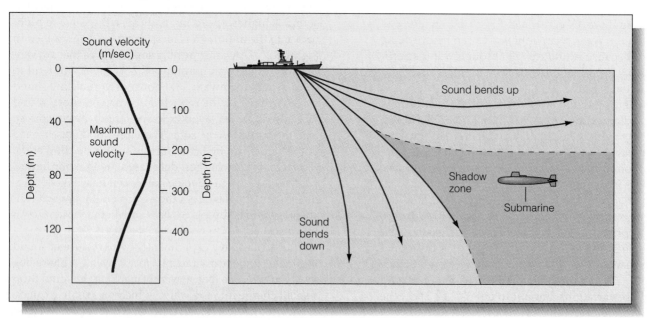

BOX 6.2

Sofar Away

Early in 1991, underwater sounds generated by a ship near Heard Island in the southern Indian Ocean were detected off the West Coast of the United States 17,960 kilometers (11,160 miles) away. The foghornlike sounds were produced by powerful underwater loudspeakers suspended 200 meters (650 feet) beneath a U.S. Navy research vessel. The ship's five transmitters emitted a booming low-frequency hum in all directions for an hour, then were silent for an hour. The pattern was repeated through most of one day.

Some of the sound was trapped in the Indian Ocean's minimum-velocity layer, the sofar channel, in which it traveled around Australia and into the Pacific. The sound was detected by a ship equipped with sensitive hydrophones lowered to the sofar depth off the Oregon coast near Coos Bay. It took about 3½ hours for the sound to travel that far, nearly halfway around the world. Similarly equipped ships and offshore stations detected the sound at other locations.

The experiment was part of a plan to measure ocean temperature within a few thousandths of a degree. Since sound travels more rapidly through warmer water, water temperature differences can be calculated by recording the time taken for sound to reach receivers around the world. The temperature of the ocean is an important measure of potential global warming, and the sound transmission experiment collected baseline data for a study designed to help settle the argument over whether carbon dioxide and other greenhouse gases are warming the planet artificially. Computer models of the Earth's climate suggest that greenhouse gases could cause the ocean to warm by about 0.005°C (0.009°F) per year at a depth of 1,000 meters (3,300 feet). Such a subtle temperature change could not be detected by conventional monitoring, but over a decade, this warming would cut a few seconds from the travel time of a sound signal. Scientists from Russia, France, Canada, New Zealand, South Africa, India, Australia, and Japan are participating in the experiment. More information on the possibilities and consequences of global warming may be found in Chapter 15.

arrival times of the signal and then compute the position of the raft. The project—since abandoned in favor of radio beacons—was called **sofar** (for *so*und *f*ixing *a*nd *r*anging). The minimum-velocity layer has come to be known as the **sofar layer**.

Sound travels slowly in the sofar layer, but it moves rapidly near the bottom of the well-mixed surface layer. Temperature and salinity conditions are homogeneous there; so they do not produce any refraction. Pressure still increases with depth, however, causing a thin high-velocity layer at around 80 meters (260 feet) just above the pycnocline (see again Figure 6.20).

Depending on the angle at which the sound waves arrive at the high-velocity layer, they will sometimes split and refract to the surface or bend into the depths. An object beyond the area of divergence may be undetectable—it would be within a **shadow zone**, a region into which very little sound energy penetrates. Shadow zones are of particular interest to submariners and the ship captains that use sound to hunt them. As depicted in **Figure 6.22**, a smart submarine captain can use the shadow zone to hide from a pursuer.

CHAPTER SUMMARY

Water, a chemical compound composed of two hydrogen atoms and one oxygen atom, is abundant on and within the Earth. The polar nature of the water molecule produces some unexpected chemical properties, one of the most important of which is water's remarkable ability to dissolve more substances than any other natural solvent. Though most solids and gases are soluble in water, the ocean is in chemical equilibrium, and neither the proportion nor amount of most dissolved substances changes significantly through time. Most of the properties of seawater are different from those of pure water because of the substances dissolved in the seawater.

The physical characteristics of the world ocean are largely determined by the physical properties of seawater. These properties include water's heat capacity, density, salinity, and its ability to transmit light and sound.

The thermal properties of water are responsible for the mild physical conditions at the Earth's surface. Liquid water is remarkably resistant to temperature change with the addition or removal of heat; and ice, with its large latent heat of fusion and low density, melts and re-freezes over large areas of the ocean to absorb or release heat with no change in temperature. These thermostatic effects, combined with the mass movement of water and water vapor, prevent large swings in Earth's surface temperature.

Changes in temperature and salinity greatly influence water density. Ocean water is usually layered by density, with the densest water on or near the bottom.

Sound and light in the sea are affected by the physical properties of water, with refraction and absorption effects playing important roles.

Terms and Concepts to Remember

absorption	heat	refraction
acid	heat budget	residence time
adhesion	heat capacity	salinity
aphotic zone	hydrogen bond	salinometer
base	latent heat of	scattering
buffer	evaporation (= of	sensible heat
calorie	vaporization)	shadow zone
chemical bond	latent heat of fusion	sofar
chemical	mixed layer	sofar layer
equilibrium	(= surface zone)	sound
chlorinity	mixing time	state
cohesion	molecule	surface zone
deep zone	Nansen bottle	(= mixed layer)
degrees	pH scale	temperature
density	photic zone	thermal
density curve	polar	equilibrium
electron	principle of	thermal inertia
excess volatiles	constant	thermocline
Forchhammer's	proportions	thermostatic
Principle	proton	properties
halocline	pycnocline	trace elements

Study Questions

1. Why is water a polar molecule? What properties of water derive from its polar nature?

2. Other than hydrogen and oxygen, what are the most abundant elements in seawater?

3. How is salinity determined? How are modern methods dependent on the principle of constant proportions?

4. Which dissolved gas is represented in the ocean in much greater proportion than in the atmosphere? Why the disparity?

5. What factors affect seawater's pH? How does the pH of seawater change with depth? Why?

6. How is heat different from temperature?

7. How does water's high heat capacity influence the ocean? Leaving aside its effect on beach parties, how do you think conditions on Earth would differ if our ocean consisted of ethyl alcohol?

8. Why does ice float? Why is this fact important to thermal conditions on Earth?

9. What factors affect the density of water? Why does cold air or water tend to sink? What is the role of salinity in water density?

10. How is the ocean stratified by density? What physical factors are involved? What names are given to the ocean's density zones?

11. What percentage of incoming sunlight reaches the ocean? What happens to that light? Is the heat budget balanced? What would happen if it were not?

12. What factors influence the intensity and color of light in the sea? What factors affect the depth of the photic zone? Could there be a "photocline" in the ocean?

For Further Study

Broecker, W. S. 1983. "The Ocean." *Scientific American*, September, 146–60. Information on the origin and nature of the ocean's dissolved solids.

Hill, M. N. 1963. *The Sea: Composition of Seawater.* New York: Wiley Interscience. A standard reference on the topic.

Horne, R. A. 1969. *Marine Chemistry: The Structure of Water and the Chemistry of the Hydrosphere.* New York: Wiley.

Kerr, R. A. 1988. "Ocean Crust's Role in Making Seawater." *Science* 239 (no. 4837): 260. The possible role of water circulation through active rift zones in reducing Mg^{2+} and increasing Ca^{2+}.

MacIntyre, F. 1970. "Why the Sea Is Salt." *Scientific American*, November, 104–15.

Munk, W. 1991. "The Heard Island Experiment." *Oceanus* (no. 1): 6–8.

Oceanus 20 (no. 2, Spring 1977) was dedicated to "Sound in the Sea."

Sverdrup, H. U. 1954. "Oceanography." In *The Earth as a Planet*, ed. C. P. Kuiper, 215–57. Chicago: University of Chicago Press. The thermostatic role of water on Earth; the importance of ocean and atmosphere in transporting heat.

7 ATMOSPHERIC CIRCULATION

Andrew

The costliest natural disaster in the history of the United States began as a cluster of rolling thunderclouds drifting westward across the coast of central Africa. The Earth's rotation had given the clouds a gentle twist, and warm oceanic air rose into their core. Like water spinning down a bathtub drain, humid air was sucked into a growing vortex of clouds. The moisture condensed to form rain, liberating heat energy that caused the storm to grow even larger. On 20 August 1992, near the center of the tropical Atlantic, the storm became Hurricane Andrew.

As the days passed, the hurricane tracked uncertainly toward the west. On two occasions, shearing winds threatened to tear the storm apart; each time it survived and strengthened. Then, on the evening of 23 August, Andrew metamorphosed into a towering black mountain of wind and rain, a rare category-5 storm. Its vast violence sucked the sea surface into a 6-meter (19-foot) dome tens of kilometers across. The wind speed near its center rose to 240 kilometers (150 miles) per hour. In the predawn hours of 24 August, the leading edge of the storm touched the Florida coast near Biscayne Bay. Eyewitnesses described the rising howl of wind—a tearing, clawing force that entered every window and door, that

Hurricane Andrew storms across southern Florida, 24 August 1992.

A neighborhood near Homestead, Florida, bears witness to the shattering force of hurricane winds that reached 350 kilometers (220 miles) per hour.

tore the roofs off their homes and children from their grasp, that hurled cars and mobile homes about like toys, and that lasted more than 90 minutes and reached speeds above 350 kilometers (220 miles) per hour. The division between ocean and land blurred beneath an onrushing wall of windblown seawater. At the height of the storm the hot, wet air glowed yellow-green and was ear-poppingly thin. Devastation was all but absolute.

Before the giant storm died later that week in a series of rattling thunderstorms across the Mississippi Valley, 160,000 people would be homeless, 43 would be dead, and 68,000 businesses would be destroyed; the damage exceeded $30 billion.[1] Scouring currents had rearranged the south Florida coast, sunk ships, shattered harbor installations, uprooted submarine cables, and inundated parts of the Everglades with salt water. Forecasters had not foreseen the ultimate violence of Andrew. Clearly, marine scientists have much to learn about the dynamics of air–ocean interactions.

[1] As this is written, damage estimates from the January 1994 earthquake in southern California are approaching $15 billion.

Figure 7.1 Steam fog over the ocean indicates rapid evaporation. Water vapor is invisible, but as water vapor rises into cool air it can condense into visible droplets.

The Earth's atmosphere and its ocean are intimately intertwined, their gases and waters freely exchanged. Gases entering the atmosphere from the ocean have important effects on climate; and gases entering the ocean from the atmosphere can influence sediment deposition, the distribution of life, and some of the physical characteristics of the seawater itself. Water evaporated from the ocean surface and moved by the winds helps minimize worldwide extremes of surface temperature and, through rain, provides moisture for agriculture. The weather that so profoundly affects our daily lives is shaped at the junction of wind and water, and winds greatly influence the movement of seawater. In this chapter we will investigate the movement of air; in the next chapter, the movement of water.

COMPOSITION AND PROPERTIES OF THE ATMOSPHERE

The lower atmosphere, which represents all but two-millionths of the planet's air, is a nearly homogenous mixture of gases, most plentifully nitrogen (78.1% by volume) and oxygen (20.9%). Air is never completely dry; **water vapor**, the gaseous form of water, can occupy as much as 4% of its volume. Sometimes liquid droplets of water are visible as clouds or fog, but more often the water is simply there but invisible, having entered the atmosphere from the ground, plants, and sea surface (see **Figure 7.1**). The residence time of water vapor in the lower atmosphere is about ten days. Water leaves the atmosphere by condensing into dew, rain, or snow.

Air has mass. A square-inch column of air (6.45 square centimeters), extending from sea level to the top of the atmosphere, weighs about 6.7 kilograms (14.7 pounds). A square-foot column of air the same height weighs more than a ton.

The temperature and water content of air greatly influence its density. Because the molecular movement associated with heat causes a mass of warm air to occupy more space than an equal mass of cold air, warm air is less dense than cold air. And contrary to what we might guess, humid air is *less* dense than dry air at the same temperature because molecules of water vapor weigh less than the nitrogen and oxygen molecules that the water vapor displaces.

Near Earth's surface, air is packed densely by its own weight. Air lifted from near sea level to a higher altitude is subjected to less pressure and will expand. As anyone knows who has felt the cool air rushing from an open tire valve, air becomes *cooler* when it *expands*. The opposite effect is also familiar: Air *compressed* in a tire pump becomes *warmer*. Air descending from high altitude toward sea level warms as it is compressed by the higher atmospheric pressure near the Earth's surface.

Warm air can hold more water vapor than cold air. Water vapor in rising, expanding, cooling air will often condense into clouds (aggregates of tiny droplets) because the cooler air can no longer hold as much water vapor. If rising and cooling continues, the droplets may coalesce into raindrops or snowflakes. The atmosphere will then lose water as **precipitation**—liquid or solid water that falls from the air to the Earth's surface. As we will see, these rising-expanding-cooling and falling-compressing-heating relationships are important in understanding atmospheric circulation, weather, and climate.

ATMOSPHERIC CIRCULATION

About half of the energy radiated toward the Earth from the sun is absorbed by the Earth, but this energy is not distributed evenly across the planet's surface. The amount of solar energy reaching Earth's surface per minute varies with the transparency of the atmosphere, the angle of the sun above the horizon, and the local reflectivity of the surface. The most important factors that affect solar angle are latitude and season.

Uneven Solar Heating and Latitude

As can be seen in **Figure 7.2**, sunlight striking polar latitudes spreads over a greater area, approaches the surface at a low angle favoring reflection, and filters through more atmosphere. Polar regions receive no sunlight at all during the depths of local winter. Contrast this to the tropical latitudes in Figure 7.2. The high solar angle in the tropics distributes the same amount of sunlight over a much smaller area. The more nearly vertical angle at which the light approaches means that it passes through less atmosphere and minimizes reflection. As you would expect, the tropics are warmer than the polar regions.

Uneven Solar Heating and the Seasons

At mid-latitudes the Northern Hemisphere receives about three times as much solar energy per day in June as it does in December. This difference is due to the 23.27° tilt of the Earth's rotational axis relative to the plane of its orbit around the sun (see **Figure 7.3**). The mounting angle you may have noticed in library reference globes indicates this tilt, or **orbital inclination**. The inclination of the axis causes the change of seasons.

The rotating Earth spins around the polar axis. This spin causes the daily rising and setting of the sun. Notice in Figure 7.3 that the axes of the Earth (extensions of the North and South poles) point toward the same places in space regardless of the season. Luckily for navigators, the north polar axis currently passes near the bright star Polaris, making Polaris the North Star. As Earth revolves around the sun, the constant tilt of its rotational axis causes the Northern Hemisphere to lean toward the sun in June, and away from the sun in December. The sun therefore

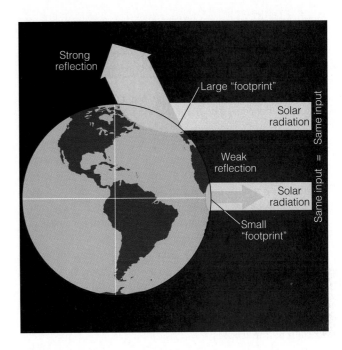

Figure 7.2 How solar energy input varies with latitude. Equal amounts of sunlight are spread over a greater surface area near the poles than in the tropics. Ice near the poles reflects much of the energy that reaches the surface there.

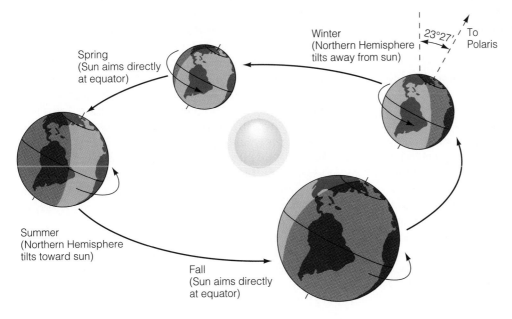

Figure 7.3 The seasons (shown here for the Northern Hemisphere only) are caused by variations in the amount of incoming solar energy as the Earth makes its annual rotation around the sun on an axis tilted by 23°27'.

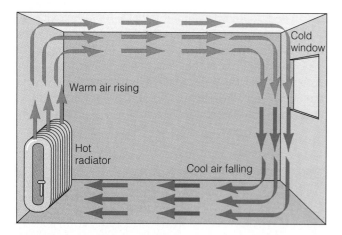

Figure 7.4 A convection current forms in a room when air flows from a hot radiator to a cold window and back.

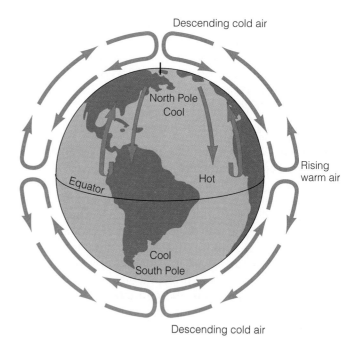

Figure 7.5 A hypothetical model of Earth's air circulation if uneven solar heating were the only factor to be considered. The thickness of the atmosphere is greatly exaggerated in this drawing.

appears higher in the sky in the summer, and lower in winter. The inclination of the Earth's axis also causes days to become longer as summer approaches, and shorter with the coming of winter. Longer days mean more time for the sun to warm the Earth's surface.

The greatest amount of solar heating is possible where the sun appears directly overhead. This never happens north of 23.27° north latitude or south of 23.27° south latitude. (Can you see why?) Most of the sun's warmth is therefore concentrated between these two latitudes, a zone known as the tropics.

Uneven Solar Heating and Atmospheric Circulation

Since the poles have such a marked deficiency of heat and the equator has such a pronounced surplus, why don't the polar oceans freeze solid and the equatorial ocean boil away? The reason is that currents in the atmosphere and ocean are moving huge amounts of heat between tropics and poles.

Water's high heat capacity makes it an ideal fluid to equalize the polar-tropical heat imbalance. Both atmosphere and ocean transfer heat by movement, but water's exceptionally high latent heat of evaporation means that water vapor transfers much more heat (per unit of mass) than liquid water. About half of the solar energy entering water results in evaporation. The solar energy required for this evaporation is later surrendered during condensation, but usually at a distance from where the initial evaporation occurred. The ocean surface near Cuba may be cooled by evaporation today, the water vapor moved north by winds, and eastern Canada warmed by condensation of the same water in a rainstorm later in the week. Atmospheric circulation accounts for about two-thirds of

the poleward transfer of heat; ocean currents directly transport the other third.

The concentration of solar energy near the equator has important effects on the atmosphere. We know that warm air rises and that cool air sinks. Think of air circulation in a room with a hot radiator opposite a cold window (see **Figure 7.4**). Air warms, expands, becomes less dense, and rises over the radiator. Air cools, contracts, becomes more dense, and falls near the cold glass window. The circular current of air in the room, a **convection** current, is caused by the difference in temperature between the ends of the room. A similar process occurs over the surface of the Earth. Surface temperatures are higher at the equator than at the poles, and air can gain heat from warm surroundings. Since air is free to move over the Earth's surface, it would be reasonable to assume that an air circulation pattern like the one shown in **Figure 7.5** would develop over the Earth. In this ideal model, air heated in the tropics would expand and become less dense, rise to high altitude, turn poleward, and "pile up" as it converged near the poles. The air would then cool and contract by radiating heat into space, sink to the surface, and turn equatorward, flowing along the surface to the tropics to complete the circuit.

But this is *not* what happens. Global circulation of air is governed by two factors: uneven solar heating *and* the rotation of the Earth. The eastward rotation of the Earth on its axis deflects the moving air or water (or any moving object having mass) away from its initial course. This deflection is called the **Coriolis effect** in honor of **Gaspard Gustave de Coriolis**, the French scientist who worked out its mathematics in 1835. An understanding of the Coriolis effect is important to an understanding of atmospheric and oceanic circulation.

The Coriolis Effect

To an earthbound observer, any object moving freely across the globe appears to curve slightly from its initial path. In the Northern Hemisphere, this curve is to the right (or **clockwise**) from the expected path; in the Southern Hemisphere, to the left (or **counterclockwise**). To earthbound observers the deflection is very real—it isn't caused by some mysterious force, and it isn't an optical illusion or other trick caused by the shape of the globe itself. *The observed deflection is caused by the observer's moving frame of reference on the spinning Earth.*

The influence of this deflection can be illustrated by performing a mental experiment involving concrete objects—in our case cities and cannonballs—and then by applying the principle to atmospheric circulation. Let's pick as examples for our experiment the equatorial city of Quito (capital of Ecuador) and Buffalo, New York. Both cities are on almost the same line of longitude (79°W), so Buffalo is almost exactly north of Quito (as **Figure 7.6** shows). Like everything else attached to the rotating Earth, both cities make one trip around the world each 24 hours. Through one day, the north-south relationship of the two cities never changes—Quito is *always* due south of Buffalo.

A complete trip around the world is 360°; so each city moves eastward at an angular rate of 15° per hour (360°/24 hours = 15°/hour). Yet even though their angular rates are the same, the two cities move eastward at different speeds. Quito is on the equator, the "fattest" part of the Earth. Buffalo is farther north at a "skinnier" part. Imagine both cities isolated on flat disks, and imagine the Earth's sphere being made of a great number of these disks strung together on a rod connecting North Pole to South Pole. From Figure 7.6, you can see that Buffalo's disk is smaller, and therefore has a smaller circumference, than Quito's. Buffalo doesn't have as far to go in one day as Quito because its disk is not as large. That means that Buffalo must move eastward *more slowly* than Quito to maintain its position due north of Quito.

Look at the Earth from above the North Pole in **Figure 7.7**. The Quito disk and the Buffalo disk must turn through 15° each hour (or Earth would rip itself apart), but the city on the equator must move faster to the east to turn its 15° each hour because its slice of the "pie" is larger. Buffalo must move at 1,260 kilometers (787 miles) per hour to go around the world in one day, while Quito

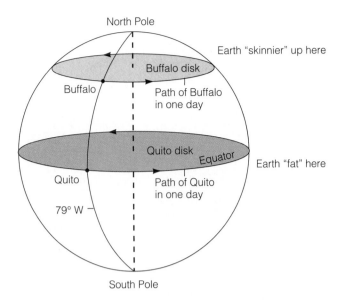

Figure 7.6 Sketch of the thought experiment in the text, showing that Buffalo travels a shorter path on the rotating Earth than Quito does each day.

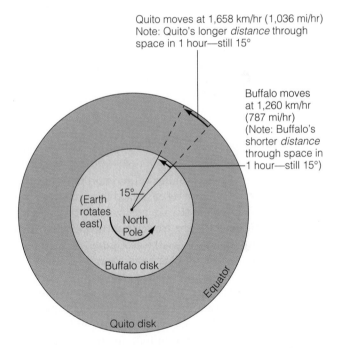

Figure 7.7 A continuation of the thought experiment. A look at the Earth from above the North Pole shows that Buffalo and Quito move at different velocities.

must move at 1,658 kilometers (1,036 miles) per hour to do so.

Now imagine a massive object moving between the two cities. A cannonball shot north from Quito toward Buffalo would carry Quito's eastward component as it

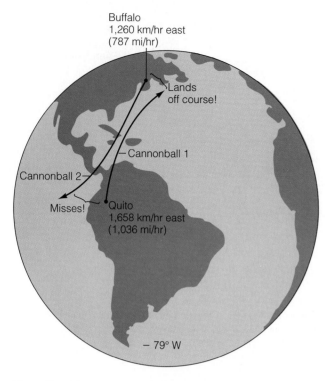

Buffalo
1,260 km/hr east
(787 mi/hr)

Lands
off course!

Cannonball 1

Cannonball 2

Misses!

Quito
1,658 km/hr east
(1,036 mi/hr)

— 79° W

Figure 7.8 The final step in the experiment. As observed from *space*, cannonball #1 (shot northward) and cannonball #2 (shot southward) move as we might expect; that is, they travel straight away from the cannons and fall to Earth. Observed from the *ground*, however, cannonball #1 veers slightly east and cannonball #2 veers slightly west of their intended targets. The effect depends on the observer's frame of reference.

goes; that is, regardless of its northward speed, the cannonball is also moving east at 1,658 kilometers (1,036 miles) per hour. The fact of being fired northward by the cannon does not change its eastward movement in the least. As the cannonball streaks north, an odd thing happens. The cannonball veers from its northward path, angling slightly to the right (east) (see **Figure 7.8**). Actually, this first cannonball is moving as an observer from space would expect, but to those of us on the ground, the cannonball "gets ahead of the Earth." As cannonball #1 moves north, the ground beneath it is no longer moving eastward at 1,658 kilometers (1,036 miles) per hour. During the ball's time of flight, Buffalo (on its smaller disk) *has not moved eastward enough to be where the ball will hit.* If the time of flight for the cannonball is one hour, a city 398 kilometers east of Buffalo (1,658 [Quito's speed] – 1,260 [Buffalo's speed] = 398) will have an unexpected surprise. Albany may be in for some excitement.

If a cannonball were fired south from Buffalo toward Quito, the situation would be reversed. This second cannonball has an eastward component of 1,260 kilometers (787 miles) per hour, even while it sits in the muzzle. Once fired and moving southward, cannonball #2 travels

over portions of the Earth that are moving increasingly faster in an eastward direction. The ball again appears to veer off course to the right (see Figure 7.8 again), falling into the Pacific to the west of Ecuador. Don't be deceived by the word *appears* in the last sentence. The cannonballs really do veer to the right, or clockwise. Only an observer in space would see them appear to go straight, and the Earth appear to move out from underneath.

The Coriolis effect is a real effect dependent on our rotating frame of reference. Part of that frame of reference involves the direction from which you view the problem. Thus in Figure 7.8, cannonball #2 looks to you as if it is veering *left*, but to the citizens of Buffalo facing south to watch the cannonball disappear, it moves to the right (west). The Coriolis deflection works counterclockwise in the Southern Hemisphere because the frame of reference there is reversed. A quick way to remember the Coriolis effect: Northern Hemisphere/clockwise/right; Southern Hemisphere/counterclockwise/left. The Coriolis effect also influences the path of objects moving from east to west, or west to east. Because the Coriolis effect influences any object with mass, as long as that object is moving, it plays a large role in the movements of air and water on the Earth.

The Coriolis Effect and Atmospheric Circulation Cells

We can now modify our original model of atmospheric circulation (Figure 7.5) into the more correct representation provided in **Figure 7.9**. Yes, air does warm, expand, and rise at the equator; and air does cool, contract, and fall at the poles. But instead of continuing all the way from equator to pole in a continuous loop in each hemisphere, air rising from the equatorial region moves poleward and is gradually deflected eastward—that is, turns to the *right* in the Northern Hemisphere and to the *left* in the Southern Hemisphere. This eastward deflection is caused by the Coriolis effect. (Note that the Coriolis effect does not *cause* the wind; it only influences the wind's direction.)

As air rises at the equator, it loses moisture by precipitation (rainfall) caused by expansion and cooling. This drier air now grows more dense in the upper atmosphere as it radiates heat to space and cools. When it has traveled about a third of the way from the equator to the pole—that is, to about 30°N and 30°S latitudes—the air becomes dense enough to fall back toward the surface. Most of the descending air turns back toward the equator when it reaches the surface. In the Northern Hemisphere the Coriolis effect again deflects this surface air to the right—the air blows across the ocean or land from the northeast. (This air is represented by the arrows labeled "NE trade winds" in Figure 7.9.) Though it has been heated by compression during its descent, this air is

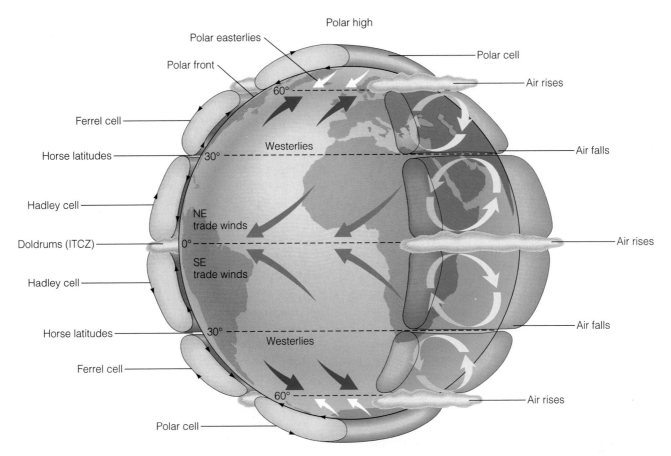

Figure 7.9 Global air circulation as described in the three-cell circulation model. As in Figure 7.5, air rises at the equator and falls at the poles. But instead of one great circuit in each hemisphere from equator to pole, there are three. Note the influence of the Coriolis effect on wind direction.

generally colder than the surface over which it flows. The air soon warms as it moves equatorward, however, evaporating surface water and becoming humid. The warm, moist, less-dense air begins to rise as it approaches the equator and completes the circuit.

Such a large circuit of air is called an **atmospheric circulation cell**. A pair of these tropical cells exists, one on each side of the equator. They are known as **Hadley cells** in honor of George Hadley, the London lawyer and philosopher who worked out an overall scheme of wind circulation in 1735. Look for them in Figure 7.9.

A more complex pair of circulation cells operates at mid-latitudes in each hemisphere. Some of the air descending at 30° latitude turns poleward rather than equatorward. Before this air descends to the surface, it is joined by air at high altitude returning from the north. As can be seen in Figure 7.9, a loop of air forms between 30° and about 50°–60° of latitude. As before, the air is driven by uneven heating and influenced by the Coriolis effect. Surface air in this circuit is again deflected to the right, this time flowing from the west to complete the cir-

cuit. (This air is represented by the arrows labeled "Westerlies" in Figure 7.9.) The mid-latitude circulation cells of each hemisphere are named **Ferrel cells** after William Ferrel, the American who discovered their inner workings in the mid-nineteenth century. They, too, can be seen in Figure 7.9.

Meanwhile, air that has grown cold over the poles begins moving toward the equator at the surface, turning to the west as it does so. At between 50° and 60° latitude in each hemisphere, this air has taken up enough heat and moisture to ascend. However, this polar air is more dense than the air in the adjacent Ferrel cell and does not mix easily with it. The unstable zone between these two cells generates most mid-latitude weather. At high altitude the ascending air from 50°–60° latitudes turns poleward to complete a third circuit. These are the **polar cells**.

Three large atmospheric circulation cells—a Hadley cell, a Ferrel cell, and a polar cell—therefore exist in each hemisphere. Air circulation within each cell is powered by uneven solar heating and influenced by the Coriolis effect.

WIND PATTERNS

The model of atmospheric circulation described above has many interesting features. At the bands between circulation cells, the air is moving *vertically*, and surface winds are weak and erratic. Such conditions exist at the equator (where air rises and atmospheric pressure is generally low) or at 30° latitude in each hemisphere (where air falls and atmospheric pressure is generally high). Places within circulation cells where air moves rapidly *horizontally* across the surface from zones of high pressure to zones of low pressure are characterized by strong, dependable winds.

Sailors have a special term for the calm equatorial areas where the surface winds of the two Hadley cells converge: the **doldrums**. The word has come to be associated with a gloomy, listless mood, perhaps reflecting the sultry air and variable breezes found there. Scientists who study the atmosphere call this area the **intertropical convergence zone**, or **ITCZ**, to reflect the influence of wind convergence on conditions near the equator. Strong heating in the ITCZ causes surface air to expand and rise. The humid, rising, expanding air loses moisture as rain, some of which contributes to the success of tropical rain forests.

Sinking air, in contrast, is generally arid. The great deserts of both hemispheres, dry bands centered around 30°, mark the intersection of the Hadley and Ferrel cells. Because evaporation is higher than precipitation in these areas, ocean surface salinity tends to be highest at these latitudes. At sea, these areas of high atmospheric pressure and little surface wind are called the **horse latitudes**. Spanish ships laden with supplies for the New World were often becalmed there, sometimes for weeks on end. When the mariners ran out of water and feed for their livestock, they were forced to throw the dead horses over the side.

Of much more interest to sailing masters were the bands of dependable surface winds *between* the zones of ascending and descending air. Most constant of these are the persistent **trade winds**, centered at about 15°N and 15°S latitude. The *trades* are the surface winds of the Hadley cells as they move from the horse latitudes to the doldrums. In the Northern Hemisphere they are the *northeast trades*; the *southeast trades* are the Southern Hemisphere counterpart.[2] The **westerlies**, surface winds of the Ferrel cells centered at about 45°N and 45°S latitude, flow between the horse latitudes and the boundaries of the polar cells in each hemisphere. Thus, the westerlies approach from the southwest in the Northern Hemisphere

and from the northwest in the Southern Hemisphere. Sailors outbound from Europe to the New World learned to drop south to catch the trades and to return home by a more northerly route to take advantage of the westerlies. Trade winds and westerlies are shown in Figure 7.9.

The three-cell model of atmospheric circulation discussed here represents an average of air flow through many years over the planet as a whole. Though the model is accurate in a general sense, local details of cell circulation vary because surface conditions are different at different longitudes. The ocean's thermostatic effect is the major factor reducing irregularities in cell circulation over water.

BREAKDOWN IN CELL CIRCULATION: EL NIÑO

Sometimes cell circulation doesn't seem to play by the rules. In three- to eight-year cycles, atmospheric circulation changes significantly from the patterns shown in Figure 7.9. The changes are caused by the **Southern Oscillation**, a reversal in the usual westward flow of air between the normally stable low-pressure area over the western Pacific north of Australia and the normally high-pressure area over the eastern Pacific near Easter Island. This reversal in the distribution of atmospheric pressure between the eastern and western Pacific causes the trade winds to weaken or reverse. The trade winds normally drag huge quantities of water westward along the ocean's surface near the equator, but without the winds these equatorial currents crawl to a stop. Warm water that has accumulated at the western side of the Pacific can return east along the equator toward the coast of Central and South America.

Normally, a current of cold water, rich in upwelled nutrients, flows north and west away from the South American continent (see **Figure 7.10a**). The Southern Oscillation replaces it by a warm, eastward- and southward-moving nutrient-poor current called **El Niño (Figure 7.10b)**. The warm water usually arrives around Christmastime; hence the current's name, "the Christ Child." Local sea level rises (sometimes by 20 centimeters, or 8 inches, in the Galápagos); water increases in temperature by up to 7°C (11°F); increased evaporation intensifies coastal storms; and rainfall inland may be much higher than normal. Marine organisms are deprived of their normal food supplies, and commercial fishing is seriously disrupted. The phenomena of the Southern Oscillation and El Niño are coupled, so the terms are often combined to form the acronym **ENSO**, "*El Niño/Southern Oscillation.*" An ENSO event lasts about a year. The effects are not only felt in the Pacific; all ocean areas at trade-wind latitudes in both hemispheres can be affected.

---

[2]Winds are named by the direction from which they blow. A west wind blows from the west toward the east; a northeast wind blows from the northeast toward the southwest.

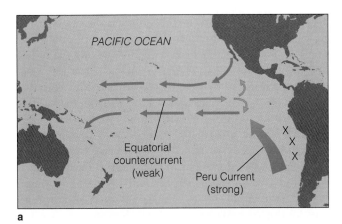

a

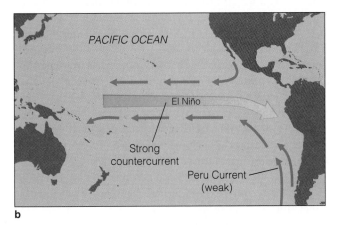

b

Figure 7.10 The relationship between the Southern Oscillation and El Niño. (a) Normally, the air and surface water flow westward, and upwelling of cold water (indicated by XXX) occurs along the west coast of South America. (b) When the Southern Oscillation develops, the trade winds diminish and then reverse, leading to an eastward movement of warm water along the equator. The surface waters of the central and eastern Pacific become warmer; upwelling ceases off the coast of South America.

The most severe ENSO event in this century occurred in 1982–83. Effects associated with El Niño were spectacular over much of the Pacific. Ecuador and Peru experienced severe flooding, and many rainfall records (and piers) were broken along the coast of California. Sea level at San Diego rose 10 centimeters (4 inches) higher than normal. Two years later, water temperatures off the West Coast of the United States were still unusually warm, and many exotic marine species were reported. Some birds that had not been seen in southern California for decades returned. Anglers along the West Coast were astonished to find fish species typical of Mexican waters dangling from their hooks. The 1985–86 El Niño was not as pronounced.

Studies of the ocean and atmosphere in 1982–83 have given researchers new insight into the behavior and effects of the Southern Oscillation. Some researchers believe

the 1982–83 event was triggered by the violent 1982 eruption of El Chichón, a Mexican volcano, which injected huge quantities of obscuring dust and sulfur-rich gases into the atmosphere. Others have recently suggested that the Southern Oscillation may be triggered by pulses of heat entering the Pacific at crustal spreading zones near the equator. Some of this heat is moved by currents to the ocean surface. The warming of surface water may cause air above the ocean to warm, atmospheric pressure to fall, and the east-to-west trade winds to slacken. Though the exact cause or causes of the Southern Oscillation are not yet understood, subtle changes in the atmosphere permit meteorologists to predict a severe El Niño nearly a year in advance of its most serious effects.

STORMS

Storms are regional atmospheric disturbances characterized by strong winds, often accompanied by precipitation. Few natural events underscore human insignificance like a great storm. When powered by stored sunlight, the combination of atmosphere and ocean can do fearful damage.

In Bangladesh, on 13 November 1970, a tropical cyclone (a hurricane) with wind speeds of more than 200 kilometers (125 miles) per hour roared up the mouths of the Ganges River, carrying with it masses of seawater up to 12 meters (40 feet) high. Water and wind clawed at the aggregation of small islands, most just above sea level, that makes up this impoverished country. In only 20 minutes at least 300,000 lives were lost, and estimates ranged to 1 million dead! Property damage was complete. Photographs taken soon after the storm showed a horizon-to-horizon morass of flooded, deep-gashed ground tortured by furious winds. There was almost no trace of human inhabitants, farms, domestic animals, or villages. Another great storm, which struck in May 1991, killed another 200,000 people. The economy of the shattered country may never fully recover.

A much different type of storm buried Buffalo, New York, in the winter of 1976–77. By 28 January snow had fallen for 40 days, and nearly a meter (39 inches) lay on the ground. On that day winds of 115 kilometers (70 miles) per hour began to buffet the city. After three days of storm, snowdrifts were piled to 9 meters (30 feet), 29 people had died, and damage was estimated at $250 million. At one point 17,000 people were trapped in offices, factories, and homes; 5,000 vehicles were abandoned.

These two large storms exemplify tropical cyclones and extratropical cyclones at their worst. As the name implies, tropical cyclones like the hurricane that struck Bangladesh are primarily a tropical phenomenon. Extratropical cyclones, the weather disturbances with which residents of Buffalo (and other mid-latitude dwellers) are most familiar, are found mainly in the Ferrel cells of each hemisphere.

Both kinds of storms are **cyclones**, huge rotating masses of low-pressure air in which winds converge and ascend. The word *cyclone*, derived from the Greek noun *kyklon* (an object moving in a circle), underscores the spinning nature of these disturbances. (Don't confuse a cyclone with a **tornado**, a much smaller funnel of fast-spinning wind associated with severe thunderstorms.)

Air Masses

Cyclonic storms form between or within air masses. An **air mass** is a large body of air with nearly uniform temperature, humidity, and therefore density throughout. Air pausing over water or land will tend to take on the characteristics of the surface below. Cold, dry land causes the mass of air above to become chilly and dry. Air above a warm ocean surface will become hot and humid. Cold, dry air masses are dense and form zones of high atmospheric pressure. Warm, humid air masses are less dense and form zones of lower atmospheric pressure.

Air masses can move within or between circulation cells. Density differences will prevent the air masses from mixing when they approach one another. Energy is required to mix air masses, and since that energy is not always available, a dense air mass may slide beneath a lighter air mass, lifting the lighter one and causing its air to expand and cool. Water vapor in the rising air may condense. All of these effects contribute to turbulence at the boundaries of the air masses.

The boundary between air masses of different density is called a **front**. The term was coined by **Vilhelm Bjerknes**, a Norwegian meteorologist, who saw a similarity between the zone where air masses meet and the violent battle fronts of World War I.

Extratropical cyclones form at a front between *two* air masses. Tropical cyclones form from disturbances within *one* warm and humid air mass.

Extratropical Cyclones

Extratropical cyclones form at the boundary between each hemisphere's polar cell and its Ferrel cell—the **polar front**. These great storms occur mainly in the winter hemisphere when temperature and density differences across the polar front are most pronounced. Remember, the cold wind poleward of the front is generally moving from the east; the warmer air equatorward of the front is generally moving from the west (see again Figure 7.9). The smooth flow of winds past each other at the front may be interrupted by zones of alternating high and low atmospheric pressure that bend the front into a series of waves. Because of the difference in wind direction of the air masses north and south of the polar front, the wave shape will enlarge, and a twist will form along the front. The different densities of the air masses prevent easy mixing; so the cold, dense air mass will slide beneath the warmer lighter one. Formation of this twist in the Northern Hemisphere, as seen from above, is shown in **Figure 7.11**. The twisting mass of air becomes an extratropical cyclone.

The twist that generates an extratropical cyclone circulates counterclockwise in the Northern Hemisphere, seemingly in opposition to the Coriolis effect, but the reasons for this paradox become clear when we consider the wind directions and the nature of interruption of the airflow between the cells. (In fact, the counterclockwise motion of the cyclone *is* Coriolis-driven because the large-scale airflow pattern at the edges of the cells is generated in part by the Coriolis effect.) Wind speed increases as the storm "wraps up" in much the same way that a spinning skater's rotation speed increases as she brings her hands to her body. Air rushing toward the center of the spinning storm rises to form a low-pressure zone at the center. Extratropical cyclones are embedded in the westerly winds and thus move eastward. They are typically 1,000–2,500 kilometers (625–1,600 miles) in diameter and last from two to five days. **Figure 7.12** provides a beautiful example.

Precipitation can begin as the circular flow develops. **Figure 7.13**, a cross section through the colliding air masses of Figure 7.11d, shows why. Precipitation is caused by the lifting, and consequent expansion and cooling, of the mass of mid-latitude air involved in the twist. As it rises and cools, this air can no longer hold all of its water vapor; so clouds and rain result. When *cold* air advances and does the lifting (as on the left side of Figure 7.13) a *cold front* occurs. A *warm front* happens when *warm* air is blown on top of the retreating edge of cold air (as on the right side of Figure 7.13). The wind and precipitation associated with these fronts are sometimes referred to as **frontal storms**.

Chains of extratropical cyclones swirl over land and ocean in the middle cells of both hemispheres. They are the principal cause of weather in the mid-latitude regions, where most of the world's people live. When a weather forecaster speaks of a cold front bringing winter snow or rain to your area, the forecaster is describing the approach of a cold south-sweeping arm of an extratropical cyclone.

Tropical Cyclones

Tropical cyclones are great masses of warm, humid rotating air. They occur in all tropical oceans except the equatorial South Atlantic. Large tropical cyclones are called **hurricanes** (*Huracan* = the Taino god of the wind)

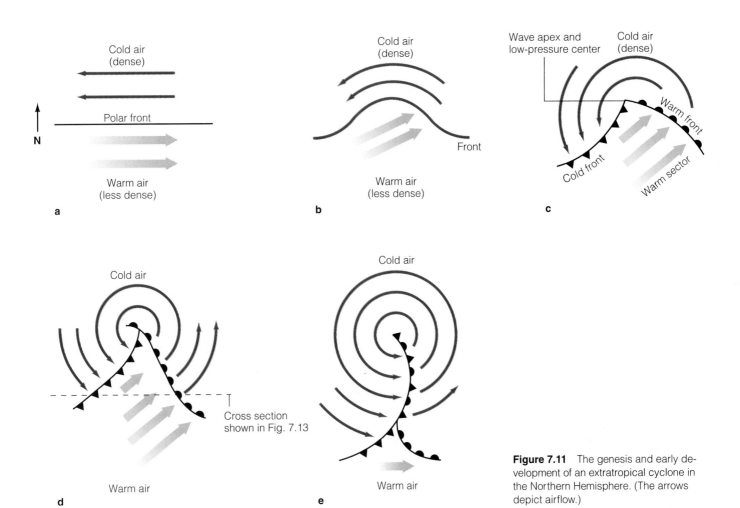

Figure 7.11 The genesis and early development of an extratropical cyclone in the Northern Hemisphere. (The arrows depict airflow.)

Figure 7.12 An extratropical cyclone over Ohio, 25 April 1991. Compare with Figure 7.11d.

BOX 7.1 ● *The Fastnet Disaster*

Centuries of effort have taught us much about atmosphere–ocean interaction, but our understanding is still incomplete. Sometimes this lack of knowledge is merely inconvenient: An incorrect forecast results in a rainy picnic or a cloudy day at the beach. At other times, though, our ignorance can contribute to the loss of hundreds or thousands of human lives and the destruction of property worth billions of dollars.

Air–ocean interaction is perhaps the most complex problem in all of physical oceanography, and weather forecasting depends on an accurate understanding of what happens at that boundary. Our best guesses aren't always good enough. Consider, for example, the 1979 Fastnet race across the shallow waters southwest of the British Isles.

The Fastnet Yacht Race is always a difficult test of seamanship and equipment. The 1979 race began on 11 August in weather so mild that the crews' major concern was sea fog and calm winds. It ended three days later in shrieking winds and towering seas generated by a storm that exploded into the North Atlantic from its birthplace over the American Great Plains. One specialist called the unpredicted storm a "meteorological bomb."

The first morning of the race, a weak storm system began moving from Nova Scotia toward the racecourse. Its actions were not unusual—it behaved as a mild summer storm is expected to behave. The next night, under the cover of darkness, it blended with a jet stream, one of a number of high-altitude bands of strong wind that blow across the Earth's surface. At around noon on 13 August the fast-moving storm suddenly "deepened": Its atmospheric pressure plummeted, and its internal winds increased. At this crucial juncture, three merchant ships that were nearest to the low-pressure center somehow reported incorrect pressure and wind-speed readings. The conflicting reports confused the forecast, but the race was already underway. That night, after dark, the "bomb" exploded amidst the Fastnet fleet.

What was supposed to be a 600-mile sail around a lighthouse off the Irish coast became every yachtsman's worst nightmare. Twenty-seven hundred men and women in 303 yachts were trapped in a tempest with winds gusting to 147 kilometers (92 miles) per hour. More than a third of the boats were knocked down by the 15-meter (50-foot) waves, their masts paralleling the ocean surface. Fully a fourth capsized. Sailors who had retreated to the safety of their cabins were battered by tins of food and pieces of equipment as their boats violently rolled and pitched. Twenty-four crews abandoned their boats, five of which sank. Fifteen people died of drowning or hypothermia. It was the greatest disaster in the century-long history of ocean yacht racing.

Were the forecasters responsible for the disaster? A board of inquiry found such ". . . explosively developing meteorological and oceanographic conditions . . . to be essentially unpredictable." Before the race started, no forecaster could have foreseen the violence of this storm. The lesson is clear: Ocean scientists need to learn more about the complex relationship between ocean and air. Far more than yacht races is at risk.

Bad day for a yacht race. *Siska* struggles in Force 9 winds in the Celtic Sea west of England.

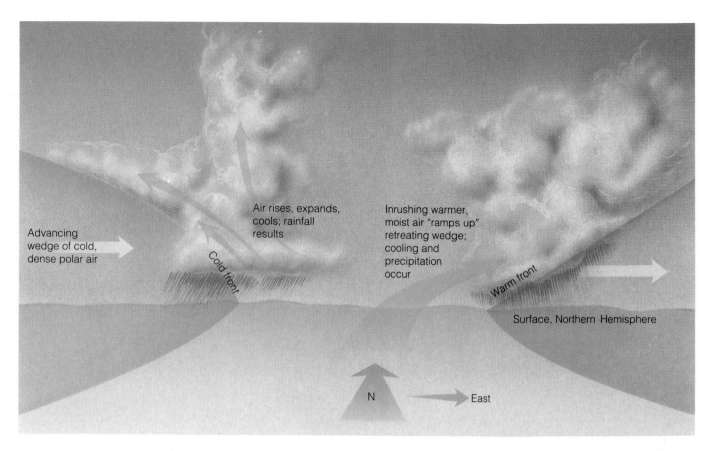

Figure 7.13 How precipitation develops in an extratropical cyclone. See Figure 7.11d to find the location of this section.

in the North Atlantic and eastern Pacific, *typhoons* (*Tai-fung* = Chinese for "great wind") in the western Pacific, *tropical cyclones* in the Indian Ocean, and *willi-willis* in the waters near Australia. To qualify formally as a hurricane or typhoon, the tropical cyclone must have winds of at least 118 kilometers (74 miles) per hour. About 100 tropical cyclones grow to hurricane status each year. A very few of these develop into superstorms, with winds near the core that may exceed 400 kilometers (250 miles) per hour! (To imagine what winds in such a storm might feel like, picture yourself clinging to the wing of a fast twin-engined private airplane in flight!) Tropical cyclones smaller than a hurricane are called *tropical storms* and *tropical depressions*.

From above, tropical cyclones appear as circular spirals, as **Figure 7.14** shows. They may be a thousand kilometers (630 miles) in diameter and 15 kilometers (9.5 miles; 50,000 feet) high. The calm center, or *eye* of the storm—a zone some 13–16 kilometers (8–10 miles) in diameter—is sometimes surrounded by clouds so high and dense that the daytime sky above looks dark. Farther out, churned by furious winds, the rainwall clouds condense huge amounts of water vapor into rain. A tropical cyclone is outlined in cross section in **Figure 7.15**.

Figure 7.14 Hurricane Elena over the Gulf of Mexico, photographed from the space shuttle *Discovery* on 2 September 1985. The eye of the storm is clearly visible. Note the spiral bands of cloud extending outward from the eye.

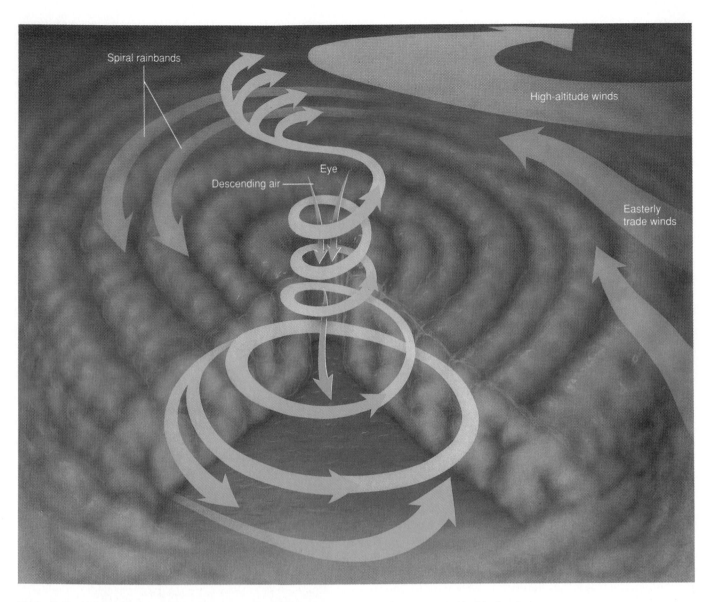

Spiral rainbands

High-altitude winds

Eye

Descending air

Easterly
trade winds

Figure 7.15 The internal structure of a hurricane. The vertical dimension is greatly exaggerated in this drawing.

Unlike extratropical cyclones, these greatest of storms form within *one* warm, humid air mass between 10° and 25° latitude in both hemispheres. (Though air conditions would be favorable, the Coriolis effect closer to the equator is too weak to initiate rotary motion.) The origins of tropical cyclones are not well understood. A tropical cyclone usually develops from a small tropical depression. Tropical depressions form in easterly waves, areas of lower pressure within the easterly trade winds. Easterly waves, sometimes accompanied by a westward-moving line of thunderstorms, are thought to originate over a large, warm land mass. When air containing the disturbance is heated by the proximity of tropical water with a temperature of about 26°C (79°F) or more, circular winds

begin to blow in the vicinity of the wave, and some of the warm humid air is forced upward. Condensation begins, and the storm takes shape. Under ideal conditions, the embryo storm reaches hurricane status (that is, with wind speeds in excess of 118 kilometers or 74 miles per hour) in two to three days.

The centers of most tropical cyclones move westward and poleward at between 5 and 40 kilometers (3–25 miles) per hour. Typical tracks of these storms are shown in **Figure 7.16**. A trio of Pacific tropical cyclones is shown in **Figure 7.17**.

Although its birth process is somewhat mysterious, the source of the storm's power is well understood. Its strength is derived from the same seemingly innocuous

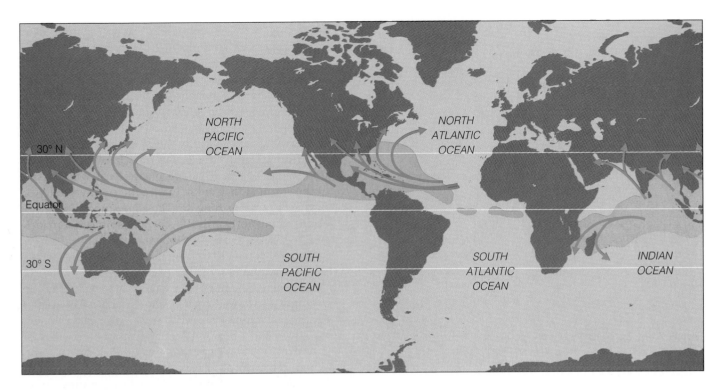

Figure 7.16 The breeding grounds of tropical cyclones, which can develop when sea surface temperature exceeds 26°C (79°F) (shaded areas). The storms follow curving paths—first moving westward with the trade winds, then either dying over land or turning eastward until they lose power over the cooler ocean of mid-latitudes. Cyclones are not spawned over the South Atlantic or the southeast Pacific because their waters are too chilly, nor in the still air—the doldrums—within 5° of the equator.

process that warms a chilled soft-drink can when water from the atmosphere condenses on its surface. As you may recall from Chapter 6, it takes quite a bit of energy to evaporate water into the atmosphere—water's latent heat of evaporation is very high. That heat energy is released when the water vapor recondenses as liquid. It tends to warm your drink very quickly, and the more humid the air, the faster the condensing and warming. In tropical cyclones the condensation energy generates air movement (wind), not more heat. Fortunately, only 2% to 4% of this energy of condensation is converted into motional energy!

A tropical cyclone is an ideal machine for "cashing in" water vapor's latent heat of evaporation. Warm, humid air forms in great quantity only over a warm ocean. As noted above, tropical cyclones originate in ocean areas having surface temperatures in excess of 26°C (79°F) (see Figure 7.16). As hot, humid tropical air rises and expands, it cools and is unable to contain the moisture it held when warm. Rainfall begins. The rainfall rate in some parts of the storm routinely exceeds 2.5 centimeters (1 inch) per hour, and 20 billion tons of water can fall from a large tropical cyclone in a day! Tremendous energy is released

Figure 7.17 Three tropical cyclones lined up between Hawaii and Central America. On the left is Hurricane Kate; in the center is a tropical depression that never reached hurricane force; and on the right is Hurricane Liza (October 1976), which within hours smashed into La Paz, Mexico, and caused millions of dollars in damage.

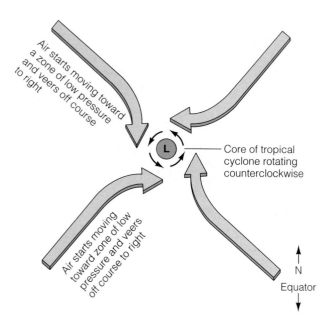

Air starts moving toward a zone of low pressure and veers off course to right

Air starts moving toward zone of low pressure and veers off course to right

Core of tropical cyclone rotating counterclockwise

L

N

Equator

Figure 7.18 The dynamics of a tropical cyclone, showing the influence of the Coriolis effect. Note that the storm turns the "wrong" way (that is, counterclockwise) in the Northern Hemisphere for the "right" reasons.

as this moisture changes from water vapor to liquid. In one day, a large tropical cyclone generates about 2,400 billion kilowatts of power, equivalent to the electrical energy needs of the entire United States for a year! Thus, solar energy ultimately powers the storm in a cycle of heat absorption, evaporation, condensation, and conversion of heat energy to kinetic energy. This energy is available as long as the storm stays over warm water and has a ready source of hot, humid air. When water conditions are right, tropical cyclones sometimes leave the tropics: one struck Montréal, Canada, in 1938.

Three aspects of a tropical cyclone can cause property damage and loss of life: wind, rain, and storm surge. The destructive force of winds to 400 kilometers (250 miles) per hour is self-evident. Rapid rainfall can cause severe flooding when the storm moves onto land. But the most danger lies in **storm surge**, a mass of water driven by the storm. The low atmospheric pressure at the storm's center produces a dome of seawater that can reach a height of 1 meter (3.3 feet) in the open sea. The water height increases when waves and strong hurricane winds ramp the water mass ashore. If a high tide coincides with the arrival of all this water at a coast, or if the coastline converges (as is the case in Bangladesh at the mouths of the Ganges), rapid and catastrophic flooding will occur. Storm surges of up to 12 meters (40 feet) were reported in Bangladesh in 1970. Much of the $3 billion in damage done by Hurricane Hugo to Charleston, South Carolina, in September 1989 was caused by a 5-meter (17-foot)

storm surge arriving at high tide. You'll learn more about storm surges in Chapter 9's discussion of large waves.

You may have noticed that tropical cyclones turn counterclockwise in the Northern Hemisphere and clockwise in the Southern Hemisphere. Does this mean that the Coriolis effect does not apply to tropical cyclones? No. This apparent anomaly is caused by the Coriolis deflection of winds approaching the center of a low-pressure area from great distances. **Figure 7.18** provides a view from above a Northern Hemisphere storm, with the core of the storm at the center. Notice the rightward deflection of the approaching air. The edge spin given by this approaching air causes the storm to spin counterclockwise in the Northern Hemisphere.

Tropical cyclones last from three hours to three weeks; most have lives of five to ten days. They eventually run down when they move over land or over water too cool to supply the humid air that sustains them. The friction of a land encounter rapidly drains a tropical cyclone of its power, and a position above ocean water cooler than 24°C (76°F) is a sure harbinger of the storm's demise. When deprived of energy, the storm "unwinds" and becomes a mass of unstable humid air pouring rain, lightning, and even tornadoes from its clouds. Tropical cyclones can be dangerous to the end: Torrential rain streaming from the remnants of Hurricane Agnes in 1976 caused more than $2 billion in damage, mostly to Pennsylvania; Chesapeake and Delaware Bays were flooded with fresh water and sediments, destroying much of the shellfish industry there.

Tropical cyclones are nature's escape valves, flinging solar energy poleward from the tropics. They are beautiful, dangerous examples of the power represented by water's latent heat of fusion.

CHAPTER SUMMARY

The water, gases, and energy at the Earth's surface are shared between the atmosphere and the ocean. The two bodies are in continuous contact, and conditions in one are certain to influence conditions in the other. The interaction of ocean and atmosphere moderates surface temperatures, shapes the Earth's weather and climate, and creates most of the sea's waves and currents.

The atmosphere responds to uneven solar heating by flowing in three great circulating cells over each hemisphere. This circulation of air is responsible for about two-thirds of the heat transfer from tropical to polar regions. The flow of air within these cells is influenced by the rotation of the Earth. To observers on the surface, the Earth's rotation causes moving air (or any moving mass) in the Northern Hemisphere to curve to the right of its initial path, and in the Southern Hemisphere to the left.

The apparent curvature of path is known as the Coriolis effect.

Uneven flow of air within cells is one cause of the atmospheric changes we call *weather*. Large storms are spinning areas of unstable air occurring between or within air masses. Extratropical cyclones originate at the boundary between air masses; tropical cyclones, the most powerful of Earth's atmospheric storms, occur within a single humid air mass. The immense power of tropical cyclones is derived from water's latent heat of evaporation.

Terms and Concepts to Remember

air mass	ENSO	polar cells
atmospheric	extratropical	polar front
circulation cell	cyclone	precipitation
Bjerknes, Vilhelm	Ferrel cells	Southern
clockwise	front	Oscillation
convection	frontal storm	storm
Coriolis, Gaspard	Hadley cells	storm surge
Gustave de	horse latitudes	tornado
Coriolis effect	hurricane	trade winds
counterclockwise	intertropical	tropical cyclone
cyclone	convergence	water vapor
doldrums	zone (ITCZ)	westerlies
El Niño	orbital inclination	

Study Questions

1. What factors contribute to the uneven heating of Earth by the sun?

2. How does the atmosphere respond to uneven solar heating? How does the rotation of the Earth affect the resultant circulation?

3. Describe the atmospheric circulation cells in the Northern Hemisphere. At what latitudes does air move vertically? Horizontally? What are the trade winds? The westerlies?

4. Describe an El Niño. Do we know what causes an ENSO event?

5. How do the two kinds of large storms differ? How are they similar? What causes an extratropical cyclone? What happens in one?

6. What triggers a tropical cyclone? From what is its great power derived? What causes the greatest loss of life and property when a tropical cyclone reaches land?

For Further Study

Anthes, R. A. 1982. *Tropical Cyclones—Evolution, Structure and Effects*. Boston, MA: American Meteorological Society. The definitive technical work on the subject.

Fisher, B. 1980. *The Fastnet Disaster and After*. London: Pelham Books. The chilling account of the lethal Fastnet yacht race of 1979. An extratropical cyclone at its nastiest.

Graham, N. E., and W. B. White. 1988. "The El Niño Cycle: A Natural Oscillator of the Pacific Ocean–Atmosphere System." *Science* 240 (no. 4857): 1293–1302. An attempt to explain the periodicity of ENSO events.

Lutgens, F. K., and E. J. Tarbuck. 1989. *The Atmosphere—An Introduction to Meteorology*. 4th ed. Englewood Cliffs, NJ: Prentice-Hall. Excellent general text.

McDonald, J. E. 1952. "The Coriolis Effect." *Scientific American*, May, 72–78. Good discussion of theory and practical application.

Oceanus 27 (no. 2, 1984). Issue dedicated to the known causes and effects of the 1982–83 El Niño event. Available from Woods Hole Oceanographic Institution, Woods Hole, MA.

Whipple, A. B. C. 1982. *Storm*. New York: Time-Life Books. Fine general overview of storms. Excellent photos, diagrams, and charts, coupled with lively writing.

8 OCEAN CIRCULATION

Going with the Flow

In July 1969, six researchers sealed themselves into a specially designed submarine for a 2,640-kilometer (1,650-mile) drift within the Gulf Stream, a great ocean current that flows northward off the east coast of the United States. Their vehicle, named the *Ben Franklin* in honor of the man who first took a scientific interest in the current, was designed by Dr. Jacques Piccard, the builder of the bathyscaphe *Trieste* (see Box 4.1).

The 15-meter (50-foot) *Ben Franklin* was towed into the Gulf Stream off West Palm Beach, Florida, and positioned at a depth of 150 meters (500 feet). The month-long program of observations was designed to track the Gulf Stream and measure physical conditions within it. The scientists aboard were in regular contact with two surface ships that followed the drifting submarine and monitored its exact position. They kept a nearly continuous record of water temperature, salinity, and density.

The research submarine *Ben Franklin* enters New York harbor in August 1969 after drifting for 31 days within the Gulf Stream.

The team made surprising discoveries right from the start. First, the submarine drifted more rapidly than expected. The speed of the surface current often exceeds 8 kilometers (about 5 miles) per hour, but oceanographers believed the current would move more slowly at greater depths. It didn't; the surface ships drifted at about the same rate as the submarine. And the drift was not always smooth. Underwater hills looming from the seabed, many of them previously uncharted, disrupted the flow of the submerged current and caused the submarine to rise and fall alarmingly. Gulf Stream eddies also caused problems. Thirteen days into the mission, a huge eddy spun the submarine 80 kilometers (50 miles) out of the Gulf Stream core, forcing it to surface in order to be towed back into the center of the Stream. These great eddies were found to reach much deeper than previously believed (we now know that they can reach all the way to the ocean bottom).

Another surprise was the general lack of marine life. The crew sighted several species of sharks, a huge jellyfish with tentacles 10 centimeters (4 inches) thick, and an aggressive broadbilled swordfish that jousted with their vessel before swimming away unharmed; but they found less life than they had expected. The biology program did make some strides, however. Tape recordings of underwater sounds of biological origin were studied on board, and some of these recordings were later used in the first detailed analyses of whale vocalizations. Dr. Piccard photographed some unusual bioluminescent organisms, a few of which were new to science. And the crew completed medical and psychological investigations on each other and analyzed the performance of the vehicle's life support systems. These studies were used in later submarine and spacecraft design.

When the *Ben Franklin* finally bobbed to the surface 510 kilometers (310 miles) south of Halifax, Nova Scotia, researchers were confident of their biggest finding: They hadn't known as much about the Gulf Stream as they thought they did.

Ocean water circulates in vast currents. Some, like the Gulf Stream, are obvious and fast, others slow and difficult to detect. Ocean currents are affected by two kinds of forces: the **primary forces** that start water moving and determine its velocity, and the **secondary forces** and factors that influence the direction and nature of its flow. Primary forces are thermal expansion and contraction of water, the stress of wind blowing over the water, and density differences between water layers. Secondary forces and factors are the Coriolis effect, gravity, friction, and the shape of the ocean basins themselves.

About 10% of the water in the world ocean is involved in **surface currents**, horizontally flowing water at the ocean's surface driven by thermal expansion and wind friction. Most surface currents move water within and above the pycnocline, the zone of rapid density change with depth. Water beneath the pycnocline also circulates, but the power for this slower, deeper circulation comes from the action of gravity on adjacent water masses of different densities. Since density is largely a function of temperature and salinity, circulation due to density differences is called thermohaline circulation. We'll investigate thermohaline circulation after a discussion of surface currents.

SURFACE CURRENTS

Solar heating causes water to expand slightly. Because of this, sea level near the equator is about 8 centimeters (3 inches) higher than sea level in temperate ocean areas. Water in the cold regions near the poles cools and contracts by a similar amount. This global difference creates a very slight slope, and warm equatorial water flows "downhill" (poleward) in response to gravity. Poleward-moving surface water tends to lag behind the Earth's eastward rotation, however, and moves toward the ocean's western boundaries. The water's slow travel is also influenced by the Coriolis effect. A sluggish circular flow therefore develops in ocean basins on either side of the equator.

But the difference in temperature between polar and tropical regions is not the major primary force responsible for surface currents. That honor must go to winds. As you read in Chapter 7, surface winds form patterns within latitude bands (see **Figure 8.1**). Most of Earth's surface-wind energy is concentrated in each hemisphere's trade winds and westerlies. Tiny irregularities in the sea surface, called *capillary waves*, transfer some of the energy from the moving air to the water by friction. This tug of wind on the ocean surface begins a more rapid mass flow of water. As a rule of thumb, the friction of wind blowing for at least ten hours will cause

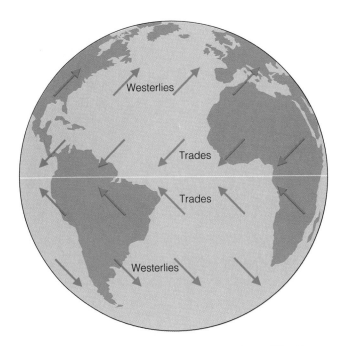

Figure 8.1 Winds driven by uneven solar heating and Earth's spin power the movement of the ocean's surface currents.

surface water to flow downwind at about 2% of wind speed. The water flowing beneath the wind forms a surface current.

Because of the Coriolis effect, Northern Hemisphere surface currents flow to the right of the wind direction. Southern Hemisphere currents flow to the left. Intervening continents and basin topography often block continuous flow and help to deflect the moving water into a circular pattern. This flow around the periphery of an ocean basin is called a **gyre** (*gyros* = circle). Two gyres are shown in **Figure 8.2**.

Flow Within a Gyre

Figure 8.3 shows the North Atlantic gyre in more detail. Though it flows continuously, without obvious places where one current ceases and another begins, oceanographers subdivide the North Atlantic gyre into four interconnected currents because each has distinct flow characteristics and temperatures. (Gyres in other ocean basins are similarly divided.) Notice that the east-west currents in the North Atlantic gyre trend to the right of the driving winds; once initiated, water flow in these currents continues in a roughly east-west direction. Where their flow is blocked by continents, the currents turn clockwise to complete the circuit.

Why does water flow around the *periphery* of the ocean basin instead of spiraling to the center? After all, the

Figure 8.2 A combination of four forces—the sun's heat, winds, the Coriolis effect, and gravity—circulates the ocean surface clockwise in the Northern Hemisphere and counterclockwise in the Southern Hemisphere. Gyres are thus formed, and two are shown here.

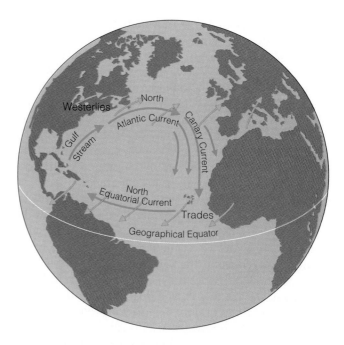

Figure 8.3 The North Atlantic gyre, a series of four interconnecting currents, each with different flow characteristics and temperatures.

Figure 8.4 Surface water blown by the winds at point A will veer to the right of its initial path and continue eastward. Water at point B veers right and continues westward.

Coriolis effect influences any moving mass *as long as it moves*; so water in a gyre might be expected to curve to the center of the North Atlantic and stop. To understand this aspect of current movement, imagine the forces acting on a surface water particle at 45° north latitude (point A in **Figure 8.4**). Here the westerlies blow from the southwest; so initially the particle will move toward the northeast. Rightward Coriolis deflection then causes the water to flow almost due east. A particle at 15° north latitude (point B) responds to the push of the trade winds from the northeast, however, and with Coriolis deflection will flow almost due west.

Research published early in this century indicates that the topmost layer of ocean water in the Northern Hemisphere, when driven by the wind, flows at about 45° to the right of the wind direction, a flow consistent with the arrows leading away from points A and B in Figure 8.4. But what about the water in the next layer down? It can't "feel" the wind at the surface; it "feels" only the movement of the water immediately above. This deeper layer of water moves *at an angle to the right* of the overlying water. The same thing happens in the layer below that, and the next layer, and so on, to a depth of about 100 meters (330 feet) at mid-latitudes. Each layer slides horizontally over the one beneath it like cards in a deck, with

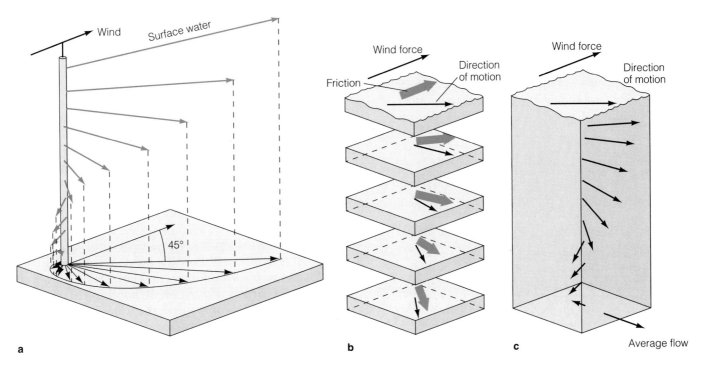

Figure 8.5 The Ekman spiral and the mechanism by which it operates. (a) The Ekman spiral model. (b) A body of water can be thought of as a set of layers. The top layer is driven forward by the wind, and each one below is moved by friction. Each succeeding layer moves with a slower speed and at an angle to the layer immediately above it—to the right in the Northern Hemisphere, to the left in the Southern Hemisphere—until friction becomes negligible. (c) Though the direction of movement is different for each layer in the stack, the theoretical *average* flow of water in the Northern Hemisphere is 90° to the right of the prevailing surface wind.

each lower card moving at an angle slightly to the right of the one above. Because of frictional losses, each lower layer also moves more slowly than the layer above. The resulting situation, portrayed in **Figure 8.5**, is known as an **Ekman spiral** after its Swedish discoverer. The term is somewhat misleading; the water does not spiral downward in a whirlpool-like motion. Rather, the spiral is a way of conceptualizing the horizontal movements in a layered water column, each layer moving in a slightly different horizontal direction. A curious result of Ekman spiral is that water at some depth will be flowing in the opposite direction from the surface current!

The *net* motion of the upper ocean—to a depth of about 100 meters—after allowance for the summed effects of Ekman spiral is known as **Ekman transport**. In theory, the direction of Ekman transport is 90° to the right of wind direction (in the Northern Hemisphere).

Armed with this information, we can look in more detail at the area around point B, which is enlarged in **Figure 8.6**. In nature, Ekman transport in gyres is less than 90°; in most cases the deflection barely reaches 45°. This deviation from theory occurs because of an interaction between the Coriolis effect and a secondary force,

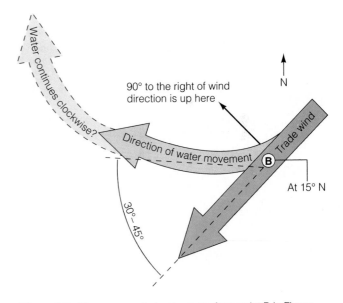

Figure 8.6 The movement of water away from point B in Figure 8.4 is influenced by the Coriolis effect and gravity.

gravity. Some flowing Atlantic water has already turned to the right to form a hill of water; it followed the rightward dotted-line arrow in Figure 8.6. Why does the rest of the water now go straight west from point B without deflecting? Because, as **Figure 8.7** shows, to turn further right the water would now have to move uphill in defi-

ance of gravity, but to turn left in response to gravity would defy the Coriolis effect. So the water continues westward, dynamically balanced between the force of gravity and Coriolis deflection.

Yes, there really *is* a hill near the middle of the North Atlantic, centered in the area of the Sargasso Sea. This hill is formed of surface water gathered at the ocean's center of circulation. It is not a steep mountain of water—its maximum height is an unspectacular 1.5 meters (5 feet)—but a gradual rise and fall from coastline to open ocean and back to opposite coastline. Its slope is so gradual you wouldn't notice it on a transatlantic crossing.

The hill is maintained by wind energy. If the winds did not continuously inject new energy into currents, friction within the fluid mass and with the surrounding ocean basins would slow the flowing water, gradually converting its motion into heat. The balance of wind energy and friction, and of Coriolis effect and gravity, propels the currents of the gyre and holds them along the outside edges of the ocean basin.

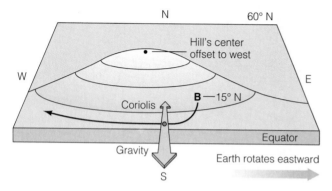

Figure 8.7 The surface of the North Atlantic Ocean is raised to form a low hill. The eastward rotation of the Earth offsets the center of the hill to the west. Water from point B (see also Figures 8.4 and 8.6) turns westward and flows along the side of this hill. The westward-moving water is balanced between the Coriolis effect (which would turn the water to the right) and gravity (which would turn it to the left). Thus water in a gyre moves along the outside edge of an ocean basin.

The Locations of Geostrophic Gyres

Gyres in balance between gravity and the Coriolis effect are called **geostrophic** gyres (*Geos* = Earth + *strophe* = turning), and their currents are geostrophic currents. Because of the patterns of driving winds and the present

Figure 8.8 Major surface currents of the world ocean.

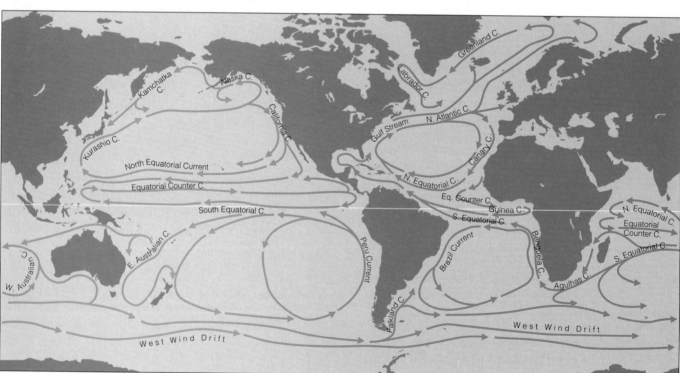

positions of continents, the geostrophic gyres are largely independent of each other in each hemisphere.

There are six great current circuits in the world ocean, two in the Northern Hemisphere and four in the Southern Hemisphere. They are shown in **Figure 8.8**. Five are geostrophic gyres: the North Atlantic gyre, the South Atlantic gyre, the North Pacific gyre, the South Pacific gyre, and the Indian Ocean gyre. Though it is a closed circuit, the sixth and largest current is technically not a gyre because it does not flow around the periphery of an ocean basin. The **West Wind Drift** (or Antarctic Circumpolar Current), as this exception is called, flows endlessly eastward around Antarctica, driven by powerful and nearly ceaseless westerly winds. This greatest of all surface ocean currents is never deflected by a continental mass.

We might expect the two gyres in the North and South Pacific (and the two gyres in the North and South Atlantic) to converge exactly at the geographic equator. However, as Figure 8.8 shows, the junction of equatorial currents lies a few degrees *north* of the geographical equator, at the meteorological equator (the imaginary line that divides the Earth into two equally warm hemispheres). The meteorological equator and the Intertropical Convergence Zone (the band at which the trade winds converge) are displaced 5°–8° northward, primarily because of the heat accumulated in the Northern Hemisphere's greater tropical land-surface area. Ocean circulation, like atmospheric circulation, is balanced around the meteorological equator.

Because of the different factors that drive and shape them, the currents comprising geostrophic gyres have different characteristics. Geostrophic currents may be classified by their position within the gyre as western boundary currents, eastern boundary currents, or transverse currents.

Currents Within Gyres

Western Boundary Currents   The fastest and deepest geostrophic currents are found at the western boundaries of ocean basins (that is, off the *east* coast of continents). These narrow, fast, deep currents move warm water poleward in each hemisphere. There are five large **western boundary currents**: The Gulf Stream (in the North Atlantic), the Japan or Kuroshio Current (in the North Pacific), the Brazil Current (in the South Atlantic), the Agulhas Current (in the Indian Ocean), and the East Australian Current (in the South Pacific).

The **Gulf Stream** is the largest of the western boundary currents. Studies of the Gulf Stream have revealed that off Miami, the Gulf Stream moves at an average speed of 2 meters per second (5 miles per hour) to a depth of over 450 meters (1,500 feet). Water in the Gulf Stream can move more than 160 kilometers (100

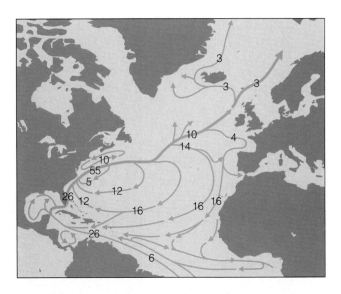

Figure 8.9   The general surface circulation of the North Atlantic. The numbers indicate flow rates in sverdrups (1 sv = 1 million cubic meters of water per second).

miles) in a day. Its average width is about 70 kilometers (43 miles).

The volume of water transported in western boundary currents is extraordinary. The unit used to express volume transport in ocean currents is the **sverdrup (sv)**, named in honor of Harald Sverdrup, one of this century's pioneering oceanographers. A sverdrup equals 1 million cubic meters per second.[1] Gulf Stream flow is at least 55 sv (55 million meters per second), about 300 times the usual flow of the Amazon, the greatest of rivers. In **Figure 8.9** the surface currents of the North Atlantic gyre are shown with their volume transport (in sverdrups) indicated.

Water in a current, especially a western boundary current, can move for surprisingly long distances within well-defined boundaries. In the Gulf Stream, the current-as-river analogy can be startlingly apt—the western edge of the current is often clearly visible. Water within the current is usually warm, clear, and blue, often depleted of nutrients and incapable of supporting much life. By contrast, water over the continental slope adjacent to the current is often cold, green, and teeming with life.

Long straight edges are the exception rather than the rule in western boundary currents, however. Unlike rivers, ocean currents lack well-defined banks, and friction with adjacent water can cause a current to form waves along its edges. Western boundary currents meander as they flow poleward. The looping meanders sometimes connect to form turbulent rings, or **eddies**, that trap cold or

[1]One million cubic meters is about one-half the volume of the Louisiana Superdome.

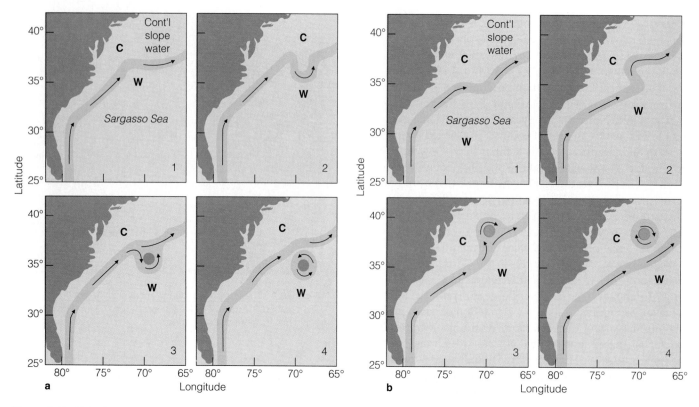

Figure 8.10 Eddy formation. (a) A meander in the Gulf Stream forms a counterclockwise cold-core eddy. (b) The eddy formed here rotates clockwise and has a warm core. (C = cold, W = warm.)

warm water in their centers and then separate from the main flow. For example, cold-core eddies form in the Gulf Stream as it meanders southward upon leaving the coast of North America off Cape Hatteras (see **Figure 8.10a**). Warm-core eddies can form north of the Gulf Stream when the warm current loops into the cold water lying to the north (see **Figure 8.10b**). Because of the mechanics of their formation, warm-core eddies rotate clockwise, and cold-core eddies rotate counterclockwise.

The slowly rotating eddies move away from the current and are distributed across the North Atlantic. Some may be 1,000 kilometers (625 miles) in diameter and retain their identity for more than three years. In mid-latitudes as much as one-fourth of the surface of the North Atlantic may consist of old cold-core ring remnants! Both cold and warm eddies are visible in the satellite image of **Figure 8.11**. Recent research suggests their influence reaches to the seafloor. Warm- and cold-core eddies may be responsible for the slowly moving *abyssal storms* that leave often-observed ripple marks in deep sediments.

Eastern Boundary Currents   There are five **eastern boundary currents** found at the eastern edge of ocean

basins (that is, off the *west* coast of continents): the Canary Current (in the North Atlantic), the Benguela Current (in the South Atlantic), the California Current (in the North Pacific), the West Australian Current (in the Indian Ocean), and the Peru or Humboldt Current (in the South Pacific).

Eastern boundary currents are the opposite of their western boundary counterparts in nearly every way: They carry cold water equatorward; they are shallow and broad (sometimes more than 1,000 kilometers or 625 miles across); their boundaries are not well defined; and they tend not to form eddies. Their total flow is less than that of their western counterparts. The Canary Current in the North Atlantic carries only 16 sv of water at about 2 kilometers per hour (1.2 mph). The current is so shallow and broad that sailors may not notice it. Contrast the flow rates of the North Atlantic's western and eastern boundary currents in Figure 8.9. **Table 8.1** summarizes the major differences between boundary currents in the Northern Hemisphere.

Westward Intensification   Why should western boundary currents be concentrated and eastern boundary currents

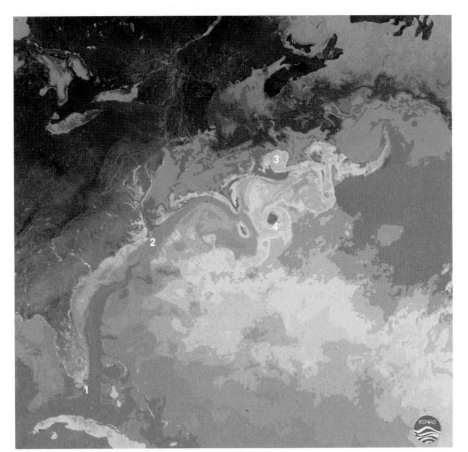

Figure 8.11 The Gulf Stream viewed from space. This image is a composite of temperature data returned from NOAA polar orbiting meteorological satellites during the first week of April 1984. The composite image is printed with an artificial color scale: Reds and oranges are warm, 24° to 28°C (76° to 84°F); yellows and greens 17° to 23°C (63° to 74°F); blues 10° to 16°C (50° to 61°F); and purples are cold, 2° to 9°C (36° to 48°F). The Gulf Stream appears like a red (warm) river as it moves from the southern tip of Florida (1) north along the East Coast. Moving offshore at Cape Hatteras (2), it begins to meander, with some meanders pinching off to form warm-core (3) and cold-core (4) eddies. As it moves northeastward, the water cools dramatically, releasing heat to the atmosphere and mixing with the cooler surrounding waters. By the time it reaches the middle of the North Atlantic, it has cooled so much that its surface temperature can no longer be distinguished from that of the surrounding waters.

Table 8.1   Boundary Currents in the Northern Hemisphere

Type of Current (example)	General Features	Speed	Transport (millions of cubic meters per second)	Special Features
Western boundary currents Gulf Stream, Japan Current	Narrow, <100 km Deep—substantial transport to depths of 2 km	Swift, hundreds of kilometers per day	Large, usually 50 sv or greater	Sharp boundary with coastal circulation system; little or no coastal upwelling; waters tend to be depleted in nutrients, unproductive; waters derived from trade wind belts
Eastern boundary currents California Current, Canary Current	Broad, ~1,000 km Shallow, <500 m	Slow, tens of kilometers per day	Small, typically 10–15 sv	Diffuse boundaries separating from coastal currents; coastal upwelling common; waters derived from mid-latitudes

be diffuse? One reason is the converging flow of the trade winds on either side of the equator. Water moved by the trades approaches the meteorological equator and is shepherded west, where it piles up at the western edge of the basin before turning swiftly poleward. This concentration of water produces the poleward-moving western boundary currents. In contrast, the westerly winds of each hemisphere do not converge, and water driven by them is not swept along a line of convergence. Coriolis deflection can therefore move some of the eastward-moving water equatorward before the basin's eastern boundary is reached.

A second reason is the rotation of the Earth itself. The hill described in Figure 8.7 is offset to the west because of the Earth's eastward rotation, so water must squeeze closer to the ocean basin's western edge to pass around the hill at the western boundary. The combined effect on current flow is known as **westward intensification**, a phenomenon clearly visible in Figure 8.9.

Transverse Currents As we have seen, most of the power for ocean currents is derived from the trade winds at the fringes of the tropics and from the mid-latitude westerlies. The stress of winds on the ocean in these bands gives rise to the **transverse currents**—currents that flow from east to west, and west to east—which link the eastern and western boundary currents.

The trade-wind-driven North and South Equatorial Currents in the Atlantic and Pacific are moderately shallow and broad, but each transports about 30 sv westward. Because of the thrust of the trades, Atlantic water at Panama usually is 20 centimeters (8 inches) higher, on average, than water across the isthmus in the Pacific. The greater expanse of water at the equator in the Pacific and the stronger trade winds there develop more powerful westward-flowing equatorial currents, and the height differential between the western and eastern Pacific is thought to approach 1 meter (3.3 feet)!

Westerly winds drive the eastward-flowing transverse currents of the mid-latitudes. Because they are not shepherded by the trade winds, eastward-flowing currents are wider and flow more slowly than their equatorial counterparts. The North Pacific and North Atlantic Currents are Northern Hemisphere examples.

As can be seen in Figure 8.8, the westward flow of the transverse currents near the equator proceeds unimpeded for great distances, but the eastward flow of transverse currents at mid- and high-latitudes in the northern ocean basins is interrupted by continents and island arcs. In the far south, however, eastward flow is almost completely free. Intense westerly winds over the southern ocean drive the greatest of all ocean currents, the unobstructed West Wind Drift (or Antarctic Circumpolar Cur-

rent). This current carries more water than any other—at least 100 sv west-to-east in the Drake Passage between the tip of South America and the adjacent Palmer Peninsula of Antarctica!

Countercurrents and Undercurrents

Equatorial currents are typically accompanied by **countercurrents** flowing on the surface in the opposite direction from the main current. As you may remember, at the meteorological equator air is rising and the trade winds do not blow across the ocean. Without persistent winds to drive water to the west, some backward flow of water occurs at the meteorological equator. Some water flows away from the main equatorial currents (which flow westward a bit north and south of the meteorological equator) to return on the surface to the east. Look for this in the Pacific in Figure 8.8.

Countercurrents can also exist *beneath* surface currents. These are sometimes referred to as **undercurrents**. The first undercurrent was discovered in 1951 in the central Pacific by Townsend Cromwell, a researcher employed by the U.S. Fish and Wildlife Service to investigate deep long-line fishing techniques pioneered by the Japanese. The Pacific Equatorial Undercurrent, or Cromwell Current, flows beneath the North Equatorial Current with an average velocity of 5 kilometers (3 miles) per hour at a depth of 100–200 meters (330–660 feet). It is about 300 kilometers (190 miles) wide and carries a volume equivalent to about half the Gulf Stream. It has been traced for over 14,000 kilometers (8,700 miles), from New Guinea to Ecuador. Countercurrents have since been found under most major currents. Their volumes are often a significant percentage of the current above.

Oceanographers are uncertain about what causes undercurrents, but near the equator (where the Coriolis effect is negligible) some water is not deflected north or south when it reaches the west side of the basin. It stays on the equator and underflows the surface current back to the east. This eastward-flowing current is kept on course by the Coriolis effect. If the current veers north from the equatorial region, the Coriolis effect bends it to the right, or south. Should the current wander too far south, the Coriolis effect in the Southern Hemisphere deflects the water to the left (north).

El Niño

El Niño, an anomaly of air and ocean circulation introduced in Chapter 7, is an exception to normal surface flow. When the propelling trade winds falter, warm equatorial water that would normally flow westward in the equatorial Pacific "backs up" to flow east (see Figure 7.10). Much of the flow is in the nature of a greatly strengthened equatorial countercurrent, but oceanographers have recently discovered that the Pacific Equatorial Undercurrent also increases greatly in volume during an ENSO event. The normal northward flow of the cold Peru Current is interrupted or overridden by the warm water. This current, rich in upwelled nutrients, is responsible for the great biological productivity of the ocean off the coasts of Peru and Chile. When the Peru Current slows, fish and seabirds dependent on the abundant life it contains die or migrate elsewhere. During major El Niño events, sea level and water temperature rise at the mid-Pacific's eastern boundary. The warmer water causes more evaporation, and an area of low atmospheric pressure develops, centered some 2,000 kilometers (1,300 miles) west of Peru. Humid air rising in this zone causes high precipitation in normally dry areas. Marine and terrestrial habitats and organisms can be significantly impacted by these changes.

Effects of Surface Currents on Climate

Surface currents distribute tropical heat worldwide. Warm water flows to higher latitudes, transfers heat to the air, cools, moves back to low latitudes, absorbs heat again, and the cycle repeats. The greatest amount of heat transfer occurs at mid-latitudes, where about 10^{15} (10 million billion) calories of heat are transferred each second—more than a million times the power consumed by all the world's human population! The combination of water flow and heat transfer (from and to water) influences climate and weather in several ways.

Commercial airliners flying to Europe from U.S. West Coast cities fly over central Ontario. In winter and spring months, the ground is hidden beneath masses of ice and snow, but passengers can often make out the frozen surface of James Bay. When those airliners land in London or Edinburgh, their passengers find a much milder climate than the barren whiteness of central Canada. Yet London's latitude of 51°N is the same as the southern tip of James Bay. Edinburgh lies at 55°N, the same latitude as Ontario's Polar Bear Provincial Park, a sanctuary for migrating polar bears. Why the difference in climate? The predominant direction of air flow is over the water eastward toward the British Isles. Even in winter Edinburgh and London are bathed in air only recently in contact with the relatively warm North Atlantic Current. Scotland and England therefore have a maritime climate, and their only polar bears are found in zoos. These cities are warmed in part by the energy of tropical sunlight transported to their high latitudes by the Gulf Stream (see again Figure 8.8).

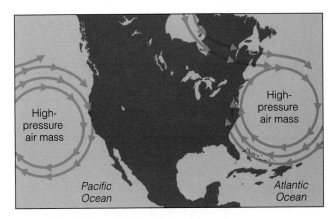

Figure 8.12 General summer air circulation patterns of the East and West coasts of the United States.

At lower latitudes on an ocean's eastern boundary the situation is often reversed. Mark Twain is supposed to have said that the coldest winter he ever spent was a summer in San Francisco. Summer months in that West Coast city are cool, foggy, and mild, while Washington, D.C., on nearly the same line of latitude (but on the western boundary of an ocean basin), is known for its August heat and humidity. Why the difference? Look at Figure 8.8 and follow the currents responsible. The California Current, carrying cold water from the north, comes close to the coast at San Francisco. As shown in **Figure 8.12**, air normally flows clockwise in summer around an offshore zone of high atmospheric pressure. Wind approaching the California coast loses heat to the cold sea and comes ashore to chill San Francisco. Summer air often flows around a similar high off the East Coast (the Bermuda High). Winds approaching Washington, D.C., therefore blow from the south and east. Heat and moisture from the Gulf Stream contribute to the capital's oppressive summers. In winter Washington, D.C., is colder than San Francisco because westerly winds approaching Washington are chilled by the cold continent they cross.

WIND-INDUCED VERTICAL CIRCULATION

Wind blowing offshore or parallel to shore can cause what is known as coastal **upwelling**. The friction of wind blowing along the ocean surface causes the water to begin moving, the Coriolis effect deflects it to the right (in the Northern Hemisphere), and the resultant Ekman transport moves it offshore. As shown in **Figure 8.13a** coastal upwelling occurs when this surface water is replaced by deep water rising along the shore.

BOX 8.1   *Studying Currents*

Traditional methods of measuring currents divide into two broad categories: the float method and the flow method. The **float method** depends on the movement of a drift bottle or other free-floating object. In the **flow method** the current is measured as it flows past a fixed object.

Surface currents can be traced with drift bottles or drift cards (**Figure a**). These tools are especially useful in determining coastal circulation but provide no information on the path the drift bottle or card may have taken between its release and collection points. More elaborate drift devices, such as the drogue arrangement in **Figure b**, can be tracked continuously by radio direction-finders or radar. Surface currents can also be tracked by noting the difference between the daily expected and observed positions of ships at sea.

Deeper currents can also be surveyed by free-floating devices. The Swallow float (named after its developer) is used to detect the drift of intermediate water masses. Adjusted to descend to a specific density (and therefore

depth), the Swallow float emits sonar "pings" as it drifts so that a tracking vessel can follow the movement of the water mass in which it is embedded. More sophisticated devices developed in the 1970s and 1980s depend on sofar channels (see Chapter 6) to transmit sound. Low-frequency tones are broadcast from autonomous submerged probes to moored listening stations (see **Figure c**). These sofar probes have accumulated more than 240 float-years of data in the North Atlantic at depths from 700 to 2,000 meters (2,300 feet to 6,600 feet). One of these drifting probes has been sending data for nine years!

Bottom currents can also be investigated. The Woodhead seabed drifter (**Figure d**), a sort of drift card for bottom currents, looks like a small weighted parasol. Released in groups, a few of these objects may eventually wash on shore to be returned by interested beach-combers.

Current meters, or flow meters, measure the speed and direction of a current from a fixed position. Most

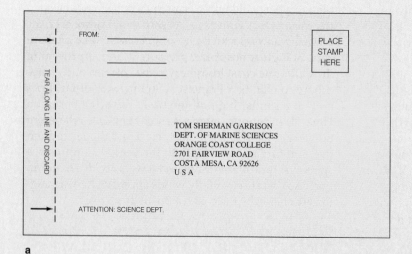

a

(a) A drift card. The card is weighted at one end with a metal rod and made buoyant by a narrow foam block at the other. The whole assembly is enclosed in a waterproof clear plastic envelope. Methods for measuring currents: (b) A drogue, which drifts with the current and is continuously tracked. (c) A sofar float being launched from the Woods Hole Oceanographic Institution's research ship *Oceanus*. The probe will drop to a depth of 3,500 meters (11,500 feet) and produce a low-frequency tone once each day for tracking. (d) A Woodhead seabed drifter, for charting bottom currents. (e) An Ekman flow meter, which measures current speed and direction at a fixed position.

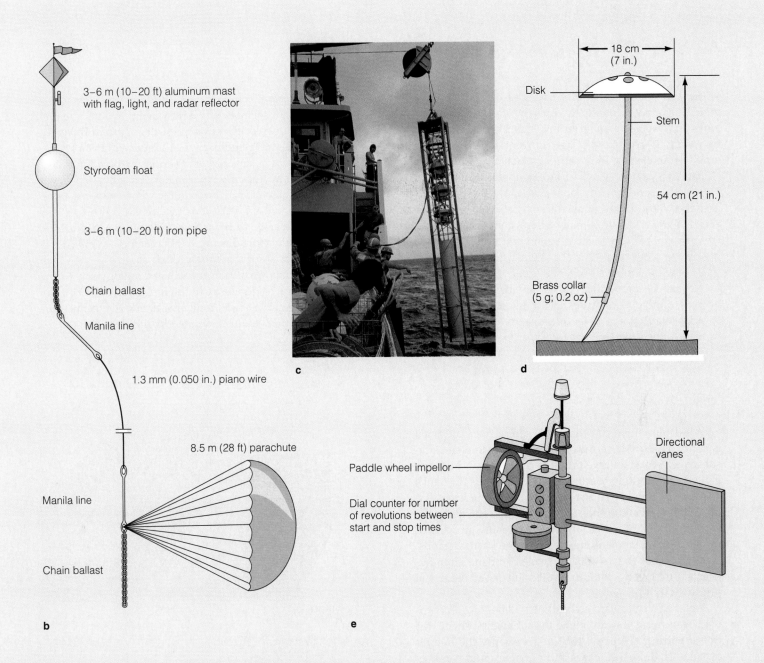

b

3–6 m (10–20 ft) aluminum mast
with flag, light, and radar reflector

Styrofoam float

3–6 m (10–20 ft) iron pipe

Chain ballast

Manila line

1.3 mm (0.050 in.) piano wire

8.5 m (28 ft) parachute

Manila line

Chain ballast

c

d

18 cm
(7 in.)

Disk

Stem

54 cm (21 in.)

Brass collar
(5 g; 0.2 oz)

e

Paddle wheel impellor

Dial counter for number
of revolutions between
start and stop times

Directional
vanes

flow meters, such as the Ekman type shown in **Figure e**, use rotating vanes to measure current speed and a recording compass to measure direction. Bottom water movements are usually too slow to be measured by flow meters.

Advances in electronics and computer design have made possible several new methods of study. One new device pioneered by the U.S. Office of Naval Research measures current speed by sensing the electromagnetic force generated by seawater as it moves in Earth's mag-

netic field. Buoys equipped with these sensors can record current speed and direction without dependence on delicate moving parts.

Yet another method, developed for study of thermohaline circulation, senses the presence in seawater of tritium, a radioactive isotope of hydrogen. A small amount of tritium is produced in the upper atmosphere when hydrogen is bombarded by cosmic rays, but most of it was produced by hydrogen bomb testing in the 1960s. Most tritium combines with oxygen to become radioactive

BOX 8.1 ● *Studying Currents (continued)*

water, and some of this water enters the ocean by precipitation or river runoff. At high latitudes this water sinks and, labeled by its tritium content, can be traced by sampling. The speed of deep currents has been measured by analysis of their tritium content.

Satellites can sense sea-surface temperature (see Figure 8.11, for example) and topography, ocean color, and even chlorophyll content (an indication of biological productivity and therefore of nutrient content), giving fresh insight into ocean circulation. So far, satellites can observe only the ocean surface, but satellite-borne lasers may someday probe the ocean to greater depths.

Acoustical tomography (*tomos* = to cut or slice + *graph* = to write) uses pulses of low-frequency sound to sense variations in water temperature, salinity, and movement beneath the surface. Experiments begun in

1981 have used multiple transmitters and receivers to investigate the formation and movement of Gulf Stream eddies and deep-ocean circulation at a depth of 760 meters (2,500 feet). Preliminary results suggest that a great deal of the ocean's energy of motion is involved in relatively small fluctuating flows analogous to atmospheric weather systems. Unlike weather systems (which generally extend across 1,000 kilometers, or 630 miles, and last less than a week), however, this midscale oceanic turbulence typically extends only 100 kilometers (63 miles) but may persist for more than 100 days.

Researchers are also using acoustic tomography to conduct a long-term study of the formation of North Atlantic Deep Water south of Greenland. The technique may revolutionize our understanding of deep sea circulation.

Wind blowing from the north along the California coast causes offshore movement of surface water and subsequent coastal upwelling. The overlying air becomes chilled, contributing to San Francisco's famous fog banks and cool summers (see again Figure 8.12). Wind-induced upwelling is also common in the Peru Current, along the west coast of Antarctica's Palmer Peninsula, in parts of the Mediterranean, and near some large Pacific islands. Again, because the new surface water is often rich in nutrients, prolonged wind can result in increased biological productivity.

Water driven toward a coastline will be forced downward, returning seaward along the continental shelf. This **downwelling (Figure 8.13b)** helps supply the deeper ocean with dissolved gases and nutrients, and it assists in the distribution of living organisms. Unlike upwelling, downwelling has no direct effect on the climate or productivity of the adjacent coast.

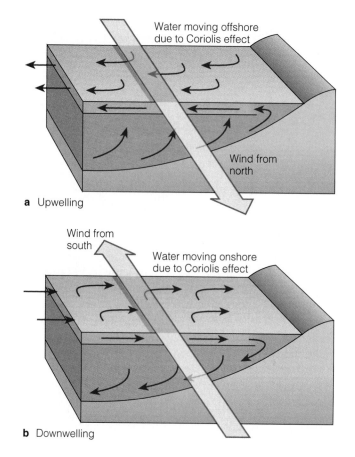

a Upwelling

b Downwelling

Figure 8.13 Coastal upwelling and downwelling. (a) In the Northern Hemisphere, coastal upwelling can be caused by winds from the north blowing along the west coast of a continent. Water moving offshore is replaced by cold, deep, nutrient-laden water. (b) A prolonged southerly wind along a Northern Hemisphere west coast can result in downwelling.

THERMOHALINE CIRCULATION

The surface currents we have discussed affect the uppermost layer of the world ocean (about 10% of its volume), but horizontal and vertical currents also exist below the pycnocline in the ocean's deeper waters. The slow circulation of water at great depths is driven by *density* differences rather than by wind energy. Because density is largely a function of water temperature and salinity, the movement of water due to differences in density is called **thermohaline circulation** (*therme* = heat + *halos* = salt). Virtually the entire ocean is involved in slow thermohaline circulation, a process responsible for most of the vertical movement of ocean water.

Water Masses

As you may recall from Chapter 6, the ocean is density-stratified, with the most dense water near the seabed and the least dense near the surface. Each water mass has specific temperature and salinity characteristics. Density stratification is most pronounced at temperate and tropical latitudes because the temperature difference between surface water and deep water is greater there than near the poles.

The water masses possess distinct, identifiable properties. Like air masses, water masses don't mix easily when they meet; instead, they flow above or beneath each other. Water masses can be remarkably persistent and will retain their identity for great distances and long periods of time. Oceanographers name water masses according to their relative position.

In temperate and tropical latitudes the common water masses are:

- *Surface water*, to a depth of about 200 meters (650 feet)
- *Central water*, to the bottom of the main thermocline (which varies with latitude)
- *Intermediate water*, to about 1,500 meters (5,000 feet)
- *Deep water*, water below intermediate water but not in contact with the bottom, to a depth of about 4,000 meters (13,000 feet)
- *Bottom water*, water in contact with the seabed

Surface currents move in the relatively warm upper environment of surface and central water. The boundary between central water and intermediate water is the most abrupt and pronounced.

No matter at what depth they are located, the characteristics of each water mass have been determined by conditions of heating, cooling, evaporation, and dilution that occurred at the ocean *surface* when the mass was formed. The heaviest (and deepest) masses were formed by surface conditions that caused water to become very cold and salty. Water masses near the surface are warmer and less saline, and they may have formed in warm areas where precipitation exceeded evaporation. Water masses at intermediate depths are intermediate in density.

The Temperature–Salinity Diagram

Perhaps the best way to visualize ocean layering is with a **temperature–salinity** (or **T-S**) **diagram** like the one in **Figure 8.14**. The S-shaped curve through the center of the figure shows the temperature and salinity of water at each depth indicated. Note that many combinations of temperature and salinity can yield the same density, and that the density of the water tends to increase with depth. The shape of the curve is governed by the position and nature of water masses. **Figure 8.15** is a typical T-S diagram for the South Pacific Ocean. It shows that salinity increases abruptly at the transition from Antarctic Intermediate Water to Deep Circumpolar Water, and that temperature drops rapidly between Deep Circumpolar Water and Antarctic Bottom Water.

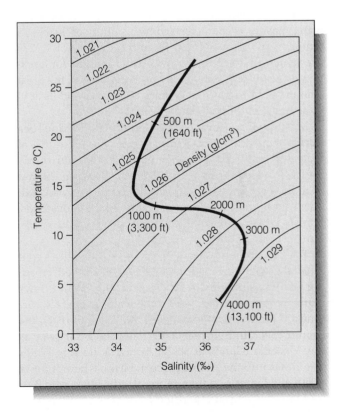

Figure 8.14 A general temperature–salinity (T-S) diagram, with density curves added. The S curve tracks the combination of temperature and salinity (and therefore density) with increasing depth.

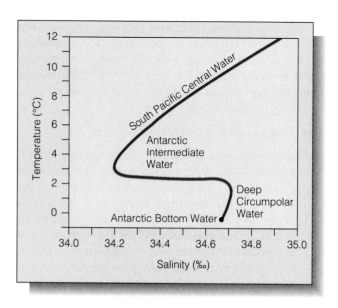

Figure 8.15 A typical T-S diagram for the South Pacific Ocean.

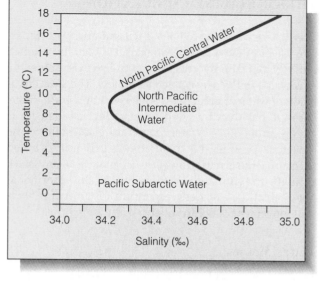

Figure 8.16 A typical T-S diagram for the North Pacific Ocean.

Formation and Downwelling of Deep Water

Antarctic Bottom Water, the most distinctive of all deep-water masses, is characterized by a salinity of 34.65‰, a temperature of –0.5°C (30°F), and a density of 1.0279 grams per cubic centimeter. This water is noted for its extreme density (the densest in the world ocean), the great amount of it produced near Antarctic coasts, and its ability to migrate north along the seafloor.

Most Antarctic Bottom Water forms in the Weddell Sea during winter. Sea ice can incorporate only about 15% of seawater's salt, and the salt remaining in the surrounding cold water forms a frigid brine. Between 20 and 50 million cubic meters of this water forms every second! The water's great density causes it to sink. As it descends, it mixes with nearly equal parts of water from the southern West Wind Drift.

The mixture settles along the edge of Antarctica's continental shelf and spreads along the deep-sea bed, creeping north in slow sheets. Antarctic Bottom Water flows many times more slowly than the water in surface currents; in the Pacific it may take 1,000 years to reach the equator. Six hundred years later it may be as far away as the Aleutian Islands at 50°N! Antarctic Bottom Water also flows into the Atlantic Ocean basin, where it flows north at a faster rate than in the Pacific. Antarctic Bottom Water has been identified as high as 40° *north* latitude on the Atlantic floor, a journey that has taken some 750 years.

Some dense bottom water also forms in the northern polar ocean, but the topography of the Arctic Ocean basin prevents most of it from escaping. Thus, a much smaller volume of cold, salty water reaches the North Pacific Ocean floor. **Figure 8.16** is a typical T-S diagram for the North Pacific. Note that the abrupt transition to a very dense layer of bottom water seen in Figure 8.15 is missing here.

Other distinct deep-water masses exist. Their positions in relation to each other are always determined by their relative densities. North Atlantic Deep Water forms when relatively warm and salty North Atlantic Central Water rises due to cold winds from northern Canada sweeping surface water aside. Exposed to the chilled air, water at the latitude of Iceland releases heat, cools from 10°C to 2°C (50°F to 36°F), and sinks. (Transferred to the air, this bonus heat helps to moderate European winters.) Similar water forms at the West Wind Drift in the South Atlantic. Pacific Deep Water also forms in the Pacific in the West Wind Drift and along the east coast of the Kamchatka Peninsula. Atlantic or Pacific Deep Water is less dense than Antarctic Bottom Water and so floats above it, out of contact with the ocean floor.

A deep-water mass also forms in the enclosed Mediterranean Sea, where surface water is made more saline by the excess of evaporation over freshwater input. About 300,000 cubic kilometers (72,000 cubic miles) more water evaporates annually from the Mediterranean than is replaced by river runoff or precipitation. In the cool winter months, Mediterranean water with a salinity of about 38‰ flows past the lip of Gibraltar and spreads into the Atlantic as Mediterranean Deep Water (**Figure 8.17**). Mediterranean Deep Water underlies much of the central water mass in the Atlantic, and some of this water can be traced as far south as the basins of the Antarctic. Though saltier than Antarctic Bottom Water or Atlantic Deep Water, Mediterranean Deep Water is considerably warmer and therefore not as dense. It will lie atop the layers of more dense water at high southern latitudes.

The age of deep water can be determined by analyzing its dissolved oxygen content. Ocean water picks up free oxygen only at or near the surface, by contact with the atmosphere or through the action of photosynthetic plants. After leaving the surface, the water gradually loses its oxygen through the respiration of organisms or by chemical reactions with sediments, rocks, or dissolved components. The dissolved oxygen content of a water mass is therefore a rough index of its age—the length of time since the water left the surface. Researchers have found that water masses slowly mix with the surrounding water as they flow away from their sources, losing their individual identity as they grow older.

Thermohaline Circulation Patterns

The great quantities of dense water sinking at ocean-basin edges must be offset by equal quantities of water rising elsewhere. **Figure 8.18** shows an idealized model of thermohaline flow. Note that water sinks relatively rapidly in a small area where the ocean is very cold, rises much more gradually across a very large area in the warmer temperate and tropical zones, and slowly returns poleward near the surface to repeat the cycle. The continual diffuse upwelling of deep water maintains the existence of the permanent thermocline found everywhere at low and middle latitudes. This slow upward movement is estimated to be about 1 centimeter (½ inch) per day over most of the ocean. If this rise were to stop, downward movement of heat would cause the thermocline to descend and would reduce its steepness. In a sense, the thermocline is "held up" by the continual slow upward movement of water.

Most features of this ideal circulation pattern exist in nature. **Figure 8.19** shows deep circulation in the Atlantic. The water masses, each of distinct density and sandwiched in layers, are slowly propelled by gravity. Masses butt against one another in **convergence zones**, and the heavier water slides beneath the lighter water. Hundreds of years may pass before water masses complete a circuit or blend to lose their identities. Remember, Antarctic Bottom Water in the Pacific retains its character for up to 1,600 years! The residence time of most

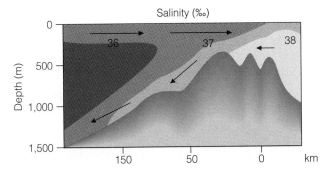

Figure 8.17 A vertical section at the Strait of Gibraltar, showing the movement of water masses of different salinity over the sill that separates the Mediterranean Sea from the Atlantic.

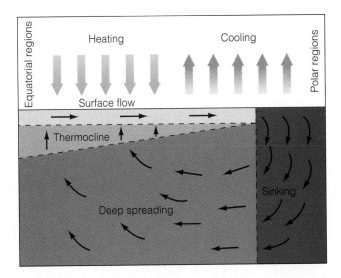

Figure 8.18 The classical model of a pure thermohaline circulation caused by heating in lower latitudes and cooling in higher latitudes.

deep water is less, however; it takes about 200–300 years to rise to the surface. (By contrast, a bit of water in the North Atlantic gyre may take only a year to complete a circuit.)

Not all thermohaline circulation is so sedate. Ripple marks in sediments, scour lines, and the erosion of rocky outcrops on deep-ocean floors provide evidence that relatively strong, localized bottom currents exist. Some currents may move as rapidly as 60 centimeters per second (2 feet per second). These relatively fast currents are strongly influenced by bottom topography, and they are sometimes called **contour currents** because their dense

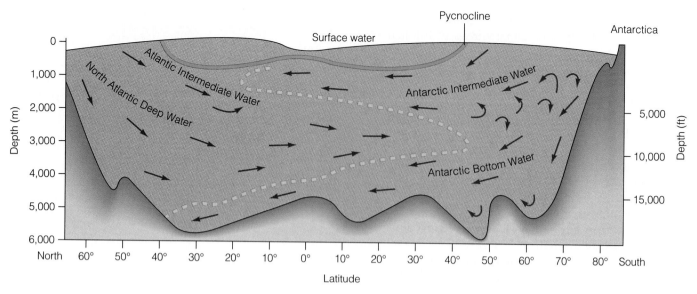

Figure 8.19 Schematic diagram of the circulation of the Atlantic Ocean. Arrows indicate the direction of water movement. Surface contour is greatly exaggerated. The boundaries between water masses are very diffuse.

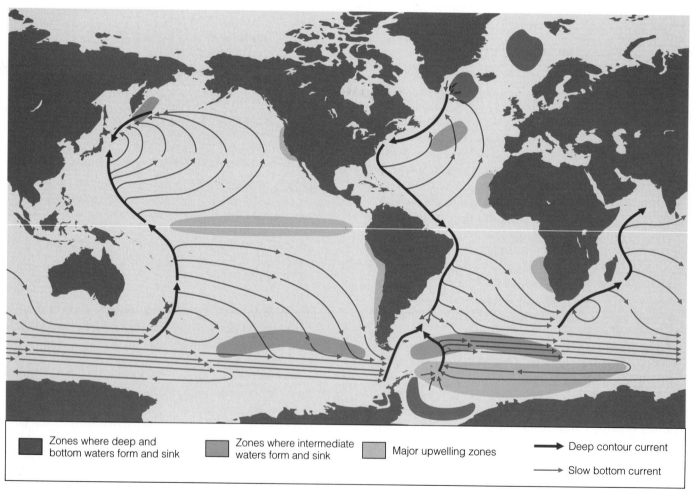

■ Zones where deep and bottom waters form and sink	■ Zones where intermediate waters form and sink	■ Major upwelling zones	→ Deep contour current
			→ Slow bottom current

Figure 8.20 An overview of thermohaline circulation.

water flows around (rather than over) seafloor projections. Bottom currents generally move equatorward at or near the western boundaries of ocean basins (below the western boundary surface currents). The subtle density differences between deep-water masses are not capable of moving water at the speed of the wind-driven surface currents. Water in some of these currents may move only 1–2 meters (3–7 feet) per day. Even at that slow speed, the Coriolis effect modifies their pattern of flow.

Vertical currents may also occur near the deep-ocean floor. Researchers are investigating the movement of plumes of hot water rising from vents on the mid-ocean ridges. Preliminary evidence suggests that this water rises a few hundred meters off the bottom, reaches a position of equilibrium in the water column, and then spreads horizontally over great distances. Its movement is also influenced by the Coriolis effect, and the overall flow pattern seems to resemble that of surface gyres.

An overview of large-scale thermohaline circulation is provided in **Figure 8.20**. The areas at which deep and bottom waters form and sink are shown, as are zones of coastal and equatorial upwelling. Deep contour currents are indicated by thick arrows, and slower bottom circulation by thin arrows. The gradual rising of water through most latitudes, evident in Figure 8.18, is not shown.

CHAPTER SUMMARY

Ocean water circulates in currents. Surface currents affect the uppermost 10% of the world ocean. The movement of surface currents is powered by the warmth of the sun and by winds. Some surface currents are rapid and riverlike, with well-defined boundaries; others are slow and diffuse. Water in surface currents tends to flow horizontally, but it can also flow vertically in response to wind blowing near coasts (upwelling or downwelling). Surface currents transfer heat from tropical to polar regions, influence weather and climate, distribute nutrients, and scatter organisms. They have contributed to the spread of humanity to remote islands, and they are important factors in maritime commerce.

Circulation of the 90% of ocean water beneath the surface zone is driven by the force of gravity, as dense water sinks and less dense water rises. Since density is largely a function of temperature and salinity, the movement of deep water due to density differences is called thermohaline circulation. Currents near the seabed flow as slow, riverlike masses in a few places, but the greatest volumes of deep water creep through the ocean at an almost imperceptible pace.

The Coriolis effect, gravity, and friction shape the direction and volume of surface currents and thermohaline circulation.

Terms and Concepts to Remember

acoustical tomography	float method	thermohaline circulation
Antarctic Bottom Water	flow method	transverse currents
	geostrophic	undercurrent
contour current	Gulf Stream	upwelling
convergence zone	gyre	West Wind Drift
countercurrent	primary forces	western boundary currents
downwelling	secondary forces	westward intensification
eastern boundary currents	surface currents	
	sverdrup (sv)	
eddy	temperature–salinity (T-S) diagram	
Ekman spiral		
Ekman transport		

Study Questions

1. What forces are responsible for the *movement* of ocean water in currents? What forces and factors influence the *direction and nature* of ocean currents?

2. What is a gyre? How many large gyres exist in the world ocean? Where are they located?

3. Why does water tend to flow around the periphery of an ocean basin? Why are western boundary currents the fastest ocean currents? How do they differ from eastern boundary currents?

4. What are countercurrents? Undercurrents? How is El Niño thought to be related to these currents?

5. What is the role of ocean currents in the transport of heat? How can ocean currents affect climate? Contrast the climate of a mid-latitude coastal city at a western ocean boundary with that of a mid-latitude coastal city at an eastern ocean boundary.

6. What are water masses? Where are distinct water masses formed? What determines their relative position in the ocean?

7. What drives the vertical movement of ocean water? What is the general pattern of thermohaline circulation?

8. What methods are used to study ocean currents?

For Further Study

Huyghe, P. 1990. "The Storm Down Below." *Discover*, November, 70–76. "Abyssal storms" may be caused by the influence of warm- and cold-core eddies.

Knauss, J. A. 1961. "The Cromwell Current." *Scientific American*, April, 105–16. A look at deep-water currents under the equatorial currents.

Kunzig, R. 1991. "Can Earth's Internal Heat Drive Ocean Circulation?" *Science* 252 (no. 5013): 1620–21. Heat coming from vents in oceanic ridges may churn abyssal waters.

MacLeish, W. H. 1989. "The Blue God." *Smithsonian*, February, 44–58. The Gulf Stream's story.

MacLeish, W. H. 1989. "Painting a Portrait of the Stream from Miles Above—and Below." *Smithsonian*, March, 42–55. Satellite imagery, wonderful pictures. (Companion piece to article above.)

Price, J., R. Weller, and R. Schudlich. 1987. "Wind-Driven Ocean Currents and Ekman Transport." *Science* 238 (no. 4833): 1534–38. Verification of theoretical Ekman transport. Observed transport is consistent with theoretical Ekman transport to within 10%.

Spinel, R. C., and P. F. Worcester. 1990. "Ocean Acoustic Tomography." *Scientific American*, October, 94–99. Recent information on the predominance of midscale circulation.

Stommel, H. 1987. *A View of the Sea*. Princeton, NJ: Princeton University Press. A famous oceanographer explains ocean currents in an imaginary dialog between himself and an inquisitive chief engineer of an oceanographic research vessel. Currents and the author's view of life are both on display here. A wonderful book.

Weibe, P. 1982. "Rings of the Gulf Stream." *Scientific American*, March, 60–79. Discussion of the physics and biology of Gulf Stream eddies.

Whitehead, J. A. 1989. "Giant Ocean Cataracts." *Scientific American*, February, 50–57. "Waterfalls" of dense water plunging down continental slopes are likened to cataracts. The role of dense water in maintaining the chemistry and climate of the deep ocean is explored.

9 WAVES

"... these blue-water hills ..."

In the spring of 1916 Sir Ernest Shackleton found himself and the members of his failed south-polar expedition marooned on a desolate island at the tip of Antarctica's Palmer Peninsula, far from any hope of rescue. Their expedition ship, aptly named *Endurance*, had been crushed by ice and lay at the bottom of the Weddell Sea. Someone had to go for help.

They had only one chance—to reach a whaling station on South Georgia Island, over 1,300 kilometers (800 miles) away. With the force of the westerlies and the thrust of the West Wind Drift, it would be impossible to return if they were blown past the small island. But Shackleton's largest and strongest boat, the *James Caird*, was only 6.85 meters (22.5 feet) long, and ahead lay the most violent stretch of ocean on Earth.

Although they had little hope of success, a crew of six set out on the morning of 24 April 1916. In his account of their desperate voyage, the captain of *Endurance*, Frank Worsley, gives us a feel for the power of the ocean encircling Antarctica:

> In the afternoon [of the fifth day] the swell settled and lengthened out—the typical deep-sea swell of these latitudes. Offspring of the westerly gales, the great unceasing westerly swell of the Southern Ocean rolls almost unchecked around this end of the world in the Roaring Forties and the Stormy Fifties.[1] The highest, broadest, and longest swells in the world, they race on their encircling course until they reach their birthplace again, and so, reinforcing themselves, sweep forward in fierce and haughty majesty, four hundred, a thousand yards, a mile apart in fine weather, silent and stately they pass along. Rising forty or fifty feet and more from crest to hollow, they rage in apparent disorder during heavy gales. . . .
>
> At times, rolling over their allotted ocean bed, in places four miles deep, they meet a shallow of thirty to a hundred fathoms. Their bases retarded by the bank, their crests sweep up in furious anger at this check, until their front forms an almost perpendicular wall of green rushing water that smashes on a ship's deck, flattening steel bulwarks, snapping two-inch steel stanchions, and crushing deckhouses and boats like eggshells. These blue-water hills in a very heavy gale move as fast as thirty-five miles an hour, but striking the banks, the madly leaping crests falling over and onward, probably attain a momentary speed of fifty or sixty miles or more. The impact of hundreds of tons of *solid* water at this speed can only fairly be imagined.
>
> So we held our way; in those valleys and on those ridges alternately. First, half becalmed—a hill of water ahead, another astern—the following hill lifts us, as the boat slides with increasing speed down the ever steepening slope, till with a sudden upward swoop, the sea boiling white around and over us, we are on the summit with a commanding view of a panorama of dark grey and indigo blue rollers, topped and broken with white horses. The crest passing leaves the boat apparently stationary, gravity now dropping us back till the next hollow reaches us, and so on *ad nauseam*.

After an all but unbearable voyage of 16 days they made a perfect landfall, reported the position of the rest of the expedition to the whalers, and assisted with the rescue. Against all odds, not a single life was lost.

[1]*Forties* and *Fifties* here refer to the ocean in the area of 40° and 50° south latitude.
Source: Worsley, F. A. 1977. *Shackleton's Boat Journey.* New York: Norton, 116–18.

Figure 9.1 A floating sea gull demonstrates that wave forms travel but that the water itself does not. In this sequence, a wave moves from left to right as the gull (and the water in which it is resting) rotates in an imaginary circle, moving slightly to the left up the front of an approaching wave, then to the crest, then sliding to the right down the back of the wave.

To most people an ocean wave in deep water appears to be a massive moving object—a ridge of water traveling across the sea surface. An ocean wave is one of several kinds of **waves**, all of which are disturbances caused by the movement of energy from a source through some medium (solid, liquid, or gas). As the energy of the disturbance travels, the medium through which it passes moves in specific ways. Sometimes this movement is visible to us as a hump in the medium. The traveling hump, or ridge, produces the appearance of movement we see in a wave. In an ocean wave, *energy* is moving at the speed of the wave, but *water* is not.

Picture a resting sea gull as it bobs on the wavy ocean surface far from shore. The gull moves in *circles*: up and forward as the tops of the waves move to his position, down and backward as the tops move past. Each circle is equal in diameter to the wave's height. As may be seen in **Figure 9.1**, energy in waves flows past the resting bird, but the gull and its patch of water move only a very short distance forward in each up-and-forward, down-and-back wave cycle. The water on which the bird rests does not move continuously across the sea surface as the wave illusion suggests.

The nearly friction-free transfer of energy from water particle to water particle in these circular paths, or **orbits**, transmits wave energy across the ocean surface and causes the wave form to move. This kind of wave is known as an **orbital wave**—a wave in which particles of the medium (water) move in closed circles as the wave passes. Orbital ocean waves occur at the boundary between two media (between air and water) or between layers of water of different densities. Because the wave form in these waves moves forward, they are known as **progressive waves**.

The progressive wave that moved the gull was caused by wind. Other forces can generate much greater progressive waves, in which water molecules move through much larger circular or elliptical orbits. Some of these waves are so large that they do not appear to us as waves at all, but rather as the slow sloshing of water in a harbor or bay, as dangerous flooding surges of water, or as rhythmic and predictable ocean tides.

Ocean waves have distinct parts. The **wave crest** is the highest part of the wave, above average water level; the **wave trough** is the valley between wave crests, below average water level. **Wave height** is the vertical distance between a wave crest and the adjacent trough, while **wavelength** is the horizontal distance between two successive crests (or troughs). The relationships among these parts are shown in **Figure 9.2**. The time it takes for two successive wave crests (or troughs) to pass a fixed point, usually measured in seconds, is known as the **wave period**. **Frequency** is the number of waves passing a fixed point per second. Frequency is the inverse of period.

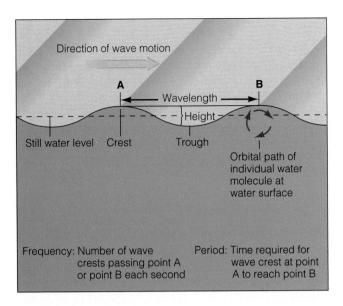

Figure 9.2 The anatomy of a progressive wave.

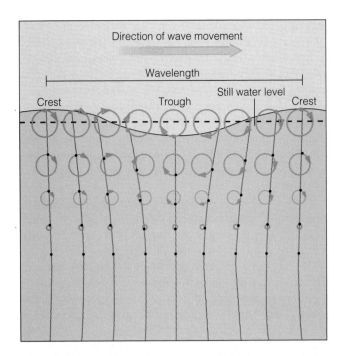

Figure 9.3 The orbital motion of water particles in a wave, which extends to a depth of about one-half of the wavelength.

The circular motion of water particles at the surface of a wave continues underwater. As **Figure 9.3** shows, the diameter of the orbits through which water particles move diminishes rapidly with depth. For all practical purposes, wave motion is negligible below a depth of one-half the wavelength, where the circles are only ⅑ the diameter of those at the surface. This means that a diver in 20 meters of water would not notice the passage of a wind wave of 30-meter wavelength, and he might barely notice the wave if he were at a depth of 15 meters. Since most ocean waves have moderate wavelengths, the circular disturbance of the ocean that propagates these waves affects only the uppermost layer of water. Note that the movement of water in circles doesn't resemble interlocking mechanical gears. Instead, there is a coordinated, uniform circular movement of water molecules in one direction as the waves pass.

CLASSIFYING WAVES

Ocean waves are classified by the disturbing force that creates them, the restoring force that tries to flatten them, and their wavelength. (Wave height is not often used for classification because it varies greatly depending on water depth, interference between waves, and other factors.)

Disturbing Force

Energy that causes ocean waves to form is called a **disturbing force**. Wind blowing across the ocean surface provides the disturbing force for wind waves. Arrival of a storm surge or seismic sea wave in an enclosed harbor or bay, or a sudden change in atmospheric pressure, are the disturbing forces for the resonant rocking of water known as a *seiche*. Landslides, volcanic eruptions, or faulting of the seafloor associated with earthquakes are the disturbing forces for seismic sea waves (also known as *tsunami*). The disturbing forces for tides are changes in the direction of gravitational forces among the Earth, moon, and sun, combined with Earth's rotation.

Restoring Force

Restoring force is the dominant force trying to return the water surface to flatness after a wave has formed in it. If the restoring force of a wave were quickly and fully successful, a disturbed sea surface would immediately become smooth, and the energy of the embryo wave would be dissipated as heat. But that isn't what happens. Waves continue after they form because the restoring force overcompensates and causes oscillation. The situation is analogous to a weight bobbing at the bottom of a very flexible spring, constantly moving up and down past its normal resting point.

The restoring force for very small water waves—those with wavelengths less than 1.73 centimeters (0.68 inch)—is surface tension (cohesion), the property that allows individual water molecules to stick to each other by means of hydrogen bonds (see Figure 6.2). These **capillary waves** are transmitted across a puddle because cohesion, the force that makes the tea creep up on the sides of a teacup, tugs the tiny wave troughs and crests toward flatness. Capillary waves are the first waves to form when the wind blows. These small ripples are important in transferring energy from air to water to drive ocean currents, but they are of little consequence in the overall picture of ocean waves because they are tiny and carry very little energy.

All waves with wavelengths greater than 1.73 centimeters depend on gravity to provide the restoring force. Like the spring weight moving up and down, gravity pulls the crests downward, but the momentum of the water causes the crests to overshoot and become troughs. The repetitive nature of this movement gives rise to the circular orbits of individual water molecules in an ocean wave. These larger waves are called **gravity waves**. Since the circular motion of water molecules in a wave is nearly friction-free, gravity waves can travel across thousands of miles of ocean surface without dissipating, eventually to break on a distant shore.

Wavelength

Wavelength is a direct measure of wave size. **Table 9.1** lists the causes and typical wavelengths of the four types of ocean waves: wind waves, seiches, seismic sea waves, and tides. **Figure 9.4** shows the relation among disturbing and restoring forces, period, and the relative amount of energy present in the ocean's surface for each wave type. Note that more energy is stored in wind waves than in any of the other wave types.

DEEP-WATER WAVES, SHALLOW-WATER WAVES

Most of the characteristics of ocean waves depend on the relationship between their wavelength and water depth. Wavelength determines the *size* of the orbits of water molecules within a wave, but water depth determines the *shape* of the orbits. The paths of water molecules in a wind wave are circular only when the wave is traveling in deep water. A wave cannot "feel" the bottom when it moves through water deeper than one-half its wave-

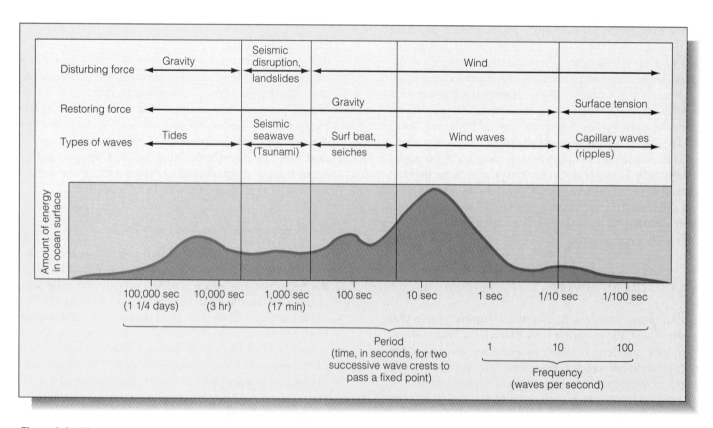

Figure 9.4 Wave energy in the ocean as a function of the wave period. As the graph shows, most wave energy is typically concentrated in wind waves. However, tsunami, rare events in the ocean, can transmit more energy than all wind waves for a brief time.

Table 9.1 Wavelengths and Disturbing Forces of Important Ocean Waves

Wave Type	Typical Wavelength	Disturbing Force
Wind wave	60–150 m (200–500 ft)	Wind over ocean
Seiche	Large, variable; a fraction of basin size	Changes in atmospheric pressure, storm surge, tsunami
Seismic sea wave (tsunami)	200 km (125 mi)	Faulting of seafloor, volcanic eruption, landslide
Tide	½ circumference of Earth	Gravitational attraction, rotation of Earth

length; too little wave energy is contained in the small circles below that depth. Waves moving through water deeper than one-half their wavelength are known as **deep-water waves**. A wave has no way of knowing how deep the water is, only that it is in water deeper than about one-half its wavelength. An example may help: A wind wave of 20-meter wavelength will act as a deep-water wave if it is passing through water more than 10 meters deep (see **Figure 9.5a**).

The situation is different for wind-generated waves close to shore. The orbits of water molecules in waves moving through shallow water are flattened by the proximity of the bottom. Water just above the seafloor cannot move in a circular path, only forward and backward. Waves in water shallower than $\frac{1}{20}$ their original wavelength are known as **shallow-water waves** (see **Figure 9.5b**). A wave with a 20-meter wavelength will act as a shallow-water wave if the water is less than 1 meter deep.

Intermediate-depth water waves (or transitional waves) travel through water deeper than $\frac{1}{20}$ their original wavelength, but shallower than one-half their original wavelength. In our example, this would be water between 1 meter and 10 meters deep. The orbital motion of water molecules in an intermediate-depth wave is shown in **Figure 9.5c**.

Of the four wave types listed in Table 9.1, only wind waves can ever be deep-water waves. To understand why, remember that most of the ocean floor is deeper than 125 meters (400 feet), one-half the wavelength of very large wind waves. The wavelengths of the larger waves are *much* longer; the wavelength of seismic sea waves, for example, usually exceeds 100 kilometers (62 miles). No ocean is 50 kilometers (31 miles) deep; so seiches, seismic sea waves, and tides are always in water that to them is shallow or intermediate in depth; their huge orbit-circles flatten against a distant bottom always less than half a wavelength away.

In general, the longer the wavelength of a wave, the faster the wave energy will move through the water. For

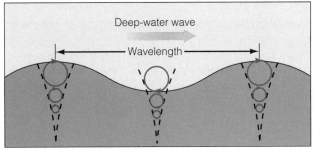

a Depth ≥ $\frac{1}{2}$ wavelength

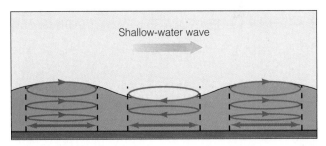

b Depth ≤ $\frac{1}{20}$ wavelength

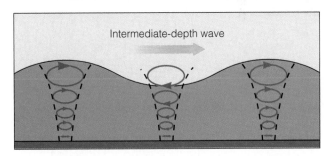

c Depth ≤ $\frac{1}{2}$ to ≥ $\frac{1}{20}$ wavelength

Figure 9.5 Progressive waves. (a) A deep-water wave; (b) a shallow-water wave; (c) an intermediate-depth wave.

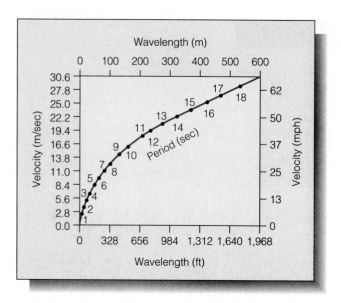

Figure 9.6 The theoretical relationship between velocity, wavelength, and period in deep-water waves. Velocity is equal to wavelength divided by period. If one characteristic of a wave can be measured, the other two can be calculated. The easiest to measure exactly is period, shown here as numbers on the curve (in seconds).

deep-water waves this relationship is shown in the formula

$$V = L/T$$

in which V represents velocity, L is wavelength, and T is time, or period (in seconds). Wavelength is difficult to determine at sea, but period is comparatively easy to find—by having an observer time the movement of waves past the bow of a stopped ship, for example. It turns out that if period (T) is known, velocity (V) can be calculated from the relation

$$V \text{ (in meters/sec)} = 1.56T$$

or, if you prefer,

$$V \text{ (in feet/sec)} = 5T$$

Figure 9.6 shows the relationship between wavelengths of deep-water waves and their velocity and period.

The action of shallow-water waves is described by a different equation, which may be written as

$$V = \sqrt{gd} \quad \text{or} \quad V = 3.1\sqrt{d}$$

where V is velocity (in meters per second), g the acceleration due to gravity (an average of 9.8 meters per second

per second), and d the depth of water in meters. The *period* of a wave remains unchanged regardless of the depth of water through which it is moving. But as deep-water waves enter the shallows and feel bottom, their velocity is reduced and their crests "bunch up"; so their wavelength shortens.

Following is a comparison of two very different kinds of ocean waves (a comparison of apples and oranges), but notice the general relationship between wavelength and wave velocity: The longer the wavelength, the greater the velocity.

Wind Waves
(Deep-Water Waves)

- Period to about 20 *seconds*
- Wavelength to perhaps 600 meters (2,000 ft.) in extreme cases
- Speed to perhaps 31 meters per second (100+ feet per second, or 70 miles per hour) in extreme cases

Seismic Sea Waves
(Shallow-Water Waves)

- Period to perhaps 20 *minutes*
- Wavelength typically 200 kilometers (125 miles)
- Speeds of 760 kilometers (470 miles) per hour

Remember that *energy*—not the water mass itself—is moving through the water at the astonishing speed of 760 kilometers per hour in seismic sea waves. (A speed of 760 kilometers per hour is the speed of a commercial jet airliner!)

WIND WAVES

Wind waves are gravity waves formed by the transfer of wind energy into water. Most wind waves are less than 3 meters (10 feet) high. Wavelengths from 60 to 150 meters (200 to 500 feet) are most common in the open ocean.

Wind waves grow from capillary waves. Capillary waves form as wind friction stretches the water surface and as surface tension tries to restore it to smoothness. These waves are nearly always present on the ocean. Capillary waves interrupt the smooth sea surface, deflect surface wind upward, slow it, and cause some of the wind's energy to be transferred into the water to drive the capillary wave crest forward. The wind may eddy briefly behind the tiny crest, creating a slight partial vacuum there. Atmospheric pressure pushes the trailing crest forward (downwind) toward the trough, adding still more energy to the water surface. The increasing energy in the water surface expands the circular orbits of water particles in the direction of the wind, enlarging the small wave's size. The capillary wave becomes a wind wave when its wave-

a

Figure 9.7 Wind waves. (a) Swell—mature, regular wind waves—off the Oregon coast. (b) Sea, an area of wind wave formation, west of San Francisco.

length exceeds 1.73 centimeters (0.68 inch), the wavelength at which gravity supersedes capillary action as the restoring force.

If the wind wave remains in water deeper than one-half its wavelength and the wind continues to blow, the wave becomes larger. Its crest is thrust higher into faster wind, extracting even more energy from the moving air. The circular orbits of water particles within the wave grow larger with more energy input; height, wavelength, and period increase proportionally. The irregular peaked waves in the area of wind wave formation are called **sea** (**Figure 9.7a**); the chaotic surface is formed by simultaneous wind waves of many wavelengths. When the wind slows or ceases, as it does away from a storm, the wave crests become rounded and regular. These mature wind waves of uniform wavelength outside their original area of generation are called **swell**. Swell form smooth undulations in the ocean surface (**Figure 9.7b**). Since waves with the longest wavelengths move fastest, swell are sorted by wavelength, with longer waves moving away from the storm most rapidly. Observers at a distance

b

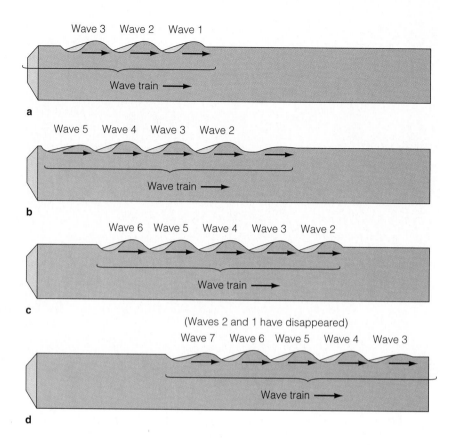

Figure 9.8 The progress of a wave train. (a) The energy in the leading waves (here, waves 1 and 2) is transferred into circular movement in undisturbed water. (b) As waves 1 and 2 are drained of energy, they gradually disappear, but the circular movement forms new waves 4 and 5 at the end of the train.

would first encounter quick-moving waves of long wavelength, then middle-sized waves, and then slow, small ones.

Progressing groups of swell with the same origin and wavelength are called **wave trains**. The leading waves in the wave train are drained of energy because they must begin the circular movement of the undisturbed water into which they are intruding. These leading waves gradually disappear, but after the wave train has passed, some energy remains behind in the circles to form new waves. New waves thus form behind as the leading waves disappear at the front of the wave train. This process is shown in **Figure 9.8**.

The implications of this detail are surprising. Though each wave moves forward with a velocity proportional to its wavelength, the wave train itself moves forward at only *half* that speed. Groups of waves therefore move ahead at half the speed of individual waves within the group. The half-speed advance of the wave train is called **group velocity**. As deep-water waves move into shallow water, the speed of individual waves within the group slows until wave speed equals group velocity.

Factors Affecting Wind Wave Development

Three factors affect the growth of wind waves. Wind must be moving faster than the wave crests for energy transfer from air to sea to continue; so the mean speed of the wind, or **wind strength**, is clearly important to wind wave development. A second factor is the length of time the wind blows, or **wind duration**. The third factor is the uninterrupted distance over which the wind blows without significant change in direction: the **fetch**.

A strong wind must blow continuously for nearly three days for the largest waves to develop fully. A **fully developed sea** is the maximum wave size theoretically possible for a wind of a specific strength, duration, and fetch. Longer exposure to wind at that speed will not increase the size of the waves. The first three columns of **Table 9.2** give examples of the combinations of wind speed, fetch, and duration required to form waves of the fully developed sea described in the remaining columns. Note that waves in a fully developed sea are small if wind speed is low. However, if the wind speed is 74 kilometers (46 miles) per hour through 1,313 kilometers (816 miles)

Table 9.2 Conditions Necessary for a Fully Developed Sea at Given Wind Speeds, and the Parameters of the Resulting Waves

Wind Conditions			Wave Size		
Wind Speed	Fetch	Wind Duration	Average Height	Average Wavelength	Average Period
19 km/hr (12 mi/hr)	19 km (12 mi)	2 hr	0.27 m (0.9 ft)	8.5 m (28 ft)	3.0 sec
37 km/hr (23 mi/hr)	139 km (86 mi)	10 hr	1.5 m (4.9 ft)	33.8 m (111 ft)	5.7 sec
56 km/hr (35 mi/hr)	518 km (322 mi)	23 hr	4.1 m (13.6 ft)	76.5 m (251 ft)	8.6 sec
74 km/hr (46 mi/hr)	1,313 km (816 mi)	42 hr	8.5 m (27.9 ft)	136 m (446 ft)	11.4 sec
92 km/hr (58 mi/hr)	2,627 km (1,633 mi)	69 hr	14.8 m (48.7 ft)	212.2 m (696 ft)	14.3 sec

Source: Data from U.S. Army Corps of Engineers Coastal Research Center, Virginia.

Figure 9.9 An immense wave rushes toward a ship.

for 42 hours—conditions not wholly unrealistic in the Pacific—waves that average 8.5 meters (27.9 feet) high can result. The highest 10% of the waves in this sea will exceed 17.2 meters (56.6 feet) in height! The combination of factors required to produce truly great seas doesn't occur very often. In the example given at the bottom of Table 9.2, it seems unlikely that a wind of 92 kilometers (58 miles) per hour would blow steadily in one direction for 69 hours over 2,627 kilometers (1,633 miles).

The greatest potential for large waves occurs beneath the strong and nearly continuous winds of the West Wind Drift surrounding Antarctica. Jules Dumont d'Urville, the early nineteenth-century French explorer of the South Seas, encountered a wave train with heights estimated "in excess" of 30 meters (100 feet) in Antarctic waters, and

in 1916 Ernest Shackleton contended with occasional waves of similar size in the West Wind Drift during a heroic voyage to remote South Georgia Island in an open boat. Satellite observations have shown that wave heights to 11 meters (36 feet) are fairly common in the West Wind Drift.

In zones of high winds, a less than fully developed sea can also attract attention. Though wind speed within cyclonic storms is often very strong, the circular motion of air doesn't allow long fetches, and fully developed seas rarely occur beneath them. Officers standing deck watches during storms rarely quibble with theoretical maximum height *vs.* observed height, however. Wind waves can be overwhelming even if they are not fully developed (**Figure 9.9**).

Wind Wave Height

Wave height is not directly represented in the deep-water wave formula $V = L/T$. During their formation, moderately sized wind waves in the open ocean exhibit a maximum 1:7 ratio of wave height to wavelength (see **Figure 9.10**); this ratio is the **wave steepness**. Waves 7 meters long will not be more than 1 meter high, and waves of 70-meter wavelength will not exceed 10 meters of height. The angle at their crest will not exceed 120°. A peaked appearance usually indicates the continuing injection of wind energy. If a wave gets any higher than the 1:7 ratio for its wavelength, it will break, and excess energy from the wind will be dissipated as turbulence; hence the "whitecaps" or "combers" associated with a fully developed sea.

The highest wave ever measured was sighted on the night of 7 February 1933 by Lieutenant Frederick Marggraff, a watch officer aboard the U.S. Navy tanker *Ramapo*. USS *Ramapo* was steaming from Manila to San Diego through a furious storm—a storm made more intense by the coalescence of three low-pressure centers. For days a steady wind had blown at 107 kilometers (67 miles) per hour, and gusts to 126 kilometers (78 miles) per hour often lashed the decks. But the wind blew persistently from one direction, and though the monstrous waves it generated dwarfed the tanker, they were surprisingly orderly in form.

"The conditions for observing the seas from the ship were ideal," wrote *Ramapo*'s executive officer. "We were running directly down the wind with the sea. There were no cross seas and therefore no peaks along wave crests. There was practically no rolling, and the pitching motion was easy because of the fact that the sides of the waves were much longer than the ship. The moon was out astern and facilitated observations during the night. The sky was partly cloudy." At about three in the morning, Lieutenant Marggraff observed a train of tremendous waves looming in the moonlight. As the trough of the first wave approached the ship, he noted that its distant crest was on a level with the crow's nest on the mainmast. At that instant the ship's stern sank into the bottom of the onrushing trough. The next two waves were about the same size. Not surprisingly, such immense waves made an indelible impression on all who witnessed them.

How big were these waves? When the executive officer had some time to spare, he did some calculations. **Figure 9.11** illustrates the attitude of the ship when the largest waves were measured. The height of the waves was determined by using a set of the ship's plans, a calculation of the height of the observer above the sea surface, and a sight to the horizon. The largest wave for which a dependable observation had been made was 34 meters (112 feet) high, still a record!

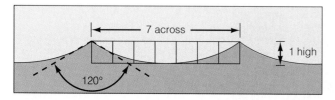

Figure 9.10 A wind wave of moderate size shown during the time of its formation. The ratio of height to wavelength, called wave steepness, is 1:7; the crest angle does not exceed 120°.

Figure 9.11 How the great wave observed from the USS *Ramapo* was measured. An officer on the bridge was looking toward the stern and saw the crow's nest in his line of sight to the crest of the wave, which had just come in line with the horizon. Wave height was later calculated based on the geometry of the situation.

Wavelength Records

But is that as big as wind waves can get? This is an interesting question, and one not dispassionately answered. Calculating the wavelength from wave period (T in $V = L/T$), Bigelow and Edmondson (1947) reported that mariners have sighted waves with wavelengths of 451 meters (1,481 feet) off the west coast of Ireland, 583 meters (1,914 feet) off the Cape of Good Hope, and 829 meters (2,719 feet) in the equatorial Atlantic. (That last monster would have had a period of 25 seconds!) The nineteenth-century French admiral J. Mottez reported a wave with a wavelength of 790 meters (2,600 feet), a period of 23 seconds, and a speed of 123 kilometers (76 miles) per hour in the equatorial Atlantic west of Africa. The wavelength of the *Ramapo* wave discussed earlier was calculated at 360 meters (1,180 feet); so these sightings are perhaps exaggerated. Much smaller waves can also add a great deal of excitement to a deck officer's day (see **Figure 9.12**)!

Interference

The real situation of wind waves at sea is not as simple as has been suggested here. The ideal vision of one set of waves moving in one direction at one speed across an otherwise smooth surface is almost never observed in the ocean.

Independent wave trains exist simultaneously in the ocean most of the time. Since long waves outrun shorter ones, wind waves from different storm systems can overtake and interfere with one another. One wave doesn't

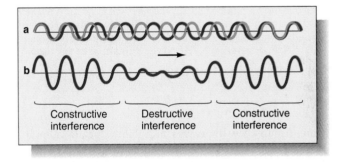

Figure 9.13 Interference. Two overlapping waves of different wavelength (a) generate both constructive and destructive interference (b).

crawl over the others when they meet; instead they add to (or subtract from) one another. Such interaction is known as **interference**. In **Figure 9.13a** a wave of one wavelength is represented as a light blue line, a slightly longer wavelength wave as a dark blue line. In the sea surface where these waves coincide (shown in **Figure 9.13b**) you can see the alternation between addition (large crests and troughs) and subtraction (almost no waves at all). The cancellation effect of subtraction is termed **destructive interference**, not because of harm to lives or property, but because wave interference destroys or cancels the waves involved. **Constructive interference** is the additive formation of large crests or deep troughs. This pattern of constructive and destructive interferences explains why the ocean surface will be relatively calm for a

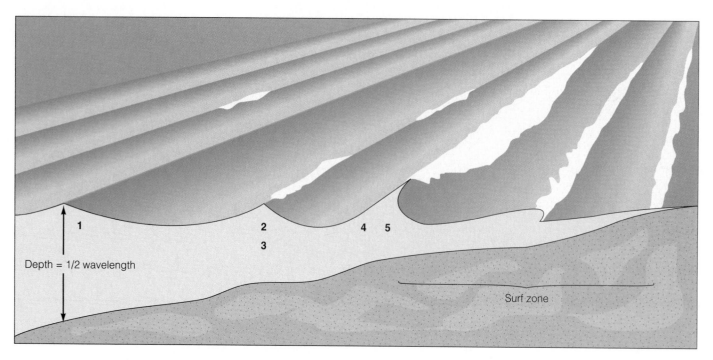

Figure 9.14 How a wave train breaks against the shore. (1) The swell "feels" bottom when the water is shallower than one-half the wavelength. (2) The wave crests become peaked because the wave's energy is packed into less water depth. (3) Friction with the bottom slows the wave, while waves behind it maintain their original rate. Therefore, wavelength shortens, but period remains unchanged. (4) The wave approaches the critical 1:7 ratio of wave height to wavelength. (5) The wave breaks when the ratio of wave height to water depth is about 3:4.

time, rise to a few big waves, and then return to calm. Interference also explains why in some instances every ninth wave might be quite large. As wavelengths change, though, every seventh wave might be large, or every fifth, or every twelfth. Contrary to folklore, there is no set ratio.

Interference can have sudden unpleasant consequences on the open sea. In or near a large storm, wind waves at many wavelengths and heights may approach a single spot from different directions. If such a rare confluence of crests occurred at your position, a huge wave crest would suddenly erupt from a moderate sea to threaten your ship. The freak wave—called a **rogue wave**—would be much larger than any noticed before or after, and it would be higher than the theoretical maximum wave capable of being sustained in a fully developed sea. In such conditions, one wave in about 1,175 is over three times average height, and one in every 300,000 is over four times average height!

Wind Waves Approaching Shore

Most wind waves eventually find their way to a shore and break, dissipating all their order and energy. The process begins when our now familiar deep-water wave becomes an intermediate-depth water wave in water less than half a wavelength deep. **Figure 9.14** outlines the events leading to the break:

1. The wave train moves toward shore. When the water is less than half the wavelength in depth, the wave "feels" bottom.

2. The circular motion of water molecules in the wave is interrupted. Circles near the bottom flatten to ellipses. The wave's energy must now be packed into less water depth; so the wave crests become peaked rather than rounded.

3. Friction with the bottom slows the wave. Waves behind it continue toward shore at the original rate. Wavelength therefore decreases, but period remains unchanged.

4. The wave becomes too high for its wavelength, approaching the critical 1:7 ratio.

5. As the water becomes even shallower, the part of the wave below average sea level slows because of friction with the bottom. When the wave was in

a

b

Figure 9.15 Types of breaking waves. (a) Plunging waves form when the bottom slopes steeply toward shore. (b) Spilling waves form when the bottom slopes gradually. (c) Surging waves form when the beach slopes too abruptly for the waves to break.

c

deep water, molecules at the top of the crest were supported by the molecules ahead (thus transferring energy forward). This is now impossible because the *water* is moving faster than the *wave*. As the crest moves ahead of its supporting base, the wave breaks. The break occurs at about a 3:4 ratio of wave height to water depth. (That is, a 3-meter wave will break in 4 meters of water.) The turbulent mass of agitated water rushing shoreward during and after the break is known as **surf**. The **surf zone** is the region between the breaking waves and the shore.

Waves break against the shore in different ways, depending in part on the slope of the bottom. The break can be violent and toppling, leaving an air-filled channel (or *tube*) between the falling crest and the foot of the wave. These **plunging waves (Figure 9.15a)** form when waves approach a steeply sloping bottom. A more gradually sloping bottom generates a milder **spilling wave** as the crest slides down the face of the wave (**Figure 9.15b**). **Surging waves (Figure 9.15c)** occur when an abrupt beach slope prevents a proper break and water simply

BOX 9.1 • *Surfing*

More than 2 million Americans surf regularly. The thrill of rushing down the face of a growing, breaking wave is exhilarating! People willingly endure cold, boredom, and some danger for a ride lasting only a few seconds. If you're a good swimmer, give it a try. You will need to paddle your board or swim vigorously to match your speed to that of the advancing wave crest. As the wave rises to break, your forward speed (and sense of timing) will place you on the leading edge of the crest, accelerating downward and forward. The technique takes time to master, but the feeling is worth the effort!

Prejudice compels me to reveal that *real* surfers are body surfers. The buoyant forces on which board surfing depends are diminished in body surfing, but forces within the wave propel the body surfer with greater acceleration. A body surfer's intimate contact with the water and close proximity to the wave surface greatly increase the sensation of movement.

The ultimate trick is to surf the wave completely submerged, as a dolphin does. You push powerfully off the bottom (or swim rapidly toward the surface) as the wave crest approaches, inserting yourself into the freshly breaking wave from below. The rapid flow of water within the toppling wave will ripple your skin, and the sudden acceleration can thrust you from the face of the wave like a wet bar of soap shooting from a fist. Altogether a wonderful oceanic experience!

rushes ashore. Along with some wind waves, storm surges and seismic sea waves form surging waves.

Slope alone does not determine the position and nature of the breaking wave. The contour and composition of the bottom can also be important. Bottoms that get shallow gradually can sap waves of their strength because of prolonged friction against the bottom of the lowest elliptical water orbits. Energy may be lost even more rapidly if the bottom is covered with loose gravel or irregular growths of coral. Masses of moving seaweeds or jostling chunks of sea ice can also extract energy from a wave. In a few rare cases the shore is configured in such a way that waves don't break at all; the waves have lost virtually all their energy by the time their remnants arrive at the beach.

Wave Refraction

What happens when a wave line approaches the shore at an *angle*, as it almost always does, as shown in **Fig-**

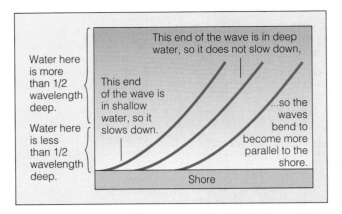

a

b

Figure 9.16 Wave refraction. (a) Diagram showing the elements that produce refraction, as seen from above. (b) Wave refraction around Maili Point, Oahu, Hawaii. Note how the wave crests bend almost 90° as they move around the point.

Figure 9.17 A "checkerboard" sea surface off La Jolla, California, the result of wave interference.

ure 9.16? The line does not break simultaneously because different parts of it are in different depths of water. The part of the wave line in shallow water slows down, but the attached segment still in deeper water continues at original speed; so the wave line bends (or refracts). The bend can be as much as 90° from the original direction of the wave train. This slowing and bending of waves in shallow water is called **wave refraction**. The refracted waves break in a line almost parallel to the shore.

Wave refraction can produce some odd surf patterns. A 19-second swell "feels" bottom at 300 meters (1,000 feet), while an 11-second swell "feels" bottom at 100 me-

ters (330 feet). You may remember the average depth at the outer continental shelves is only about 150 meters (500 feet). As these two wave trains approach shore, the longer swell will be slowed by the shelf for perhaps 60 kilometers (38 miles) before the shorter swell is slowed by the bottom. With one wave bending and the other heading in its original direction, the potential for complex interference is great. Thus, waves may break for a time at one spot, cease, and then begin breaking at another spot a few hundred meters down the beach. The ocean surface might have a checkerboard appearance from the constructive and destructive interference of these crests (see **Figure 9.17**).

Figure 9.18 A storm surge. The low pressure and high onshore winds generated by a hurricane can produce a storm surge up to 9 meters (30 feet) high. Hurricane Camille, which struck the Mississippi coast in 1969, produced a storm surge 7 meters (23 feet) high. The massive inundation was responsible for a substantial fraction of the $1.4 billion in damage suffered in the area. In August 1992, Hurricane Andrew struck the coasts of South Florida and Louisiana with sustained winds of 140 miles per hour. A 4-meter (13-foot) storm surge was responsible for heavy damage to harbors, coastal towns, and wildlife refuges. Losses from the storm have been estimated at $30 billion.

LONG WAVES

We now turn our attention to the longer waves generated by the low atmospheric pressure of large storms, by the sloshing of water in enclosed spaces, and by the sudden displacement of ocean water.

A Word About "Tidal Waves"

The popular media and general public tend to label *any* unusually large wave a "tidal wave" regardless of its origin, and the waves described below are prime candidates for this error. Press accounts of storms at sea usually list a rogue wave as a tidal wave, and very large sets of wind waves are called tidal waves by some yachtsmen or surfers. The sea waves associated with earthquakes are almost always called tidal waves in media damage reports. The waves caused by the approach of a tropical cyclone to land may also incorrectly be termed tidal waves. The term *tidal wave* is *not* synonymous with *large wave*, however. And, as we will see in the next chapter, the only true tidal waves are relatively harmless waves associated with the tides themselves.

STORM SURGES

The abrupt bulge of water driven ashore by a tropical cyclone (hurricane) or frontal storm is called a **storm surge**.

Its crest can temporarily add up to 7.5 meters (25 feet) to coastal sea level. Water can reach even greater heights when the surge is funneled into a confined bay or estuary.

Many factors contribute to the severity of a storm surge. The most important factor is the strength of the storm generating the surge. The low atmospheric pressure associated with a great storm will draw the ocean surface into a broad dome as much as 1 meter (about 3 feet) higher than average sea level. This dome of water accompanies the storm to shore, becoming much higher as the water gets shallower at the coast. There the water ramps ashore, driven forward by large storm-generated wind waves (**Figure 9.18**). A storm surge is a short-lived phenomenon. Technically it is not a progressive wave because it is only a crest; wavelength and period cannot be assigned to it.

Water in a storm surge does not come ashore as a single breaking wave, but rushes inland in what looks like a sudden, very high, wind-blown tide. Indeed, storm surges are sometimes called *storm tides* because the volume of water they force on shore is greatly increased if the surge arrives at the same time as a high tide. The wrong combination of low atmospheric pressure, strong onshore winds, high tide, and bottom contour can be especially dangerous if estuaries in the area have been swollen by heavy rainfall preceding the storm.

Storm surges have had catastrophic consequences. The frightful tropical storm of November 1970 in Bangladesh (described in Chapter 7) generated a storm surge up to

Figure 9.19 The Thames tidal barrier in London. The funnel shape of the Thames estuary concentrates storm surges as they near London. The barrier sections can be raised (a) to prevent flooding upstream. Each barrier section is controlled by hydraulic arms within towers. The great size of the $1.3 billion project can be seen in (b).

12 meters (40 feet) high, which caused the death of more than 300,000 people. Storm surges associated with extra-tropical cyclones (frontal storms) can also do tremendous damage. On 1 February 1953, a storm surge and high tide arrived simultaneously against the Dutch coast. Wind waves breached the dikes and flooded the low country, covering more than 3,200 square kilometers (800,000 acres) and drowning 1,783 people. The Dutch anticipate this coincidence of events will occur only about once in 400 years. The dikes have been rebuilt and the land reclaimed from the North Sea. On the opposite side of the North Sea, Londoners have spent $1.3 billion on a flood defense system at the mouth of the Thames River. The centerpiece of the project is an immense barrier against storm surges (see **Figure 9.19**). Experts expect the barrier to prevent a devastating flood on an average of once every 50 years.

The U.S. coast is also at risk. In 1900 a storm surge topped the Galveston seawall, swept into the city, and killed more than 6,000 area residents. (In contrast, no lives were lost during a similar Texas storm in 1961 because coastal barriers had been constructed and advance warning permitted preparation and evacuation.) In August 1954, the combined effects of a high tide and Hurricane Carol caused water to rise almost 5 meters (16 feet) above normal to flood much of Providence, Rhode Island. Damage exceeded $41 million. City officials have since built a protective surge barrier across upper Narragansett Bay. And as we saw in Chapter 7, Hurricane Andrew's 1992 assault on the Florida coast was made even more lethal by its storm surge. Anyone living in a low-lying coastal area frequented by violent storms should be aware of and prepared for the potential danger of storm surge.

SEICHES

When disturbed, water confined to a small space (such as a bucket, a bathtub, or a bay) will slosh back and forth at a specific resonant frequency. The frequency changes with different amounts of water, or with different sizes or shapes of containers. If you carry a shallow container of water (like an ice cube tray) from one place to another, you're careful not to move the tray at its resonant frequency—to avoid a spill. Most of the water's random motion quickly settles down after you place the tray on a table top, but the water in the tray may rock gently for some seconds at this one resonant frequency. That rocking is a **seiche** (pronounced "saysh").

The seiche phenomenon was first studied in Switzerland's Lake Geneva by eighteenth-century researchers curious about why the water level at the ends of the long, narrow lake rises and falls at regular intervals after windstorms. They found that constant breezes tend to push water into the downwind end of the lake. When the wind stops, the water is released to rock slowly back and forth at the lake's resonant frequency, completing a crest-trough-crest cycle in a little more than an hour. At the ends of the lake the water rises and falls a foot or two; at the center it moves back and forth without changing height (**Figure 9.20**). This kind of wave is called a **standing wave** because it oscillates vertically with no forward movement. The point (or line) of no vertical wave action in a standing wave—the place in the lake where the water moves only back and forth—is called a **node**. In Lake Geneva, the wavelength of the seiche is twice the length of the lake itself; the node lies at the center. The lake acts like a large version of the ice cube tray.

This kind of activity can occur in confined areas of ocean ranging in size from bays and harbors to entire ocean basins. As we have seen, a drop in atmospheric pressure beneath a storm can draw water to a slightly higher level. When the storm moves inland, the water is released and the oscillation begins. The wavelength of the oscillation is a multiple of the length of the ocean basin, often thousands of miles. The rhythmic oscillation of tides can also cause seiches in bays or harbors. Seiche periods may range from a few minutes to more than a day. In large areas, the wavelength of these standing waves may rival that of tsunami.

Lakeside property may occasionally be threatened, but damage from seiches along ocean coasts is rare. The wavelength may be tremendous, but seiche wave height in the ocean rarely exceeds a few inches. Seiches may disturb shipping schedules by interfering with the predicted arrival times of tides; or they may cause currents in harbors, which could snap mooring lines. In rare instances, seiches contribute to the peril of people on shore, but only

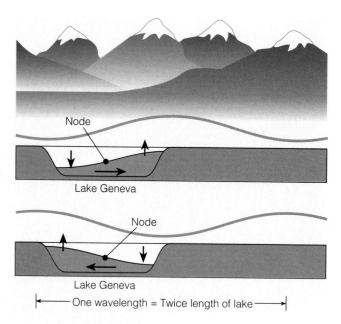

Figure 9.20 Seiche in Lake Geneva. After a windstorm, the lake sloshes back and forth, forming a standing wave, or seiche, with a wavelength of 220 kilometers (132 miles), which is twice the length of the lake. The period of the seiche is about 72 minutes, and its height is about 0.5 meter (1.5 feet).

because the seiche is coincidentally accompanied by wind waves of considerable height, or by a tsunami.

TSUNAMI AND SEISMIC SEA WAVES

Long-wavelength, shallow-water, progressive waves caused by the rapid displacement of ocean water are called **tsunami**, a descriptive Japanese term combining *tsu* (harbor) with *nami* (wave). The word is both singular and plural. Tsunami caused by the sudden vertical movement of the Earth along faults (the same forces that cause earthquakes) are properly called **seismic sea waves**. Tsunami can also be caused by landslides, icebergs falling from glaciers, volcanic eruptions, and other direct displacements of the water surface. Note that all seismic sea waves are tsunami, but not all tsunami are seismic sea waves.

Origins

"Small" tsunami are caused by the displacement of surface water. Although less energy is released by landslides than by most seismic fractures, the resulting sea waves are still very destructive for people or structures near their point of origin. This is especially true if the wave is

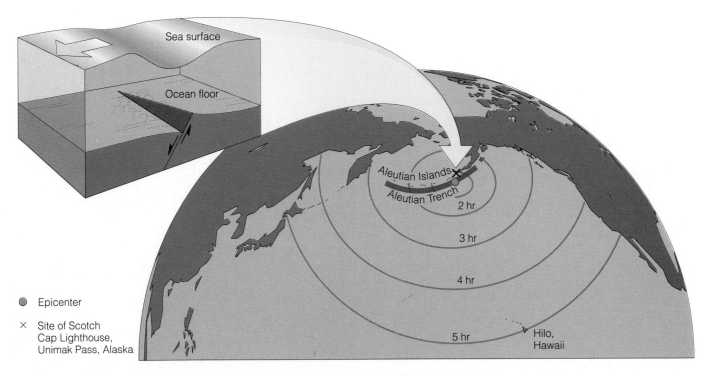

○ Epicenter

✕ Site of Scotch
Cap Lighthouse,
Unimak Pass, Alaska

Figure 9.21 The tsunami of 1 April 1946 began when a rupture along a submerged fault in the Aleutian Trench lifted the sea surface above. The wave moved outward at a speed of about 212 meters per second (472 miles per hour). At this speed, it took only about 5 hours for the wave to travel to the Hawaiian Islands.

formed within a confined area, as in the Lituya Bay examples cited in **Box 9.2**.

Seismic sea waves originate on the ocean floor when Earth movement along faults displaces ocean water. **Figure 9.21** shows the birth of a seismic sea wave in the Aleutians. Rupture along a submerged fault lifts the sea surface above. Gravity pulls the crest downward, but the momentum of the water causes the crest to overshoot and become a trough. The oscillating ocean surface generates progressive waves that radiate from the epicenter in all directions. Waves would also form if the fault movement were downward. In that case a depression in the water surface would propagate outward as a trough. The trough would be followed by smaller crests and troughs caused by surface oscillation.

It seems strange to refer to tsunami caused by movement along deeply submerged faults—waves with wavelengths of up to 200 kilometers (125 miles)—as shallow-water waves. Yet half their wavelength would be 100 kilometers (62 miles), and even the deepest ocean trenches do not exceed 11 kilometers (7 miles) in depth. These immense waves therefore never find themselves in water deeper than half their wavelength. As any intermediate-depth wave or shallow-water wave would be, seismic sea waves are affected by the contour of the bottom and are commonly refracted, sometimes in unexpected ways.

Speed

The velocity of a tsunami is given by the formula for the speed of a shallow water wave:

$$V = \sqrt{gd}$$

Since the acceleration due to gravity (g) is 9.8 meters (32.2 feet) per second per second, and a typical Pacific abyssal depth (d) is 4,600 meters (15,000 feet), solving for velocity (V) shows that the wave would move at 212 meters per second (472 miles per hour). At this speed a seismic sea wave will take only about five hours to travel from Alaska's seismically active Aleutians to the Hawaiian Islands.

Encountering Tsunami

We are familiar with the steepness of a wind wave and the short period of a few seconds between its crests. Tsunami are much different. Once a tsunami is generated, its steepness (ratio of height to wavelength) is

BOX 9.2 ● *The Gods of Lituya Bay*

The largest wave ever observed on Earth formed on 9 July 1958. Four of the six eyewitnesses lived to tell about it.

The great wave formed in Lituya Bay, a narrow glacier-fed inlet on the northeast shore of the Gulf of Alaska. The area was explored in 1788 by the French explorer Jean-François La Pérouse, but the only notable oceanic feature mentioned in his account was the strong tidal current encountered at the bay's narrow entrance. Had he inquired, the local inhabitants could have told him of the awesome magic contained within the bay: Waves "as big as gods."

Three pleasure boats were anchored in the bay on the calm evening the great wave appeared. At 10:16 P.M. the crew of the U.S. Geologic Survey ship *Stephen R. Capps*, lying at anchor in another bay about 100 kilometers (63 miles) to the east, felt a strong earthquake along the Fairweather Fault, a fault that angles west along the steep mountains rimming the innermost reaches of Lituya Bay. The earthquake dislodged 61 million cubic meters (80 million cubic yards) of rock, mud, and glacial ice, which plummeted into the upper bay. Water displaced by this enormous mass of material generated a huge surge that raced to the bay's opposite wall, where it climbed 530 meters (1,740 feet) above sea level.[2] The displacement also generated a gravity wave that roared out into the bay at a speed of between 155 and 210 kilometers per hour (97 and 130 miles per hour). This wave was at least 49 meters (160 feet) high, roughly the height of a 15-story building.

Mr. Howard Ulrich, aboard the small boat *Edrie*, was probably the first to see the wave in the gathering darkness. Ulrich and his seven-year-old son had been awakened by a deafening roar and had gone on deck to investigate. Ulrich did not comprehend the wave at first; it looked like a low cloud or fogbank extending from shore to shore. The monster began to break as it rounded the north side of an island in the center of the bay, but on the south side, where his boat rested at anchor, the wave had not started to curl. Ulrich was unable to free the an-

[2]The Sears Tower in Chicago, the world's tallest inhabited structure, is 87 meters (286 feet) shorter than the height of this "splash"!

a A panoramic view of Lituya Bay, Alaska, showing destruction caused by a landslide-generated tsunami on 9 July 1958. The surge and giant wave of water generated by a rock slide at the far end of the bay destroyed the forest in the light areas along the shores of the bay.

chor, but the line snapped as the boat rose to meet the wave. *Edrie* accelerated upward to the crest, was carried forward over a submerged peninsula, and was deposited into a different section of the now wildly churning bay. Amazingly, the two were able to leave the bay under their boat's own power the next day.

Mr. and Mrs. William Swanson, aboard their fishing trawler *Badger*, anchored on the north side of the bay, were not so fortunate. The Swansons were awakened by a violent vibration. At first, William Swanson saw nothing unusual; then a blue wall moved past the northern end of the island toward their boat, traversing the distance in about 90 seconds. The trawler's anchor chain parted, and she ascended the wave bow-first at a terrifying angle. Then the top of the wave broke, engulfing the boat. *Badger* shuddered, then rushed stern-first down the face of the shattered wave, wind howling through her rigging, as she was carried at a height of 25 meters (80 feet) over the tops of the highest trees, over the bay mouth bar, and out to sea. The Swansons' boat was fatally broken. A deserted skiff was floating nearby and they clambered aboard as *Badger* sank. They were rescued by a fishing boat two hours later.

The two-person crew of the third boat was unluckiest of all; they perished within the wave. William Swanson last saw the *Sunmore* accelerating in the trough of the wave toward a rocky cliff. No sign of crew or boat was ever found.

The 49-meter height of the wave was established by survey crews analyzing the line of tree devastation around the periphery of the bay (clearly visible in **Figure a**). It gives one pause to learn that this wave was comparatively *small*. Though they have not been directly observed, much larger waves have left trails of smashed timber even farther inland. Evidence of a 61-meter (200-foot) wave was reported in 1899; an 1853 wave reached 120 meters (395 feet) in height. And on 27 October 1936 a landslide-generated wave rose to 150 meters (490 feet) in Lituya Bay!

The "gods" of Lituya Bay are tsunami, members of the class of long waves described in this chapter.

extremely low. This lack of steepness, combined with the wave's very long period (5 to 20 minutes), enables it to pass unnoticed beneath ships at sea. A ship on the open ocean that encounters a tsunami with a 16-minute period would rise slowly and imperceptibly for about 8 minutes to a crest only 0.3 to 0.6 meter (1 or 2 feet) above average sea level and then ease into the following trough 8 minutes later. With all the wind waves around, such a movement would not be noticed.

As the seismic sea wave crest approaches shore, however, the situation changes rapidly and often dramatically. The period of the wave remains constant, velocity drops, and wave height greatly increases. As the crest arrives at the coast, observers see water surge ashore in the manner of a very high, very fast tide. In confined coastal waters relatively close to their point of origin, tsunami can reach a height of 30 meters (100 feet). The wave is a fast on-rushing flood of water, not the huge plunging breaker of popular folklore.

The wave energy spreads through an enlarging circumference as a tsunami expands from its point of origin. People on shore near the generating shock have reason to be concerned because the energy will not have dissipated very much. On 1 April 1946, a fracture along the Aleutian Trench generated a seismic sea wave that quickly engulfed the Scotch Cap lighthouse on Unimak Island in the Aleutians. The lighthouse was completely destroyed, and the five coastguardsmen operating the lighthouse died (**Figure 9.22**).

The same seismic sea wave reached the Hawaiian Islands about five hours later. By this time the wave circumference was enormous and its energy more dispersed, but even so, successive waves surged onto Hawaiian beaches at 15-minute intervals for more than two hours. One 9-meter (30-foot) wave struck the town of Hilo, and water rose to a height of 17 meters (56 feet) in an exposed valley near Polulu! At least 150 people were killed in Hawaii that morning, and property damage was in excess of $25 million. A photograph taken near Hilo pier during the tsunami is included in **Figure 9.23**.

Note that the destruction in Hawaii was not caused by one wave (as at Scotch Cap), but by a series of waves following one another at regular intervals. Some energy from the main tsunami wave was distributed into smaller waves ahead of or behind the main wave as it moved. If the epicenter of the displacement responsible for a tsunami is very far away, sea level at shore will rise and fall as these waves arrive. The interval between crests (the wave period) is usually about 15 minutes. Coastal residents far from a tsunami's origin can be lulled into thinking the waves are over; they return to the coastline only to be injured or killed by the next crest. This behavior contributed to loss of life in Hawaii.

a

b

Figure 9.22 Tsunami destruction. Scotch Cap lighthouse guards Unimak Pass in Alaska's Aleutian Islands. (a) The lighthouse before 1 April 1946. Its foundation is 15 meters (49 feet) above sea level, the upper plateau 36 meters (118 feet) above sea level. (b) The site after the tsunami. The lighthouse, support buildings, and radio masts are gone, and the slopes are heavily washed almost to the plateau. Five coastguardsmen died in the wave.

a

Figure 9.23 Tsunami in progress. (a) The wave front rushes up the Wailuku River in Hilo, Hawaii, during the tsunami of 1 April 1946. Note the steep front, the turbulence of the water behind it, and the placidity of the water ahead of the wave. The right-hand span of the bridge was destroyed by an earlier wave of the same tsunami. (b) The 1 April 1946 tsunami rushes ashore in Hilo, Hawaii, as terrified residents run for their lives.

b

Tsunami in History

The greatest recent tsunami was associated with a massive earthquake along the Peru–Chile trench on 22 May 1960. The earthquake and associated landslides killed more than 4,000 people. Waves reaching Japan, 14,500 kilometers (9,000 miles) away, killed 180 people and caused $50 million in structural damage, and Los Angeles and San Diego harbors were badly disrupted by seiches excited by the tsunami.

Half the time a tsunami's trough will arrive first. This occurred with catastrophic results in Lisbon's Tagus estuary on 1 November 1755. Many curious people were attracted to the shore when sea level dropped several meters over about 15 minutes. People ventured out to pick up fish, inspect grounded vessels, and look around. Thousands were killed when the first large crest arrived, and would-be rescuers were drowned by later crests, some of which exceeded 10 meters (33 feet) in height! In all, more than 50,000 people died in and around Lisbon.

There have been even greater tsunami disasters. In 1703, at Awa, Japan, 100,000 people died. On 27 August 1883, the explosive volcanic eruption of Krakatoa in Indonesia generated 35-meter (116-foot) waves that destroyed 163 villages and killed more than 36,000 people. Evidence of even larger waves has been discovered. Researchers have found signs of a huge wave, perhaps as high as 91 meters (300 feet), that crashed against the Texas coast 66 million years ago. It may have been caused by a comet or asteroid striking the Gulf of Mexico near Yucatán. The wave scoured the floor of the Gulf, picked up sand, gravel, and shark's teeth, and deposited the material in what is now central Texas.

The Tsunami Warning Network

Since 1948, an international tsunami warning network has been in operation around the seismically active Pacific to alert coastal residents to possible danger. Warnings must be issued rapidly because of the speed of these waves through the water. Telephone books in coastal Hawaiian towns contain maps and evacuation instructions for use when the warning siren sounds.

The tsunami warning system was responsible for averting the loss of many lives after the great 27 March 1964 earthquake in Alaska (see Chapter 3). A 3.7-meter (12-foot) wave, probably the fourth crest to reach the coast, swept into Crescent City, California, about six hours after the quake. Though more than 300 buildings were destroyed or damaged, five gasoline storage tanks set ablaze, and 27 blocks of the city demolished, there were relatively few casualties. But sometimes there is not enough time to provide a warning. In July 1993 an earthquake in the Sea of Japan generated one of the largest tsunami ever to strike Japan. Waves washed over areas 30 meters (97 feet) above sea level; 120 people were drowned or crushed by shattered buildings.

CHAPTER SUMMARY

Waves transmit energy, not water mass, across the ocean's surface. The speed of ocean waves usually depends on their wavelength, with long waves moving fastest. Arranged from short to long wavelengths (and therefore from slowest to fastest), ocean waves are generated by very small disturbances (capillary waves), wind (wind waves), rocking of water in enclosed spaces (seiches), seismic and volcanic activity or other sudden displacements (tsunami), and gravitational attraction (tides). The behavior of waves depends largely on the relation between a wave's size and the depth of water through which it is moving. Waves can refract and reflect, break, and interfere with one another

Unlike wind waves, the waves of very long wavelengths are always in "shallow water" (water less than half their wavelength deep). These long waves travel at high speeds. Some of the waves can be destructive, but their ability to cause damage is fortunately not proportional to their wavelengths. The waves with the longest wavelengths of all—the tides—are covered in the next chapter.

Terms and Concepts to Remember

capillary wave
constructive
  interference
deep-water wave
destructive
  interference
disturbing force
fetch
frequency
fully developed sea
gravity wave
group velocity
interference
intermediate-depth
  water wave
node
orbit

orbital wave
plunging wave
progressive wave
restoring force
rogue wave
sea
seiche
seismic sea wave
shallow-water
  wave
spilling wave
standing wave
storm surge
surf
surf zone
surging wave
swell

tsunami
$V = L/T$
$V = \sqrt{gd}$
wave
wave crest
wave height
wave period
wave refraction
wave steepness
wave train
wave trough
wavelength
wind duration
wind strength
wind wave

Study Questions

1. How is an ocean wave different from a wave in a spring or rope? How is it similar? How does it relate to a "stadium wave"—a waveform made by sports fans in a circular arena?

2. Draw a deep-water progressive ocean wave and label its parts. Show the orbits of water particles. Include a definition of *wave period*. How would you measure wave frequency?

3. What factors influence the growth of a wind wave? What is a fully developed sea? Where would we regularly expect to find the largest waves? Are waves in fully developed seas always huge?

4. What happens when a wind wave breaks? What factors affect the break? How are plunging waves different from spilling waves?

5. Though they move through all the ocean, seiches and tsunami are referred to as intermediate-depth waves or shallow-water waves rather than deep-water waves. How can this be?

6. What causes tsunami? Are all seismic sea waves tsunami? Are all tsunami seismic sea waves? How fast do tsunami travel? Do they move in the same way or at the same speed in a confining bay as they would in the open ocean?

7. Are tsunami ever dangerous if encountered in the open sea? What happens when they reach shore?

For Further Study

Bascom, W. 1959. "Ocean Waves." *Scientific American*, August, 89–97. A classic reference on the subject, well known to every oceanography professor and almost every student. Excellent illustrations.

Bascom, W. 1980. *Waves and Beaches*. Rev. ed. New York: Anchor/ Doubleday. One of the best general references available. Clear and lively writing.

Dudley, W. C., and M. Lee. 1988. *Tsunami!* Honolulu: University of Hawaii Press. Combines first-hand accounts of survivors with scientific information on tsunami.

Kampion, D. 1989. *The Book of Waves*. Santa Barbara, CA: Arpel Graphics. The definitive coffee-table book of waves. Magnificent photography.

Shepard, F. P. 1969. *The Earth Beneath the Sea*. New York: Atheneum. A widely admired and wonderfully written book for the general reader. Includes an eyewitness account of a tsunami.

Sverdrup, H., et al. 1942. *The Oceans: Their Physics, Chemistry, and General Biology*. Englewood Cliffs, NJ: Prentice-Hall. Thorough mathematical account of waves.

10 TIDES

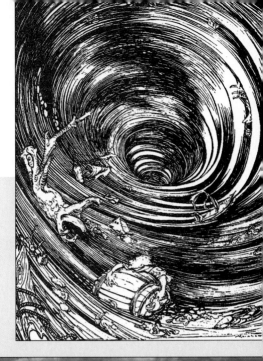

Maelstrom

Along some coasts the rhythmic daily rise and fall of tides is barely observable; in others it is impossible to miss. When geological features restrict the free flow of tidal water, the force of the tide can generate a rushing current and occasionally violent whirlpools.

The American writer Edgar Allan Poe wrote about a whirlpool called the maelstrom, which lies among the southern Lofoten Islands off Norway's west coast. The rapid spinning of water in the maelstrom raises its outer edge and depresses the central core. Here is Poe's description, from his 1841 story, *A Descent into the Maelstrom*:

> The edge of the whirl was represented by a broad belt of gleaming spray; but no particle of this slipped into the mouth of the terrific funnel, whose interior, as far as the eye could fathom it, was a smooth, shining, and jet-black wall of water, inclining to the horizontal at an angle of some forty-five degrees, speeding dizzily round and round with swaying and sweltering motion, and sending forth to the winds an appalling voice, half shriek, half roar, such as not even the mighty cataract of Niagara ever lifts up in its agony to Heaven. . . .

Do whirlpools genuinely inspire the dread that Poe's character expressed, or is their reputation largely a figment of artistic imagination (see **Figure a**)?

Whirlpools arise in shallow, restricted straits connecting two large bodies of deep water. They are caused by the tides. If the large bodies of water move to different tidal cycles, high tide in one will not occur at the same time as high tide in the other. Turbulence develops when one mass of water tries to pass the other during changes in tide. Under ideal conditions, bottom contours and the rush of opposing tidal currents cause seawater in the confined area to spin vigorously. The larger the opposing bodies of water and the smaller the passage through which the confined water must pass, the greater the vortex caused by the tidal currents.

Maelstrom is a Norwegian word derived from the Dutch *malen* (to grind in a circle, as a millstone grinds

a Maelstrom: fiction. A representation by the early twentieth-century English illustrator Arthur Rackham.

b Maelstrom: fact. This maelstrom, the world's strongest, is located 33 kilometers (21 miles) north of the city of Bodø, Norway.

grain) and *strøm*, or stream. Today's maelstrom is a heaving mass of active tidal water connecting Vestfjord and the Norwegian Sea between the islands of Moskenesøya and Mosken. Look for the area on a chart at 67°48′N, 12°50′E, about 165 kilometers (100 miles) north of the Arctic Circle. Currents in the passage can reach a speed of 13 kilometers (8 miles) per hour when the tides change, and strong local winds make the passage even more dangerous. The shallow "saddle" separating the fjord and the sea is only about 20 meters (70 feet) deep, and water rushing over its rocky bottom is driven to spin.

But is the chaos as horrifying as Poe suggested? Well, no. Fishermen avoid the strait during the times of maximum tidal difference, but as can be seen in **Figure b**, the sucking, roaring, cavernous maw of the maelstrom is more fiction than fact.

Tides are periodic short-term changes in the height of the ocean surface at a particular place, caused by a combination of the gravitational force of the moon and sun and the motion of the Earth. With a wavelength that can equal half Earth's circumference, tides are the longest of all waves. Unlike the other waves we have met, these huge shallow-water waves are never free of the forces that cause them, and so they act in unusual but generally predictable ways.

Around 300 B.C. the Greek navigator and astronomer Pytheas first wrote of the connection between the position of the moon and the height of a tide, but full understanding of tides had to await Newton's analysis of gravitation. We now know that the main cause of tides is the gravity of the moon acting on the ocean.

TIDAL BULGES

The pull of gravity between two bodies is proportional to the masses of the bodies and inversely proportional to the square of the distance between them. This means that gravitational attraction drops off quickly with distance. Gravity tends to pull the Earth and moon together, but inertia—the tendency of moving objects to continue in a straight line—keeps them apart. The Earth and the moon don't smash into each other (or fly apart) because they are in a stable orbit; their mutual gravitational attraction is exactly offset by their inertia. They rotate around a common center of mass.

Though their sums are equal and opposite, the inward pull of gravity and the outward-moving tendency of inertia don't always act in exactly the same balanced way on each particle of Earth and moon. In **Figure 10.1**, the five numbered points represent particles at different places on Earth. Notice that points 2 and 4 are in line with the moon. Point 2 is closer to the moon, however; so gravitational attraction (arrow labeled F_g) slightly exceeds the outward-moving tendency of inertia (arrow labeled F_i). Water there tends to be pulled toward the moon. At point 4, farther from the moon, inertia exceeds gravitational force; so water there tends to be flung away from the moon. The moon is equally far from points 1 and 3; so inertia and gravitational force are balanced, but gravity and inertia at those points do not pull in exactly opposite directions. The resultant forces (represented by dashed arrows that show their direction and strength) tend to squeeze the Earth at points 1 and 3 and stretch it at points 2 and 4. Only at point 5 are the inward pull of gravity and the outward-moving tendency of inertia exactly equal and opposite. Note that point 5 is not located at Earth's center. This is because the moon does not really revolve around the Earth; both the Earth and moon revolve around a common center of mass located 4,671 kilometers (2,903 miles)

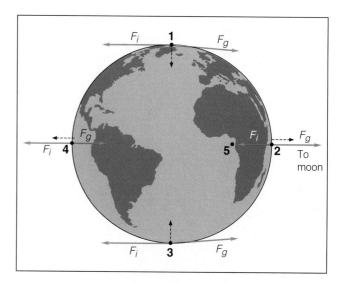

Figure 10.1 The actions of gravity and inertia on particles at five different locations on the Earth. The arrows labeled F_i represent the outward-moving tendency of inertia, and those labeled F_g represent the force of gravity. (See text for details.)

from Earth's center, about 1,700 kilometers (1,060 miles) inside the Earth.

The solid Earth cannot move much in response to these forces, but the fluid atmosphere and ocean can.[1] We don't notice the changes in the height of the atmosphere, but changes in water level are visible to coastal observers. Water is pulled away from points 1 and 3 and toward points 2 and 4. The moon's gravity pulls water toward point 2, and inertia flings water toward point 4, so that those areas "bulge" with water. **Figure 10.2** shows the result (greatly exaggerated in the figure). Bulges have formed beneath the moon's position relative to Earth—at a point on Earth directly opposite that position (though you can't see it in the figure). The bulges are the crests of the planet-sized waves that cause **high tides**. **Low tides** correspond to the troughs. The wave crests and troughs that cause high and low tides are actually very small: A 2-meter (7-foot) rise or fall in sea level is insignificant in comparison to the ocean's great size. The bulges (the tide wave crests) travel about 1,600 kilometers (about 1,000 miles) per hour at the equator in an attempt to keep up

---

[1]The Earth isn't stiff enough to resist the tidal pulls of the moon and sun. Bulges occur in the solid Earth just as they appear in the ocean or the atmosphere. The crests (bulges) of the Earth are, of course, much smaller—25 to 30 centimeters (10 to 12 inches) is about average. They pass unnoticed beneath us twice a day. On the other hand, tidal variability in the height of the atmosphere has been measured in miles.

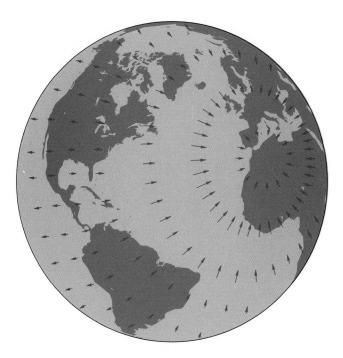

Figure 10.2 The high tide bulges. A bulge of water forms below the moon's position over the Earth as a result of the moon's gravitational pull, and a corresponding bulge forms on the opposite side of the Earth because of inertia.

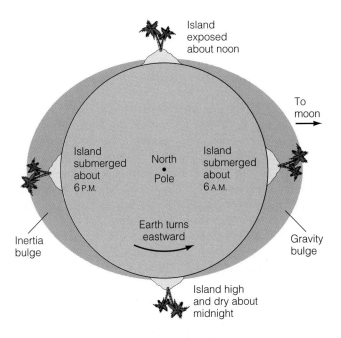

Figure 10.3 How the Earth's rotation beneath the tidal bulges produces high and low tides. Notice that the tidal cycle is 24 hours 50 minutes long because the moon rises 50 minutes later each day.

with the moon. Theoretically, the wavelength of these tide waves is as long as 20,000 kilometers (12,500 miles)!

The bulges tend to stay aligned with the moon as the Earth spins around its north-south axis. **Figure 10.3** shows this situation as seen from above the North Pole. As the Earth turns eastward, an island on the equator is seen to move in and out of these bulges through one rotation (one day). Starting at 0000 (midnight) we see the island in shallow water at low tide. Around six hours later, at 0613 (6:13 A.M.), the island is submerged in the lunar bulge at high tide. At 1226 (about noon) the island is within the tide wave trough at low tide. At 1838 (6:38 P.M.) the island is again submerged, this time in the opposite crest caused by inertia. About an hour after midnight on the next day, the island is back in shallow water, where it began. The key to understanding tides is to see Earth turning beneath these bulges.

There are complications, of course. For example, the **lunar tides**, tides caused by gravitational and inertial interaction of moon and Earth, complete their cycle in a tidal day (also called a lunar day.) A complete tidal day is 24 hours 50 minutes long because the moon, which exerts the greatest tidal influence, rises 50 minutes later each day. Thus, the highest tide also arrives 50 minutes later each day.

As shown in **Figure 10.4**, another complication arises from the fact that the moon does not stay right over the

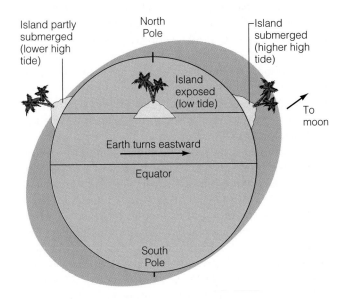

Figure 10.4 How the changing position of the moon relative to the Earth produces higher and lower high tides.

equator but, each month, moves from a position as high as 28°30' above Earth's equator to 28°30' below. When the moon is 28°30' north of the equator, an island at 28°30' north latitude will pass through the bulge on one side of the Earth but miss the bulge on the other side. During one day the island passes through a very high tide, a low tide, a lower high tide, and another low tide.

The Sun's Role

The sun's gravity also attracts particles on Earth. Remember, closeness counts for much in determining the strength of gravitational attraction. The sun is about 27 million times more massive than the moon, but the moon is about 387 times closer than the sun; so the sun's influence on the tides is only 46% that of the moon's. Smaller solar bulges tend to follow the sun through the day. These are the **solar tides**, tides caused by the gravitational and inertial interaction of the sun and Earth.

Like the moon, the sun also appears to move above and below the equator (23°27′N to 23°27′S, as you may recall from Chapter 7); so the position of the solar bulges varies like that of the lunar bulges. The Earth revolves around the sun only once a year, however; so the position of the solar bulges above or below the equator changes much more slowly than the position of the lunar bulges.

Sun and Moon Together

The ocean responds simultaneously to inertia and to the gravitational force of both sun and moon. If Earth, moon, and sun are all in a line (as shown in **Figure 10.5a**), the lunar and solar tides will be additive and extreme. This results in higher high tides and lower low tides. But if the Earth, moon, and sun form a right angle (as shown in **Figure 10.5b**), the solar tide will tend to diminish the lunar tide. Because the moon's contribution is more than twice that of the sun, however, the solar tide will not completely cancel the lunar tide.

The large tides caused by the linear alignment of sun, Earth, and moon are called **spring tides** (*springen* = to move quickly); these times of very high high tides and very low low tides occur at two-week intervals, corresponding to the new and full moons. (Please note that spring tides don't happen only in the spring of the year!) **Neap tides** (*naepa* = hardly disturbed) alternate at two-week intervals; they occur when the Earth, moon, and sun form a right angle. During times of neap tides, high tides are not very high and low tides are not very low. **Figure 10.6** plots tides at two coastal sites through spring and neap cycles.

Because their orbits are not perfect circles, the moon and the sun are closer to the Earth at some times than at others. The difference is significant—some 25,000 kilometers (16,000 miles) in the moon's case. Because the intensity of gravitational attraction is inversely proportional to the square of the distance between the bodies, the closer moon raises a higher tidal crest. If the moon

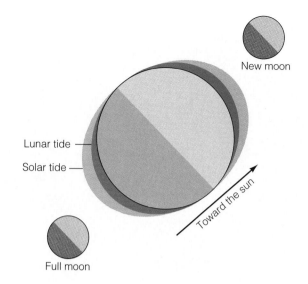

a Spring tides

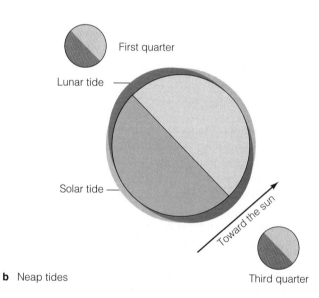

b Neap tides

Figure 10.5 Relative positions of the sun, moon, and Earth during spring and neap tides. (a) Spring tides: At the new and full moons, the solar and lunar tides reinforce each other, making the highest high and lowest low tides. (b) Neap tides: At the first and third quarter moons, the sun, Earth, and moon form a right angle, creating the lowest high and the highest low tides.

and sun are over nearly the same latitude, and if the Earth is also close to the sun, extreme tides will result.

TIDAL DATUM, TIDAL RANGE

The reference level to which tidal height is compared is called **tidal datum**. Tidal datum is the zero point seen in

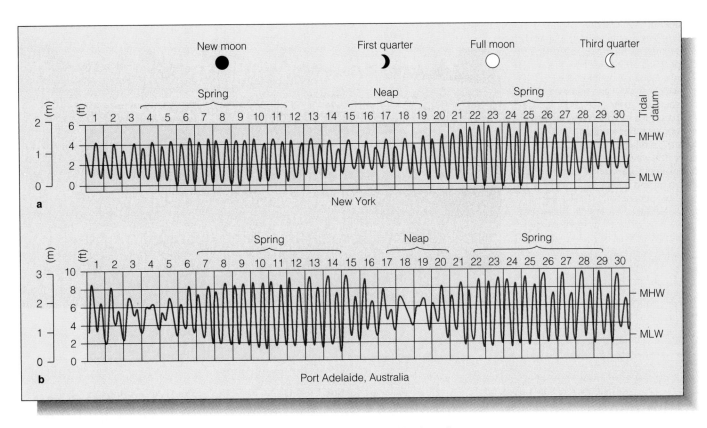

Figure 10.6 Tidal records for a typical month at (a) New York and (b) Port Adelaide, Australia. Note the relationship of spring and neap tides to the phases of the moon. MHW = mean high water; MLW = mean low water.

tide graphs such as Figures 10.6 and 10.8. This reference plane is not always set at **mean sea level**, the height of the ocean surface averaged over a few years' time. For the U.S. West Coast, tidal datum is *mean lower low water* (MLLW), an average of the *lower* of the two low tides arriving on shore most days. The U.S. East and Gulf coasts experience a different tidal pattern, and tidal datum there is usually *mean low water* (MLW). Some countries use other measures to set the datum.

The **tidal range** (high-to-low-water height difference) varies with basin configuration. In small areas such as lakes, the tidal range is very small. In larger enclosed areas, such as the Baltic or Mediterranean seas, tidal range is also moderate. Tidal range is not the same over a whole ocean basin; it varies from the coast to the centers of oceans. The largest tidal ranges occur at the edges of the largest ocean basins, especially in bays or inlets that concentrate tidal energy because of their shape. Tidal range at the apex of a funnel-shaped sea, gulf, or bay can often be extreme. In the Bay of Fundy, near Moncton, New Brunswick (Canada), tidal range is especially wide: up to 15 meters (50 feet) from highs to lows! The northern reaches of the Sea of Cortez east of Baja California have a tidal range of about 9 meters (30 feet). Tide waves

sweeping toward the narrow southern end of the North Sea can build to great heights along the southeast coast of England and the north coast of France.

A **tidal bore** (*bara* = wave) will form in some inlets (and their associated rivers) exposed to great tidal fluctuation. Here, at last, is a true **tidal wave**—a steep wave moving upstream generated by the action of the tide crest in the enclosed area of the river mouth (see **Figure 10.7**). Bores may be up to 3 meters (10 feet) high and move from 5 to 7.5 meters per second (11 to 17 miles per hour). Their potential danger is lessened by their predictability. Accurately predicting the arrival of tidal bores is essential to safe navigation. In addition to the Bay of Fundy, tidal bores are common in the Amazon, in the Ganges Delta, and on England's Severn River. Some rivers in southern China may have three or four simultaneous bores at different places along the river's length.

TIDAL PATTERNS

The wavelength of tides rarely reaches the theoretical maximum of half the circumference of the Earth. Land masses obstruct the tidal crests, diverting, slowing, and otherwise

Figure 10.7 A tidal bore on the Severn River in southwestern England. A tide crest sweeping up the funnel-shaped Bristol Channel from the Atlantic Ocean encounters the increasingly shallow and narrow river passage and rushes forward to cause the bore. Like all tidal bores, the Severn bore reaches its greatest height during the days surrounding spring tides. The waves that form the bore in this example are nearly 1 meter (3.3 feet) high and are traveling at about 5 meters per second (11 miles per hour). Note that the smooth waves are followed by turbulent water also moving rapidly upriver.

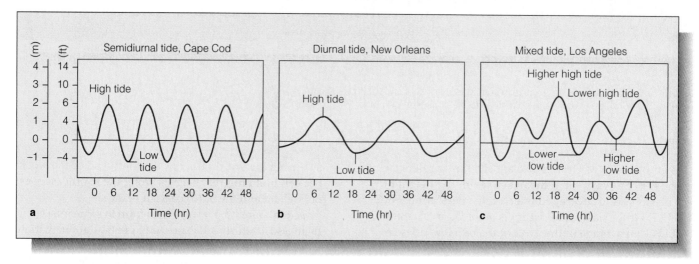

Figure 10.8 Tide curves for the three common types of tides. (a) A semidiurnal tide pattern at Cape Cod, Massachusetts. (b) A diurnal tide pattern at New Orleans, Louisiana. (c) A mixed tide pattern at Los Angeles, California.

complicating their movements. This interference produces different patterns in the arrival of tide crests at different places. In some places tides disappear altogether.

The shape of the basin itself has a strong influence on the patterns and heights of tides. As we have seen, water in large basins can rock rhythmically back and forth in seiches. Tide crests can stimulate this oscillation, and the configuration of coasts around a basin can alter its rhythm. Because tides are shallow-water waves, frictional forces also play an important role. For these and other reasons, some coastlines experience **semidiurnal** (twice daily) tides: two high tides and two low tides of nearly equal level each lunar day. Others have **diurnal** (daily) tides: one high and one low. The tidal pattern is called **mixed** if successive high tides or low tides are of significantly different heights through the cycle.

Figure 10.8 gives an example of each tidal pattern. At Cape Cod two tide crests arrive per lunar day, a semidiurnal pattern (**Figure 10.8a**). The natural tendency of water in an enclosed ocean basin to rock at a specific frequency modifies the tidal pattern in the Gulf of Mexico; so New Orleans sees one crest per lunar day, a diurnal pattern (**Figure 10.8b**). The West Coast of the United States has a mixed tidal pattern: often a *higher* high tide,

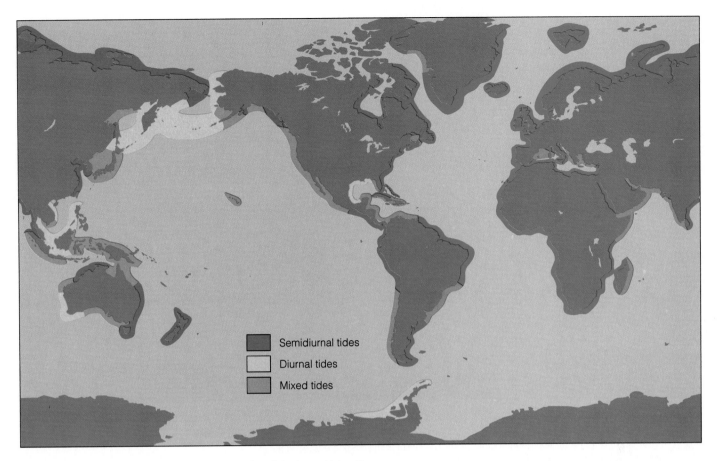

Figure 10.9 The worldwide geographic distribution of the three tidal patterns.

followed by a *lower* low tide, a *lower* high tide, and a *higher* low tide each lunar day. (That's not as confusing as it first seems—see **Figure 10.8c**.)

The Pacific has a unique pattern of diurnal, semidiurnal, and mixed tides. As **Figure 10.9** shows, it has the most complex of all tidal patterns. The east coast of Australia, all of New Zealand, and much of the west coasts of Central and South America have a semidiurnal tidal pattern. The Aleutians have diurnal tides. The Pacific coasts of North America and some of South America have mixed tides. Why the differences?

Remember the surface of Lake Geneva in our discussion of seiches? The water level at the center of the lake remains at the same height, while water at the ends rises and falls (see again Figure 9.20). The node at the center of the lake has an analog in ocean basins called an **amphidromic point** (*amphi* = around + *dromas* = running). An amphidromic point is a no-tide point in the ocean, around which the tide crest rotates through one tidal cycle. Because of the shape and placement of land masses around ocean basins, the tide crests and troughs cancel each other at these points. The crests sweep around amphidromic points like wheel spokes from a rotating hub,

radiating crests toward distant shores. The waves are influenced by the Coriolis effect because a large volume of water moves with the tide wave. The tide waves move counterclockwise around the amphidromic point in the Northern Hemisphere, and clockwise in the Southern Hemisphere. Tide ranges increase with distance from an amphidromic point.

Why do tides move counterclockwise in the Northern Hemisphere? Imagine Lake Geneva again. The long axis of the lake stretches east and west. Because of the Coriolis effect, water moving east at the center of the lake is deflected slightly to the right (to the south). If the lake were larger and the flow of water greater (and thus the Coriolis effect stronger), water would hug the southern shore as it traveled eastward. When the water begins to rock the other way—back to the west—the Coriolis effect would move the water to the right toward (and along) the northern shore. Note that the *overall* movement of water would be counterclockwise. Water moving in a tide wave tends to stay to the right of an ocean basin for the same reason. As water moves north in a Northern Hemisphere ocean, it moves toward the eastern boundary of the basin; as it moves south, it moves toward the western

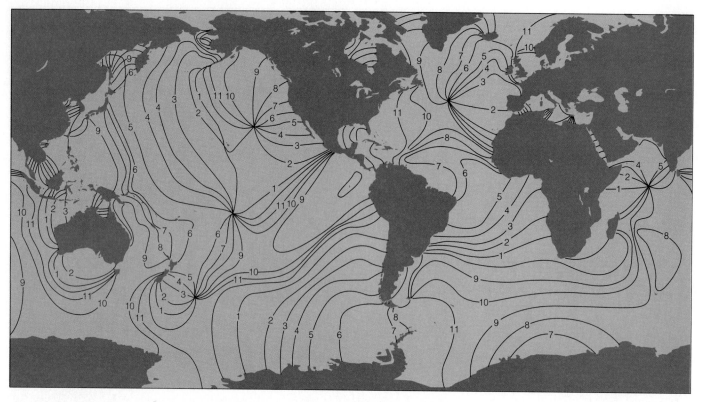

Figure 10.10 Amphidromic points in the world ocean. Tidal ranges generally increase with increasing distance from amphidromic points. Lines radiating from the points indicate tide waves moving around these points, counterclockwise in the Northern Hemisphere and clockwise in the Southern Hemisphere. (See text for details.)

boundary. A wave crest moving counterclockwise will develop around a node if this motion continues to be stimulated by tidal forces.

About a dozen amphidromic points exist in the world ocean; **Figure 10.10** shows their locations. Notice the complexity of the Pacific, which contains five such points. It's no wonder that the arrival of tide wave crests at the Pacific's edge produces such a complex mixture of tide patterns, depending on shoreline location.

OTHER ASPECTS OF TIDES

Tidal Currents

The rise or fall in sea level as a tide crest approaches and passes will cause a **tidal current** of water to flow into or out of bays and harbors. Water rushing into an enclosed area because of the rise in sea level as a tide crest approaches is called a **flood current**. Water rushing out because of the fall in sea level as the tide trough approaches is called an **ebb current**. (The terms *ebb tide* and *flood tide* have no technical meaning.)

Anyone who has stood at the narrow mouth of a large

bay or harbor cannot help but be impressed with the speed and volume of the tidal current that occurs between tidal extremes. Midway between high and low spring tides, the ebb current rushing from San Francisco Bay strikes the base of the south tower of the Golden Gate Bridge with such force that a bow wave is formed, giving the illusion that the bridge itself is moving rapidly. Tidal currents at the Golden Gate can reach 3 meters per second (about 7 miles per hour) because of the volume of enclosed water and the narrowness of the channel through which it must escape. Navigators must know the times of tidal currents to safely negotiate any harbor entrance or other narrow strait. In some places, this knowledge may save their lives.

Tidal currents become more complex in the open sea. One's position relative to an amphidromic point, the shape of the basin, and the magnitude of gravitational forces and inertia must all be considered to calculate the speed and direction of tidal currents over a deep bottom. The velocity of tidal currents is less in the open sea because the water is not confined as it is in a harbor. Open-sea tidal currents have been measured at a few centimeters per second, and their velocity tends to decrease with depth.

Figure 10.11 Tidal power installation at the Rance estuary in western France.

Tidal Friction

The daily rise and fall of the tides consumes a very large amount of energy, energy ultimately dissipated as heat. Most of this energy comes directly from the rotation of the Earth itself, and tidal friction is gradually slowing Earth's rotation by a few hundredths of a second per century. Even such a small change has long-term planetary effects. Geologists studying the daily growth rings of fossil corals and clams estimate that the length of the day has grown longer; so the number of days in a year has become less as planetary rotation has slowed. Evidence suggests that 350 million years ago a year contained between 400 and 410 days, with each day being about 22 hours long; and 280 million years ago, there were about 390 22½-hour days in a year.

Tidal friction affects other bodies. Tidal forces have locked the rotation of the moon to that of the Earth. As a result, the same side of the moon is always facing the Earth, and a day on the moon is a month long.

Predicting Tides

There are at least 150 tide-generating and tide-altering forces and factors. The seven most important of these—such as the position of the moon and sun, their distance, the rocking tendency of the basin in question, and so on—must be considered if one wishes to predict tides mathematically. But the interactions of all the forces and factors are so complex that if a previously unknown continent were discovered on Earth, the coastal tide times and ranges on its shores could not be accurately predicted.

It is the study of past records that allows tide tables to be projected into the future. This experience permits prediction of tidal height to an accuracy of about 3 centimeters (1.2 inches) for years in advance.

Even so, extraneous factors can affect the estimates. For example, the arrival of a storm surge will greatly affect the height or timing of a tide—as will gentle, atmospherically induced seiching of the basin or excitement of large-scale resonances by a tsunami. Even a strong, steady onshore or offshore wind will affect tidal height and the arrival time of the crest. Weather-related alterations are sometimes called **meteorological tides** after their origin.

TIDES AND ELECTRICITY GENERATION

Humans have found ways to use the tides. Ships sail to sea and return to port with the tides. Intentional grounding of a ship with the fall of a tide can provide a convenient, if temporary, drydock. To these traditional uses has been added a potential alternate to our growing dependence on fossil fuels: taking advantage of trapped high-tide water to generate electricity.

Tidal power is the only marine energy source that has been successfully exploited on a large scale. The first major tidal power station was opened in 1966 in France, on the estuary of the river Rance (**Figure 10.11**), where tidal range reached a maximum of 13.4 meters (44 feet). Built at a cost of $75 million, this 850-meter- (2,800-foot-) long dam contains 24 turbo-alternators, capable of generating 544 million kilowatt-hours of electricity annually.

At high tide, seawater flows from the ocean through the generators into the estuary. At low tide the seawater and river water from the estuary flow out through the same generators. Power is generated in both directions. A similar installation has recently started generating power at Passamaquoddy Bay, a part of the Bay of Fundy between Maine and New Brunswick, Canada.

Tidal power has many advantages: operating costs are low, the source of power is free, and no carbon dioxide or other pollutants are added to the atmosphere. But even if tidal power stations were built at every appropriate site worldwide, the power generated would amount to less than 1% of current world needs. And, of course, this method of power generation is not free of trade-offs. The dam and electrical generators can be damaged by storms, and the large, intricate metal valves and vanes at the heart of the plant are easily corroded by seawater. Computer simulations have suggested that installing a dam would change the resonance modes of a bay or estuary and therefore the height of the tide wave. Studies also suggest that sensitive planktonic and benthic marine life would be disrupted and even that increased tidal friction would cause a minute decrease in the rate of Earth's rotation.

We humans are not alone in benefiting from the tides.

The daily ebb and flood sweeps pollutants from the shallows, moves the juvenile stages of animals from intertidal nurseries to the deeper ocean, and generates currents that mix nutrients and distribute sediments. But we may owe a greater debt to tides than we realize. By periodically exposing marine organisms to drying, the rhythmic rise and fall of the tides may have played a crucial role in life's colonization of the dry land.

CHAPTER SUMMARY

Tides have the longest wavelengths of the ocean's waves. They are caused by a combination of the gravitational force of the moon and the sun, the motion of the Earth, and the tendency of water in enclosed ocean basins to rock at a specific frequency. Unlike the other waves, these huge shallow-water waves are never free of the forces that cause them; so they act in unusual but generally predictable ways. Basin resonances and other factors combine to cause different tidal patterns on different coasts. The rise and fall of the tides can be used to generate electrical power, and they are important in many physical and biological coastal processes.

Terms and Concepts to Remember

amphidromic point	mean sea level	tidal bore
diurnal tide	meteorological tide	tidal current
ebb current	mixed tide	tidal datum
flood current	neap tide	tidal range
high tide	semidiurnal tide	tidal wave
low tide	solar tide	tides
lunar tide	spring tide	

Study Questions

1. What causes the rise and fall of the tides? What celestial bodies are most important in determining tides?

2. What is a high tide and a low tide? A spring tide and a neap tide?

3. What are the most important factors influencing the heights and times of tides? What tidal patterns are observed? Are there tides in the open ocean? If so, how do they behave?

4. How are tides important?

5. From what you learned about tides in this chapter, where would you locate a plant that generated electricity from tidal power? What would be some advantages and disadvantages of using tides as an energy source?

For Further Study

Bowditch, N. 1966. *American Practical Navigator*. Washington, DC: U.S. Naval Hydrographic Office. A one-volume nautical reference of unexcelled brevity and precision. The chapters on waves are clear and direct.

Defant, A. 1958. *Ebb and Flow: The Tides of Earth, Air and Water*. 1958. Ann Arbor: University of Michigan Press.

Von Arx, W. S. 1962. *An Introduction to Physical Oceanography*. New York: Addison-Wesley. Excellent technical discussion of tides, good illustrations.

11 COASTS

Dynamic, Changeable Places

Coasts are dynamic, changeable places, especially when viewed on a human scale. Henry David Thoreau wrote that, to the people of Massachusetts, the sea is their garden, and the dog that growls at their door is the Atlantic Ocean. Few stretches of offshore water are more dangerous and turbulent than the winter Atlantic off New England, and this particular dog has an expensive appetite for the same shores favored by thousands of tourists and longtime seasonal residents. The resort town of Scituate, south and east of Boston, illustrates the ephemeral nature of some coasts. In February 1978, a particularly strong Atlantic storm combined with exceptionally high tides wreaked havoc on the town's sandy shore. Waves washed over the coastal zone, heavily built with vacation homes and businesses. The private property loss from the 36-hour storm was great, but in this one storm the cost to public structures like seawalls, groins, and piers amounted to millions of dollars *more* than the value of the private property. In the 350 years since English settlers began recording the history of this stretch of shore, it has been inundated by storms and severely damaged at least 84 times. Since 1938 the Scituate seawall has been rebuilt 18 times at public expense.

The West Coast is also subject to coastal damage. An interesting example involved Ellen Scripps, heiress to a publishing fortune and benefactor of (among other good works) the land and first buildings of what is now the Scripps Institution of Oceanography. After World War I, Scripps developed land along the San Diego shore into Sunset Cliffs Park—a property landscaped with gazebos, walkways, and gardens—which she donated to the citizens of California. Unfortunately, the sandstone that formed the dramatic shore cliffs and adjacent park was so soft that a determined artist could scratch his initials into the rock using only his fingernails. The cliffs eroded at a rapid rate, and 55 years after its completion, almost nothing remained of the once-beautiful park. The coast in the area continues to erode, threatening a highway and a number of expensive homes.

Beach erosion has toppled this home on Nantucket Island off the coast of Massachusetts.

Coastal areas join land and sea. The place where ocean meets land is usually called the **shore**, while the term **coast** refers to the larger zone affected by the processes occurring at this boundary. A sandy beach might form the shore in an area, but the coast (or coastal zone) includes the marshes, sand dunes, and cliffs just inland of the beach, as well as the sand bars and troughs immediately offshore. The world ocean is bounded by about 440,000 kilometers (273,000 miles) of shore.

Because of its proximity to both ocean and land, a coast is subject to natural events and processes common to both realms. A coast is an active place. Here is the battleground on which wind waves break and expend their energy. Tides sweep water on and off the rim of land, rivers drop most of their sediments at the coasts, and ocean storms pound the continents. The *shape* of a coast is a product of many processes: uplift and subsidence, the wearing-down of land by erosion, and the redistribution of material by sediment transport and deposition. The *location* of a coast depends primarily on global tectonic activity.

CLASSIFYING COASTS

As we saw in Chapter 3, no area of geology was left undisturbed by the revelations of plate tectonics. In the 1960s, geologists began to classify coasts according to tectonic position. *Active* coasts, near the leading edge of moving continental masses, were found to be fundamentally different from the more *passive* shores near trailing edges. The ages and characteristics of coasts are better understood by taking plate movements into account, as are the coasts' composition and physical shapes. But the slow forces of plate movement are frequently obscured by the more rapid action of waves, by erosion of the land, and by the transport of sediments. Coasts are modified by small-scale (erosional/depositional) as well as large-scale (tectonic) processes.

Another factor that must be considered in the classification of coasts is long-term change in sea level. Five factors can cause sea level changes:

- The amount of water in the world ocean can vary. Sea level is lower during periods of global glaciation (ice ages) because there is less water in the ocean. Periods of abundant volcanic outgassing or an increase in the number of icy comets striking the Earth add water to the ocean and raise sea level.

- The volume of the ocean's "container" may vary. High rates of seafloor spreading are associated with the expansion in volume of the mid-ocean ridges. This displaces the ocean's water, which

climbs higher on the edges of the continents. Sediments shed by the continents during periods of rapid erosion can also decrease the volume of ocean basins and raise sea level.

- The water itself may occupy more or less volume as its temperature and salinity vary. During times of global warming, seawater expands and occupies more space, raising sea level.

These three factors are responsible for **eustatic change**—variations in sea level that can be measured all over the world ocean. Of course, the continents rarely stay still as sea level rises and falls. Local changes are bound to occur, and two other factors produce variations in *local* sea level:

- Tectonic motion and isostatic adjustment can change the shape of the boundary of the container.

- Wind and currents, seiches, storm surges, and other effects of water in motion can force water against a shore or draw it away.

Sea level has been at its current elevation (give or take 0.5 meter, 1.5 feet) for only about 3,500 years. Over the past million years, worldwide sea level has varied from about 120 meters (400 feet) below to about 10 meters (33 feet) above its present position. These changes in sea level have produced major differences in the position and nature of coastlines, especially in areas where the edge of the continent slopes gradually or where the coast is rising or sinking. **Figure 11.1** shows an estimate of previous shore positions along the U.S. East Coast in the geologically recent past, and a prediction for the future should the present warming trend cause the polar ice to melt more rapidly.

Perhaps the most useful scheme for classifying a coast is based on the predominant events occurring there: *erosion* (from *rodere* = to gnaw) and *deposition*. **Erosional coasts** are new coasts in which the dominant processes are those *removing* coastal material. **Depositional coasts** are those coasts that are *steady or growing* because of their rate of sediment accumulation. The material being deposited may be worn from a nearby erosional segment of coast, or, as is more likely, transported from adjacent land to the shore by the action of wind or rivers. The rocky shores of Maine are erosional because erosion exceeds deposition; the sandy coastline from New Jersey to Florida is typically depositional because deposits of sediment tend to protect the shore from new erosion. The rocky central California coast is erosional, while the broad beaches of southern California are depositional. About 30% of the U.S. coastline is depositional, 70% erosional. We will use the erosional-depositional classification scheme in the rest of this chapter.

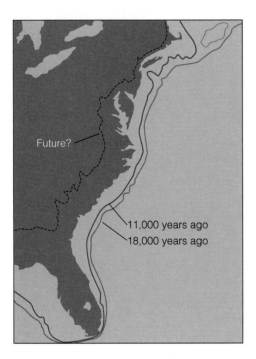

Figure 11.1 Because of changes in sea level within the last 18,000 years, the position of the coast in the eastern United States has been as much as 200 kilometers (125 miles) from the present shoreline, leaving much of the continental shelf exposed. In the future, if the ocean were to expand and the polar ice caps were to melt because of global warming, sea level could rise perhaps 100 meters (330 feet), driving the coast inland as much as 250 kilometers (160 miles).

Figure 11.2 Waves pound an exposed rocky coast.

EROSIONAL COASTS

Land erosion and marine erosion both work to modify the nature of a rocky coast. Erosional coasts are shaped and attacked from the land by stream erosion; the abrasion of wind-driven grit; the alternate freezing and thawing of water in rock cracks; the probing of plant roots; glacial activity; rainfall; dissolution by acids from soil; and slumping. As anyone who hikes in the mountains knows, these factors are not limited to coasts.

Attack from the sea is by waves and currents. The continual onslaught of waves does most of the work, with currents distributing the results of their labor. Some indication of the violence of this activity may be inferred from **Figure 11.2**.

Large storm waves routinely generate forces of 27,000 kilograms per square meter (25 tons per square yard). The crashing waves push air and water into tiny rock crevices. The repeated buildup and release of pressure within these crevices can weaken and fracture the rock. But it is not the hydraulic pressure of moving water alone that abrades the coasts; tiny pieces of sand, bits of gravel, or stones hurled by waves toward the shore are even more effective at erosion. **Dissolution**, the dissolving of minerals in the rocks by water, contributes to the erosion of easily soluble coastal rocks such as limestone. Even the burrowing and wearing activities of marine organisms have an effect.

The rate at which a shore erodes is proportional to the hardness and resistance of the material of which it is composed, the violence of the wave shock to which it is exposed, and the local range of tides. Hard rock resists wear. Coasts made of dense granite or basalt may retreat an insignificant amount over a human lifetime. The granite coast of Maine erodes only a few centimeters per decade. Coasts of soft sandstone or other weak (or soluble)

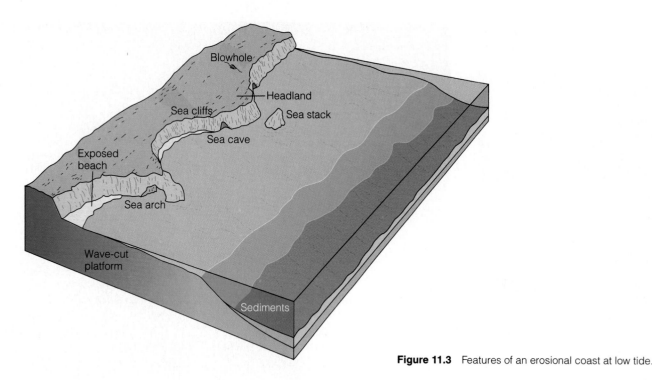

Figure 11.3 Features of an erosional coast at low tide.

materials, however, may disappear at a rate of a few meters per year. In an extreme case, some soft shore cliffs facing the North Sea in East Suffolk, England, eroded more than 11 meters (36 feet) during two hours of a severe storm!

Marine erosion is usually most rapid on **high-energy coasts**, areas frequently battered by large waves. High-energy coasts are most common adjacent to stormy ocean areas of great fetch and along the eastern edges of continents exposed to tropical storms. The coasts of Maine and British Columbia, and the southern tips of South America and South Africa, are typical high-energy coasts. **Low-energy coasts** are only infrequently attacked by large waves. Because of their generally protected location in the Gulf of Mexico, the U.S. Gulf states share a low-energy coast—at least between hurricanes!

Waves can affect the coast only where they strike; so erosional activity is concentrated near average sea level. A shore with little tidal variation erodes quickly because the wave action is concentrated near one level for longer times. Low-energy coasts protected by offshore islands usually erode slowly, as do areas below the low-tide line. Some erosion does occur below the surface because of the orbital motion of water in waves, but even the largest waves have little erosive effect at depths greater than about 15 meters (50 feet) below average sea level. Cliffs above shore are subject to pounding, either directly from waves or by rocks hurled by waves. In one case, windows of a Scottish lighthouse 100 meters (328 feet) above sea level were broken by wave-thrown rocks during an Atlantic winter storm.

Features of Erosional Coasts

Erosive forces can produce a wave-cut shore showing some or all of the features illustrated in **Figure 11.3**. Note the complex small-scale irregularities of this rocky erosional coastline. **Sea cliffs** fall abruptly from land into the ocean, their steepness usually resulting from the collapse of undercut notches. The position of the sea cliffs marks the shoreward limit of marine erosion on a coast. The parade of waves cuts **sea caves** into the cliffs at local zones of weakness in the rocks. Most sea caves are accessible only at low tide. A blowhole can form if erosion follows a zone of weakness upward to the top of the cliff. When the tide is at just the right height, spray can blast from the fissure as waves crash into the cliff. Offshore features of rocky coasts can include natural arches, stacks, and a smooth, level **wave-cut platform** just offshore, which marks the submerged limit of rapid marine erosion. Much of the debris removed from cliffs during the formation of these structures is deposited in the quieter water farther offshore, but some can rest at the bottom of the cliffs as exposed beaches. Indeed, broad beaches are sometimes found on erosional coasts.

Shore Straightening

The *first* effect of marine erosion on a newly exposed shore is to intensify the irregularity of the coastline. This happens because coastal rocks are usually not uniform in composition over long horizontal distances. Some hard

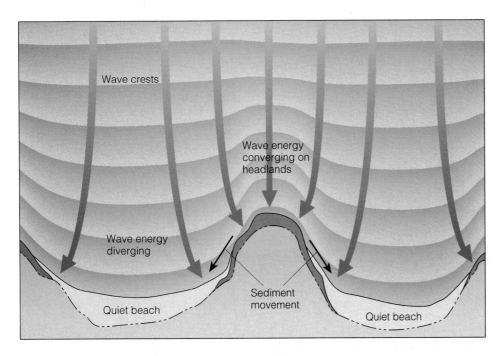

Figure 11.4 Wave energy converges on headlands and diverges in the adjoining bays. The accumulation of sediment in the tranquil bays eventually smoothes and straightens the contours of the shore.

rocks will resist erosion well, while softer rocks on the same coast may disappear almost overnight. (This explains the uneven character of the stacks, arches, and sea cliffs.)

Eventually, however, coastal erosion tends to produce a smooth shoreline. Wave energy is focused onto headlands and away from bays by wave refraction (**Figure 11.4**). Sediment eroded from the headlands tends to collect as beaches in the relatively calm bays. As erosion continues, the deposits may eventually protect the base of the shore cliffs from the waves. Coastal irregularities are thus smoothed with the passage of time. As you might expect, straightening occurs most rapidly on high-energy coasts.

Transition from Erosional to Depositional Shore

If the submerged slope of the seafloor is steep, the eroded sediments will quickly drain out to sea. If the slope is not too steep, sediments will be transported along the coast by wave and current action.

The net amount of sediment (usually sand) that moves along the coast, driven by wave action, is referred to as **longshore drift**. Longshore drift occurs in two ways: the wave-driven movement of sand along the exposed beach, and the current-driven movement of sand in the surf zone just offshore (**Figure 11.5**).

As we have seen, wind waves arriving from distant storms do not always approach the coast straight on.

Most wind waves approach at an angle, then refract in shallow water to break almost parallel to shore. Refraction is usually incomplete, however, and some angle remains when the waves break. If sediments have accumulated to form a beach, water from the breaking wave will rush up the beach at a slight angle, but return to the ocean by running straight downhill under the influence of gravity. The millions of sand grains disturbed by the wave will follow the water's path, moving up the beach at an angle but retreating down the beach straight down the slope. Net transport of the grains is *longshore*, parallel to the coast, *away* from the direction of the approaching waves. Net sand flow along the U.S. East and West coasts is usually to the south because the waves that drive the transport system usually approach from the north, where storms most commonly occur.

Sediment is also transported in the surf zone in a **longshore current**. The waves breaking at a slight angle distribute a portion of their energy away from their direction of approach. This energy propels a narrow current in which sediment already suspended by wave action can be transported downcoast. The speed of the longshore current sometimes reaches almost 4 kilometers (about 2.5 miles) per hour.

Sand moving in the wash of waves along the beach and sediments propelled in the longshore current just offshore are often joined by much greater loads of sediments brought to the coast by rivers. Net southward transport of all this material along the central California

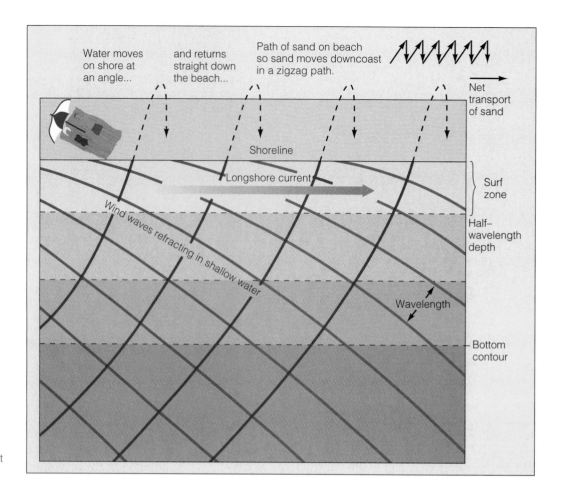

Figure 11.5 The wind wave refraction and longshore current components of longshore drift.

Within the figure:
- Water moves on shore at an angle...
- and returns straight down the beach...
- Path of sand on beach so sand moves downcoast in a zigzag path.
- Net transport of sand
- Shoreline
- Longshore current
- Surf zone
- Half-wavelength depth
- Wind waves refracting in shallow water
- Wavelength
- Bottom contour

coast exceeds 230,000 cubic meters (300,000 cubic yards) per year. Typical East Coast figures are about two-thirds of this value.

Accumulation and distribution of a layer of protective sediments along a coast can insulate that coast from rapid erosion because wave energy expended in churning overlying sediment particles cannot erode the underlying rock. Thus, with time, erosional shorelines can evolve into depositional ones. Unless the coast is rapidly rising or sinking, or unless other large-scale geological processes interfere, the inexorable process of erosion will tend to change the character of any coast from erosional to depositional.

DEPOSITIONAL COASTS

The features found on depositional coasts are usually composed of sediments rather than solid rock. As noted above, most sediment reaches the coast in rivers. Not all the incoming sediment joins the longshore drift; some of the fine particles moved by rivers will stay in suspension

long enough to be transported to the outer continental shelf and beyond. If the rate of deposition of larger particles exceeds the ability of the longshore transport system to remove and distribute the material along the coast, the sediments may build up at the river mouth to form a fan called a **delta**. The term, first used by Herodotus in the fifth century B.C. to describe the triangular mass of sediment at the terminus of the Nile, is derived from the triangular shape of the capital Greek letter delta (Δ). Deltas are most common on the low-energy shores of enclosed seas, where the tidal range is not extreme. Two very different deltas are shown in **Figure 11.6**.

Other sources of sediments for depositional coasts include material deposited by glaciers, dust and sand driven to the shore by wind, pulverized lava and volcanic ash, and the skeletal remains of tiny marine organisms.

BEACHES

The most familiar feature of a depositional coast is the beach. A **beach** is a zone of unconsolidated (loose) parti-

a

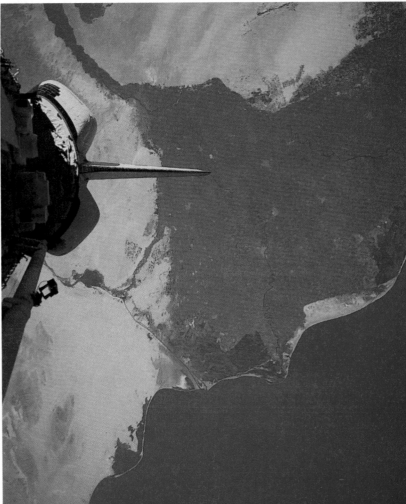

b

Figure 11.6 River deltas form where sediment-rich rivers enter enclosed or semienclosed seas where wave energy is limited. (a) The "bird foot" shape of the Mississippi Delta is clearly seen in this photograph. Submerged overlapping lobes of the Mississippi Delta extend completely across a wide continental shelf, and sediment from the river is now being deposited directly on the continental slope. Lobed and "bird-foot" deltas form where depositional processes overwhelm the processes of erosion and sediment transportation. (b) The Nile Delta, which has been smoothed into a more rounded shape by erosion and sediment transportation. The tail of the space shuttle can be seen in the upper left of the photo.

cles that covers part or all of a shore. The landward limit of a beach may be vegetation, a sea cliff, relatively permanent sand dunes, or construction such as a seawall. The seaward limit occurs where sediment movement on- and offshore ceases—a depth of about 10 meters (33 feet) at low tide. The United States has 17,762 kilometers (10,983 miles) of beaches, about 30% of the country's total shoreline. Beaches result when sediment, usually sand, is transported to places suitable for deposition. Such places include the calm spots between headlands, shores sheltered by offshore islands, and regions with usually quiet surf. Sometimes the sediment is transported a very short distance; grit may simply fall from the cliff above and accumulate at the shoreline. More often, however, the sediment on a beach has been moved for long distances to its present location.

Wherever they are found, beaches are in a constant state of change. They may be thought of as rivers of sand—zones of continuous sediment transport.

The Composition and Slope of Beaches

The material of beaches can range from boulders, through cobbles and pebbles and gravel, to very fine silt. The rare black-sand beaches of Hawaii are made of finely fragmented lava. Some beaches consist of shells and shell debris, or fragments of coral. Unfortunately, some also include large quantities of human junk—glass or metal or plastic beaches are not unknown. Cobble beaches can be very steep (occasionally with slopes in excess of 20°), but wide beaches of fine sand are sometimes flatter than parking lots.

Table 11.1	The Relationship Between the Particle Size of Beach Material and the Average Slope of the Beach	
Type of Beach Material	Size (mm)	Average Slope of Beach
Very fine sand	.0625–.125	1°
Fine sand	.125–.25	3°
Medium sand	.25–.50	5°
Coarse sand	.50–1.0	7°
Very coarse sand	1–2	9°
Granules	2–4	11°
Pebbles	4–64	17°
Cobbles	64–256	24°

Source: Shepard, 1973.

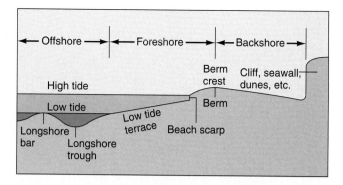

Figure 11.7 A typical beach profile.

In general, the flatter the beach, the finer the material from which it is made (**Table 11.1**). The relation between particle size and beach slope depends on wave energy, particle shape, and the porosity of the packed sediments. Water from waves washing onto a beach—the **swash**—carries particles onshore, increasing the beach's slope. If water returning to the ocean—the **backwash**—carries back the same amount of material as it delivered, the beach slope will be in equilibrium, and the beach will not become larger or steeper.

On fine-grain beaches, the ability of small, sharp-edged particles to interlock discourages water from percolating down into the beach itself; so water from waves runs quickly back down the beach, carrying surface particles toward the ocean. This process results in a very gradual slope. Broad, flat beaches also have a large area on which to dissipate wave energy and can provide a calm environment for the settling of fine sediment particles. In contrast, coarse particles (gravel, pebbles) do not fit together well and so readily allow water to drain between them. Onrushing water disappears into a beach made of coarse particles; so little water is left to rush down the slope, thereby minimizing the transport of sediments back to the ocean. Thus larger particles tend to build up at the back of the beach, increasing its steepness.

Beach Shape

Figure 11.7 shows a beach profile, or cross section, affected by small-to-moderate wave and tidal action. The scale is exaggerated vertically to show detail. The key feature of any beach is the **berm**, an accumulation of sediment that runs parallel to shore and marks the normal limit of sand deposition by wave action. The peaked top

of the berm, called the **berm crest**, is usually the highest point on a beach. It corresponds to the shoreward limit of wave action during most high tides. Inland of the berm crest, extending to the farthest point where beach sand has been deposited, is the **backshore**. The backshore is the relatively inactive portion of the beach, which may include windblown dunes and grasses. The **foreshore**, seaward of the berm crest, is the active zone of the beach, washed by waves during the daily rise and fall of the tides. It extends from the base of the berm—where a **beach scarp**, a vertical wall of variable height, is often carved by wave action at high tide—to the low tide mark where the offshore zone begins. The **low tide terrace** is the smooth area between the beach scarp and low tide mark on which waves expend most of their energy. The low tide terrace is the part of the beach favored by runners because of its constant slope and hard, water-packed sand. Below the low tide mark, wave action, turbulent backwash, and longshore currents excavate a **longshore trough** parallel to shore. Irregular **longshore bars** (submerged or exposed accumulations of sand) complete the seaward profile.

How Beach Contours Form and Change

This beach profile is only temporary, generated by the interplay of sediments, waves, and tides. Great storm waves can rearrange a beach in a day, transporting thousands of tons of sediment from the beach to hidden sand bars offshore. **Figure 11.8** shows this transition in progress during a period of high wave action on a beach in southern California. The beach scarp has grown to enormous proportions as sand is gouged from a thick berm with each new wave on the incoming tide. Later in the year, smaller waves will move the sand back onto the beach.

Indeed, most temperate-climate beaches undergo a seasonal transformation. Beaches are cut to a lower level in winter than in summer because higher waves accompany winter storms. **Figure 11.9** traces the changes in a

Figure 11.8 A high beach scarp in southern California. Rising tides and large late-summer waves have removed huge amounts of sand from a thick berm. The sand was deposited through the spring and summer months when wave energy was low.

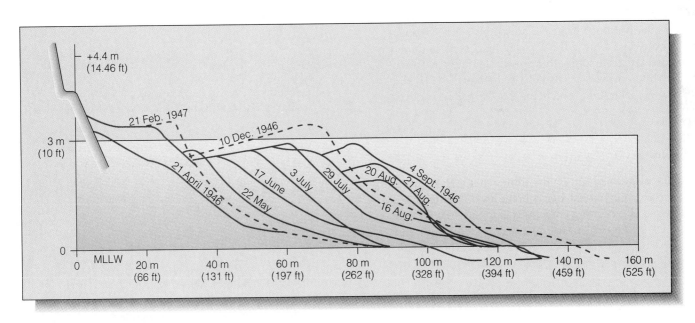

Figure 11.9 Seasonal changes in a beach profile. The berm at Carmel, California, builds 60 meters (200 feet) seaward during summer, when the waves are small. The beach then almost completely disappears when struck by the large waves of winter storms.

Figure 11.10 The movement of sand on Carlsbad State Beach in California. (a) The beach in the summer of 1978. (b) The same beach in the spring of 1983 after a winter of severe storms. The sand has moved offshore, and coastal structures have been damaged.

beach profile through most of a year. Notice that the beach builds by deposition during the summer, from its low point in April to its height in December, and then is torn down again by winter storms. Changes from summer to winter on a southern California beach are shown in **Figure 11.10**.

Though the coasts on which they are found are classified as depositional, beaches can temporarily disappear altogether and expose the underlying basement to new erosion by marine forces. Beaches containing particles of many different sizes are usually in the midst of this kind of change, with finer sand being lost (or gained) above a cobble or rocky floor.

Minor Beach Features

Small features are among the things that make beaches such interesting places. *Ripples* in the sand (**Figure 11.11a**) are caused by rushing currents. Diamond-shaped *backwash marks* (**Figure 11.11b**) form when projecting shells, pebbles, or animals interrupt the backwash along the low-tide terrace and cause uneven deposition of very fine

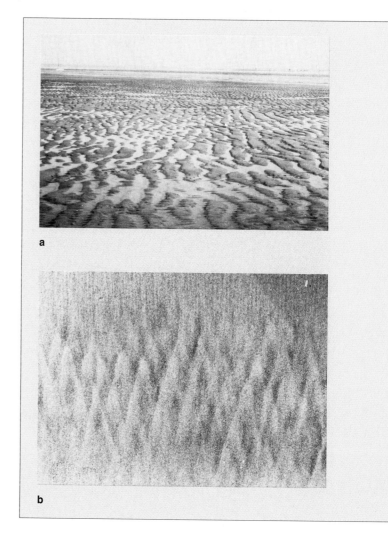

Figure 11.11 Minor features of beach sand. (a) Ripple marks. (b) Backwash marks. (c) The streamlike pattern of rill marks.

sediment of a contrasting color. Another result of uneven deposition can be seen in *beach layering*. A vertical slice into a calm beach will often reveal layer upon layer of sediments with different colors and textures. Structures similar to ripples can be seen in dry river bottoms or stream beds. *Rills* (**Figure 11.11c**) are small branching surface depressions that channel water back to the ocean from a saturated beach during a falling tide.

On a somewhat larger scale, the beach shoreline is sometimes scalloped by *cusps* (*cuspis* = point) (**Figure 11.12**). These repeating points, interrupted by miniature bays, are evenly spaced every few meters. Cusps tend to form at neap-tide periods and are destroyed during the greater range of spring tides. Cusp size appears to be correlated with wave energy. The origin of cusps is not well understood, but they are probably caused by interference between approaching waves. Offshore sand bars

Figure 11.12 Beach cusps at San Simeon, California, formed in a gently sloping fine-sand beach. The horizontal beach berm can be seen to the right.

and troughs are also more pronounced during periods of moderate tides. The variability of cusps and bars reflects the transitory nature of beach processes.

LARGE-SCALE FEATURES OF DEPOSITIONAL COASTS

We have already noted the presence of small longshore bars off beaches (see Figure 11.7), but mature depositional coasts—especially the coasts along a subsiding irregular continental margin—also exhibit much larger features. Some of these features are illustrated in **Figure 11.13**.

Sand Spits and Bay Mouth Bars

Sand spits are among the most common of these features. A sand spit forms where the longshore current slows as it clears a headland and approaches a quiet bay. The slower current is unable to carry as much sediment; so sand and gravel are deposited in a line downcurrent of the headland. As can be seen in Figure 11.13, sand spits often have a curl at the tip. This is caused by the current-generating waves being refracted around the tip of the spit.

A **bay mouth bar** forms when a sand spit closes off a bay by attaching to a headland adjacent to the bay. The bay mouth bar protects the bay from waves and turbulence and encourages the accumulation of sediments there. An **inlet**—a passage to the ocean—may be cut

c

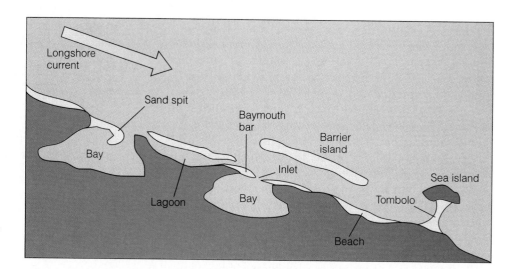

Figure 11.13 A composite diagram of the large-scale features of depositional coasts. Not all of these features would be found in such close proximity or in this order on a real coast.

Figure 11.14 A bay mouth bar.

through a bay mouth bar by tidal action, by water flowing from a river emptying into the bay, or by heavy storm rains. A bay mouth bar is shown in **Figure 11.14**.

Barrier Islands and Sea Islands

Depositional coasts can also develop narrow, exposed sand bars that are parallel to but separated from land. These are known as **barrier islands** (see **Figure 11.15**). About 13% of the world's coasts are fringed with barrier islands.

Barrier islands can form when sediments accumulate on submerged rises paralleling the shoreline. Some islands off the Mississippi–Alabama coast developed in this way. Larger barrier islands are thought to form in a different way, however. Near the end of the last major rise in sea level, about 6,000 years ago, coastal plains near the edge of the continental shelf were fronted by lines of

sand dunes. Rising sea level caused the ocean to break through the dunes and form a **lagoon**, a long, shallow body of seawater isolated from the ocean. The high lines of coastal dunes became islands. As sea level continued to rise, wave action caused the islands and lagoons to migrate landward. Most of the barrier islands off the southeast coast of the United States originated in this way. They are still migrating slowly landward as sea level continues to rise.

There are 295 barrier islands along the East and Gulf coasts of the United States, with a combined length of 2,591 kilometers (1,610 miles). About 30 times a year, severe storms generate waves intense enough to erode barrier island beaches. The largest of these storms can generate waves that overwash the low islands. Runoff from rivers swollen by rains, coupled with water driven by wind waves and storm surge, can rapidly flood a lagoon and cut new inlets through barrier islands.

Figure 11.15 Barrier islands off the Texas Gulf Coast.

Despite these dangers, about 70 of the barrier islands have been commercially developed, and millions of people live on them. The most famous barrier islands include Atlantic City, New Jersey; Ocean City, Maryland; Miami Beach and Palm Beach, Florida; and Galveston, Texas. About once every 100 years a winter storm has catastrophic effects on populated areas of Atlantic barrier islands, and the southeastern Atlantic and Gulf coasts must contend with occasional large hurricanes. The continuing subsidence of these passive coasts (combined with changes resulting from commercial development and the ongoing rise in sea level) will cost lives and property. **Figure 11.16** suggests the extent of the threat.

Unlike barrier islands, **sea islands** contain a firm central core that was part of the mainland when sea level was lower. The rising ocean separated these high points from land, and sedimentary processes surrounded them with beaches. Hilton Head, South Carolina, and Cumberland Island, Georgia, are sea islands. If the island is close to shore, a bridge of sediments called a **tombolo** may accumulate to connect the island to the mainland. Tombolos can also connect offshore rocky outcrops or volcanoes to the mainland.

a

b

Figure 11.16 Barrier island modification. (a) The beach extending along the Matagorda Peninsula (Texas) barrier in September 1960. (b) The same area six days after the passage of Hurricane Carla in September 1961. The beach and barrier island have been breached, and washover deltas are clearly seen.

ESTUARIES

An **estuary** (*æstus* = tide) is a body of water partially surrounded by land where fresh water from a river mixes with ocean water. Estuaries are often areas of remarkable biological productivity and diversity. Chesapeake Bay, San Francisco Bay, and the mouth of the Columbia River are all estuaries.

Three factors determine the characteristics of estuaries: the shape of the estuary, the volume of river flow at the head of the estuary, and the range of tides at the estuary's mouth. The mingling of waters of different densities, the rise and fall of the tide, and the variations in river flow along with the actions of wind, ice, and the Coriolis effect guarantee that patterns of water circulation in an estuary will be complex.

Estuaries are categorized by their circulation patterns. The simplest circulation patterns are found in **salt wedge estuaries**, which form where a rapidly flowing, large river enters the ocean in an area where tidal range is low or moderate. The exiting fresh water holds back a wedge of intruding seawater (**Figure 11.17a**). Note that density differences cause fresh water to flow over salt water. The seawater wedge moves seaward at times of low tide or strong river flow, and it returns landward as the tide rises or when river flow diminishes. Some seawater from the wedge joins the seaward-flowing fresh water at the steeply sloped upper boundary of the wedge, and new seawater from the ocean replaces it. Nutrients and sediments from the ocean can enter the estuary in this way. Examples of salt wedge estuaries are the mouths of the Hudson and Mississippi rivers.

A different pattern occurs where the river flows more slowly and tidal range is moderate to high. As their name implies, **well-mixed estuaries** contain differing mixtures of fresh and salt water through most of their length. Tidal turbulence stirs the waters together as river runoff pushes the mixtures to sea. A well-mixed estuary is illustrated in **Figure 11.17b**. The mouth of the Columbia River is an example.

Deeper estuaries exposed to similar tidal conditions but greater river flow become **partially mixed estuaries**. Partially mixed estuaries share some of the properties of salt wedge and well-mixed estuaries. Note in **Figure 11.17c** the influx of seawater beneath a surface layer of fresh water flowing seaward; mixing occurs along the junction. Energy for mixing comes from both tidal turbulence and river flow. London's Thames River, San Francisco Bay, and Chesapeake Bay are examples.

In well-mixed and partially mixed estuaries in the Northern Hemisphere, the incoming seawater will trend toward the right side of the estuary because of the Coriolis effect. Outflowing river water will also trend to the right of its direction of travel. This rightward drift can be seen in the contour of lines representing surface salinity in **Figure 11.18**.

Reverse estuaries can form along arid coasts when rivers cease to flow. The evaporation of seawater in the uppermost reaches of these estuaries will cause water to flow from the ocean into the estuary, producing a gradient of *increasing* salinity from the ocean to the estuary's upper reaches (see **Figure 11.17d**). Reverse estuaries—sometimes called lagoons—are common on the Pacific coast of Mexico's Baja Peninsula and along the U.S. Gulf Coast.

The coasts of the United States contain about 15,150 square kilometers (5,850 square miles) of estuarine waters. Many of the estuaries along the East Coast have been formed by the combined effects of rising sea level and tectonic subsidence. Chesapeake Bay and the Hudson

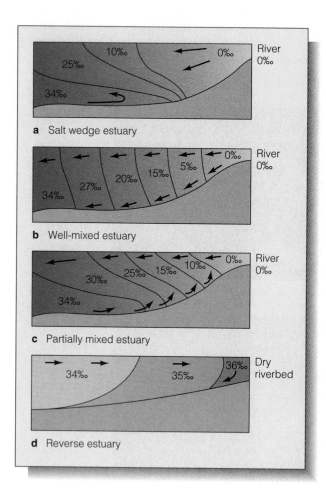

Figure 11.17 Types of estuaries. The salinity values show the amount of mixing between fresh water (0‰) and seawater (33‰) in the various types. (a) Salt wedge estuary. (b) Well-mixed estuary. (c) Partially mixed estuary. (d) Reverse estuary, in which evaporation plays a major role.

River Valley are examples of "drowned" river mouths. Some West Coast estuaries were formed by tectonic forces; San Francisco Bay has filled with water because of faulting and local subsidence. Deep, narrow estuaries known as **fjords** were originally cut by glaciers slicing deep U-shaped indentations at the mouths of high-latitude rivers. Fjords are found in British Columbia, Alaska, Norway, and New Zealand. Estuaries can also form in river-fed bays behind bay mouth bars or within river deltas. Estuaries have become much more common as sea level has risen over the last 6,000 years.

As we will see in Chapter 14, estuaries often support a tremendous number of living organisms. The easy availability of nutrients and sunlight, protection from wave shock, and the presence of many habitats permit the growth of many species and individuals. Estuaries are frequently nurseries for marine animals; several species of perch, anchovy, and Pacific herring take advantage of the abundant food in estuaries during their first weeks of life. Unfortunately for their inhabitants, estuaries are in high demand for development into recreational resources and harbors. Estuaries have become the most polluted of all marine environments.

CHARACTERISTICS OF U.S. COASTS

Plate tectonic forces have had immense influence on the margins of continents, and the edges of the United States mainland are no exception. The results of plate movement on the West Coast differ greatly from those on the East and Gulf coasts, primarily because the West Coast is near an active plate margin while the East and Gulf coasts are not.

The West Coast

The West Coast is an actively rising margin on which volcanoes, earthquakes, and other indications of recent tectonic activity are easily observed. West Coast beaches are typically interrupted by jagged rocky headlands, volcanic intrusions, or the effects of submarine canyons. Wave-cut terraces (**Figure 11.19**) are found as much as 400 meters (1,300 feet) above sea level in a number of places, evidence that uplift has exceeded the general rise in sea level through the past million years.

Most of the sediments on the West Coast originated from erosion of relatively young granitic rocks of the coastal mountains. The particles of quartz and feldspar that make up most of the sand were transported to the shore by flowing rivers. The volume of sedimentary material transported to West Coast beaches from inland areas greatly exceeds the amount originating at the coast. Because West Coast beaches are usually high in wave

Figure 11.18 The Chesapeake Bay estuary, an example of a partially mixed estuary. The typical distribution of surface salinity in the estuary ranges from 28‰ at the mouth to 1‰ near the upper reaches. The Coriolis effect forces the inflowing salt water against the right bank; notice how the 20‰ contour trends toward the right bank.

Figure 11.19 Wave-cut terraces on San Clemente Island.

energy, deltas tend not to form at river mouths. The predominant direction of longshore drift is to the south.

The East Coast

The East Coast is a passive margin, tectonically calm and subsiding because of its trailing central position on the North American Plate. Subsidence along the coast has been considerable—3,000 meters (10,000 feet) over the last 150 million years. A deep layer of sediment built up offshore—material that produced the ancestors of today's barrier islands. Relatively recent subsidence has been more important in shaping the present-day coast, however. Coastal sinking and rising sea level have combined to submerge some parts of the East Coast at a rate of about 0.3 meter (1 foot) per century. This process has formed the huge flooded valleys of Chesapeake and Delaware bays, the landward-migrating barrier islands, and the shrinking lowlands of Florida and Georgia.

Rocks to the north (in Maine, for example) are among the hardest and most resistant to erosion of any on the continent; so beaches are uncommon in Maine. But from New Jersey southward the rocks are more easily fragmented and weathered, and beaches are much more common. As on the West Coast, sediments are transported coastward by rivers from eroding inland mountains, but the transported material is trapped in sunken estuaries and therefore plays a less important role on beaches. Eastern beaches are typically formed of sediments from nearby erosional shores, or from the shoreward movement of offshore deposits laid down when the sea level was lower. The amount of sand in an area thus depends in part on the resistance or susceptibility of nearby shores to erosion. Sand moves generally south on these beaches just as it does on the West Coast, but the volume of moving sand in the East is less.

Glaciers have also contributed to the shaping of the northern part of the East Coast. Long Island and Cape Cod are remnants of debris moved into position by glaciers.

The Gulf Coast

The Gulf Coast experiences a smaller tidal variance and—hurricanes excepted—a smaller average wave size than either the West or East coast. Reduced longshore drift and an absence of interrupting submarine canyons allow the great volume of accumulated sediments from the Mississippi and other rivers to form large deltas, barrier islands, and a long, raised "super berm" that prevents the ocean from inundating much of this sinking coast.

These are fortunate conditions, because the rate of subsidence in the Gulf Coast is greater than that for most of the East Coast. Human activity can make the subsidence even worse. Pumping from oil and water wells, combined with sediment starvation and dredging, have exacerbated the situation around some large cities. At Galveston, Texas, for example, sea level is nearly 64 centimeters (25 inches) higher than it was a century ago, and most of New Orleans is now 2 meters (6.6 feet) below sea level. As we have seen, the results of hurricanes at such places can be tragic. The protective natural berm can easily be breached, and flood waters can drive far inland.

Figure 11.20 Growth of a beach protected by a breakwater: Santa Monica, California.

HUMAN INTERFERENCE IN COASTAL PROCESSES

Beaches exist in a tenuous balance between accumulation and destruction. Human activity can tip the balance one way or the other. For example, consider the rocky **breakwater** shown in **Figure 11.20**. The breakwater interrupts the progress of waves to the beach, shielding it from the longshore current and allowing sand to accumulate there. Without dredging, the beach will eventually reach the breakwater and fill the small boat anchorage the breakwater was built to provide. This is a minor example of human alteration to a beach, yet it serves to introduce the growing problem of human influences on coastal processes.

We often divert rivers, build harbors, and develop property with surprisingly little understanding of the impact our actions will have on the adjacent coast. Our role then becomes that of powerless observers. Residents of erosional coasts can only accept the inexorable loss of their property to the attack of natural forces, but residents of depositional coasts are sometimes presented with alternatives. The choices are almost never simple. For example, should rivers be dammed to control devastating floods? If the dams are built, they will trap sediments on their way from mountains to coast. Beaches within the coastal cell fed by the dammed river will shrink because the sand on which they depend to replenish losses at the shore is blocked. Alarmed coastal residents will then take steps to hang onto whatever sand remains. They may try to trap "their" beaches by erecting

Figure 11.21 Groins change the shape of a coastline by interfering with longshore transport. Sand drifting eastward (toward the right) is trapped against the groins, and the beach beyond is much narrower. Erosion downcoast may be accelerated.

groins, short extensions of rock or other material placed at right angles to longshore drift to stop the longshore transport of sediments. This temporary expedient usually accelerates erosion downcoast (**Figure 11.21**). Diminished beaches then expose shore cliffs to accelerated erosion; wind wave energy that would have harmlessly

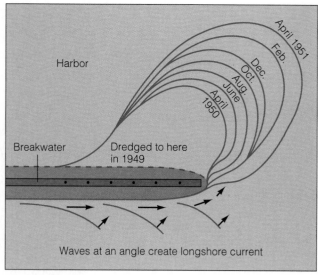

Harbor

Breakwater

Dredged to here
in 1949

April 1951

Feb.

Dec.

Oct.

Aug.

June

April
1950

Waves at an angle create longshore current

a

b

Figure 11.22 (a) A sand spit grows into the small-boat anchorage at Santa Barbara, California. (b) Sand accumulates at the tip of the breakwater when longshore transport is interrupted.

churned sand grains now speeds the destruction of natural and artificial structures. Was protection from periodic flooding worth the loss of the beach? I suppose it depends on where your property is situated!

Less frequently, the deposition of sand is an unwanted result of human constructions. Interruption of longshore transport caused a new beach to form against what once were bare cliffs upcoast from the breakwater at Santa Barbara, California. Downcoast, calm water at the breakwater's end interfered with longshore drift, and sand accumulated in the harbor (**Figure 11.22**). Without continuous dredging of about 600 cubic meters (800 cubic yards) per day, the harbor would soon fill with sand. The dredging cost to residents of Santa Barbara is considerable.

What are the implications of these unlooked-for sand movements? Douglas Inman, director of the Center for Coastal Studies at Scripps Institution of Oceanography, feels that at least 20% of the U.S. beach-bounded coastline is in danger of serious or catastrophic alteration. On the West Coast, a 30-year period of relatively mild weather may be ending. During this time people felt it was safe to build close to the shore. Increased dam building and breakwater, jetty, and groin construction have made southern California's beaches more vulnerable. In coastal California alone, 3,666 homes and 1,020 businesses were damaged during the stormy winter of 1983; losses exceeded $100 million. A large storm in January 1988 did similar damage along the same coast, a clear indication that coastal resi-

dents do not always learn by example. The barrier islands of the East and Gulf coasts are at least as vulnerable, as the repeated cycles of commercial development followed by destructive hurricanes indicate.

Shores that look permanent through the short perspective of a human lifetime are in fact among the most ephemeral of all marine structures. Let's enjoy them in their present stages.

CHAPTER SUMMARY

Our personal experience with the ocean usually begins at the coast. These impermanent, often beautiful junctions of land and sea are subject to rearrangement by waves and tides, by gradual changes in sea level, and by tectonic activity. Coasts may be classified in many ways, but one of the most useful schemes is based on the balance of coastal erosion and deposition. Erosional coasts are typically rocky but may contain beaches. Depositional coasts are usually characterized by broad beaches. About 30% of the U.S. coast is depositional, 70% erosional. The shape and composition of a typical beach change through the year as sediments, driven by waves and currents, move on- and offshore. All coasts are temporary structures. Human interference with coastal processes can accelerate erosion and subject commercial developments to destruction.

Terms and Concepts to Remember

backshore	estuary	partially mixed
backwash	eustatic change	estuary
barrier island	fjord	reverse estuary
bay mouth bar	foreshore	salt wedge estuary
beach	groin	sand spit
beach scarp	high-energy coast	sea cave
berm	inlet	sea cliff
berm crest	lagoon	sea island
breakwater	longshore bars	shore
coast	longshore current	swash
delta	longshore drift	tombolo
depositional coast	longshore trough	wave-cut platform
dissolution	low-energy coast	well-mixed estuary
erosional coast	low tide terrace	

Study Questions

1. What wears down an erosional coast? What agents are most effective? How does a high-energy coast differ from a low-energy coast?

2. What are some of the features of an erosional coast? Are they temporary or permanent? What causes shore straightening?

3. What two processes contribute to longshore drift? What powers longshore drift? What is the predominant direction of drift on U.S. coasts? Why?

4. What are some of the features of a depositional coast? Are they temporary or permanent? Is there a relationship between wave energy on a depositional coast and the size (or slope, or grain size) of the beaches found there?

5. Compare and contrast the U.S. West, East, and Gulf coasts.

6. How do human activities interfere with coastal processes? What steps can be taken to minimize loss of life and property along U.S. coasts?

For Further Study

Bascom, W. 1980. *Waves and Beaches*. Rev. ed. New York: Anchor/Doubleday. A valuable reference, well and clearly written, nicely illustrated.

Burk, K. 1979. "The Edges of the Ocean: An Introduction." *Oceanus* 22 (no. 3): 2–9. Terminology and overview.

Dolan, R., and H. Lins. 1987. "Beaches and Barrier Islands." *Scientific American*, July, 68–77. Excellent summary of the evolution of these ephemeral structures.

Gornits, V., et al. 1982. "Global Sea Level Trend in the Past Century." *Science* 215 (no. 4540): 1611–14. Mean sea level rise has been 12 centimeters (4.7 inches) in the past century. Most of this rise has been due to ocean warming and the subsidence of coasts.

Oceans Magazine 20 (no. 2, March–April 1987). An entire issue dedicated to coasts and coastal processes.

Oceanus 36 (nos. 2 and 3, spring and summer 1993). Dedicated to coastal science and policy.

Sackett, R. 1983. *The Edge of the Sea*. Chicago: Time–Life Books. Part of the "Planet Earth" series. Excellent photos of coasts and coastal processes.

Shepard, F. P. 1977. *Geological Oceanography: Evolution of Coasts, Continental Margins, and the Deep Ocean Floor*. New York: Crane, Russack & Co., Inc.

Shepard, F. P., and H. R. Wanless. 1971. *Our Changing Coastlines*. New York: McGraw-Hill.

12 LIFE IN THE OCEAN

The Forest Beneath the Waves

The ocean often moves placidly in a kelp forest—the long seaweeds interfere with the circular movement of water molecules as the waves pass. Because the plants secrete a lubricant that smooths a diver's way, visitors can glide easily between the closely entwined seaweeds, gently nudging the strands from their paths. Sunlight filters between the fronds, sending illuminating shafts flickering into the deeper water below. Schools of fishes often move among the kelp, and countless animals—many too small to be seen—nestle around their dark bases. The feeling is like that of being in a great grove of redwoods, and a diver's relative weightlessness allows nearly all parts of the area to be explored. It is usually quiet here, peaceful and calm.

A biologist sees even more beauty in the scene and notes the rapid growth; the seaweeds are longer on this visit than a few weeks ago, and their stalks are thicker. Some worms have begun to make spiral tracings on the kelp surfaces. The animals on the seabed are different now, and arrayed in new patterns. A family of otters has moved to the area, attracted by the many nutritious sea urchins scattered across the gravel-covered bottom. Schools of anchovetta flash overhead, swimming between the kelp blades in long follow-the-leader schools, efficiently sieving the ocean for food. Importantly, there are many more plantlike organisms living here than the big seaweeds swaying in the currents: a living haze crowds the water immediately ahead of the diver's faceplate. A flashlight beam reveals swarms of dust-sized organisms in every direction. Countless millions of organisms drift unobserved, too small to reflect the light. Life abounds.

The ocean here fairly hums with productivity. Food is being produced rapidly and in great quantity. Big plants and microscopic ones are harnessing light from the sun to assemble carbohydrates, as their forebears have done since the ocean was young. The temperate coastal ocean brims with life, much of it dependent on the subtle light-driven biochemistry proceeding in this lovely, gracefully moving place.

Sunlight penetrates a kelp forest, one of the ocean's most productive habitats.

To the weekend sailor, the ocean may seem interesting mainly because of its winds and currents and waves, but the seemingly empty seawater next to a small sailboat may support millions of invisible plantlike organisms. The waters beneath the hull can conceal wonderful worms and colorful crustaceans, the gently swaying kelp forests just described, whole communities of microscopic creatures drifting with the currents, schools of fishes, maybe even whales. On the dark distant bottom, animals locate one another with glowing lures or jostle for food near volcanic vents. The volume of living flesh in the ocean greatly exceeds that on dry land. The great wide sea is the ideal habitat for life.

The ocean can support such a bewildering array of life forms because of water's unique physical properties—its dissolving power, its density, its ability to absorb large quantities of heat yet rise very little in temperature. All life on our planet is water-based; indeed, life began in the ocean. Life and the Earth have changed together, generation by generation, over about 4 billion years. The ability of living things to change through time to fit the physical and chemical environment, to colonize virtually every location capable of sustaining them, and finally to investigate themselves by scientific logic appears to have come about through **evolution**—the process of organic change through the natural selection of favorable traits.

There are at least 50 million different species (kinds) of living things on Earth. Yet despite their stupefying diversity in form and life-style, all species share the same underlying mechanisms for capturing and storing energy, manufacturing proteins, and transmitting information between generations. Biologists know that there is nothing special about the atoms or energy of life, no way to distinguish the physical components of life from their nonliving counterparts. What *does* distinguish life from nonlife is the ability of living things to capture, store, and transmit energy—and the ability to reproduce.

ENERGY AND MARINE LIFE

Nearly all the energy marine organisms need to function (and to evolve) comes directly or indirectly from the sun. The sun produces enormous quantities of energy—some in the form of visible light, a tiny portion of which strikes the Earth. Only about one part in 2,000 of the light that reaches the Earth's surface is captured by organisms, but that "small" input of energy powers all the growth of living things. Light energy from the sun is trapped by chlorophyll in organisms called *producers* (certain bacteria, algae, and green plants) and changed into chemical energy. The chemical energy is used to build simple carbohydrates and other organic molecules—**food**—which are then used by the plant itself or are eaten by animals

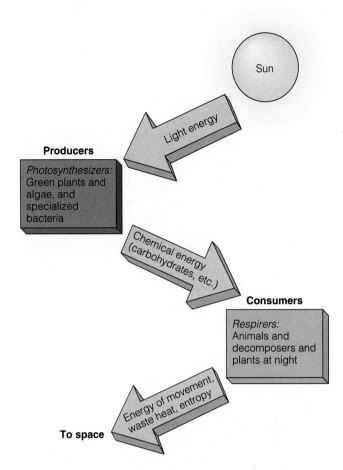

Figure 12.1 The flow of energy through living systems. At each step, energy is degraded (transformed into a less useful form).

(or other organisms) called *consumers*. Because light energy is used to synthesize molecules rich in stored energy, the process is called **photosynthesis** (*photos* = light, *syn* + *tithenai* = to place together). Here is a general formula for photosynthesis:

> **carbon dioxide + water – (sunlight) → glucose + oxygen**

The energy is released, or **metabolized**, when food such as glucose (a carbohydrate) is used for growth, repair, movement, reproduction, and the other functions of organisms. The metabolism of food eventually produces waste heat, which flows away from Earth into the coldness of space. This one-way flow of energy is shown in **Figure 12.1**.

Photosynthesis is the usual method of binding energy into carbohydrates, but there is another. **Chemosynthesis**, employed by a few relatively simple forms of life, is the production of usable energy directly from energy-rich

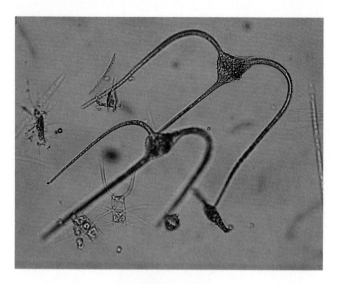

Figure 12.2 *Ceratium*, a common primary producer in the plankton. In a very bright light these two single-celled plantlike organisms would barely be visible to the unaided eye.

inorganic molecules available in the environment, rather than from the sun. As we will see in Chapter 14, some unusual forms of marine life depend on chemosynthesis. Overall, chemosynthetic production of food is very small in comparison to photosynthetic production.

Primary Productivity

The synthesis of organic materials from inorganic substances by photosynthesis or chemosynthesis is called **primary productivity**. Primary productivity is expressed in grams of carbon bound into organic material per square meter of ocean surface area per year ($gC/m^2/yr$). The organic material produced is usually glucose. The source of carbon for glucose is dissolved CO_2. *Phytoplankton*—minute, drifting photosynthetic organisms described in Chapter 13—produce between 90% and 98% of oceanic carbohydrates. Seaweeds—larger marine plants that we will meet in Chapter 14—contribute from 2% to 10% of the ocean's primary productivity; chemosynthetic organisms probably account for less than 1% of the total. Though estimates vary widely, recent studies suggest that total ocean productivity ranges from 50 to 100 grams of carbon bound into carbohydrates per square meter of ocean surface per year (50 to 100 $gC/m^2/yr$). (For comparison, a well-tended alfalfa field produces about 1,600 $gC/m^2/yr$).

The total weight of a **primary producer** is assumed to be about ten times the mass of the carbon it has bound into carbohydrates (see **Figure 12.2**). Thus, a primary productivity of 100 $gC/m^2/yr$ represents the yearly growth

of about 1,000 grams of primary producers for each square meter of ocean surface. Between 20 and 25 billion metric tons (22 to 28 billion tons) of carbon is believed to be bound into carbohydrates in the ocean each year; so between 200 and 250 billion metric tons (220 to 280 billion tons) of marine plants and plantlike organisms are produced annually! Each year this vast bulk is consumed by the metabolic activity of the producers themselves and by the consumers that graze on them. The component atoms are then reassembled by photosynthesis into carbohydrates in a continuous cycle.

Feeding (Trophic) Relationships

Photosynthetic and chemosynthetic organisms can be called either primary producers or **autotrophs** (*auto* = self + *trophe* = nourishment) because they make their own food. The bodies of autotrophs are rich sources of chemical energy for any organisms capable of consuming them. **Heterotrophs** (*hetero* = other, different) are organisms such as animals that must consume other organisms because they are unable to synthesize their own food molecules. Some heterotrophs consume autotrophs, and some consume other heterotrophs.

We can label organisms by their positions in a "who eats whom" feeding hierarchy called a **trophic pyramid** (*trophos* = one who feeds). The primary producers shown at the bottom of the pyramid in **Figure 12.3** are mostly chlorophyll-containing photosynthesizers. The animal heterotrophs that eat the photosynthesizers are called **primary consumers** (or herbivores); the animals that eat the primary consumers are called secondary consumers; and so on to the **top consumer** (or top carnivore).

Note that the mass of consumers becomes smaller as energy flows toward the top of the pyramid. There are many small primary producers at the base and a very few large top consumers at the apex. Only about 10% of the energy from the organisms consumed is stored in the consumers as flesh; so each level is about one-tenth the mass of the level directly below. The rest of the energy is lost as waste heat as organisms live and work to maintain themselves.

Pyramids such as these can lead to the misconception that one kind of fish eats only one other kind of fish, and so on. Real communities are more accurately described as food webs, an example of which is included as **Figure 12.4** on page 208. A **food web** is a group of organisms linked by complex feeding relationships in which the flow of energy can be followed from primary producers through consumers. Organisms in a food web almost always have some choices of food species.

So organisms interact with each other, feed on one another, and transfer energy as food from producing autotrophs through a web of consuming heterotrophs.

Trophic level

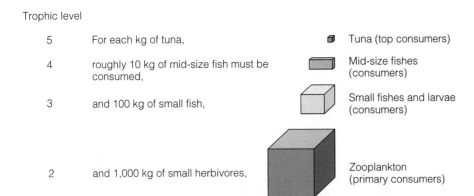

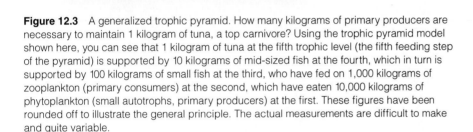

5	For each kg of tuna,	Tuna (top consumers)
4	roughly 10 kg of mid-size fish must be consumed,	Mid-size fishes (consumers)
3	and 100 kg of small fish,	Small fishes and larvae (consumers)
2	and 1,000 kg of small herbivores,	Zooplankton (primary consumers)
1	and 10,000 kg of primary producers.	Phytoplankton (primary producers)

Figure 12.3 A generalized trophic pyramid. How many kilograms of primary producers are necessary to maintain 1 kilogram of tuna, a top carnivore? Using the trophic pyramid model shown here, you can see that 1 kilogram of tuna at the fifth trophic level (the fifth feeding step of the pyramid) is supported by 10 kilograms of mid-sized fish at the fourth, which in turn is supported by 100 kilograms of small fish at the third, who have fed on 1,000 kilograms of zooplankton (primary consumers) at the second, which have eaten 10,000 kilograms of phytoplankton (small autotrophs, primary producers) at the first. These figures have been rounded off to illustrate the general principle. The actual measurements are difficult to make and quite variable.

Nearly all are ultimately dependent on sunlight and photosynthesis. What is the physical role of the ocean in this?

PHYSICAL FACTORS AFFECTING MARINE LIFE

Marine organisms depend on the ocean's chemical composition and physical characteristics for life support. Any aspect of the physical environment that affects living organisms is called a **physical factor**. Living in the ocean has advantages over living on land; physical conditions in the sea are usually milder and less variable than physical conditions on land. The most important physical factors for marine organisms are water transparency, temperature, dissolved nutrients, salinity, dissolved gases, acid–base balance, and hydrostatic pressure.

Transparency

On land, most photosynthesis proceeds at or just above ground level. But seawater is relatively transparent, which allows photosynthesis to proceed for some distance below the ocean surface. Incoming sunlight must run a gauntlet of difficulties, however, before it can be absorbed by the chlorophyll in marine autotrophs.

One important factor is that water is more transparent to some colors of light than to others. In clear water, blue light penetrates to the greatest depth, while red light is absorbed near the surface. **Figure 12.5** shows the depths attained by light of various wavelengths (colors) in clear ocean water. Light energy absorbed by water turns to heat.

The depth to which light penetrates is also limited by the number and characteristics of particles in the water. These particles, which may include suspended sediments, dustlike bits of once-living tissue, or the organisms themselves, scatter and absorb light. High concentrations of particles quickly absorb most blue and ultraviolet light. This absorption, combined with the reflection of green light by chlorophyll within the producers, changes the color of productive coastal waters to green.

How far down does light penetrate? Near the coasts, conditions become uncomfortably dark for divers at depths around 30 to 40 meters (100 to 130 feet). The human eye is most sensitive to the blue wavelengths; in very clear tropical waters with a smooth surface and high solar angle, human observers in submersibles have seen

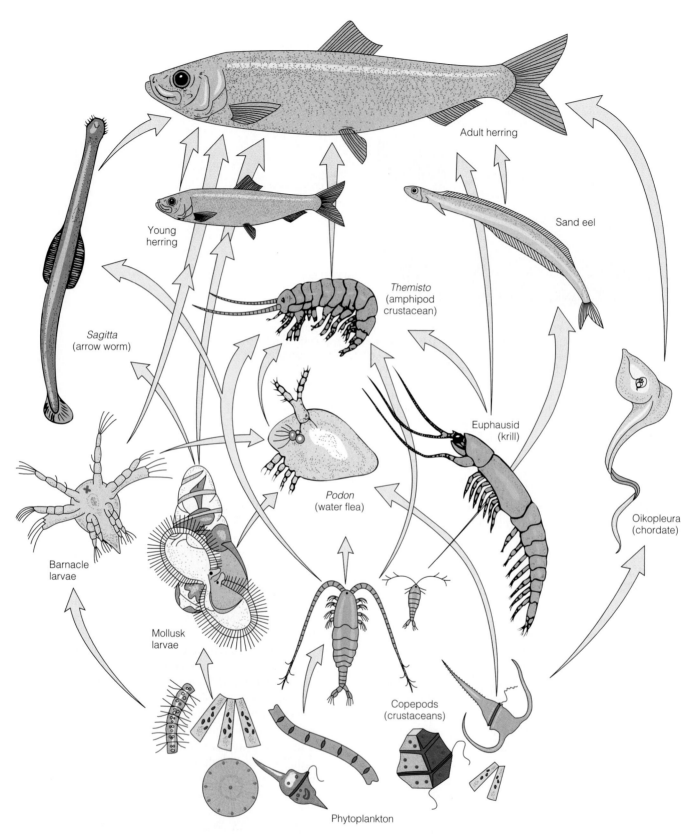

Adult herring

Young herring

Sand eel

Sagitta
(arrow worm)

Themisto
(amphipod
crustacean)

Podon
(water flea)

Euphausid
(krill)

Oikopleura
(chordate)

Barnacle
larvae

Mollusk
larvae

Copepods
(crustaceans)

Phytoplankton

Figure 12.4 A food web, illustrating the major trophic relationships leading to an adult herring. The arrows show the direction of energy flow. Note that feeding relationships are not as simple as one might assume from Figure 12.3.

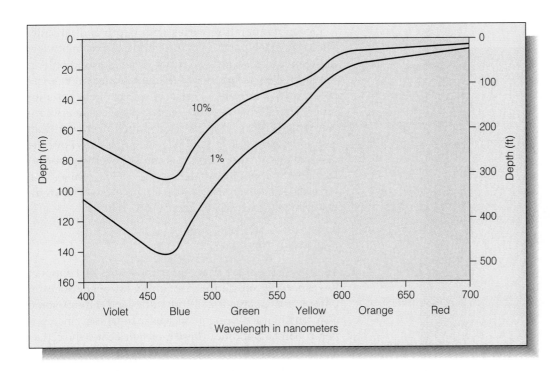

Figure 12.5 Depths at which the surface radiation of light is reduced to 10% and 1% for various colors in clear ocean water.

dark blue light at about 200 meters (660 feet). There is evidence to suggest that some deep-water fish use even this dim light for body orientation, feeding, and predator avoidance. Photometers much more sensitive than the human eye have detected light at even greater depths; the present record is 520 meters (1,706 feet) in the tropical Atlantic!

Photosynthesis proceeds slowly at low light levels. Most of the biological productivity of the ocean occurs in an area near the surface called the **euphotic zone** (*eu* = good + *photos* = light). This is where marine plants trap more energy than they use. Though it is difficult to generalize for the ocean as a whole, the euphotic zone typically extends to a depth of approximately 40 meters (130 feet) in mid-latitudes. The upper productive layer of ocean is a very thin skin indeed. The water within this zone amounts to less than 2% of world ocean volume, yet nearly all marine life depends on this fine illuminated band.

Temperature

The rate at which chemical reactions occur in a living organism is largely dependent on the molecular vibration we call *heat*. Since agitation brings reactants together, warmer temperatures increase the rate at which chemical reactions occur. Thus, an organism's **metabolic rate**, the rate at which energy-releasing reactions proceed within an organism, increases with temperature. The metabolic rate approximately doubles with a 10°C (18°F) temperature rise. The interior temperature of an organism is directly related to the rate at which it moves, reacts, and lives.

The great majority of marine organisms are "cold-blooded," or **ectothermic** (*ektos* = outside + *therme* = heat), having an internal temperature that stays very close to that of their surroundings. A few complex animals—mammals and birds and some of the larger, faster fishes—are "warm-blooded," or **endothermic** (*endon* = within), meaning that they have a stable, high internal temperature.

In general, the warmer the environment of an ectotherm within its tolerance range, the more rapidly its metabolic processes will proceed. Tropical fish in a heated aquarium will therefore eat more food and require more oxygen than goldfish of the same size living in an unheated but otherwise identical aquarium. The tropical fish will generally grow faster, have a faster heartbeat, reproduce more rapidly, swim more swiftly, and live shorter lives. But you can't just crank the heater up another notch for even faster fish; eventually the little fellows will cook as their proteins distort. The upper limit of temperature that an ectotherm can tolerate is often not much higher than its optimum temperature. The lower limit is usually more forgiving because molecules are merely slowed.

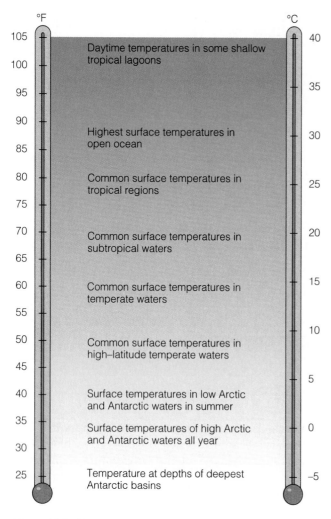

°F   °C

105 — 40

Daytime temperatures in some shallow
tropical lagoons

100

95 — 35

90

Highest surface temperatures in — 30
open ocean

85

80 — 25

Common surface temperatures in
tropical regions

75

70 — 20

Common surface temperatures in
subtropical waters

65

60 — 15

Common surface temperatures in
temperate waters

55

50 — 10

Common surface temperatures in
high–latitude temperate waters

45

— 5

40

Surface temperatures in low Arctic
and Antarctic waters in summer

35

Surface temperatures of high Arctic — 0
and Antarctic waters all year

30

25

Temperature at depths of deepest — −5
Antarctic basins

Figure 12.6   Temperatures of marine waters.

Do endotherms have narrow temperature requirements? Yes and no. Endotherms can tolerate a tremendous range of *external* temperature compared to ectotherms; think of a whale migrating from polar waters to the tropics, or an emperor penguin incubating an egg at −51°C (−60°F). Their *internal* temperatures, however, vary only slightly. In our own case, consider the temperatures of places inhabited by humans in contrast to the narrow internal temperature range physicians consider normal. Sophisticated thermal regulation mechanisms make it possible for endotherms to live in a variety of habitats, but they pay a price. Their high metabolic rates make proportionally high demands on food supply and gas transport, but the benefit of having a biochemistry fine-tuned to a single efficient temperature is worth the regulatory difficulties involved.

Ocean temperature varies with depth and latitude. The average temperature of the world ocean is only a few degrees above freezing; warmer water is found only in the lighted surface zones of the temperate and tropical ocean, and in rare, deep, warm chemosynthetic communities. Though temperature ranges of the ocean are considerable (**Figure 12.6**), they are much narrower than comparable ranges on land. Marine organisms have the definite advantage: They are almost never exposed to sustained temperatures above 30°C (86°F), while some terrestrial species must tolerate long periods when temperature reaches or even exceeds 60°C (140°F).

Dissolved Nutrients

The chemistry of life depends on molecules that support it—especially substances containing nitrogen, calcium, phosphorus, potassium, silicon, sulfur, and sodium; some also contain trace elements such as manganese, zinc, copper, cobalt, and iodine. These molecules enter the cell as **nutrients**, compounds that autotrophs require for the production of organic matter. Some nutrients help form the structural parts of organisms, some make up the chemicals that directly manipulate energy, and some have other functions. A few of these necessary nutrients are always present in seawater, but most are not readily available.

The main inorganic nutrients required in primary productivity include nitrogen (as nitrate, NO_3^-) and phosphorus (as phosphate, PO_4^{3-}). As any gardener knows, plants require fertilizer—mainly nitrates and phosphates—for success. Ocean gardeners would have more trouble raising crops than their terrestrial counterparts, though, because the most fertile ocean water contains only about 1/10,000 the available nitrogen of topsoil. Phosphorus is even more scarce in the ocean, but fortunately less of it is required by living things, which have about 1 atom of phosphorus for every 16 atoms of nitrogen.

Nitrogen and phosphorus are often depleted by autotrophs during times of high productivity and rapid reproduction. Also in short supply during rapid growth are dissolved silicates and calcium compounds (used for shells and other hard parts) and trace elements such as iron, copper, and magnesium (used in enzymes, vitamins, and other large molecules). Marine plants have no choice but to recycle these nutrients; nutrient cycles move nutrient molecules rapidly and repeatedly into and out of living systems. Though primary productivity may be very high when light is available, the total mass of living material cannot increase until more inorganic nutrients are made available by recycling, upwelling, runoff from land, or other means.

Salinity

Cell membranes are greatly affected by the salinity of surrounding water. The salinity of seawater (see Chapter 6) can vary in places because of rainfall, evaporation, runoff of water and salts from land, and other factors. Surface salinity varies most, with lows of 6‰ or less along the coast of the outer Baltic Sea in early summer, to year-round highs exceeding 40‰ in the Red Sea. Salinity is less variable with increasing depth, with the ocean typically becoming slightly saltier with depth.

Fluctuating salinity can physically damage membranes, and concentrated salts can alter protein structure. Salinity can affect specific gravity, density, and therefore the buoyancy of an organism. Salinity is also important because it can cause water to enter or leave a cell through the membrane, changing the cell's overall water balance. Seawater is nearly identical in salinity to the interior of all but the most advanced forms of marine life; this means that maintaining salt balance, and therefore water balance, is easy for most marine species.

Dissolved Gases

Virtually all marine organisms require dissolved gases—in particular carbon dioxide and oxygen—to stay alive. Oxygen does not dissolve easily in water, and as a result there is about 100 times more gaseous oxygen in the atmosphere than in the ocean. But CO_2, essential to primary productivity, is much more soluble and reactive in seawater than oxygen (as **Table 12.1** shows). Although as much as 1,000 times more carbon dioxide than oxygen can dissolve in water, normal values at the ocean surface average around 50 milliliters per liter for CO_2 and around 6 milliliters per liter for oxygen. At present the ocean holds about 60 times as much carbon dioxide as the atmosphere. Because of this abundance, marine plants almost never run out of CO_2.

Deep water tends to contain more carbon dioxide than surface water. Why should this be? Table 12.1 also shows the relationship between water temperature and its ability to dissolve gases. Note that colder water contains more gas at saturation. You may recall that the deepest and densest seawater masses are formed at the surface in the cold polar regions, and, as we have seen, more CO_2 can dissolve in that low-temperature environment. The dense water sinks, taking its large load of CO_2 to the bottom, and the pressure at depth helps to keep it in solution. CO_2 also builds in deep water because only heterotrophs (animals) live and metabolize there, and because CO_2 is produced as decomposers consume falling organic matter. No photosynthetic primary producers are present in the dark depths to use this excess CO_2.

Table 12.1   Solubility of Gases in Seawater as a Function of Temperature (Salinity = 33‰)

Temperature	Solubility (ml/l at atmospheric pressure)[a]		
	N_2	O_2	CO_2
0°C (32°F)	14.47	8.14	8,700
10°C (50°F)	11.59	6.42	8,030
20°C (68°F)	9.65	5.26	7,350
30°C (86°F)	8.26	4.41	6,660

Source: F. G. Walton-Smith. *CRC Handbook of Marine Science* (Cleveland, OH: CRC Press, 1974).

[a]Figures are given at *saturation*, the maximum amount of gas held in solution before bubbling begins.

Rapid photosynthesis at the surface lowers CO_2 concentrations and increases the quantity of dissolved oxygen. Oxygen is least plentiful just below the limit of photosynthesis because of respiration by many small animals at middle depths. These relationships were shown in Figure 6.6.

Low oxygen levels can sometimes be a problem at the ocean surface. Plants produce more oxygen than they use, but they produce it only during daylight hours. The continuing respiration of plants at night will sometimes remove much of the oxygen from the surrounding water. In extreme cases this oxygen depletion may lead to the death of the plants and animals in the area, a phenomenon most noticeable in enclosed coastal waters during spring and fall plankton blooms.

The greatest variability in levels of dissolved gas is found at the surface near shore. Less dramatic changes occur in the open sea.

Acid–Base Balance

The complex chemistry of Earth's life-forms depends on precisely shaped enzymes, large protein molecules that speed up the rate of chemical reactions. Like heat, strong acids or bases distort the shapes of these vital proteins, and they lose their ability to function normally. As we saw in Chapter 6, seawater is slightly alkaline; its average pH is about 7.8. The dissolved substances in seawater act to buffer pH changes, preventing broad swings of pH when acids or bases are introduced. The normal pH range of seawater is much less variable than that of soil (terrestrial organisms are sometimes limited by the presence of harsh alkali soils that damage cell components).

Though seawater remains slightly alkaline, it is subject to some variation. When dissolved in water, some

CO_2 becomes carbonic acid. In areas of rapid plant growth, pH will rise because CO_2 is used by the plants for photosynthesis. And because temperatures are generally warmer at the surface, less CO_2 can dissolve in the first place. Thus, surface pH in warm, productive water is usually around 8.5.[1]

At middle depths and in deep water, more CO_2 may be present. Its source is the respiration of animals and bacteria. With cold temperatures, high pressure, and no photosynthetic plants to remove it, this CO_2 will lower the pH of water, making it more acid with depth. Thus, deep, cold seawater below 4,500 meters (15,000 feet) has a pH of around 7.5. This lower pH can dissolve calcium-containing marine sediments (you may recall from Chapter 5 that sediments containing calcium carbonate are rarely found in deep water). A drop to pH 7 can occur at the deep-ocean floor when bottom bacteria consume oxygen and produce hydrogen sulfide.

Hydrostatic Pressure

Marine organisms are often subject to great pressure from the constant weight of water above them, but this so-called **hydrostatic pressure** presents very little difficulty to them. In fact, the situation in the ocean is parallel to that on land. Land animals live in air pressurized by the weight of the atmosphere above them (1 kilogram per square centimeter, or 14.7 pounds per square inch, at sea level) without experiencing any problems. Indeed, atmospheric pressure is necessary for breathing, flight, and some other physical necessities of life.

Pressures inside and outside an organism are virtually the same, both in the ocean and at the bottom of the atmosphere. Thus marine organisms do not need heavy shells to keep from being crushed by hydrostatic pressure. Great pressure does have some chemical effects: Gases become more soluble at high pressure, some enzymes are inactivated, and metabolic rates for a given temperature tend to be slightly higher. These effects are felt only at great depth, though. Unless marine organisms have gas-filled spaces in their bodies, a moderate change in pressure has little effect.

Limiting Factors

Often, too much or too little of a single physical factor can adversely affect the function of an organism. We call that factor a **limiting factor**, a physical or biological necessity whose presence in inappropriate amounts limits the normal action of the organism. Imagine, for example, an ocean area in which everything is perfect for photosyn-

[1]For a review of the pH scale, please see Figure 6.7.

thesis—warmth, nutrients, adequate CO_2—everything *except light*. In that circumstance no photosynthesis would occur; light is the limiting factor. If light were present but nitrates were absent, nitrate nutrients would be the limiting factor. Sometimes *too much* of something can be limiting—heat, for instance.

CLASSIFICATIONS OF THE MARINE ENVIRONMENT

Scientists have found it useful to divide the marine environment into *zones*, areas with homogeneous physical features. Attempts to classify the oceanic realm began in France in the 1830s, but they were hampered by lack of knowledge about the physical and biological features of the ocean. The situation changed rapidly near the end of the 1800s with the invention of tools for measuring physical factors and taking samples at various depths. Now these divisions can be made on the basis of light, temperature, salinity, depth, latitude, water density, or almost any of the other physical dimensions we have discussed.

One of the most commonly used divisions—in this book and in general—is classification by light availability. The **photic zone** (*photos* = light), the sunlit top layer of the ocean, extends in the tropics to a depth of approximately 200 meters (660 feet) and in productive mid-latitude water down to about 100 meters (330 feet). The upper half of the photic zone—the layer in which most biological productivity occurs—is called the **euphotic zone** (*eu* = good). Below the photic zone is the **aphotic zone** (*a* = without), the dark zone that extends to the bottom.

When the great English and American oceanographic institutions were founded, the need for a common, detailed nomenclature of position became urgent. Using a model from Harald Sverdrup as a point of departure, the scheme shown in **Figure 12.7** was devised by the National Academy of Sciences in the mid-1950s. It has withstood the test of time and is in common use today.

The primary division is between water and ocean bottom. Open water is called the **pelagic zone** (*pelagius* = of the sea), and it is divided into two subsections, the **neritic zone** (*neritos* = shallow) near shore over the continental shelf and the deep-water **oceanic zone** beyond the continental shelf. The oceanic zone is further divided by depth into realms. The *epipelagic zone* (*epi* = atop) corresponds to the lighted photic zone. In the aphotic depths are layered the *mesopelagic* (*mesos* = in the middle), *bathypelagic* (*bathos* = depth), and *abyssopelagic* (*a* = without + *byssos* = bottom) zones. Abyssopelagic water is the water in the deep trenches.

Divisions of the bottom are labeled **benthic** (*benthos* = bottom) and begin with the intertidal **littoral zone** (*lit-*

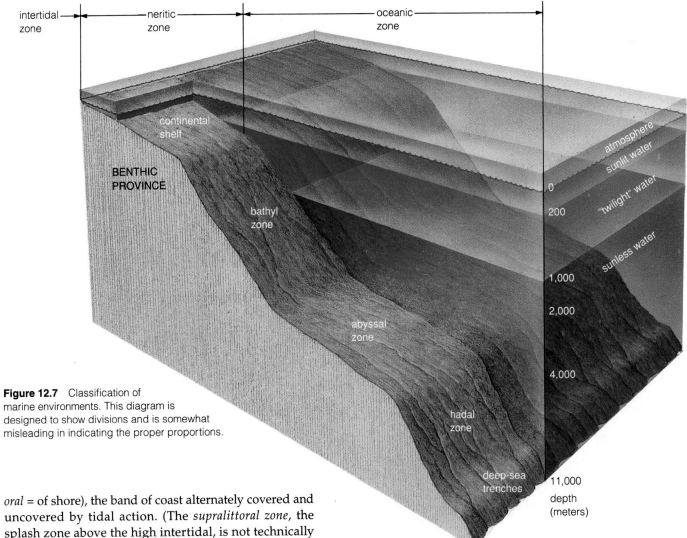

intertidal zone — neritic zone — oceanic zone

continental shelf

BENTHIC PROVINCE

bathyl zone

abyssal zone

hadal zone

deep-sea trenches

atmosphere

sunlit water

"twilight" water

sunless water

0

200

1,000

2,000

4,000

11,000

depth (meters)

Figure 12.7 Classification of marine environments. This diagram is designed to show divisions and is somewhat misleading in indicating the proper proportions.

oral = of shore), the band of coast alternately covered and uncovered by tidal action. (The *supralittoral zone*, the splash zone above the high intertidal, is not technically part of the ocean bottom.) Past the littoral is the *sublittoral zone* (*sub* = below), which is further divided into inner and outer segments. The *inner sublittoral* is ocean bottom near shore, the *outer sublittoral* is ocean floor out to the edge of the continental shelf.[2] The *bathyal zone* covers seabed on the slopes and down to great depths, where the **abyssal zone** begins. The **hadal zone** (*Hades* = underworld) is the deepest seabed of all, the trench walls and floors.

CLASSIFICATION OF OCEANIC LIFE

Just as oceanographers found it necessary to develop a standard classification and naming system for position in the oceanic realm, biologists centuries earlier had realized the value of being able to classify living things into categories and give them universally understood names.

[2]Many workers use the words *supertidal, intertidal,* and *subtidal* instead of the littoral terms.

Systems of Classification

The study of biological classification is called **taxonomy** (*taxo* = to put in order + *onoma* = name). Classification schemes have been around for as long as people have looked at living things. Putting animals in one category and plants in another is an ancient distinction, for example.

The Greek philosopher Aristotle proposed a system of classifying animals based on their exterior similarities, but his results were not very useful. Using his system we would place airline pilots, gliding squirrels, flying fish, and grasshoppers into the same group because each can fly! Such a system is an **artificial system of classification**. (Another artificial system of classification is the arrangement of books by jacket color, page size, or typeface.) By contrast, the **natural system of classification** for living

BOX 12.1 ● *Sea Serpents*

On 4 October 1848 the British frigate *Daedalus* arrived at Plymouth from the East Indies. Rumors immediately swept the port that she had sighted a 100-foot-long sea monster, at four o'clock in the afternoon of 6 August, between the Cape of Good Hope and St. Helena. The captain and most of the ship's officers and crew observed it for some 20 minutes (**Figure a**).

When this startling news reached the Lord Commissioners of the Admiralty in London, they immediately requested a formal report from Peter M'Quahe, captain of *Daedalus*. He wrote:

> . . . the object was discovered to be an enormous serpent, with head and shoulders kept about four feet constantly above the surface of the sea, and as nearly as we could approximate by comparing it with the length of what our main-topsail yard would show in the water, there was at the very least 60 feet of the animal [above the water], no portion of which was, to our perception, used in propelling it through the water. . . . It passed rapidly, but so close under our lee quarter, that had it been a man of my acquaintance, I should easily have recognised his features with the naked eye; and it did not, either in approaching the ship or after it had passed our wake, deviate in the slightest degree from its course to the [southwest],

which it held on at the pace of from 12 to 15 miles per hour, apparently on some determined purpose.

> The diameter of the serpent was about 15 or 16 inches behind the head, which was, without any doubt, that of a snake, and it was never, during the 20 minutes that it continued in sight of our glasses, once below the surface of the water; its colour a dark brown, with yellowish white about the throat. It had no fins, but something like a mane of a horse, or rather a bunch of seaweed, washed about its back.

This letter was published in *The Times* of 13 October. The British public were greatly excited by the report. After all, naval officers do not tell lies to the Lords of the Admiralty!

Nearly 60 years later, Sir Arthur Rostron, Commodore of the Cunard Line and, as captain of *Carpathia*, the man responsible for rescuing more than 700 survivors of the *Titanic*, reported a similar experience. Rostron was chief officer on the liner *Campania* when, on 26 April 1907, he saw something curious. "It was a sea monster!" he wrote in his memoirs.

> It was no more than fifty feet from the ship's side when we passed it and so both I and the junior officer of the watch had a good sight of it. So strange an animal was it that I remember crying out: "It's alive!" One has heard such yarns about these monsters and cocked a speculative eye at the teller, that I wished as never before that I had a camera in my hands. . . . The thing was turning its head from side to side for all the world as a bird will on a lawn between its pecks.

> I was unable to get a clear view of the monster's "features" but we were close enough to realize its head rose eight or nine feet out of the water while the trunk of the neck was fully twelve inches wide.

What were these competent, sea-wise men seeing? Almost certainly they *were* seeing something, an object rare enough to excite comment and disbelief. Perhaps they sighted the giant squid, a huge mollusk known to reach a total length of 17 meters (57 feet). Its two longest tentacles could conceivably extend from the water,

a "Appearance of the sea serpent when first seen from HBM ship *Daedalus*," according to the engraving that was first published in *The Times* along with Captain M'Quahe's letter.

periscope fashion, in a passable imitation of a serpent. But squid tentacles (and the squid to which they are attached) couldn't move consistently along the ocean surface "at the pace of from 12 to 15 miles per hour." A large seal might, but even the largest seals and sea lions do not approach the sizes reported in these and other sightings. More than one large seal could be mistaken for a single individual in poor observing conditions, however. Whale sharks certainly qualify in size (they can reach lengths to 18 meters, 60 feet), and although they don't often stick parts of their bodies out of water they can move purposefully in one direction for long periods of time. And what of masses of seaweed caught in a current? Mariners have reported possible "sea monsters" that upon closer inspection turn out to be clumps of seaweed rolled into curious shapes by storm waves. It's interesting that Captain M'Quahe reported that "something like a mane of a horse, or rather a bunch of seaweed, washed about its back." Lines of dolphins rising regularly to take air could also produce a long serpentlike effect. And don't forget upright-floating tree trunks, flocks of birds flying close to the water, or flotsam and other debris.

As we have seen, scientists never say never. Marine life is wonderfully varied: Many new species await discovery in the open ocean, and large oceanic reptiles may be among them. Beautiful plesiosaur-like animals (see **Figure b**) now thought to be extinct may someday be found, but the odds are overwhelming against such a dramatic discovery. No reptile could live undiscovered in the greatest depths because it would have to come to the surface to feed and breathe. Vertical movement from the surface to the abyss would expose the animal to terrific environmental stresses; no known animal is adapted to survive at all depths. Except for a few deep-diving marine mammals, air-breathers are limited to the uppermost layers. Any true sea serpents would probably live near the surface, and we would almost certainly be familiar with them. And no fresh plesiosaur carcasses or bones have washed ashore anywhere.

Does this mean there are no sea serpents in the ocean? Not necessarily, but it *does* suggest that such monsters are much more often found in the popular press and public imagination than in the ocean itself. Still, wouldn't it be wonderful if . . .

Source: J. B. Sweeney. *A Pictorial History of Sea Monsters and Other Dangerous Marine Life* (New York: Crown, 1972).

b An artist's impression of plesiosaurs, thought to be extinct for 60 million years.

Table 12.2	Classification of Organisms into Five Kingdoms	
Kingdom	Characteristics	Examples
Monera	Single-celled organisms lacking a nucleus and other internal structural subdivisions; feed by absorption, photosynthesis, chemosynthesis	Bacteria, cyanobacteria (blue-green algae)
Protista	Single-celled organisms possessing a nucleus and other internal structural subdivisions; feed by absorption, photosynthesis, or ingestion of particles	Diatoms, dinoflagellates, protozoa (foraminiferans, radiolarians, amoebas, etc.)
Fungi	Multicellular organisms lacking photosynthetic ability; nutrition by absorption	Mushrooms, molds
Plantae	Multicellular photosynthetic autotrophs	Large algae, mosses, ferns, flowering plants
Animalia	Multicellular heterotrophs	Invertebrates, vertebrates

Figure 12.8 Carolus Linnaeus—the father of modern taxonomy—in Laplander costume. (He went on a scientific expedition to Lapland in 1732.)

organisms, which biologists use today, relies on an organism's structural and biochemical similarities. We place all insects together regardless of their flying ability, just as we place all books by Melville together, all compositions of Schubert together, and all sea stars together because each group has a common underlying natural origin. The groups are arranged *systematically*—that is, in some order that makes structural and evolutionary sense.

One of the first persons to classify groups of organisms into natural categories was the eighteenth-century Swedish naturalist Carl von Linné, or as he called himself, **Linnaeus** (**Figure 12.8**). In his zeal to classify every aspect of the natural world, Linnaeus invented three supreme categories, or **kingdoms**: animal, vegetable, and mineral. Today's biologists leave the mineral kingdom to the geologists and have expanded Linnaeus's two living kingdoms to five. The names and characteristics of these five kingdoms are listed in **Table 12.2**.

Linnaeus's great contribution was a system of classification based on **hierarchy**, a grouping of objects by degrees of complexity, grade, or class. In this boxes-within-boxes approach, sets of small categories are nested within larger categories. Linnaeus devised names for the categories, starting with *kingdom* (the largest category) and passing down through *phylum, class, order, family,* and *genus*, to *species* (the smallest category). In 1758 he published a catalog of all animals then known, his monumental *Systema Naturæ* (The System of Nature). **Figure 12.9** shows the classification of a familiar sea gull using the Linnaean method. Note the nested arrangement of category-within-category, each category becoming more specific with every downward step.

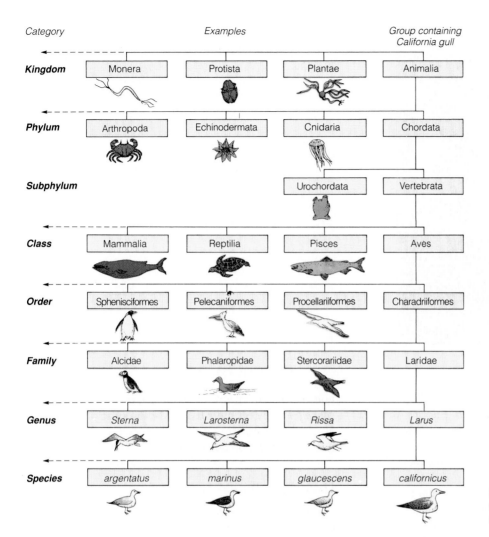

Category	Examples			Group containing California gull
Kingdom	Monera	Protista	Plantae	Animalia
Phylum	Arthropoda	Echinodermata	Cnidaria	Chordata
Subphylum			Urochordata	Vertebrata
Class	Mammalia	Reptilia	Pisces	Aves
Order	Sphenisciformes	Pelecaniformes	Procellariiformes	Charadriiformes
Family	Alcidae	Phalaropidae	Stercorariidae	Laridae
Genus	*Sterna*	*Larosterna*	*Rissa*	*Larus*
Species	*argentatus*	*marinus*	*glaucescens*	*californicus*

Figure 12.9 The modern system of biological classification, using the California gull (*Larus californicus*) as an example.

Names

Linnaeus also perfected the technique of naming animals. The *genus* and *species* names—the names of the last two nested categories—constitute an organism's **scientific name**. *Octopus bimaculatus* is the scientific name of a common West Coast octopus: *Octopus* is the generic name, *bimaculatus* the specific name. A closely related species, *Octopus dofleini*, is a larger animal that ranges to Alaska. *Octopus bimaculatus* and *Octopus dofleini* are not interfertile (they're not the same species), but, as their shared generic name suggests, they *are* closely related.

The advantage of a scientific name over a common name is immediately apparent to anyone trying to identify a shell found on the beach. The same shell may have many different common names in many different languages, but it will have *only one scientific name*. When you discover that name in a good guide to shells, you can use it to find references that will tell you what is known about the animal, its lifestyle, its range, and its evolutionary history.

MARINE COMMUNITIES

Organisms are distributed throughout the marine environment in specific groups of interacting producers, consumers, and recyclers that share a common living space. These groups are called *communities* (see **Figure 12.10**). A **community** is made up of the many populations of organisms that interact with one another at a particular location. A **population** is a group of organisms of the same species occupying a specific area. The location of a community, and the populations that make it up, depend on the physical and biological characteristics of that living space. In the next two chapters we will survey the

Figure 12.10 Students inspect a rocky intertidal community. In spite of wave shock, periodic exposure to drying wind and sun, and a broad temperature range, intertidal communities can be among the ocean's most diverse and productive. Nutrients are usually available here, and there are a large number of niches and habitats to be occupied.

organisms in the ocean's two great realms: the pelagic and benthic environments. Pelagic organisms (*pelagios* = of the sea) live suspended in the water; benthic organisms (*benthos* = seabed) live on or in the ocean bottom.

The largest marine community in area—and the most sparsely populated—is the pelagic community lying within the uniform mass of permanently dark water between the sunlit surface and the deep bottom. Few animals live there because so little food is available, but those organisms that survive are among the strangest in the ocean. Opportunities for feeding in the deep open-ocean community are few and far between, so some animals are able to consume prey larger than themselves should the occasion arise. Because so few animals are present, mating is also a rare event; in a few species, males and females become permanently bonded during their first encounter, the male burrowing into the female's body for a life-long free ride.

In contrast, the smallest obvious marine communities may be those benthic communities established against solitary rocks on an otherwise flat, featureless seabed. Drifting larvae will colonize the place. The established community can seem an oasis of life and activity in an otherwise static sedimentary desert. Seaweeds will grow, worms will burrow, snails will scrape algae from the hard surfaces, and small fishes will nestle among crevices. Hundreds of small plants and animals can live their lives within a meter of each other, interacting in a compact solitary community with no similar environment available for thousands of meters. The larvae of the next generation drift away with little chance of finding a suitable place to carry on their lives. Microscopic communities also exist: Interacting populations can exist on a single grain of sand or on one decomposing fish scale.

Organisms Within Communities

There are many different places to live and many different "jobs" for organisms within even a simple community. A **habitat** is an organism's "address" within its community, its physical *location*. Each habitat has a degree of environmental uniformity. An organism's **niche** (*nidus* = nest) is its "occupation" within that habitat, its relationship to food and enemies, an expression of what the organism is *doing*. For example, the small fishes living among the coral heads in a coral reef community share the same habitat, but each species has a slightly different niche. Each population in the community has a different "job" for which its shape, size, color, behavior, feeding habits, and other characteristics particularly suit it.

Competition

The availability of resources such as food, light, and space within a community determines the number and composition of the populations of organisms within that community. Competition for the necessities of life may occur within the community between members of the *same* population or between members of *different* populations. Subtle swings in physical or biological factors may give one population the advantage for a time but then shift to favor another.

When *members of the same population* (all members of the same species) compete with each other, some individuals will be larger, stronger, or more adept at gathering food, avoiding enemies, or mating. These animals tend to prosper, forcing their less successful relatives to emigrate, fight, or die in the course of competition. This kind of competition continually fine-tunes a population to its environment.

When *members of different populations* compete, one population may be so successful in its "job" that it eliminates competing populations. In a stable community, two populations cannot occupy the same niche for long. Eventually the more effective competitor overwhelms the less effective one. For example, the little barnacle *Chthamalus* lives on the uppermost rocks in many intertidal communities, while the larger barnacle *Balanus* lives on the lower rocks (see **Figure 12.11**). Planktonic larvae of both species can attach themselves to rocks anywhere in the intertidal zone and begin to grow. In the lower zone the faster-growing *Balanus* push the weaker *Chthamalus* off the rocks, while at higher positions *Balanus* cannot survive because it is not so resistant to drying and exposure as the tough little *Chthamalus*. At the top and bottom of their distribution the two species do not compete for food or space. The competition at the intersection of their ranges prevents each species from occupying as much of the habitat as might otherwise be possible.

As we have seen, physical and biological factors affect the number and positions of organisms in a community. The number of individuals per unit area (or volume) is known as the **population density**. Rare individuals have a much lower population density than dominant ones. In general, *more* different species exist in benign habitats where physical factors stay near optimal values (like a coral reef or rain forest), and *fewer* species exist in rigorous habitats where physical factors range to extremes (like a beach or a desert). That is, "easy" habitats typically have high **species diversity** (they contain more species in more niches within a given area), and harsh habitats usually have a lower species diversity. Relatively few species can cope with the stressful environment of the polar ocean, for example, but many species have adapted to the relatively benevolent environment of a tropical reef.

CHANGE IN MARINE COMMUNITIES

Like the organisms that compose them, communities change through time, but marine communities do not evolve as rapidly as terrestrial communities. The slow

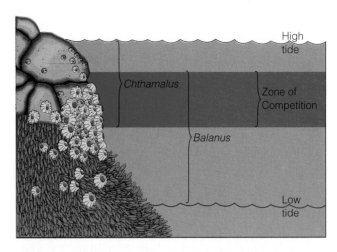

Figure 12.11 Competition between two species of barnacles prevents either from occupying as much of the intertidal zone as possible. The central zone of overlap indicates the area where *Chthamalus* and *Balanus* compete for food and space.

changes associated with seafloor spreading, climate cycles, atmospheric composition, or newly evolved species have shaped this generally slow evolution. As on land, the species, community composition, and location of a marine community are changed by the environmental factors to which members of the community are exposed. Communities themselves can gradually modify the physical aspects of their environment. A coral reef is an extreme example: The massive accumulation of coral and sediments within the reef can alter current patterns, influence ocean temperature, and change the proportions of dissolved gases.

But rapid changes can occur in marine communities. A natural catastrophe—a volcano erupting, a landslide that blocks a river, or the collision of an asteroid with Earth, for example—can disrupt a community. Similarly, human activities—such as altering an estuary by damming a river, dumping excess nutrients into a nearshore area, or stressing organisms with toxic wastes—can cause rapid, disruptive changes. The composition of offshore communities changes abruptly near new sewage outfalls, for example.

A stable, long-established community is known as a **climax community**. This self-perpetuating aggregation of species tends not to change unless disrupted by severe external forces such as violent storms, significant changes in current patterns, epidemic diseases, or influx of great amounts of fresh water or pollutants. A disrupted climax community can be reestablished through the process of **succession**, the orderly changes of a community's species composition from temporary inhabitants to long-term inhabitants. Disruption makes the environment more hostile to the original species, but destruction of species in the original community leaves open habitats and niches. A few highly tolerant species will move into the area, eventually drawing in other species that depend on them. If the environment is permanently changed by the disruption, a different climax community will be established than was previously present. In the next chapters we turn our attention to the climax communities of the ocean's vast pelagic and benthic domains.

CHAPTER SUMMARY

Life on Earth is notable for both unity and diversity: *diversity* because there are at least 50 million different species (kinds) of living things on Earth; *unity* because each species shares the same underlying mechanisms for capturing and storing energy, manufacturing proteins, and transmitting information between generations. The atoms in living things are no different from the atoms in nonliving things, and the energy that powers living things is the same energy found in inanimate objects.

Because of its watery nature and origin, in a sense all life on this planet is marine. The oceanic environment is a relatively easy place for cells to capture, transform, and store the energy they require to maintain their complex organization and to grow and reproduce. In part at least, life in the ocean is so successful—total marine productivity is so high—because of the ocean's physical charac-

teristics, but these characteristics may also be limiting factors for an organism.

Evidence suggests that life has diversified and adapted to varying environments by the process of natural selection. Life has changed through time, and life and the Earth have grown old together. Primary producers—autotrophs—are organisms that synthesize food from inorganic substances. Autotrophic marine organisms transform energy from the sun (or from certain inorganic molecules) into chemical energy to power their own growth, and they are in turn consumed by heterotrophic organisms. The feeding relationships in a community resemble complex webs.

A variety of physical factors affects the density, variety, and success of the life-forms in each marine habitat. These factors include water's transparency, temperature, dissolved nutrients, salinity, dissolved gases, hydrostatic pressure, acid–base balance, and others.

Marine organisms are classified by their physical characteristics and by the degree to which they resemble other organisms. The various marine environments populated by marine life may be classified by physical characteristics. Organisms are distributed throughout the marine environment in specific communities—groups of interacting producers, consumers, and recyclers that share a common living space. The location of a community, and the variety of organisms that constitute it, depends on the physical and biological characteristics of that living space.

Terms and Concepts to Remember

abyssal zone	heterotroph	pelagic zone
aphotic zone	hierarchy	photic zone
artificial system of classification	hydrostatic pressure	photosynthesis
autotroph	kingdom	physical factor
benthic	limiting factor	population
chemosynthesis	Linnaeus (Carl von Linné)	population density
climax community		primary consumer
community	littoral zone	primary producer
ectothermic	metabolic rate	primary productivity
endothermic	metabolism	scientific name
euphotic zone	natural system of classification	species diversity
evolution		succession
food	neritic zone	taxonomy
food web	niche	top consumer
habitat	nutrient	trophic pyramid
hadal zone	oceanic zone	

Study Questions

1. What is the ultimate source of the energy used by most living things?

2. What do primary producers produce? How is productivity expressed?

3. What is an autotroph? A heterotroph? How are they similar? How are they different?

4. What is a trophic pyramid? What is the relationship of organisms in a trophic pyramid? Does this have anything to do with food webs?

5. Name and briefly discuss five physical factors of the marine environment that impact living organisms. How is each different in the ocean from on the land?

6. What is a limiting factor? Can you think of some examples not given in the text?

7. How is the marine environment classified? Which scheme is most useful? Justify your answer.

8. How does a natural system of classification differ from an artificial system? Can you give an example of each? Was the hierarchy-based system invented by Linnaeus natural or artificial? What *is* a hierarchy-based system?

9. What are communities? How are marine organisms distributed in communities, and what factors influence who lives where? What is a climax community?

For Further Study

Andrewartha, H. G. 1961. *Introduction to the Study of Animal Populations*. Chicago: University of Chicago Press. An important classic.

Cloud, P. 1983. "The Biosphere." *Scientific American*, September, 176–89. Earth and life have evolved together.

Eiseley, L. 1956. "Charles Darwin." *Scientific American*, February, 62–72. A wonderfully written short biography worth searching for.

Friedrich, H. 1969. *Marine Biology, An Introduction to Its Problems and Results*. London: Sedgwick & Jackson. English translation from German by G. Vevers. Fine exposition on the influence of physical factors on life in the sea.

Isaacs, J. 1969. "The Nature of Oceanic Life." *Scientific American*, September, 146–62. A valuable overview, part of the now-classic issue of *Scientific American* dedicated to the revolution in marine science.

Kozloff, E. N. 1990. *Invertebrates*. Philadelphia: W. B. Saunders. Outstanding new text, very well written by an expert marine biologist.

Margulis, L,. and K. Schwarz. 1988. *Five Kingdoms*. 2d ed. San Francisco: Freeman. Excellent overview of the difficulties in classifying organisms into phyla.

Milne, David H. 1995. *Marine Life and the Sea*. Belmont, CA: Wadsworth Publishing Company.

Schmidt-Nielsen, K. 1972. *How Animals Work*. London: Cambridge University Press. A simplified look at structure, function, and biochemistry.

Stanley, S. M. 1986. *Earth and Life Through Time*. New York: Freeman. Excellent standard text.

Starr, C. 1994. *Biology: Concepts and Applications*. 2d ed. Belmont, CA: Wadsworth. Perhaps the best general biology text available. The principles of unity and evolution serve as organizers. Gracefully written, thorough.

Sumich, J. L. 1991. *An Introduction to the Biology of Marine Life*. 5th ed. Dubuque, IA: W. C. Brown.

13

PELAGIC COMMUNITIES

". . . silver-shining, swift, strong, streamlined . . ."

Among the ranks of marine drifters and swimmers, members of the tuna family are the ocean's fastest and widest-ranging animals. The body of a tuna is dedicated to speed. Its fins retract into slots, its eyes form a smooth surface with the rest of the head, and it may consume as much as 25% of its weight in food each day. Indeed, a tuna uses so much energy that one of its greatest physiological problems is to avoid overheating! The biological equivalent of the legendary Flying Dutchman, these powerful fishes are fated to travel continuously. If they ever stopped they would suffocate, and their massive bodies would fall to the depths.

Tuna and their relatives swim enormous distances and exhibit astonishing bursts of speed. Studies have shown that albacore tuna regularly migrate from the coast of California to Japan and back, a one-way trip of 8,500 kilometers (5,300 miles), moving at an average speed of not less than 26 kilometers (16 miles) per day. Tagged bluefin tuna (**a**) have traveled at least 7,770 kilometers (4,830 miles) across the North Atlantic in 119 days, that is, over 65 kilometers (40 miles) each day. In reality, the bluefins must have traveled much farther, continually detouring from a straight-line course to hunt for food. The fastest tuna can maintain speeds of over 75 kilometers (47 miles) per hour , and the sailfish, a close relative, can rocket to 110 kilometers (68 miles) per hour for a short time.

These magnificent fishes are valued for their meat—especially in Japan, where they are highly prized for sashimi, the raw fish component of sushi. In the Tokyo fish market, old and fat bluefin tuna have sold for as much as $70,000 a ton! In 1989 a school of huge bluefin appeared off the California Channel Islands. The largest weighed 458 kilograms (1,008 pounds). Many lucky fishermen paid off home mortgages and boat loans in a week of heroic fishing. But it is the living animal that provides the greatest inspiration: silver-shining, swift, strong and streamlined, these silent nomads slip through the ocean more than a million miles in a lifetime.

The bluefin tuna, the widest-ranging and one of the fastest of the ocean's pelagic animals. Sometimes called a *horse mackerel* because of its enormous size and strength, the record bluefin weighed 458 kilograms (1,008 pounds)!

The organisms that live suspended in seawater are immensely varied, but all have common problems of maintaining their vertical position in the water, obtaining food, and surviving long enough to reproduce. They can be divided into two broad groups based on their lifestyle: The **plankton** drift or swim weakly, going where the ocean goes, unable to move consistently against waves or current flow. The **nekton** are pelagic organisms that actively swim.

PLANKTON

The pelagic organisms that constitute plankton are as important as they are inconspicuous. The word is derived from the Greek word *planktos*, meaning "wandering."

The diversity of planktonic organisms is astonishing: There are giant drifting jellyfish with tentacles 8 meters (25 feet) long, small but voracious arrowworms, many single-celled creatures that glow brightly when disturbed, mollusks with slowly beating flaps that resemble butterfly wings, crustaceans that look like microscopic shrimp, miniature jet-propelled animals that live in jelly-like houses and filter food from water, and shimmering, crystal-shelled algae. The only feature common to all plankton is their inability to make consistent lateral headway, though many can and do move vertically in the water column. A photograph of concentrated plankton is shown in **Figure 13.1**.

The plankton contain many different plantlike species and virtually every major group of animals. Thus, the term *plankton* is not a collective natural category like mollusks or algae, which would imply an ancestral relationship between the organisms; instead it describes a basic ecological connection. Members of the plankton community, informally referred to as *plankters*, can and do interact with one another: There is grazing, predation, parasitism, and competition among members of this dynamic group.

Collecting and Studying Plankton

The first researcher to appreciate the uniqueness of plankton was the Danish biologist Anders S. Oersted. He was prompted to look for tiny drifters in the ocean after noticing microscopic algae that discolored water in an aquarium. His observations, made during a voyage from Denmark to the West Indies in 1847, revealed that the open ocean was not devoid of plant life. He correctly concluded that these single-celled autotrophs were the primary food source for most animal life in the ocean. His brief report, written in Danish, was generally overlooked by the scientific community for almost 50 years. The first large-scale systematic study of plankton was carried out by biologists aboard the research vessel *Meteor* during the German Atlantic oceanographic expedition of 1925–26. Many of the tools and techniques they pioneered are still in use today.

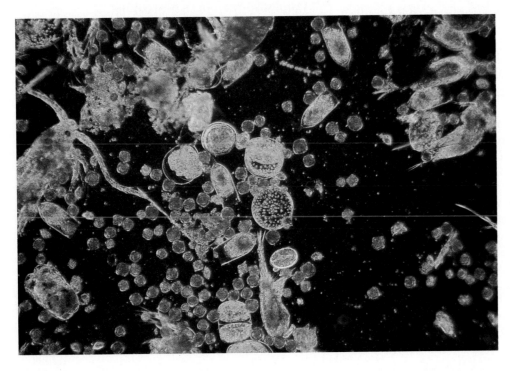

Figure 13.1 Concentrated plankton collected with a plankton net. Both plant and animal species are shown. In bright light the largest of these organisms would barely be visible to the unaided eye. The most numerous small round objects are dinoflagellates; the larger round organisms at the center of the photograph are diatoms; and the single-eyed shrimplike animal at the far left is a copepod.

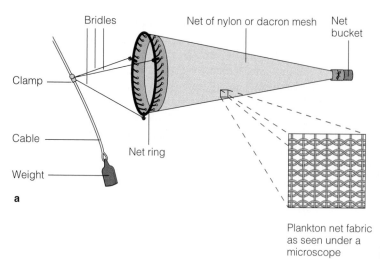

Plankton net fabric
as seen under a
microscope

Figure 13.2 Plankton nets. (a) The standard conical net is made of fine mesh and has a mouth up to 1 meter (3.3 feet) in diameter. The net is towed behind a ship for a set distance. The number of organisms present in the water can be estimated if the trapped organisms are counted and the volume of sampled water is known. A plankton net's filtering efficiency (the effectiveness with which it removes organisms from the intended water sample) must be taken into account when calculating the volume of water sampled. An abundance of large organisms or great numbers of individuals can sometimes clog the net, reducing its filtering efficiency. (b) The net shown here has a somewhat coarser mesh because its target organisms, small shrimplike crustaceans known as krill, are relatively large.

Plankton nets (Figure 13.2) of the type perfected for the *Meteor* expedition are essential to plankton studies. These conical nets are customarily made of nylon or Dacron cloth woven in a fine interlocking pattern to assure consistent spacing between threads. The net is hauled slowly for a known distance behind a ship, or cast to a set depth, and then reeled in. Trapped organisms are flushed to the net's pointed end and carefully removed for analysis. Quantitative analysis of plankton requires the organisms and an estimate of the sampled volume of water.

Very small plankton can slip through a plankton net. Their capture and study require concentration by centrifuge, or entrapment by a plankton filter through which water is drawn. The filter is later disassembled and the plankton studied in place.

Measurements of physical ocean conditions—such as dissolved carbon dioxide and oxygen content, pH, temperature, and light intensity at the time and place of sampling—are of special importance in interpreting the samples. **Synoptic sampling** (*syn* = together + *optikos* = for sight; to see together), simultaneous sampling at many locations, can be useful in pinpointing the often subtle interplay of species and physical conditions that complicate our understanding of plankton biology.

Phytoplankton: The Autotrophs

Autotrophic plankton is generally called *phytoplankton*, a term derived from the Greek word *phyton*, meaning "plant." A huge, nearly invisible mass of phytoplankton drifts within the sunlit surface layer of the world ocean. Phytoplankton is critical to all life on Earth because of its great contribution to food webs and its generation of large amounts of atmospheric oxygen through photosynthesis. Planktonic autotrophs are thought to bind at least 20 billion metric tons of carbon into carbohydrates each year, at least 25% of the food made by photosynthesis on Earth! These easily overlooked, mostly single-celled drifting photosynthesizers are much more important to marine productivity than the larger and more conspicuous seaweeds.

There are eight major types of phytoplankton, of which the most important are the diatoms, the dinoflagellates, and the coccolithophores.

Diatoms The dominant photosynthetic organisms in the plankton—and in the world—are the **diatoms**. More than

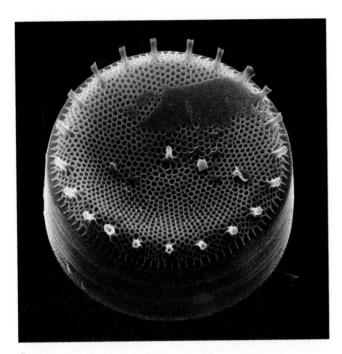

a

b

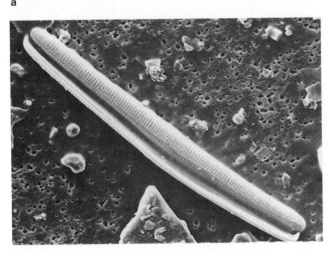

c

Figure 13.3 (a) The valve, or "shell," of the diatom *Coscinodiscus*. A scanning electron micrograph of the beautiful silica frustule. (b) A closer view, showing the intricate perforations from which diatoms take their name. (c) A scanning electron micrograph of *Nitzschia*, a pennate (elongated) diatom. (d) *Chaetoceros*, a chain-forming centric (round) diatom. Each of the cells in this figure is about 15 micrometers (300 millionths of an inch) across.

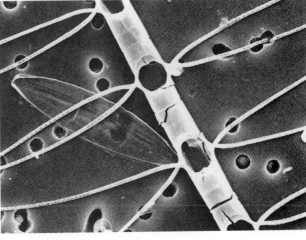

d

5,600 species of diatoms are known to exist. The larger species are barely visible to the unaided eye. Most are round, but some are elongated or branched or triangular.

Typical diatoms are shown in **Figure 13.3**. The name means "cut through" (*dia* = through + *tomos* = to cut), a reference to the patterns of perforations through the diatom's rigid cell wall, or **frustule** (*frustulum* = a small piece). As much as 95% of the mass of the frustule consists of silica (SiO_2), giving this beautiful covering the optical, physical, and chemical characteristics of glass—an ideal protective window for a photosynthesizer. Magnification reveals that the frustule consists of two closely fitting halves, or **valves**, which fit together like a well-made gift box, the top valve adhering tightly over the lip of the bottom one. The pattern of perforations, slits, striations, dots, and lines on the surface of the valves is different for each diatom species (**Figure 13.3b**). When diatoms die, their valves can drift to the seafloor to accumulate as layers of siliceous ooze (see Chapter 5.)

Inside the diatom's tailored valves lies the most nearly perfect photosynthetic machine yet to evolve on the planet. Fully 55% of the energy of sunlight absorbed by a diatom can be converted into the energy of carbohydrate chemical bonds, one of the most efficient energy conversion rates known. Excess oxygen not needed in the cell's respiration is released through the perforations in the frustule into the water. Some oxygen is absorbed by marine animals, some is incorporated into bottom sediments, and some diffuses into the atmosphere. Most of the oxygen we breathe has moved recently through the many glistening pores of diatoms.

Diatoms store energy as fatty acids and oils, compounds that are lighter than their equivalent volume of water and thus assist in flotation. As you might guess, flotation is a potential problem for diatoms because the weight of their heavy silica frustule seems at odds with their requirement for staying near the sunlit ocean surface. Oil floats, glass sinks, and a balanced amount of both produces neutral buoyancy. Not all diatoms need to float, however. Many nonplanktonic species lie on shallow bottoms where light and nutrients are able to support photosynthesis. These benthic species are nearly always elongated (or *pennate*) in shape.

Dinoflagellates   As a child I remember walking along a dark California beach where the sand in my footprints glowed after each step. Handfuls of sand thrown out over the wet beach surface burst into fans of blue light strong enough to cast a shadow, and swimmers emerged from the warm water with thousands of glowing points of light clinging to their bodies. The light I saw was caused by dinoflagellates that had multiplied rapidly in the nearshore waters off southern California. In the late 1950s, wastewater treatment was not as thorough as it is today, and the effluent was not pumped as far out to sea. The combination of warm, calm water and abundant nutrients often triggered such blooms. These displays are rare today, but on warm September or October nights, after the northeast winds have blown the surface water out to sea and upwelling has brought nutrients to the shore, the sand sometimes glows, and waves can flash blue as they break. These dinoflagellates were bioluminescent (*bios* = life + *luminis* = light). Waste oxygen, left

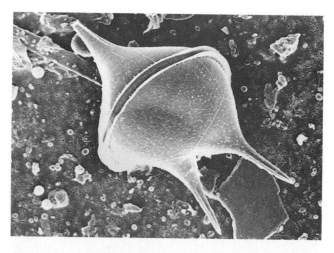

Figure 13.4   A dinoflagellate, genus *Protoperidinium*. Two flagella (not visible in this scanning electron micrograph) beat in opposing grooves in the armor-plated body. One flagellum beats in the equatorial groove, another extends from a slit in the organism's base. In a strong light, this dinoflagellate would barely be visible to the unaided eye.

over from photosynthetic reactions, combines with enzymes and an imaginatively named protein, *luciferin*, causing the organisms to glow.

Most **dinoflagellates** (**Figure 13.4**) are single-celled autotrophs. A few species live within the tissues of other organisms (the zooxanthellae of coral animals discussed in Chapter 14, for example), but the great majority of dinoflagellates live free in the water. Most have two whiplike projections called **flagella** in channels grooved in their protective outer covering of cellulose. One flagellum drives the organism forward while the other causes it to rotate in the water (hence the name: *dino* = whirling + *flagellum* = whip). Their flagella allow dinoflagellates to adjust orientation and vertical position to make the best photosynthetic use of available light.

Some species of dinoflagellates can become so numerous that the water turns a rusty red as light reflects from the accessory pigments within each cell. These species are responsible for the phenomenon of *red tide*. During times of such rapid growth (usually in springtime), concentration of these microscopic organisms may briefly reach 6 million per liter (23 million per gallon)!

Red tide can be dangerous because some dinoflagellate species synthesize potent toxins as byproducts of metabolism. Among the most effective poisons known, these toxins may affect nearby marine life or (if the dinoflagellates are dried on the beach and blown inland), even humans. Some of the toxins are similar in chemical structure to the muscle relaxant curare, but they are tens of times more powerful. You should avoid eating certain species of clams, mussels, and other filter feeders during summer months when toxin-producing dinoflagellates are abundant in the plankton. If shellfish from a particular area are unsafe, a state governmental agency will issue an advisory that may remain in effect for six weeks or more until the danger is past.

Coccolithophores and Other Phytoplankton Most other types of phytoplankton are extraordinarily small and so are called **nanoplankton** (*nanus* = dwarf). The **coccolithophores**, for example, are tiny single cells covered with discs of calcium carbonate (coccoliths) fixed to the outside of their cell walls (**Figure 13.5**). Coccolithophores live near the ocean surface in brightly lighted areas. The translucent covering of coccoliths may act as a sunshade to prevent absorption of too much light. In areas of high coccolithophore productivity, most notably in the Mediterranean and Sargasso seas, their numbers occasionally become so great that the water appears milky or chalky. Coccoliths can also build seabed deposits of ooze. The famous White Cliffs of Dover in England consist largely of fossil coccolith ooze deposits uplifted by geological forces.

Other very small phytoplankters, the **silicoflagellates**, possess filamentous internal supporting structures of silica, the same material from which diatoms construct their valves. One or two flagella are always present. Compared to diatoms, however, the overall structure and biochemistry of silicoflagellates seem primitive. Little is known about their worldwide distribution or abundance. As **Box 13.1** suggests, recent evidence suggests that silicoflagellates, coccolithophores, cyanobacteria, and other, even tinier photosynthesizers may contribute much more to overall plankton productivity than previously thought.

Phytoplankton Productivity

As we saw in Chapter 12, phytoplankton are responsible for the majority of marine primary productivity. Where is phytoplankton productivity the greatest? This question can be examined in two parts: (1) Where in the water column is productivity highest, and (2) where across the world ocean are the areas of greatest primary productivity?

At what depth is productivity the greatest? Phytoplankters are adapted to operate optimally using less light than is usually found at the ocean surface; this permits greater productivity through the euphotic zone as a whole. Though it is difficult to generalize for all the ocean, the depth at which phytoplankton productivity is often greatest is typically around 20 meters (66 feet) at local noon, when illumination is greatest and the depth of maximum productivity the deepest. Figures averaged for a whole day (see **Figure 13.6**) indicate that the depth of

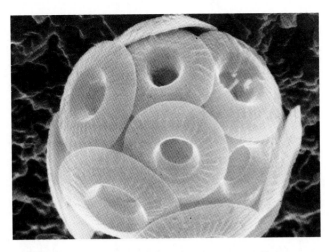

Figure 13.5 Calcium carbonate plates in place on *Umbilicosphaera*, a coccolithophore. The tiny plates are 6 micrometers (120 millionths of an inch) across, about four times larger than large ultraplankton.

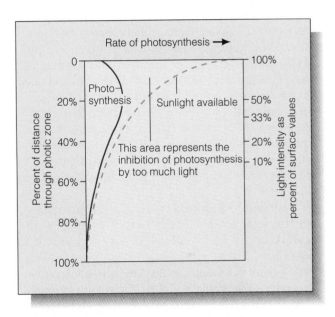

Figure 13.6 The rate of photosynthesis as a function of depth. The photic zone may extend to a depth of 100 meters (330 feet) in clear oceanic waters (as shown here) or to less than 20 meters (66 feet) in turbid nearshore waters. Note that too much light inhibits photosynthesis near the surface. As the depth increases, the available sunlight decreases, but the plants become more efficient at using it. Maximum productivity (the highest net rate of photosynthesis) takes place about 20% of the way through the photic zone.

BOX 13.1 • *Ultraplankton!*

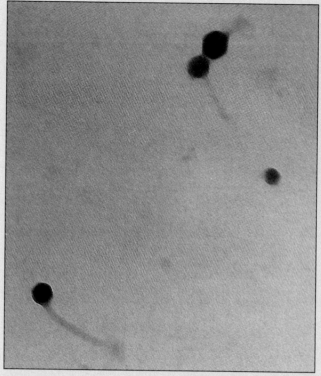

a

In the mid-1970s marine biologists began to appreciate the contribution to oceanic productivity of a new class of phytoplankton they termed *ultraplankton*. **Ultraplankton** is the smallest of the small—diminutive nanoplankters so tiny that they were long mistaken for fine particles of suspended sediments. These organisms are too small to be resolved by light microscopes and slip undetected through all but the very finest filters. Their size, typically about 2 micrometers (40 millionths of an inch) across, is made up for by their abundance: an astonishing 100,000 cells per milliliter (3 million cells per ounce) in tropical and subtropical waters previously thought to be all but sterile.

Synechococcus is typical of this newly recognized type of organism. It was discovered when individual cells—little more than naked photosynthetic machines—fluoresced a bright orange when struck by ultraviolet light. Analysis of the fluorescence spectrum (living examples of *Synechococcus* are too small to study directly) revealed the presence of an odd chlorophyll variant previously known only in a rare mutant strain of corn. Structural details revealed by electron micrographs suggest that *Synechococcus* and its relatives may be cyanobacteria. Even smaller photosynthesizers have been discovered, some less than 1.5 micrometers (30 millionths of an inch) in diameter.

Recent estimates suggest that ultraplankton may account for up to 70% of all the photosynthetic activity of the world ocean! How could such a huge contribution to oceanic productivity go unnoticed for so long? In part because these autotrophs are so extremely small, and in part because the products of their photosynthetic activity are closely cycled to even smaller heterotrophic bacteria in the immediate vicinity! Here is a complete microecosystem—a community operating on the smallest possible scale—that manufactures and consumes carbohydrates and carbon dioxide in amounts almost beyond comprehension, a sort of ecological black market below the "official economy" of the relatively huge diatoms and dinoflagellates. As if they weren't busy enough, these bacteria also decompose organic material (spilled into the water when phytoplankton are eaten by zooplankton), clean up the liquid excretions of zooplankton dissolved in seawater, and process the small amounts of cytoplasm that phytoplankton exude into the ocean as they age. Many marine biologists now believe the greatest fraction of organic particles in the water column of the open ocean is composed of these metabolically active bacterial cells—the ultraplankton.

And what happens to these bacteria? In a 1989 article in the British journal *Nature*[a], biologists announced the discovery of *still smaller* organisms, viruses less than 0.2 micrometer (4 millionths of an inch) in diameter. These viruses—a class of simple creatures known as *bacteriophages*—inject their genetic material into a bacterium and trick it into manufacturing more viruses. The bacteria die in the process. Researchers reported finding viruses in numbers that stagger the imagination: up to 100 million viruses per milliliter (nearly 3 billion per ounce). Ultraplankton, indeed!

[a]Bergh, O., K. Y. Borshem, F. Bratbak, and G. Heldal. 1989. "High Abundance of Viruses Found in Aquatic Environments." *Nature* 340 (no. 6233): 467–68.

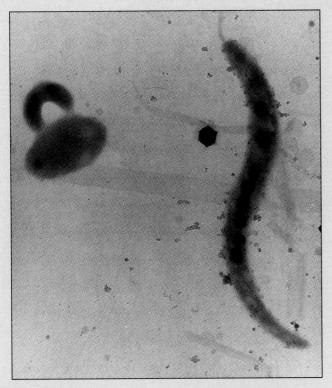

b

Ultraplankton. (a) Bacteriophages. These viruses inject their genetic material into a bacterium and trick it into manufacturing more viruses. (b) Three bacteria—ellipsoidal, **S**-shaped, and **C**-shaped—from the Gulf of Bothnia (east of Sweden). The dark dots within the bacteria are viruses. The large hexagonal structure near the **S**-shaped bacterium is a relatively large virus with a diameter of 160 nanometers. The bacteria and viruses were photographed using a transmission electron microscope.

greatest productivity is between 5 and 10 meters (17 and 33 feet) below the surface, about one-fifth the way through the photic zone.

Where are the areas of greatest productivity? Because of coastal upwelling and land runoff, nutrient levels are highest near the continents. Plankton is most abundant there, and productivity is highest. The water above some continental shelves sustains productivity in excess of 1 gC/m²/*day*! **Figure 13.7** shows the relation between shore proximity, nutrient levels, light, and productivity off the U.S. East Coast. But what of the open ocean? Where is productivity greatest away from land?

In the tropics? The open tropical oceans have abundant sunlight and CO_2 but are generally deficient in surface nutrients because the strong thermocline discourages the vertical mixing necessary to bring nutrients from the lower depths. The tropical oceans away from land are therefore oceanic deserts nearly devoid of visible plankton. The typical clarity of tropical water underscores this point. In most of the tropics productivity rarely exceeds 30 gC/m²/yr, and seasonal fluctuation in productivity is low.

Tropical coral reefs—benthic habitats—are exceptions to this general rule. Reef areas, which account for less than 2% of the tropical ocean surface, are productive places because autotrophic dinoflagellates live *within* the tissues of coral animals and don't drift randomly as plankton. Nutrients are made available by coastal upwelling and by the coral's own metabolism. These nutrients are cycled tightly through the reef and not lost to sinking.

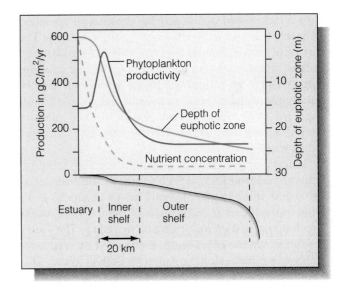

Figure 13.7 Geographical variation in the euphotic depth, nutrient concentration, and productivity on a continental shelf. The data were collected from the coast of Georgia to the edge of the continental shelf.

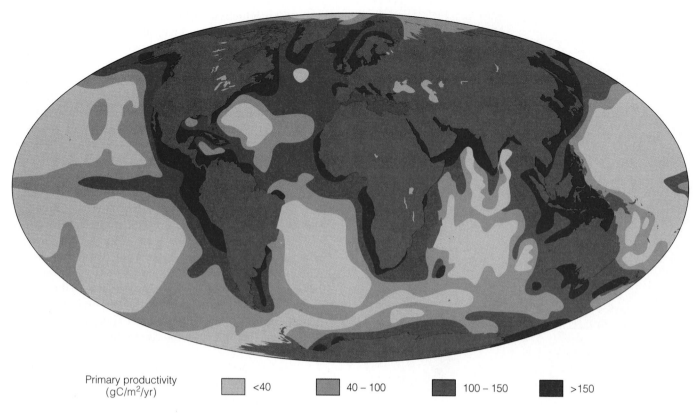

Primary productivity
(gC/m²/yr)

| | <40 | | 40 – 100 | | 100 – 150 | | >150 |

Figure 13.8 Distribution of primary productivity in the world ocean.

In the polar regions? At very high latitudes the low sun angle, reduced light penetration due to ice cover, and weeks or months of darkness in winter severely limit productivity. At the height of summer, however, 24-hour daylight, a lack of surface ice, and the presence of upwelled nutrients can lead to spectacular plankton blooms. The surface of some sheltered bays can look like tomato soup because dinoflagellates and other plankton are so abundant. This bloom cannot last because nutrients are not quickly recycled and because the sun is above the critical angle for a few weeks at best. The short-lived summer peak does not compensate for the long, unproductive winter months.

Average productivity at very high northern latitudes tends to be lower than at high southern latitudes, averaging less than 25 gC/m²/yr. The Arctic Ocean is almost surrounded by land masses, which limit water circulation and, therefore, nutrient upwelling. The southern ocean, on the other hand, is enriched by water upwelling to replace sinking Antarctic Bottom Water. This rich mixture is stirred by the West Wind Drift. The Antarctic accounts for a much greater share of high-latitude production than the Arctic because more nutrients are available.

In the temperate and subpolar zones? With the tropics generally out of the running for reasons of nutrient de-

ficiency and the north polar ocean suffering from slow nutrient turnover and low illumination, the overall productivity prize is left to the temperate and southern subpolar zones. Thanks to the dependable light and the moderate nutrient supply, annual production in the nearshore temperate and southern subpolar ocean areas is the greatest of any open-ocean area. Typical productivity in the temperate zone is about 120 gC/m²/yr. In ideal conditions southern subpolar productivity can approach 250 gC/m²/yr!

Figure 13.8 shows the levels of productivity in tropical, temperate, and polar ocean areas. Note that *nearshore* productivity is nearly always higher than *open-ocean* productivity, even in the relatively productive temperate and south subpolar zones.

Curiously, the open-ocean area with the greatest annual productivity is an exception to the general picture developed in this section. The slender, cold finger of high productivity pointing west from South America along the equator is a result of wind-propelled upwelling due to Ekman transport on either side of the geographic equator. Look for this area in Figure 13.8.

Does productivity change with the seasons? **Figure 13.9** shows the relationship of phytoplankton biomass to season. The low, flat line representing annual tropical productivity contrasts with the high, thin peak representing the Arctic summer. The higher of the two peaks for the temperate zone indicates the plankton bloom of northern spring caused by increasing illumination, while the smaller peak representing the northern fall is caused by nutrients returning to the surface.

Zooplankton: The Heterotrophs

Heterotrophic plankton—the planktonic organisms that eat the primary producers—are collectively called **zooplankton** (*zoion* = animal). Zooplankters are the most numerous primary consumers of the ocean. They graze on the diatoms, dinoflagellates, and other phytoplankton at the bottom of the trophic pyramid in a way analogous to cows grazing on grass. The variety of zooplankton is surprising; nearly every major animal group is represented. Each is expert at painstakingly concentrating food from a dilute supply in the water. The most abundant zooplankters, accounting for about 70% of individuals, are tiny shrimplike animals called copepods (**Figure 13.10**). Copepods are crustaceans, a group that also includes crabs, lobsters, and shrimp.

Not all zooplankters are small, however. Many are in the 1 to 2 centimeter (½ to 1 inch) size range. The largest drifters are giant jellyfish of genus *Cyanea*; their bells may be more than 3.5 meters (12 feet) in diameter! We have a special term for plankton larger than about 1 centimeter (½ inch) across: **macroplankton** (*macros* = large).

Most zooplankton spend their whole lives in the plankton community; so we call them **holoplankton** (*holos* = entirely). But some planktonic animals are the juvenile stages of crabs, barnacles, clams, sea stars, and other organisms that will later adopt a benthic or nektonic lifestyle. These temporary visitors are **meroplankton** (*meros* = mixed). Most animal groups are represented in the meroplankton; even the powerful tuna serves a brief

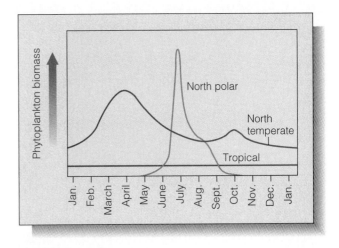

Figure 13.9 Variation in the biomass of phytoplankton by season and latitude. (Note that the total area under each curve represents productivity.)

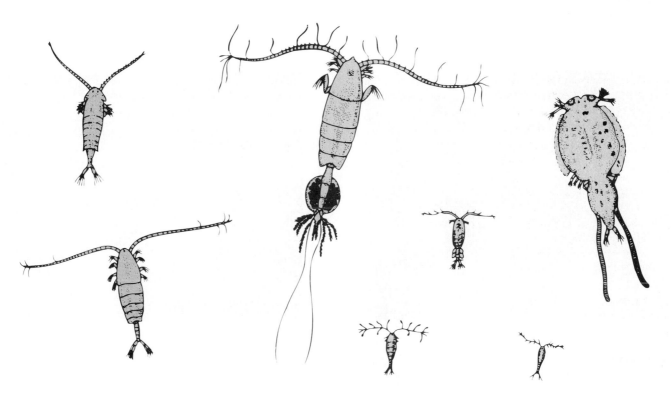

Figure 13.10 Assorted marine copepods, the most common type of zooplankton.

planktonic apprenticeship. These useful categories can be applied to phytoplankton as well as zooplankton. Holoplanktonic organisms are by far the more numerous forms of both phytoplankton and zooplankton.

Particularly lovely holoplanktonic organisms are the **radiolarians**, which are relatives of the amoeba. Like the amoeba they are able to move their protoplasm in streams,

but unlike the common amoeba they possess spikelike silica projections on which the flowing occurs. Patterns of light refracting through the protoplasm and silica needles can be beautiful when viewed with a microscope (**Figure 13.11**).

Other forms related to amoebas are **foraminiferans**, which also extend long protoplasmic filaments (**Figure 13.12**). Most foraminifera have calcium carbonate shells. As we saw in Chapter 5, extensive white deposits of calcareous ooze have been built on the seabed from their skeletons. As was the case with some phytoplankton sediments, some of these layers have been uplifted and can be found on land.

Plankton and Food Webs

Zooplankton and other animals eat phytoplankton, and still other animals eat these primary consumers. As we saw in Chapter 12, the mass of zooplankton is typically about 10% that of phytoplankton, which is reasonable because of the "harvesting" relationship that exists between them.

The zooplankters depend on phytoplankton for both food and oxygen. The activity of zooplankton often creates an **oxygen minimum zone** below the well-lighted surface zone: Oxygen is depleted by the animals there and not replaced by phytoplankton. Some species of zooplankton and a number of kinds of small swimming animals make nightly migrations from the oxygen minimum zone toward the darkened surface layer to feed on the smaller organisms drifting there.

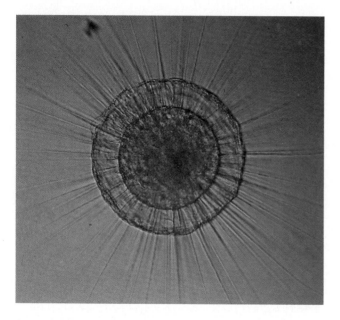

Figure 13.11 A type of radiolarian with needlelike pseudopods radiating from its body like the rays of the sun.

a

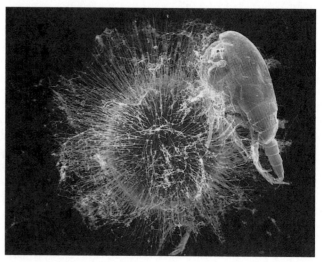

b

Figure 13.12 Foraminiferans. The word means "bearer of windows." (a) Light streams through thin parts of the shells. (b) *Orbulina* feeding on a copepod (compare with Figure 13.10). The foram's spines are made of calcium carbonate and have a sticky layer on their surfaces. Zooplankton that bump into the spines stick to them and are subsequently digested.

Figure 13.13 A chambered nautilus (*Nautilus*), from the only cephalopod group with an external shell. The animal occupies the outermost chamber; the other chambers are partially filled with gas to provide buoyancy. Chambers are added as the animal grows.

It is interesting to note that the largest marine animals, such as whale sharks (fish) and baleen whales (mammals) do not expend their energy tracking down and attacking big animals. Instead these largest of all feeders concentrate zooplankton from the water and consume it in vast quantities. The zooplankton they eat are not usually the primary consumers but the somewhat larger secondary consumers, usually crustaceans such as krill that have themselves fed on the microscopic primary consumers. In this way whales and other large filter feeders get around the inevitable energy losses of a long food web by harvesting energy closer to the source, thus gaining the advantages of efficiency and quantity.

NEKTON

Pelagic animals that swim actively are known as nekton (*nektos* = swimming). Most nektonic animals are **vertebrates** (animals with backbones, such as fishes, reptiles, marine birds, and marine mammals), but a few representatives are **invertebrates** (animals without backbones, such as squid, nautiluses, and some species of shrimp-like arthropods).

Squids and Nautiluses

The most highly evolved of the **mollusks**—a category of animals that includes clams and snails—are the magnificent **cephalopods** (*cephalon* = head + *pod* = foot), a group of marine predators containing squids, nautiluses, and octopuses. These well-named animals have a head surrounded by a foot divided into tentacles. The nautiluses retain a large coiled external shell, but squids have only a thin vestige of the shell within their bodies, and octopuses (nearly all of which are benthic organisms) have none at all. Pelagic cephalopods move by swimming with special fins or by squirting jets of water from an interior cavity.

Most cephalopods catch prey with stiff adhesive discs on their tentacles that function as suction cups, and they tear or bite the flesh with horny beaks. Nautiluses (**Figure 13.13**) hunt at considerable depths, their strong shells buoyed with gas-filled chambers. Squids are more advanced cephalopods that swim in groups and usually prey on small fishes. Their skin is embedded with tiny extensible pigment cells called *chromatophores* (*chromos* = color + *ferre* = to bear), which can rapidly expand or contract to change the animal's color and pattern. Squids can also confuse predators with clouds of ink. Some kinds of squids eject a dummy of coagulated ink that's a rough duplicate of their size and shape. The squid is long gone by the time the attacker discovers the deception! At least one species of squid living below the euphotic zone produces a sparkling luminous ink instead of black ink (which, of course, would be ineffective in the darkness). Squids can grow to surprising sizes (**Figure 13.14**). The record length, including tentacles, is 18 meters (59 feet)! Most are much smaller.

Shrimps

Arthropoda (*arthron* = joint + *pod* = foot), the animal category that includes the lobsters, shrimp, crabs, krill, and barnacles, is the most successful animal group on Earth. They occupy the greatest variety of habitats, consume the

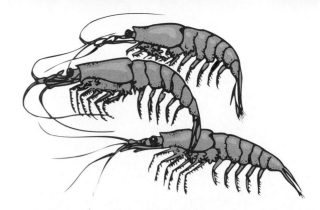

Figure 13.15 Krill (*Euphausia superba*). These shrimplike crustaceans, shown here approximately life-size, occur throughout the world ocean; they are particularly numerous in Antarctic seas.

Figure 13.14 A large squid (*Dosidicus*) and a diver inspect one another. The diver later wrote, "I felt the cold embrace of tentacles with their sharp, toothed suction cups digging into my bare skin. It was like somebody was throwing cactus on my neck." He returned to the boat minus his dive lights and decompression meter and with "nasty lesions" from the sharp protrusions on the squid's suction cups.

greatest quantities of food, and exist in almost unimaginable numbers. Their most important innovation is an **exoskeleton**. Unlike the often cumbersome shells of some other marine animals, the exoskeleton of an arthropod fits and articulates like a finely crafted suit of armor. Its three layers serve to waterproof the covering, tint it a protective color, and make it resilient and strong. Muscles within the animal are attached to the exoskeleton to move the appendages.

The 2,000-plus species of shrimps, most of which are pelagic, range in size from about 1 centimeter to more than 20 centimeters (½ inch to more than 8 inches) in length. Larger species are often called *prawns*. Their semitransparent bodies are flattened from side to side, and swimming species normally move by the rhythmic waving of their legs. Rapid movement in an emergency is accomplished by a quick flex of the abdomen and tail. Shrimping is an important "fishery" in the Gulf of Mex-

ico and up the Atlantic Coast as far as North Carolina. The annual catch of shrimp in the United States is about 100,000 metric tons (110,000 tons).

Another important pelagic arthropod is the **krill** (**Figure 13.15**), the central species of the Antarctic ecosystem. This thumbsized, shrimplike crustacean, primary consumer *extraordinaire*, grazes on the abundant diatoms of the southern polar ocean. In turn, krill are eaten in tremendous numbers by seabirds, squids, fishes, and whales. Some 500 to 750 million metric tons (550 to 825 million tons) of krill inhabit the Antarctic Ocean, with the greatest concentrations in the productive upwelling currents of the Weddell Sea. Krill travel in great schools that can extend over several square miles. They behave more like schooling fish than planktonic crustaceans; their primary swimming mode is horizontal, not vertical. Japanese researchers using two ships equipped with side-scan sonar tracked a large school of swimming krill for 14 days across 278 kilometers (172 miles); this new finding threatens krill's usual classification as zooplankton.

Fishes

Fishes are vertebrates that usually live in water and possess gills for breathing and fins for swimming. There are more species of fishes—and more individuals—than species and individuals of all other vertebrates *combined*. This is not a surprising fact, considering the vast oceanic habitat the planet provides. Fishes range in adult length from less than 10 millimeters to over 20 meters (0.4 inch to 60 feet) and weigh from about 0.1 gram to about 41,000 kilograms (0.004 ounce to 45 tons). Some fish are capable of short bursts of speed in excess of 113 kilometers (70 miles) per hour; some species hardly ever move.

Fishes live near the surface and at great depth, in warm water and cold, and even frozen within ice or dried in balls of mud. Like other ectothermic (cold-blooded) organisms, the great majority of fishes are incapable of generating and maintaining a steady internal temperature from metabolic heat; thus the internal body temperature of a fish is usually the same as that of the surrounding environment. About 40% of the 30,000-plus fish species live at least part of their lives in fresh water; 60% live exclusively in seawater. Fishes have evolved to fit almost every conceivable watery habitat but are most numerous on the bottom or in productive seawater over the continental shelves. Some species have a "sixth sense," an ability to detect small changes in the electrical field surrounding their bodies that assists in the detection of prey or avoidance of predators. Some electric eels, catfish, and rays can use internally generated electricity for defense and offense.

Fishes are divided into two major groups—the cartilaginous fishes and the bony fishes—based on the material forming their skeletons.

The Cartilaginous Fishes

All members of the class **Chondrichthyes** (*chondros* = cartilage + *ichthys* = fish)—the group that includes sharks, skates, rays, and chimaeras—have a skeleton made of a tough, elastic tissue called **cartilage**. Though there is some calcification in the cartilaginous skeleton, true bone is entirely absent from this group. These fish have paired fins, jaws with teeth, and often active life-styles. Sharks and rays tend to be larger than bony fishes, and except for some whales, sharks are the largest living vertebrates.

About 350 species of sharks and 320 species of rays are known to exist. Nearly all are marine, although a few species inhabit estuaries and a very few are permanent inhabitants of fresh water. Although there are many exceptions, sharks tend to favor swimming through open water, while rays tend to be found on or near the bottom.

Sharks have an undeservedly bad reputation. More than 80% of shark species are less than 2 meters (6.6 feet) long as adults, and only a few of the remaining 20% are aggressive toward humans. Like other cartilaginous fishes, sharks are not very intelligent and certainly don't hold grudges or behave in the malignant ways so vividly portrayed in recent popular novels and movies. Still, some sharks are indeed dangerous to humans, and the great white sharks in the genus *Carcharodon* (*karcharos* = sharp + *odontos* = tooth) (**Figure 13.16**) are perhaps the most dangerous of all. These swimmer's nightmares attain lengths of 7 meters (23 feet) and weigh up to 1,400 kilograms (3,000 pounds). (Reports of even larger great whites probably have resulted from misidentification of larger harmless species.) Great whites are not white, but grayish-brown or blue above and creamy on the lower half. A dangerous relative, the mako shark, reaches lengths of 4 meters (13 feet) and is also known to attack small boats. These and other predatory species, such as tiger sharks and hammerhead sharks, are attracted to prey by vibrations in the water, which they detect with sensitive organs arrayed in lines beneath the surface of their skin. Smell also plays an important role in hunting their prey, which is usually composed of fish and marine mammals.

Though most famous, the man-eaters are not the largest of shark species. This honor goes to the immense warm-water whale sharks in the genus *Rhineodon* (*rhine* = a file, rasp; a reference to the fish's rough skin), which reach sizes in excess of 18 meters (60 feet) and 41,000 kilograms (90,000 pounds) (**Figure 13.17**). Whale sharks and their somewhat smaller relatives the basking sharks are docile and present little threat to people. These greatest of fishes swim slowly near the surface with their huge mouths open, feeding on plankton. They may filter as much as 2,200 cubic meters (2,000 tons) of water per hour through a fine mesh of gill rakers. Accumulated plankton is periodically backflushed into the mouth, where it is concentrated for swallowing.

The Bony Fishes

The 27,000-plus species of bony fishes, members of class **Osteichthyes** (*osteum* = bone + *ichthys* = fish), owe much of their great success to the hard, strong, lightweight skeleton that supports them. These most numerous of fishes—and most numerous, diverse, and successful of all vertebrates—are found in almost every marine habitat from tidepools to the abyssal depths. Their numbers include the air-breathing lungfishes and lobe-finned coelacanths, whose ancient relatives broke from the path of fish evolution to establish the dynasties of land vertebrates.

About 90% of all living fishes are contained within the osteichthyan order **Teleostei** (*teleos* = perfect + *osteon* = bone), which contains the cod, tuna, halibut, perch, and other familiar species. Within this large category (see **Figure 13.18**) are a variety of fishes with gas-filled **swim bladders** (which assist in maintaining neutral buoyancy), independently movable fins for well-controlled swimming, great speed for pursuit or avoidance of predators, highly effective camouflage, social organization, orderly patterns of migration, and other advanced features. Their economic importance is great: Some 70 million metric tons (77 million tons) of bony fishes are taken annually from the ocean to help satisfy the human demand for protein.

The Problems of Fishes

Seawater may seem to be an ideal habitat, but living in it does present difficulties. Water is about 1,000 times more dense than air and 100 times more viscous, and it impedes motion effectively at low speeds. How can a fish best move through it? How can a fish maintain its vertical

a   b

Figure 13.16   The great white shark (*Carcharodon*). (a) Cruising for a meal. (b) Jaws.

Figure 13.17   A diver hitches a ride on a whale shark. Unless he is struck by the fish's tail as he dismounts, the diver is in no danger. Large animals (like seals and divers) are not part of the diet of this type of shark.

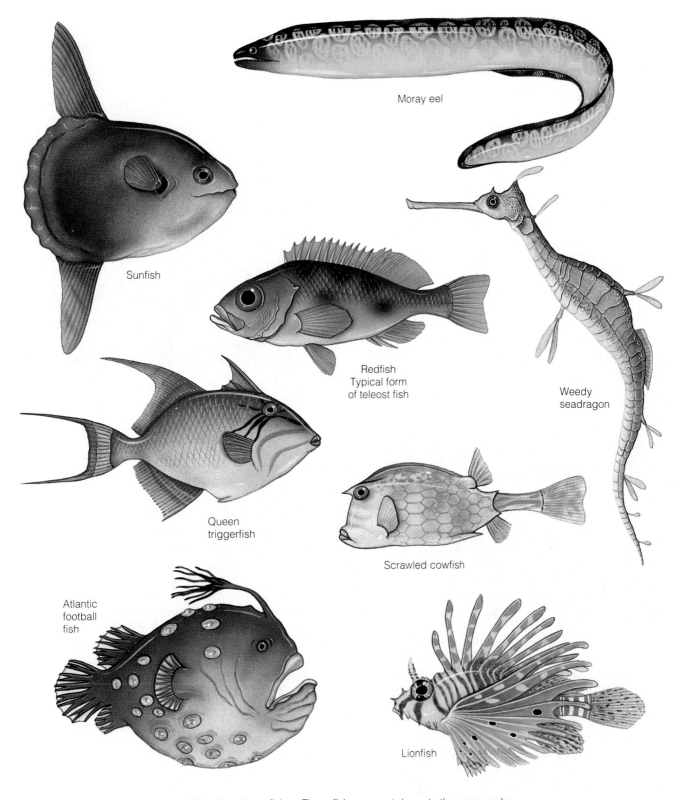

Moray eel

Sunfish

Redfish
Typical form
of teleost fish

Weedy
seadragon

Queen
triggerfish

Scrawled cowfish

Atlantic
football
fish

Lionfish

Figure 13.18 Some of the diversity exhibited by teleost fishes. These fishes are not drawn to the same scale.

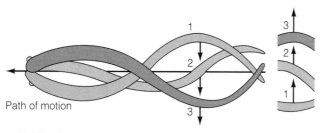

Path of motion

a Eel-like fishes

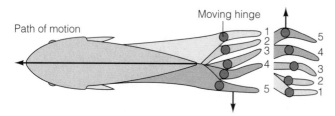

Moving hinge

Path of motion

b Advanced fishes

Figure 13.19 How a fish's body shape affects its efficiency of movement. The undulating movement of eel-like fishes (a) requires more energy to generate forward motion than the hinged-tail movement of the more advanced fishes (b).

Figure 13.20 A tunny, a pelagic teleost adapted for swimming at high sustained speeds.

position in the water column? Must a fish swim constantly to offset the weight of muscle and bone? And how about breathing? Can oxygen and carbon dioxide be exchanged efficiently under water? How can predators be thwarted? There would seem to be many problems, but these most successful vertebrates have structures and behaviors to cope.

Movement, Shape, and Propulsion Active fishes usually have streamlined shapes that make their propulsive efforts more effective. A fish's resistance to movement, or **drag**, is determined by frontal area, body contour, and surface texture. Drag increases geometrically with increasing speed. Faster-swimming fish are therefore more highly modified to minimize the slowing effects of the dense, relatively sticky medium in which they live. The most effective antidrag shape is the tapering, torpedolike body plan (as may be seen in the tuna shown at the beginning of this chapter).

A fish's forward thrust comes from the combined effort of body and fins. Muscles within slender flexible fish (such as eels) cause the body to undulate in S-shaped waves that pass down the body from head to tail in a snakelike motion. The eel pushes forward against the water much as a snake pushes against the ground (**Figure 13.19a**). This type of movement is not very efficient, however. The body must wave back and forth across a considerable distance, exposing a large frontal area to

the water; and the long body length (relative to width) needed to propagate the wave requires increased surface area, which increases drag. A more efficient swimming mechanism is shown in **Figure 13.19b**. More advanced fishes have a relatively inflexible body, which undulates rapidly through a shorter distance, and a hinged scythe-like tail to couple muscular energy to the water. The fish's body can be shorter and can face more squarely in the direction of travel; the drag losses are lower.

How efficient are the best swimmers, and how fast can they swim? Estimates vary, but it is thought that in the fastest fish between 60% and 80% of muscle force delivered to the tail fin results in forward motion. Swordfish and marlin can reach 33 meters per second (75 miles per hour) in short bursts! Some of the fastest tuna (and a few swift sharks) sustain high speeds by maintaining their internal body temperature a few degrees above that of the surrounding water. This stratagem permits them to oxidize food more rapidly, and it generates greater muscle power per unit of weight. The muscular tunny shown in **Figure 13.20** incorporates all these sustained-speed advances; few other animals are capable of such prodigious power output for such a long time. The bluefin tuna, a close relative, swims at a more or less constant 14 kilometers (9 miles) per hour and, as we have seen, may travel more than a million miles in a lifetime!

Maintenance of Level The density of fish tissue is typically greater than that of the surrounding water; so fishes will sink unless their weight is offset by propulsive forces or by buoyant gas- or fat-filled bladders. Cartilaginous fishes have no swim bladders and must swim continuously to maintain their position in the water column. Sharks generate lift with an asymmetrical tail and fins that provide lift like airplane wings. Bony fishes that appear to hover motionless in the water usually have well-developed swim bladders just below their spinal columns. The volume of gas in these structures provides enough buoyancy to offset the animal's weight. The quantity of gas is controlled by secretion and absorption of gas from the blood, and by muscular contraction of the swim bladder to compensate for temporary changes in depth.

A swim bladder may make life easier for a slow-moving teleost, but the fastest bony fishes lack swim bladders. Why? Fast, powerful predators such as tuna, mackerel, and swordfish must be able to chase prey be-

tween depths. The expansion and compression of gas in the bladder would vary rapidly with changing depth, and the risk of rupture would be too great. These speedy predators have power to spare and don't seem inconvenienced by their slight negative buoyancy.

Gas Exchange How can fish breathe underwater? **Gas exchange**, the process of bringing oxygen into the body and eliminating carbon dioxide (CO_2), is essential to all animals. At first glance the task may seem more difficult for water breathers than for air breathers, but air-breathing animals add an extra step. We air breathers must first dissolve gases in a thin film of water in our lungs before they can diffuse across a membrane. (The ease with which most gases dissolve in water was discussed in Chapter 6.)

Fish take in water containing dissolved oxygen at the mouth, pump it past fine **gill membranes**, and exhaust it through rear-facing slots (**Figure 13.21**). The higher

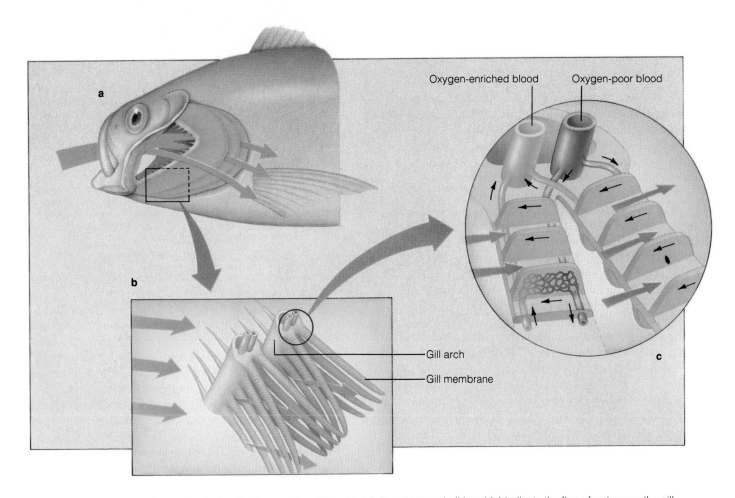

Figure 13.21 Cutaway of a mackerel, showing the position of the gills (a). Broad arrows in (b) and (c) indicate the flow of water over the gill membranes of a single gill arch. Small arrows in (c) indicate the direction of blood flow through the capillaries of the gill filament in a direction opposite to that of the incoming water. This mechanism is called countercurrent flow.

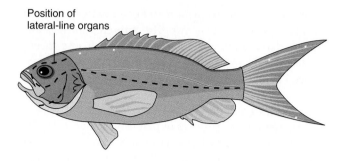

Figure 13.22 A fish's lateral-line system, which detects low-frequency vibrations. Lines of tiny pores in the fish's skin are external evidence of the lateral-line system.

concentration of free oxygen dissolved in the water causes oxygen to diffuse through the gill membranes into the animal; the higher concentration of CO_2 dissolved in the blood causes CO_2 to diffuse through the gill membranes to the outside. The gill membranes themselves are arranged in thin filaments and plates efficiently packaged into a very small space. Water and blood circulate in opposite directions—in a countercurrent flow—which increases transfer efficiency.

An active fish like a mackerel requires so much oxygen and generates so much waste CO_2 that its gill surface area must be ten times its body surface area! (Sedentary fish have proportionally less gill area.) With their large gill area and countercurrent flow, active fish extract about 85% of the dissolved oxygen in water flowing past their gills. Air-breathing vertebrates, by contrast, extract only about 25% of the oxygen from air entering their lungs.

Feeding and Defense Competitive pressure among the large number of fish species has caused a wonderful variety of feeding and defense tactics to evolve. Sight is very important to most fishes, enabling them to see their prey or avoid being eaten. Even some deep-water fishes that live below the photic zone have excellent eyesight for seeing luminous cues from potential mates or meals. Hearing is also well developed, as is the ability to detect low-frequency vibrations with the **lateral-line system** (**Figure 13.22**). This mechanism consists of a series of small canals in the skin and bones around the eyes, over the head, and down the sides of the body. The canals are richly supplied with nerves and connect to the surface through tiny pores in the skin. The nerves report changes in current direction, water pressure, or sonic environment to the brain. Predatory sharks use their lateral-line systems to detect prey.

Smell and taste also play a role, with some fishes having taste organs located on probelike extensions around their mouths. These assist in the search for food. The

a

b

Figure 13.23 Passive cryptic coloration in a kelp fish. (a) The fish resembles a seaweed blade in shape, color, and pattern. (b) The fish nestles within the seaweed blade, nearly disappearing from view, moving with the blades as they sway with the water.

bizarre, flattened crossbar that gives the hammerhead shark its name may provide a kind of "stereo smell," to sense differing amounts of interesting substances in the water. Salmon smell their way to their home streams after years at sea by detecting faint chemical traces characteristic of the water from the stream in which they hatched.

Defense measures are well advanced in fishes. Sea horses (and their relatives, the box fishes) depend on the simple expedient of armor plating for protection. Others, such as the puffer fish, inflate with water and erect bristly spines to become less attractive as a snack. More subtle means of offense or defense depend on trickery—looking like something you're not, or changing color to blend with the background. These kinds of **cryptic coloration** (*kryptos* = hidden) or camouflage may be active or passive. **Figure 13.23** shows an example of passive cryptic

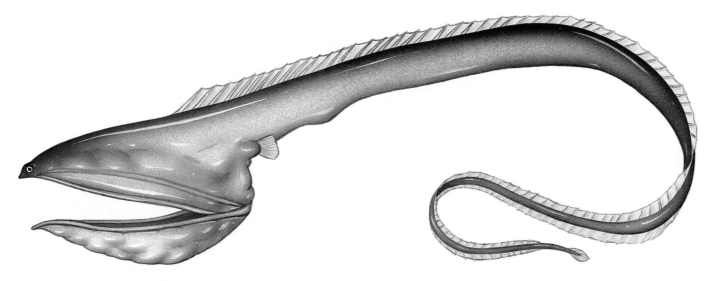

Figure 13.24 The deep-sea gulper (*Eurypharynx*), a bathypelagic species with a worldwide distribution beneath tropical waters. Its length is about 60 centimeters (24 inches).

coloration: The kelp fish closely resembles its seaweed habitat. Small animals foraging nearby run the risk of being eaten, and the fish itself may escape the notice of a larger predator. Actively swimming fish employ a different method of color blending called *countershading*. Their dark tops and silvery bottoms make the fish less obvious to predators above or below.

About a quarter of all bony fish species exhibit **schooling** behavior at some time during their life cycle. A fish school is a massed group of individuals of a single species and size class, packed closely together and moving as a unit. There is no leadership in fish schools, and the movement of fish within them seems to be controlled automatically by direct interaction between lateral-line sensors and the locomotor muscles themselves. I can personally attest to the effectiveness of schooling as a means of defense. On a few diving trips I've noticed a large moving mass just beyond the limit of clear visibility. Is it a fish school, or is it a single large animal? Many predators might not stay around long enough to find out! Schools have the added benefits of reducing chance detection by a predator, providing ready mates at the appropriate time, and increasing feeding efficiency.

Perhaps the most effective way to avoid being eaten is simply to disappear. The surface-feeding flying fishes do this by accelerating rapidly, leaping into the air, and spreading their fins to act as wings. Refractive differences between air and water prevent the surprised predator from seeing where its intended meal has gone!

Fishes living in the twilight world at the bottom of the photic zone use bioluminescence in feeding, avoiding being eaten, and mate attraction. Some members of this sparsely populated community have built-in luminescent organs that cast dim blue light downward; this light masks their own shadows, so they have less chance of being detected and eaten. Others are among the ocean's most bizarre creatures. Gulper eels (**Figure 13.24**) have extendable jaws and a stomach capable of consuming prey larger than the eels themselves, an adaptation of great importance when one considers that a gulper eel may not encounter a feeding opportunity more often than once or twice a year! Some deep-swimming organisms attract their infrequent meals with a luminous lure (**Figure 13.25**). These animals also use patterns of glowing spots or lines to identify themselves to members of the same species, a necessary first step in mating. Some use flashes of light to dazzle or frighten potential predators.

Marine Reptiles

Each of the three main groups of reptiles has marine representatives: turtles, sea snakes and marine lizards (iguanas), and marine crocodiles. Like all reptiles, marine reptiles are ectothermic, breathe air with lungs, are covered with scales and a relatively impermeable skin, and are equipped with special **salt glands** to concentrate and excrete excess salts from body fluids. Except for one widely ranging species of turtle, all marine reptiles require the warmth of tropical or subtropical waters.

The best-known and most successful living marine reptiles are the eight species of sea turtles. Unlike land turtles, sea turtles have relatively small streamlined shells without enough interior space to retract head or limbs. The shell provides an effective passive defense, and adult

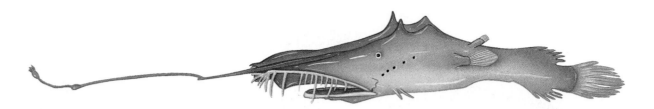

a

Figure 13.25 Two species of deep-sea anglers with bioluminescent lures. (a) *Lasiognathus saccostoma*, known only from the western Atlantic, has fishing equipment that consists of an extensible rod, filament, "float," and illuminated lure with little hooks. Its body is about 20 centimeters (8 inches) long. (b) Opportunities for meeting the opposite sex can be rare in the depths; so a female *Melanocoetus johnsoni*, about 10 centimeters (4 inches) long, carries a permanently attached male.

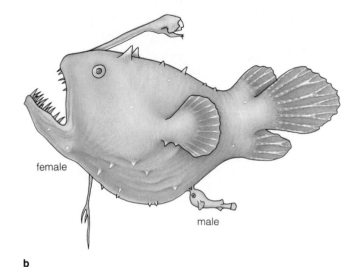

female

male

b

sea turtles have no predators except humans. Their forelimbs are modified as flippers and provide propulsive power; their hind limbs act as rudders. The two species of green sea turtles (genus *Chelonia*) (**Figure 13.26**) are the most abundant and widespread sea turtle species. Green turtles range over great distances looking for the marine algae, turtle grass, and other plants on which they feed. The largest living turtle is the carnivorous Atlantic leatherback, a streamlined animal with a soft skin-covered "shell"; it reaches lengths in excess of 2 meters (6.5 feet) and may weigh more than 600 kilograms (1,300 pounds).

Sea turtles return at two-, three-, or four-year intervals to lay eggs on the beaches at which they were hatched. Homing behavior can be a great advantage to any animal: If the parent survived its earliest childhood at this location, it will probably be a suitable place for hatching the next generation. The navigation of green turtles to tiny Ascension Island, an emergent point of the mid-Atlantic Ridge between Brazil and Africa, has been extensively studied. Researchers have found that the turtles use solar angle (to derive latitude), wave direction, smell, and visual cues—first to find the island and then to discover the spot on the beach where they hatched perhaps 20 years before!

All marine turtles are in danger of extinction. Though protected and no longer used extensively for human food, their breeding beaches are being developed or invaded by noisy recreational pursuits, their eggs and shells are in great demand, they are drowned in fishing and shrimping nets, and their feeding areas are increasingly damaged by pollutants. Much of the tropical turtle grass has disappeared because of human interference. In addition, some species mistake floating plastic bags and other

Figure 13.26 A green sea turtle, *Chelonia mydas*.

debris for the jellyfish on which they normally feed; the plastic clogs their digestive tract, and they starve to death. The outlook for these creatures is not bright.

Marine Birds

Birds probably evolved from small, fast-running dinosaurs about 160 million years ago. Their reptilian heritage is clearly visible in their scaly legs and claws, and in the configuration of their internal organs and skeletons. The success of the 8,600 living species of birds is due in large part to the evolution of feathers (derivatives of reptilian scales), used to insulate the body and to provide aerodynamic surfaces for flight. Birds (and mammals) are *endothermic*: They generate and regulate metabolic heat to maintain a constant internal temperature that is generally higher than their surroundings.

Flying birds have light, thin, hollow bones without fatty insulation; they have forsaken the heavy teeth and jaws of reptiles for a lightweight beak. Their highly efficient respiratory system can accept great quantities of oxygen, and their large four-chambered heart circulates blood under high pressure. All birds lay eggs on land, and most incubate them and provide care for the young. Some seabirds may stay at sea for years, but all must eventually return to land to breed.

Only about 270 kinds of birds, about 3% of known bird species, qualify as seabirds. Most seabirds live in the Southern Hemisphere. Like the marine reptiles, seabirds have special salt-excreting glands in their heads to eliminate the excess salt taken in with their food. Salty brine from these glands may sometimes be seen dripping from the tips of their beaks. Marine birds are voracious feeders, and the ocean will usually be teeming with life wherever they are found. True seabirds generally avoid land unless they are breeding; they obtain virtually all their food from the sea and seek isolated areas for reproduction.

Of the four groups of seabirds, the gulls and the pelicans may be most familiar to us because there are many Northern Hemisphere species and because they spend much of their time near shore. But the groups best adapted to the pelagic world are the tubenoses (albatrosses, petrels) and the penguins.

The Tubenoses
The 100 species of tubenoses of order *Procellariiformes* (*procella* = storm) are the world's most oceanic birds. Their prosaic common name does not convey any sense of their beauty and grace; it refers to the plumbing in their beak that is responsible for sensing air speed, detecting smells, and ducting saline water from the salt glands. Foraging for months across the ocean in conditions of strong winds and high waves, seeking no shelter during storms, being exposed to tropical heat and

Figure 13.27 A wandering albatross (*Diomedea exulans*). The largest of these birds has a wingspan of 3.6 meters (12 feet).

polar sleet, soaring continually through air with a gliding efficiency exceeding that of the most perfectly built human sailplane, the beautiful albatrosses, petrels, and shearwaters are true masters of the sky.

The largest of the tubenoses are the magnificent wandering albatrosses of genus *Diomedea*, which reach a wingspan of 3.6 meters (12 feet) and weight of 10 kilograms (22 pounds). (Ancestors of these great birds were even larger; some had wingspans in excess of 5.5 meters, or 18 feet!) The key to the albatross's success lies in its aerodynamically efficient wing (**Figure 13.27**), a very long, thin, narrow, cupped, and pointed structure ideal for high-speed gliding and soaring. This beautiful wing allows albatrosses to cover great distances in search of food with very little expense of energy, flying continuously for weeks at a time. They use the uplift from wind deflected by ocean waves to stay aloft and soar in long, looping arcs. Albatrosses are almost never seen to flap their wings.

Satellite tracking data indicate that wandering albatrosses routinely cover 15,000 kilometers (9,300 miles) on foraging trips and reach speeds of 80 kilometers (50 miles) per hour. High speeds over long distances are their specialty: One bird was observed to travel 808 kilometers (502 miles) at an average speed of 56 kilometers (35 miles) per hour.

Albatrosses were once thought to locate food exclusively by sight, but recent research suggests that their extraordinarily acute sense of smell also plays a role in feeding. They can evidently find schools of fish by the odor of fish oil wafted tens of kilometers downwind. Tubenoses catch fish (and squids) by dipping their bill into the water during flight or during brief stops on the surface. Albatrosses take shore leave only during breeding periods. Chicks are hatched and raised on remote islands to

Figure 13.28 A rock hopper penguin, one of the smaller species.

Penguins are native only to the Southern Hemisphere and range from the size of a large duck (see **Figure 13.28**) to a height of more than a meter (3.3 feet) and weight exceeding 36 kilograms (80 pounds). They are thought to consume about 86% of all food taken by birds in the southern ocean—about 34 million metric tons (37 million tons) per year, mostly larger zooplanktonic crustaceans. The small Galápagos penguin lives a comparatively easy life fishing the cold, nutrient-rich Humboldt Current at the equator, but its Antarctic relatives lead what must surely be the most rigorous existence of any seabird. For example, emperor penguins breed and incubate during the bitterly cold Antarctic winter. Unlike most other birds, emperors do not establish a territory, instead huddling together in tight crowds to conserve heat. The mass of penguins moves slowly around the breeding area, as warm penguins from the center of the mob circulate to the outside to be replaced at the core by their chilled peripheral friends.

Marine Mammals

The class **Mammalia** (*mamma* = breast), to which humans belong, is the most advanced vertebrate group. About 4,300 species of mammals are known. The three living groups of marine mammals are the porpoises, dolphins, and whales of order **Cetacea**; the seals, sea lions, walruses, and sea otters of order **Carnivora**; and the manatees and dugongs of order **Sirenia**.

Each of these orders arose independently from land ancestors. They exhibit the mammalian traits of being endothermic, breathing air, giving birth to living young that they suckle with milk from mammary glands, and having hair at some time in their lives. Unlike other mammals, however, these extraordinary creatures have become adapted to life in the ocean.

All marine mammals share four common features:

1. Their *streamlined body shape*, with limbs adapted for swimming, makes an aquatic life-style possible. Efficient locomotion depends on minimum drag and maximum ability to transfer propulsive energy from the muscles to the water. Thin, stiff flippers and tail flukes situated at the rear of the animal drive it forward, and similarly shaped forelimbs act as rudders for directional control. Drag is reduced by a slippery skin or hair covering.

2. They *generate internal body heat* from a high metabolic rate and *conserve* this heat with layers of insulating fat and, in some cases, fur. Their large size gives them a favorable surface-to-volume ratio; with less surface area per unit of volume, they lose less heat through the skin. This is why there are no marine mammals smaller than a sea otter; a small

which albatrosses regularly migrate from virtually any point over the world ocean. They are astonishing animals; no one who has seen one sweeping over the sea surface soon forgets the experience.

The Penguins   Penguins have completely lost the ability to fly, but they use their reduced wings to swim for long distances and with great maneuverability. The name of their order, *Sphenisciformes* (*spheniskos* = little wedge), refers to the shortness and shape of their wings. Their flightlessness makes it practical to have fatty insulation, greasy peglike feathers, stubby appendages, and large size and weight; indeed, such heat-conserving adaptations are critical to the survival of aquatic organisms in very cold climates. Their neutral buoyancy is an advantage as they forage for food underwater. Emperor penguins, the largest of the living penguin species, may dive to depths of 265 meters (875 feet) and stay submerged for 10 minutes or more. Penguins feed on fish, large zooplankters, bottom-dwelling mollusks or crustaceans, and squid. A few of the 18 species spend two uninterrupted years at sea between breedings.

mammal would lose body heat too rapidly. These adaptations are critical to animals living in cold water where food is readily available.

3. The *respiratory system is modified* to collect and retain large quantities of oxygen. The air duct "plumbing" of marine mammals is typically much different from that of land mammals, and the lungs can be more thoroughly emptied before drawing a fresh breath. The biochemistry of blood and muscle is optimized for the retention of oxygen during deep, prolonged dives. Some whales can stay submerged for 90 minutes!

4. A number of *osmotic adaptations* free marine mammals from any requirement for fresh water. Unlike other marine vertebrates, the marine mammals do not have salt-excreting glands or tissues. They swallow little water during feeding (or at any other time), and their skin is impervious to water. This minimal seawater intake, coupled with their kidneys' abilities to excrete a concentrated and highly saline urine, permits them to meet their water needs with the metabolic water derived from the oxidation of food.

Order Cetacea   The 90-plus living species of cetaceans (*ketos* = whale) are thought to have evolved from an early line of ungulates—hooved land mammals related to today's horses and sheep—whose descendants spent more and more time in productive shallow waters searching for food. Modern whales range in size from 1.8 meters (6 feet) to 33 meters (110 feet) in length and weigh up to 100,000 kilograms (110 tons). Their paddle-shaped forelimbs are used primarily for steering, and their hind limbs are reduced to vestigial bones that do not protrude from the body. They are propelled mainly by horizontal tail flukes, moved up and down by powerful muscles at the animal's posterior end. A thick layer of oily blubber provides insulation, buoyancy, and energy storage. One or two nostrils are located at the top of the head and have special valves to prevent intake of water when submerged. Whales have large, deeply convoluted brains and are thought to form complex family and social groupings.

Modern cetaceans are further divided into two suborders. **Figure 13.29** (pages 246 and 247) shows representative whales in each division. Suborder **Odontoceti** (*odontos* = tooth), the toothed whales, are active predators and possess teeth to subdue their prey. Toothed whales have a high brain-weight-to-body-weight ratio. Although much of their "extra" brain tissue is involved in formulating and receiving the sounds on which they depend for feeding and socializing, many researchers believe them to be quite intelligent. Smaller whales in this group include the killer whale and the familiar dolphins and porpoises of oceanarium shows. The largest toothed whale is the 18-meter (60-foot) sperm whale, which can dive to at least 1,140 meters (3,740 feet) in search of the large squids that provide much of its diet.

Toothed whales search for prey using **echolocation**, the biological equivalent of sonar; they generate sharp clicks and other sounds that bounce off prey species and return to be recognized (**Figure 13.30**). Odontocete whales are now thought to use sound offensively as well. Recent research indicates that some odontocetes can generate sounds loud enough to stun, debilitate, or even kill their prey. In one experiment, dolphins produced clicks as loud as 229 decibels, equivalent to a blasting cap exploding close to the target organism. Sperm whales, it has been calculated, may generate sounds exceeding 260 decibels! (The decibel scale is logarithmic; compare this

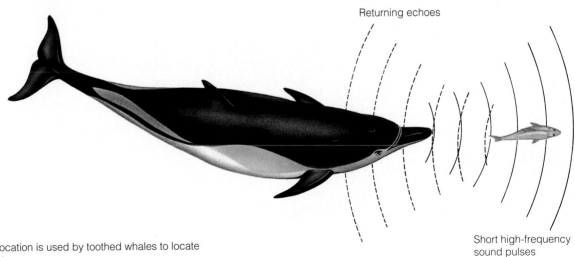

Returning echoes

Short high-frequency sound pulses

Figure 13.30   Echolocation is used by toothed whales to locate and perhaps stun their prey.

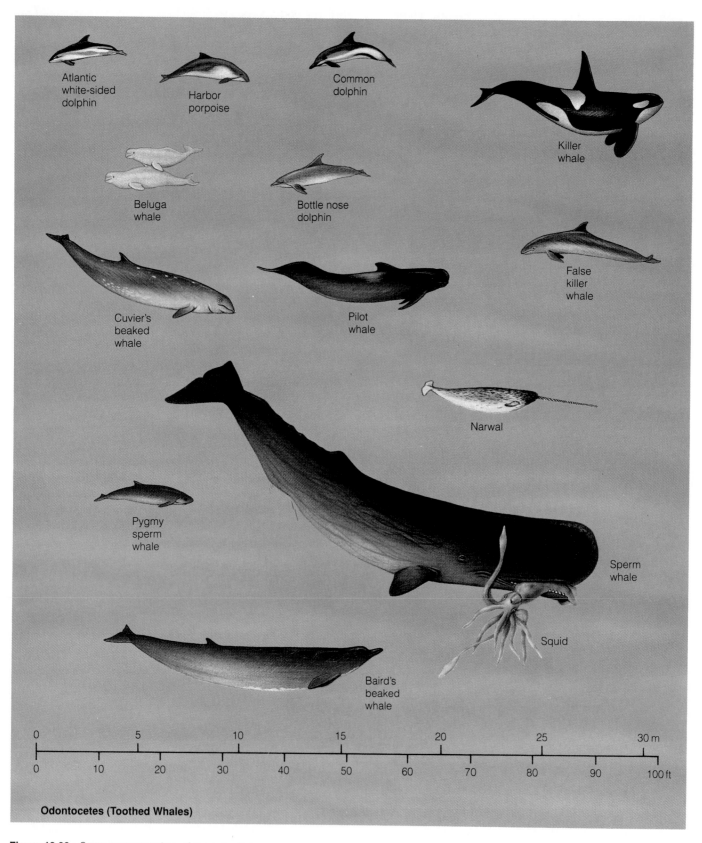

Atlantic white-sided dolphin

Harbor porpoise

Common dolphin

Killer whale

Beluga whale

Bottle nose dolphin

Cuvier's beaked whale

Pilot whale

False killer whale

Narwal

Pygmy sperm whale

Sperm whale

Squid

Baird's beaked whale

| 0 | 5 | 10 | 15 | 20 | 25 | 30 m |

| 0 | 10 | 20 | 30 | 40 | 50 | 60 | 70 | 80 | 90 | 100 ft |

Odontocetes (Toothed Whales)

Figure 13.29 Some representatives of the order Cetacea.

Humpback whale

Bowhead whale

Right whale

Minke whale

Blue whale

Fin whale

Feeding on krill

Sei whale

Gray whale

Mysticetes (Baleen Whales)

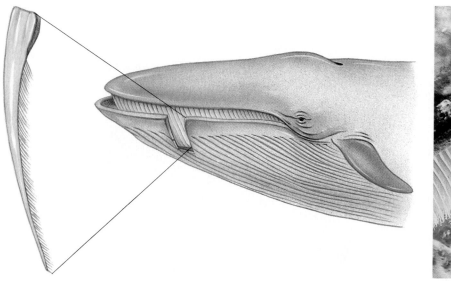

a

b

Figure 13.31 (a) A plate of baleen and its position in the jaw of a baleen whale. For clarity, the illustration shows part of the mouth cut away. (b) A student and a whale, up close and personal. This gray whale uses its stiff, coarse baleen plates to sieve crustaceans from shallow bottom mud.

figure to the 130-decibel noise of a military jet engine at full power 20 feet away!) How this prodigious noise is generated is not yet known, nor do we know how such energy is radiated from the whale without damaging the organs that produce and focus it.

Suborder **Mysticeti** (*mystidos* = unknowable), the whalebone or baleen whales, have no teeth. Filter feeders rather than active predators, these whales subsist primarily on krill, a relatively large, shrimplike crustacean zooplankter obtained in productive polar or subpolar waters. They do not dive deep but commonly feed a few meters below the surface. Their mouths contain interleaving triangular plates of bristly hornlike **baleen** (**Figure 13.31**) used to filter the zooplankton from great mouthfuls of water. The plankton is concentrated as water is expelled, swept from the baleen plates by the whale's tongue, compressed to wring out as much seawater as possible, and swallowed through a throat not much larger in diameter than a grapefruit. A great blue whale, largest of all animals, requires about 3 metric tons (6,600 pounds) of krill each day during the feeding season. The short, efficient food chain from phytoplankton to zooplankton to whale provides the vast quantity of food required for their survival. Humpback whales, another species of baleen whale, are shown in **Figure 13.32** at the end of a vertical feeding sweep.

Mysticeti is an excellent name for these odd and wonderful animals. We know comparatively little of their social structure, intelligence, sound-producing abilities, navigational skills, or physiology. We do know that humpback whales use complex songs in group com-

Figure 13.32 Humpback whales feeding at the surface.

munication and that blue whales may use very low-frequency sound to communicate over tremendous distances. Some species migrate annually from polar to tropical waters and back. Until recently our primary response to all whales has been to slaughter them in countless numbers for meat and oil, with little thought for their extraordinary abilities and assets. More of this depress-

a

b

c

Figure 13.33 Some representatives of the suborder Pinnipedia. (a) Young harbor seals anticipate lunch at a care facility. (b) The California sea lion, the "seal" of seal shows. (c) A walrus.

ing history may be found in the discussion of marine resources in Chapter 15.

Order Carnivora The order Carnivora (*carnis* = flesh + *vorare* = to devour) includes a wide variety of land predators ranging from dogs and cats to bears and weasels, but the members of the carnivoran suborder **Pinnipedia**—the seals, sea lions, and walruses (see **Figure 13.33**)—are almost exclusively marine. Unlike the cetaceans, the gregarious pinnipeds (*pinna* = wing + *pedalis* = foot) leave the ocean for varying periods of time to mate and raise their young.

True seals have a smooth head with no external ear flaps, the external part of the ear having been sacrificed to further streamline the body. They are covered with a short coarse hair without soft underfur. Seals are graceful swimmers that pursue small fish, their usual prey, with powerful side-to-side strokes of their hind limbs.

These rear appendages are partially fused and always point back from the hind end of the body; thus, they are of very little use for locomotion on land. The elephant seal, named for its long snout and large size, holds the diving depth record for all air-breathing vertebrates: 1,560 meters (5,120 feet). Sea lions, familiar to many as the performers in "seal" shows, have hind limbs with a greater range of motion and thus are more mobile on land. They have a streamlined head with small external ears and a pelt with soft underfur; unlike seals, they use their front flippers for propulsion. Walruses are much larger than either seals or sea lions and may reach weights of 1,800 kilograms (2 tons).

Walruses use their tusks like sled runners to guide their sensitive whiskers just above the sediment surface looking for clam siphons at depths up to 90 meters (300 feet). They dig up the clams with their mouths, crush the clam shell and remove the meat, and then eject the inedible

Figure 13.34 Two sea otters in a California kelp bed.

Figure 13.35 A manatee, or sea cow.

fragments. The large tusks are also useful for hauling their heavy bodies onto ice floes.

The suborder **Fissipedia** (*fissus* = split + *pedalis* = foot) has many members (including cats, dogs, raccoons, and bears) but only one truly marine representative, the sea otter (**Figure 13.34**), a relative newcomer to the marine environment. This, the smallest of marine mammals, is an active and pleasing creature. Human demand for the fur, the densest and warmest of any animal, caused its near-extermination. The modern population of the Pacific sea otter descends from a very few individuals accidentally overlooked by fur hunters of the late nineteenth and early twentieth centuries. Playful and intelligent, otters rarely exceed 1.2 meters (4 feet) in length; they eat voraciously, consuming up to 20% of their body weight in mollusks, crustaceans, and echinoderms each day. Sometimes they lie on their back in the water, balance a rock on their chest, and hammer the shell of the prey against it until it cracks. Morsels of food are extracted with small nimble fingers, and rolling over in the water cleans away the debris. A morning spent watching sea otters in their coastal habitat is a morning well spent!

Order Sirenia The bulky, lethargic, small-brained dugongs and manatees, collectively called *sirenians* (*siricis* = a mermaid) (**Figure 13.35**), are the only herbivorous marine mammals. Like the cetaceans, they appear to have evolved from the same ancestors as modern ungulates. They make their living grazing on sea grasses, marine algae, and estuarine plants in coastal temperate and tropical waters of North America, Asia, and Africa. Some species live in fresh water. The largest sirenians reach 4.5 meters (15 feet) in length and weigh 680 kilograms (1,500 pounds). They were first compared to mermaids by early Greeks, who noted the manatee's habit of resting in an upright position in the water and holding a suckling calf to her breast. Sirenians have been hunted extensively, and only about 10,000 individuals are thought to exist worldwide. Even though protected now, many are killed or wounded each year in Florida by the propellers of power boats.

CHAPTER SUMMARY

The organisms of the pelagic world drift or swim in the ocean. (Animals and plants that are associated with the bottom are known as benthic organisms.)

The organisms that drift in the ocean are known collectively as plankton. The plantlike organisms that make up phytoplankton are responsible for most of the ocean's primary productivity. Phytoplankton—and zooplankton, the small drifting or weakly swimming animals that consume them—are usually the first links in oceanic food webs. Plankton are most common along the coasts, in the upper sunlit layers of the temperate zone, in areas of equatorial upwelling, and in the southern subpolar ocean. Marine scientists have been inspired by the beauty and variety of plankton since first observing them under the microscope in the nineteenth century.

Actively swimming animals make up the nekton. Nektonic organisms include invertebrates (such as the squid, nautiluses, and shrimps) and vertebrates (such as fishes, reptiles, birds, and mammals). Each organism has a continuously evolving suite of adaptations that has brought it through the rigors of food finding, predator avoidance, salt balance, and temperature regulation—all of the challenges of the marine environment—time and time again.

Terms and Concepts to Remember

Arthropoda	gas exchange	plankton
baleen	gill membrane	plankton net
Carnivora	holoplankton	radiolarian
cartilage	invertebrate	salt gland
cephalopod	krill	schooling
Cetacea	lateral-line system	silicoflagellate
Chondrichthyes	macroplankton	Sirenia
coccolithophore	Mammalia	swim bladder
cryptic coloration	meroplankton	synoptic sampling
diatom	mollusk	Teleostei
dinoflagellate	Mysticeti	ultraplankton
drag	nanoplankton	valve
echolocation	nekton	vertebrate
exoskeleton	Odontoceti	zooplankton
Fissipedia	Osteichthyes	
flagella	oxygen minimum	
foraminiferan	zone	
frustule	Pinnipedia	

Study Questions

1. What are plankton? How are plankton collected? How are zooplankton different from phytoplankton?

2. Describe the most important phytoplankters. Which are most efficient in converting solar energy to energy in chemical bonds? By what means is this conversion achieved?

3. Where in the ocean is plankton productivity the greatest? Why?

4. Describe five nektonic organisms that are *not* fishes.

5. What are the major categories of fishes? What problems have fishes overcome to be successful in the pelagic world?

6. How does a marine bird differ from a terrestrial bird—a pigeon, for example?

7. What characteristics are shared by all marine mammals?

8. Compare and contrast the groups of living marine mammals.

9. How are odontocete (toothed) whales different from mysticete (baleen) whales? Which are the better known and studied? Why?

For Further Study

Anderson, D. M. 1994. "Red Tides." *Scientific American*, August, 62–68. Many experts believe these blooms of algae have been stimulated by pollution. Toxic species have recently become more prevalent, posing a greater threat to human and marine health.

Davis, L. S., and J. T. Darby, eds. 1990. *Penguin Biology*. San Diego: Academic Press. The first collection of biological and ecological studies of penguins since the mid-1970s.

Hardy, A. 1956. *The Open Sea*. Vol.1, *The World of Plankton*. London: Collins. A classic, readable and enthusiastic. (American printing by Houghton Mifflin, Boston.)

Isaacs, J. D. 1969. "The Nature of Oceanic Life." *Scientific American*, September, 146–62. General discussion of marine trophic relationships.

Kozloff, E. N. 1990. *Invertebrates*. Philadelphia: W. B. Saunders. Outstanding new text, very well written by an expert marine biologist.

Lagler, K. F., et al. 1977. *Ichthyology*. 2d ed. New York: Wiley. An excellent and readable standard text.

Lanting, F. 1986. "Riders on the Wind." *Oceans* 19 (Sept.–Oct.): 24–31. Excellent illustrated article on Pacific albatrosses.

Minasian, S. S. 1984. *The World's Whales—The Complete Illustrated Guide*. Washington, DC: Smithsonian Books. The title is not hyperbole. Extraordinary photographs and clear text, taxonomically arranged.

Oceanus 21 (no. 2, 1978). An entire issue devoted to marine mammals.

O'Shea, T. J. 1994. "Manatees." *Scientific American*, July, 66–72.

Pough, F. H., et al. 1989. *Vertebrate Life*. 3d ed. New York: Macmillan. Excellent and up-to-date general text containing a thorough discussion of the evolution of vertebrates.

Stevens, J. D., ed. 1987. *Sharks*. New York: Facts on File. Perhaps the best general reference on sharks available to the interested amateur.

Wimpenny, R. S. 1966. *The Plankton of the Sea*. New York: Elsevier. Good companion volume to Hardy, and confined to plankton. A complete treatise for the layman. Highly recommended.

Würsig, Bernd. 1989. "Cetaceans." *Science* 244 (no. 4912): 1550–57. Excellent review article with extensive bibliography.

14 BENTHIC COMMUNITIES

The Resourceful Hermit

Benthic marine organisms live on or in the ocean floor. There is an enormous variety of benthic creatures, but let's spend a moment with one of the more entertaining ones, a hermit crab. These small, pleasant relatives of edible crabs and lobsters have engaged the attention of generations of seaside visitors. The hermits rush around sandswept rocks or the floors of tidal pools, withdrawing into their borrowed shells at the slightest sign of danger. Their activity appears random, but their fighting, snooping, hiding, probing, and scuffling are purposeful. Like all animals, hermit crabs must struggle to eat, avoid predators, and mate. Hundreds of structural and behavioral adaptations contribute to their success. Sensors on the hermit's antennae and mouth parts alert him to the presence of food; good eyesight, muscular coordination, and a tough form-fitting covering usually foil fast-moving predators; and brilliant blue bands around the tips of his legs may signal his availability for mating.

One unique behavioral adaptation shared by all hermit crabs involves the selection of a temporary home. The front parts of a hermit crab—mostly pincers, antennae, and mouth parts—are formidable, but its hindquarters are delicate and subject to attack. To protect his flank, a hermit searches for any enclosed portable object to climb into, usually an unoccupied snail shell. A hermit crab's borrowed or stolen shell is a source of both inordinate pride and perpetual difficulty. No shell is ever completely satisfactory. A hermit crab will carefully inspect any substitute dwelling, occupied or not, and consider whether to abandon his current digs in favor of the new candidate. House hopping proceeds fairly smoothly until the supply of suitably sized shells is exceeded by the number of potential occupants (as might happen when the number of crabs grows rapidly in times of abundant food). Then things get more serious. Snails can be evicted from their self-made homes even before they're through with them; the hermit may get a house *and* a meal in a single transaction. Two crabs may fight for hours or even days over one shell. Two others might simultaneously occupy the opposite ends of an abandoned worm tube, thus spending most of their time pulling each other in different directions. Sometimes two crabs swap shells at a moment's notice. Renter's remorse sets in almost immediately, and they're off to see if they can find something even more suitable. At times human observers can't resist laughing at the all-too-human goings-on.

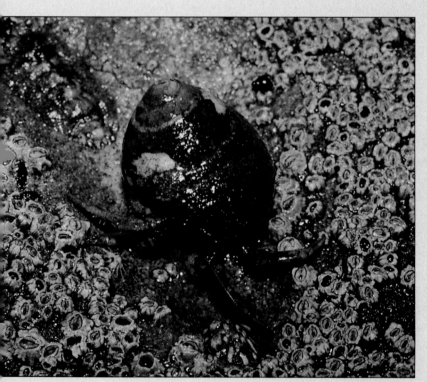

A hermit crab surveys his domain.

Unlike the drifting or actively swimming pelagic organisms discussed in the last chapter, **benthic** organisms live on or in the ocean bottom. A benthic (*benthos* = bottom) habitat may be shallow or deep, warm or cold, brimming with food and life or nearly sterile. Some benthic creatures spend their lives buried in sediment, while others rarely touch the solid seabed. Most attach to, crawl over, swim next to, or otherwise interact with the ocean bottom continuously throughout their lives.

The diversity of benthic habitats (and the diversity of organisms within them) is astonishing. Kelp forests are benthic communities, as are rocky intertidal zones, sand beaches, salt marshes, and the strangely populated areas around deep vents. Coral reefs are also benthic habitats, with communities that may include a greater number of species than any other habitat on our planet.

THE DISTRIBUTION OF BENTHIC ORGANISMS

Before looking at individual benthic communities, we will investigate ways that organisms are distributed within a community. Individual benthic organisms are almost never distributed randomly through their habitats. A **random distribution** implies that the position of *one* organism in a bottom community in no way influences the position of *other* organisms in the same community. Further, a truly random distribution (such as that shown in **Figure 14.1a**) indicates that conditions are precisely the same throughout the habitat, an extremely unlikely situation except possibly in the unvarying benthic communities of abyssal plains.

The most common pattern for distribution of benthic organisms is small patchy aggregations, or clumps. **Clumped distribution (Figure 14.1b)** occurs when conditions for growth are optimal in small areas because of physical protection (in cracks in an intertidal rock), nutrient concentration (near a dead body lying on the bottom), initial dispersal (near the position of a parent), or social interaction.

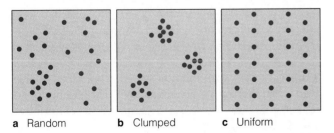

a Random   **b** Clumped   **c** Uniform

Figure 14.1 Random, clumped, and uniform population distribution patterns. The clumped pattern is most common in nature.

Uniform distribution with equal space between individuals (**Figure 14.1c**), such as the arrangement of trees we see in orchards, is the rarest natural pattern of all. The distribution of some garden eels through their territories becomes almost uniform because each eel can extend from its burrow just far enough to hassle neighbor eels spaced at equal distances. But there is still a break in the order of position; they don't line up row-upon-row like apple trees.

SEAWEEDS

When benthic communities are mentioned, many people living near a temperate shore will think first of the magnificent kelp forests. Although most of the ocean's primary productivity is generated by single-celled phytoplankton, between 2% and 10% is carried out by the seaweeds, large marine **multicellular algae**. (**Algae** is a collective term for autotrophs possessing chlorophyll and capable of photosynthesis but lacking vessels to conduct sap. The single-celled diatoms and dinoflagellates are classified as **unicellular algae**.)

Multicellular marine algae occur in a great variety of sizes and shapes. The largest can reach 62 meters (205 feet) in length, while the smallest appear as smears of cells on the surfaces of rocks. Some types of algae form underwater forests, while others grow in isolation. Lifeless seaweeds drying on shore communicate none of their natural beauty and grace to the beachcomber, but to a diver the marine forest from which they came reveals extraordinary grace and form, discloses sheltering nurseries, conceals complex interrelationships, and even provides nourishment. None of these plants grows below the euphotic zone because all depend on photosynthesis to produce the energy-rich compounds necessary for life. Nearly 7,000 species of multicellular marine algae have been identified.

The Problems of Large Marine Plants

At first glance marine plants appear to have an easy life, but living near the ocean surface is not without hazards. Intertidal plants may find themselves exposed to the drying effects of air and sunlight when the tide is out, and they may be lashed against the rocks by waves when the tide is in. The physical nature of the plants themselves provides some defense against these difficulties. Their bodies are flexible, easily able to absorb shock, resistant to abrasion, streamlined to reduce water drag, and very strong. Their surfaces are often covered by a slick mucilaginous material, which lubricates them as they move, retards drying, and deters grazing animals.

Warmth and a lack of nutrients often limit the success of seaweeds. Higher temperatures lead to higher metabolic rates, and the oxygen level in warm seawater may not be high enough to support the respiratory needs of the plants at night. Warmer temperatures can also shatter the delicate accessory pigments and proteins required for algal photosynthesis and respiration. Small increases in temperature may prevent reproduction in some species of algae. Large algae are rare in warm nutrient-poor waters, and divers visiting the tropics are usually surprised to find no sign of kelp forests. Chilly temperate and subpolar zones of nutrient upwelling often support thick algal mats and dense marine forests.

Yet another difficulty is the location of adequate anchorage or substrate for these large plants. Attached seaweeds require a stable footing. A sandy or muddy bottom is unsuitable for colonization by most large algae, and less than 2% of the ocean floor is shallow enough and solid enough to permit the growth of these attached plants.

Of course the marine lifestyle also offers some advantages. Marine plants suffer no droughts and nearly always have enough carbon dioxide for photosynthesis. Assuming suitable nutrient levels and a good foothold, only sunlight is required for productivity. Being submerged in seawater brings the additional advantage of lightweight construction. A seaweed doesn't require strong support structures because it has nearly the same density as the surrounding seawater. More of its bulk can thus be dedicated to photosynthesis. Indeed, productivity in some seaweed beds may be the highest of *any* autotrophic community on Earth.

Structure of Seaweeds

Common terms like *leaf*, *stem*, and *root* are inappropriate for seaweeds because the definitions of those parts assume the presence of vessels and flowing sap. The structures in nonvascular plants that superficially resemble leaves are called **blades** (or fronds), the stemlike structures are termed **stipes**, and the root-shaped jumble at the plant's base is appropriately named a **holdfast**. **Gas bladders** assist many species in reaching strongly illuminated surface water. Blades, stipes, and holdfast comprise the body of the plant, the **thallus**. These parts are labeled in **Figure 14.2**.

An algal thallus may be large or small, branching or tufted, in sheet form or filamentous, encrusting or elongated, rounded or pointed; variety of form within the algae is tremendous. Algal blades are symmetrically equipped with photosynthesizing tissue; they absorb gases across their entire surface and even participate in reproduction. Stipes are strong, photosynthesizing, shock-absorbing links tying the blades (or sheets or filaments) into a unit. The holdfast does not take up water and nu-

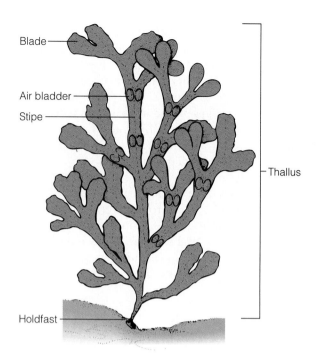

Figure 14.2 The parts of a multicellular alga.

trients from the substrate like the vascular root it superficially resembles, but it does anchor the plant in place and may provide incidental shelter for a rich variety of animal life. The gas bladders range in size from tiny grapelike bunches to single volleyball-sized floats. The flotation gas is usually nitrogen, oxygen, and argon in proportions similar to those in air.

Classification of Seaweeds

Seaweeds are classified by the presence of colored compounds in their tissues. These **accessory pigments** (or masking pigments) are light-absorbing compounds closely associated with chlorophyll molecules. They don't resemble chlorophyll chemically, but they bind loosely with it. Their presence in plants greatly enhances photosynthesis because they absorb the dim blue light at depth and transfer its energy to the adjacent chlorophyll molecules. Accessory pigments may be brown, tan, olive green, or red; they are what give most marine autotrophs, especially seaweeds, their characteristic color. The masking effect of accessory pigments is often so complete that an observer is unaware the plant contains any chlorophyll whatever, even when the seaweed is transparent. The absence of accessory pigments allows the bright green of chlorophyll to shine through.

Multicellular marine algae are segregated into three divisions based on their observable color. The green algae, with their unmasked chlorophyll, are the **Chlorophyta**

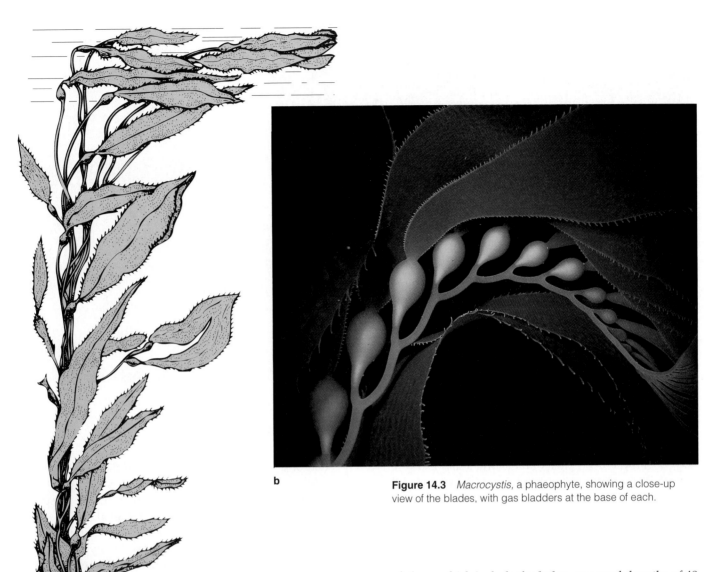

a

b

Figure 14.3 *Macrocystis*, a phaeophyte, showing a close-up view of the blades, with gas bladders at the base of each.

(*chloros* = green + *phyton* = plant), the brown algae **Phaeo-phyta** (*phae* = tan, dusky), and the red algae **Rhodophyta** (*rhodon* = rosy red). Phaeophytes are most familiar to beachcombers, and rhodophytes the most numerous.

The Phaeophytes Virtually all of the 1,500 living species of phaeophytes are marine. Some species of these largest of algae, which include the **kelps**, can reach lengths of 40 meters (132 feet); the record length exceeds 60 meters (200 feet). To attain these dimensions the plant can grow at the spectacular rate of 50 cm (20 inches) per day! Part of the strategy of rapid growth is to reach bright surface water as soon as possible. Some brown algae are annuals; others live for up to seven years. The tan or brown color of phaeophytes comes from an accessory pigment, which permits photosynthesis to proceed at greater depths than is possible for the unmasked chlorophytes. In ideal circumstances, some larger brown algae can grow in water up to about 35 meters (115 feet) deep.

The Pacific's giant kelp forests, the world's largest, consist mostly of the magnificent genus *Macrocystis* (*macro* = large + *cyst* = bladder). The size and shape of the plant and its dense canopy are shown in **Figure 14.3**. Most brown algae live in temperate and polar habitats poleward of the 30° latitude lines, but a few live in the tropics. The worldwide distribution of kelp is shown in **Figure 14.4**.

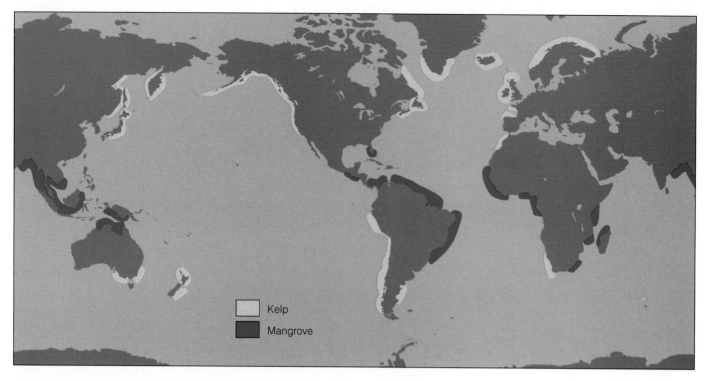

Figure 14.4 The distribution of kelp beds and mangrove communities.

	Kelp
	Mangrove

The Rhodophytes Most of the world's seaweeds are red algae; there are more rhodophyte species than all other major groups of algae combined, about 4,000 species. Rhodophytes tend to be smaller and more anatomically and biochemically complex than phaeophytes. Red algae do inhabit high latitudes, but they thrive in warmer mid- and low-latitude areas, where they usually outnumber all other algal species. Rhodophytes excel in dim light because of their sophisticated accessory pigments. These compounds absorb and transfer enough light energy to power photosynthetic activity at depths where human eyes cannot see light. The record depth for a photosynthesizer is held by a small rhodophyte discovered in 1984 at a depth of 268 meters (879 feet) on a previously undiscovered seamount in the clear tropical Caribbean. The deepest rhodophytes grow very slowly and may be tens or even hundreds of years old.

Paradoxically, many red algae also live on rocks right at the water's surface. Surface light contains all colors of the spectrum, so the color-shifting capabilities of accessory pigments are not needed there. It is believed the dark accessory pigments may act to shade the productive machinery from brilliant light. Most of the common surface-dwelling "reds" grow as a purple or pink film on rocks, on shells, on other seaweeds, and even on glass or plastic. These coralline, or calcareous, algae look like a brilliant ceramic glaze, a bit of melting raspberry sherbet, or a knobby plastic membrane thrown carelessly over the surface. The name of one very common form, genus *Lithophyllum* (named from the Latin terms for "stony" and "leaf") aptly describes an alga able to deposit calcium carbonate within its tissues (**Figure 14.5a**). Others grow in erect branching or bushy forms (**Figure 14.5b**). The gritty calcium in these plants may deter grazing animals.

Another group of shallow rhodophytes is very important in the life of coral reefs. Like coral animals, encrusting coralline algae of genus *Lithothamnium* (**Figure 14.5c**) also remove large quantities of dissolved calcium carbonate from seawater and deposit it within their tissues. This activity helps cement the reef into a mass capable of resisting heavy surf. Some reefs in the Indian Ocean are not, as was once thought, of coral origin, but instead were formed almost exclusively by the activity of coralline algae.

Seaweed Zonation

Zones, or bands of dominant seaweeds, are plainly visible on the sloping nearshore seafloor to the limit of light penetration. The vertical distribution of large marine plants depends in part on the pigments they possess. Chlorophytes are rarely found more than 10 meters (33 feet) below the surface; phaeophytes can operate from the sur-

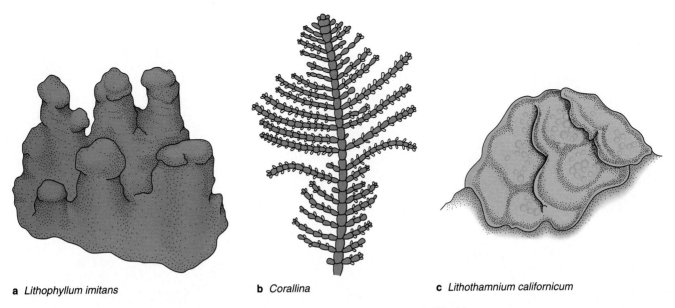

a *Lithophyllum imitans*   **b** *Corallina*   **c** *Lithothamnium californicum*

Figure 14.5   Various rhodophytes (red algae). *Corallina* (b) is an erect species, and *Lithophyllum* (a) and *Lithothamnium* (c) are encrusting species.

face down to about 30 meters (100 feet); and rhodophytes, as we have seen, can function from the surface to the photosynthetic record depth of 268 meters (879 feet).

Other factors are also important in determining where these algae will prosper, especially in the intertidal zone. Wave shock, varying salinity and pH, the presence of grazers from land and sea, the drying effects of sun and wind, abrasion by waterborne sand—the list of physical and biological factors affecting the location of intertidal plants seems almost endless. Any observant beachcomber can see the effects of these factors on marine plants (and animals) during low tide. Look for bands of different marine algae next time you go to the shore (see Figure 14.7b).

ROCKY INTERTIDAL COMMUNITIES

Anyone who spends time at the shore, especially a rocky shore, is soon struck by a curious contradiction. Although the rocky shore looks like a very difficult place for organisms to make a living, the **intertidal zone**—the band between the highest high tide and lowest low tide marks—is one of Earth's most densely populated areas. Hundreds of species and individuals crowd this junction of land and sea.

The problems of living in the intertidal zone are formidable. The tide rises and falls, alternately drenching and drying out the animals and plants. **Wave shock**, the powerful force of crashing waves, tears at the structures and underpinnings of the residents. Temperature can change

rapidly as cold water hits warm shells, or as the sun shines directly on newly exposed organisms. In high latitudes, ice grinds against the shoreline, and in the tropics intense sunlight bakes the rocks. Predators and grazers from the ocean visit the area at high tide, and those from land have access at low tide. Too much fresh water can shock the occupants during storms. Annual movement of sediment onshore and offshore can cover and uncover habitats. Yet, astonishingly, the richness, productivity, and diversity of the intertidal rocky community—especially in the world's temperate zones—is matched by very few other places. There is intense competition for space. Life abounds.

One reason for the great diversity and success of organisms in the rocky intertidal zone is the large quantity of food available. The junction between land and ocean is a natural sink for living and once-living material. The crashing of surf and strong tidal currents keep nutrients stirred and ensure a high concentration of dissolved gases to support a rich population of autotrophs. Minerals dissolved in water running off the land serve as nutrients for the inhabitants of the intertidal zone as well as for plankton in the area. Many of the larval forms and adult organisms of the intertidal community depend on plankton as their primary food source.

Another reason for the success of organisms here is the large number of habitats and niches available for occupation (see **Figure 14.6**). The habitats of intertidal animals and plants vary from hot, high, salty splash pools to cool, dark crevices. These spaces provide hiding places, quiet places to rest, attachment sites, jumping-off spots, cracks from which to peer to obtain a surprise meal, footing

a

Figure 14.6 A Pacific coast tide pool and intertidal shore. (a) A diagrammatic view. (b) Key.

1 *bushy red algae,* Endocladia
2 *sea lettuce, green algae,* Ulva
3 *rockweed, brown algae,* Fucus
4 *iridescent red algae,* Iridea
5 *encrusting green algae,* Codium
6 *bladderlike red algae,* Halosaccion
7 *kelp, brown algae,* Laminaria
8 *Western gull,* Larus
9 *intrepid marine biologist,* Homo
10 *California mussels,* Mytilus
11 *acorn barnacles,* Balanus
12 *red barnacles,* Tetraclita
13 *goose barnacles,* Pollicipes
14 *fixed snails,* Aletes
15 *periwinkles,* Littorina
16 *black turban snails,* Tegula
17 *lined chiton,* Tonicella
18 *shield limpets,* Collisella pelta
19 *ribbed limpet,* Collisella scabra
20 *volcano shell limpet,* Fissurella
21 *black abalone,* Haliotis
22 *nudibranch,* Diaulula
23 *solitary coral,* Balanophyllia
24 *giant green anemones,* Anthopleura
25 *coralline algae,* Corallina
26 *red encrusting sponges,* Plocamia
27 *brittle star,* Amphiodia
28 *common starfish,* Pisaster
29 *purple sea urchins,* Strongylocentrotus
30 *purple shore crab,* Hemigrapsus
31 *isopod or pil bug,* Ligia
32 *transparent shrimp,* Spirontocaris
33 *hermit crab,* Pagurus, *in turban snail shell*
34 *tide pool sculpin,* Clinocottus

b

from which to launch a sneak attack, secluded mating nooks, or darkness to shield a retreat. The niches of the creatures in this community are varied and numerous: Encrusting algae produce carbohydrates, snails scrape algae from rocks, hermit crabs scavenge for tidbits, octopuses wait to surprise likely meals, sea stars pry open mussels, and barnacles sweep bits of food from the water.

The most obvious and important physical factor in intertidal communities is the rise and fall of the tides (see Chapter 10). Organisms living between the high and low tide marks experience very different conditions from those residing below the low tide line. Within the intertidal zone itself, organisms are exposed to varying amounts of emergence and submergence. For example, **Figure 14.7a** plots the number of hours of exposure to air in a California intertidal zone through six months' time *vs.* the tidal height (in feet). Because some organisms can tolerate many hours of exposure while others are able to tolerate only a very few hours per week or month, the animals and plants sort themselves into three or more horizontal bands, or subzones, within the intertidal zone. Each distinct zone is an aggregation of animals and plants best adapted to the conditions within that particular narrow habitat. The zones are often strikingly different in appearance, even to a person unfamiliar with shore-

line characteristics. This zonation is clearly evident in the rocky shore of **Figure 14.7b**.

For intertidal areas exposed to the open sea, wave shock is a challenging physical factor. Fist-sized rocks have been thrown 100 meters (330 feet) into the air by the force of breaking waves. Large intertidal plants must be immensely strong, elastic, and slippery to avoid being shredded by wave energy. **Motile** (*motus* = moving) animals move to protective overhangs and crevices, where they cower during intense wave activity. Attached, or **sessile** (*sessilis* = sitting), animals hang on tightly, often gaining assistance from rounded or very low-profile shells that deflect the violent forces of rushing water around their bodies. Some sessile animals have a flexible foot that wedges into small cracks to provide a good hold; others (like mussels) form shock-absorbing cables that attach to something solid.

Desiccation (drying) by exposure to air and sunlight is another source of intertidal stress. Again, motile organisms have an advantage because they can move toward water left in tidal pools or muddy depressions by the retreating ocean. Attached animals and plants must await the water's return, huddled in low spots, moist pockets, or cracks in the rocks, or within tightly closed shells. Water trapped within a shell can keep gills moist for the

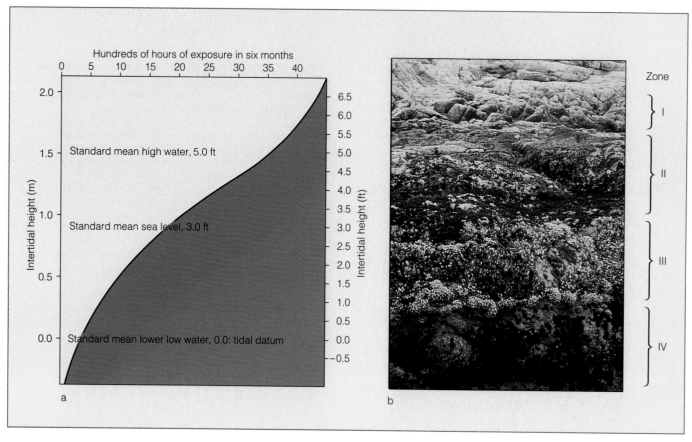

Figure 14.7 The relationship between amount of exposure and vertical zonation in a rocky intertidal community. (a) A graph showing intertidal height versus hours of exposure. The 0.0 point on the graph, the tidal datum, is the height of mean lower low water. (b) Vertical zonation, showing four distinct zones. The uppermost zone (I) is darkened by lichens and cyanobacteria; the middle zone (II) is dominated by a dark band of the red alga *Endocladia*; the low zone (III) contains mussels and gooseneck barnacles; and the bottom zone (IV) is home to sea stars (*Pisaster*) and anemones (*Anthopleura*). The bands in the photograph correspond approximately to the heights shown in the graph.

needed exchange of gases. A protective mucous coating can retard evaporative water loss from exposed soft animal body parts or blades of seaweed.

SAND BEACH AND COBBLE BEACH COMMUNITIES

Not all intertidal areas are composed of firm rock; some are sandy, some are muddy, and others consist of gravel or cobbles. (A few shores combine these elements within a small area.) The usual rigors of the intertidal zone are intensified for organisms surviving on loose substrates. Indeed, it may surprise you to learn that in spite of its generally benign conditions, the ocean contains what may well be the most hostile, rigorous, and dangerous environments for small living things on Earth: high-energy sand and cobble beaches.

As environments go, sand beaches don't seem particularly nasty places to us humans; many people consider the beach to be about the finest habitat around. Seals and sea lions spend a lot of time at the beach and seem to enjoy the experience as much as people do. In short, for organisms of about our size, the problems of living on a beach are manageable.

But for smaller organisms a beach is a forbidding place. Sand itself is the key problem. Many sand grains have sharp, pointed edges, so rushing water turns the beach surface into a blizzard of abrasive particles. Jagged grit works its way into soft tissues and wears away protective shells. A small organism's only real protection is to burrow below the surface, but burrowing is difficult without a firm footing. When the grain size of the beach is small, capillary forces can pin down small animals and prevent them from moving at all. If these organisms are trapped near the sand surface, they may be exposed to predation, to overheating or freezing, to osmotic shock from rain, or to crushing as heavy animals walk or slide on the beach.

As if this weren't enough, those that survive must contend with the difficulty of separating food from swirling sand and the dangers of leaving telltale signs of their position for predators or of being excavated by crashing waves. A few can run for their lives; some larger beach-dwelling crabs depend on their good eyesight and sprinting ability to outrace onrushing waves.

To these horrors must be added the usual problems of intertidal life already discussed. Not surprisingly, very few species have adapted to wave-swept sandy beaches! The few that have done so—mostly small, fast-burrowing clams and sand crabs and sturdy worms—consume a rich harvest of plankton and organic particles washed onto the beach and filtered from the water by the uppermost layer of sand.

Cobble beaches are even more uninviting (and they're murder on bare feet). The rounded rocks clack and bump together as waves pound the shore; most small animals are crushed. Except for nimble, insectlike "beach hoppers" and a few species of scavenging terrestrial insects, most loose, rock-strewn shores are understandably sterile of anything much larger than microscopic organisms.

Surely the most difficult of all are the black sand beaches, derived from pulverized lava on tropical volcanic islands such as Hawaii. Besides all the other difficulties mentioned, the lava on these beaches can store solar heat until temperatures approach 71°C (160°F) just below the surface of the sand. Almost nothing can tolerate these beaches for more than just a few minutes—including human feet!

SALT MARSHES AND ESTUARIES

Muddy-bottomed salt marshes are among the most interesting intertidal shores. Much of the high primary productivity of a salt marsh comes from sea grasses, mangroves, and other vascular plants that can prosper in a marine (or partly marine) environment.

As you may recall from Chapter 11, salt marshes often form in an **estuary**, a broad, shallow river mouth where fresh water and salt water mix (see Figure 11.17). A typical characteristic of estuaries is the reduction of wave shock: Surf is blocked from estuaries by longshore bars or by twisting passages connecting to the ocean. The salinity of water within an estuary may vary with tidal fluctuations, from seawater through brackish water (mixed salt and fresh water) to fresh water. Many of the organisms living in estuaries are able to tolerate varying salinities; but in areas near the river entrance the water may be almost fresh, while near the outlet it may be of oceanic salinity. These different salinities often lead to distinct horizontal zonation of organisms. Temperature range is also potentially extreme, especially in the tropics or during the temperate-zone summer, when a receding tide abandons residents to the heat of the sun. Strong currents may move in estuaries as the tide rises and falls and the river flows. Flowing water takes the place of waves in mixing nutrients and gases in the intertidal estuary community.

Estuarine marshes (such as the one shown in **Figure 14.8**) are richer and exhibit greater species diversity than marshes exposed only to seawater. Primary productivity in estuaries is often extraordinarily high because of the availability of nutrients, the great variety of organisms present, strong sunlight, and the large number of

Figure 14.8 An estuarine marsh. Urban developers often destroy coastal marshes to build marinas and homes, but citizens near this marsh in Orange County, California, have recognized its natural value and have set it aside as a marine preserve.

Benthic Communities **261**

niches. Decomposition of fast-growing, salt-tolerant plants provides the raw material for the large and complex food webs and rapid nutrient turnover characteristic of these communities. The standing biomass (mass of living matter per unit area or volume) in a typical estuary is among the highest of any marine community.

Estuarine organisms show unique adaptations to their rich and variable environment. Some estuarine plants trap fine silt particles at their roots, thus countering the erosive action of current flow. Small plants are often filamentous, bristling with tiny projections used to anchor themselves to the substratum. Larger plants have extensive root systems to hold themselves in place and to colonize new areas. Most of the resident animals burrow into the muck, scurry rapidly across the surface, or hide in the vegetation. Clams and snails work their way through the substratum, obtaining food and shelter at the same time. Polychaete worms dig for targets of opportunity, and crabs dart for any interesting morsels. Since planktonic larvae would be washed out to sea, most estuarine organisms produce nonplanktonic larvae, lay eggs on firm objects, or carry eggs on their bodies.

Estuaries are sometimes called marine nurseries because so many juvenile organisms are found there. This is especially true for fishes. Many pelagic species spend their larval lives in the protective confines of an estuary, taking advantage of the many feeding opportunities available. Most of the commercially exploited fish species on the U.S. Atlantic Coast utilize estuaries as juvenile feeding grounds. The human pressures of development and pollution are thus doubly stressful in estuaries, affecting both permanent residents and the sensitive larval stages of open-water animals.

MANGROVE FORESTS

Low, muddy coasts in tropical and some subtropical areas are often home to tangled masses of trees known as **mangroves**. These large flowering plants are never completely submerged, but because of their intimate association with the ocean they are considered marine plants (**Figure 14.9**). They thrive in the sediment-rich lagoons, bays, and estuaries of the Indo-Pacific, tropical Africa, and the tropical Americas. Their distribution, shown in Figure 14.4, depends on temperature, currents, and rainfall.

The sediment in which mangrove trees live must be covered with brackish or salt water for part or all of the day. Many mangroves avoid taking up salt ions from seawater or selectively remove salt from sap with salt-excreting cells. The fine coastal mud they colonize doesn't provide firm footing for these substantial plants; so an intricate network of arching prop roots is required for support. The strutlike prop roots are supplemented by many smaller roots equipped with breathing pores and air passages. Atmospheric oxygen is conducted by these passages to the portions of the plant submerged in oxygen-deficient mud. The root system also traps and holds sediments around the plant by interfering with the transport of suspended particles by currents. The root complex

Figure 14.9 A mangrove (*Rhizophora*) growing in salt water in Everglades National Park, Florida. The tangled roots descending from the main branch are called *prop roots* or *stilt roots*. They provide anchorage for the mangrove, trap sediment, and provide protection for small organisms.

forms an impenetrable barrier and safe haven for organisms around the base of the trees.

CORAL REEF COMMUNITIES

The tropical ocean is blue, brilliantly transparent, relatively high in salinity, and notable for its deep and abrupt thermocline and relatively low concentrations of dissolved nutrients and gases. It supports surprisingly little life.

But what of the travel-poster view of the tropical ocean? What about the thousands of brightly colored fish, strange invertebrates, and breathtaking scenes of divers swimming through living reef formations? Such scenes, photographed on reefs near islands and continents, can be found in less than 2% of the tropical ocean. The key to the difference between the open tropical ocean and the tropical reefs lies in the productivity of the **reefs** themselves, wave-resistant structures dominated by strong and rigid masses of living (or once-living) organisms. Not all reefs are built of coral; other reef builders include red and green algae, cyanobacteria, worms, and even oysters. Nonetheless, we think first of coral reefs when the words *reef* and *tropics* are mentioned together.

Coral

Although they look like flowers, **corals** are related to sea anemones and jellyfish. Some corals are solitary animals with bodies up to 30 centimeters (12 inches) in diameter, but most of the more than 500 species are ant-sized organisms crowded into colonies called **coral reefs** (**Figure 14.10**). The coral animals themselves construct the reefs by secreting hard skeletons of aragonite, a fibrous crystalline form of calcium carbonate. The matrix of cup-shaped individual skeletons secreted by coral animals gives the colony its characteristic shape.

An individual coral animal, or **polyp**, feeds by capturing and eating plankton that drift within reach of its rosette of tentacles. Victims are entrapped by stinging cells on each tentacle, transported to a central gastric cavity, and rapidly digested. Tropical corals feed at night; at dawn the polyps retract into their skeletal cups to withstand drying should the colony be exposed to air at low tide. Coral polyps can also feed by direct absorption, a process in which they simply transport dissolved food molecules through their body walls. The anatomy of a coral polyp is shown in **Figure 14.11**.

Tropical reef-building corals are **hermatypic**, a term derived from the Greek word *hermatos*, mound-builder.

Figure 14.10 Close-up of hermatypic coral, showing expanded polyps.

Figure 14.11 Anatomy of a coral reef polyp, with a blowup detail showing a cross section of the outer covering and tissue.

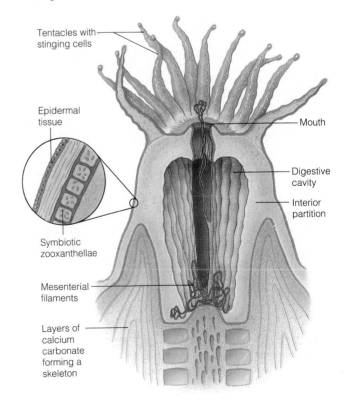

Tentacles with stinging cells

Epidermal tissue

Mouth

Digestive cavity

Interior partition

Symbiotic zooxanthellae

Mesenterial filaments

Layers of calcium carbonate forming a skeleton

Their bodies contain masses of tiny dinoflagellates called **zooxanthellae** (*xanthos* = yellow). These single-celled plantlike organisms facilitate the rapid biochemical deposition of calcium carbonate into the coral skeleton. Coral's success in the nutrient-poor water of the tropics depends upon its intimate biological partnership with zooxanthellae. The microscopic zooxanthellae carry on photosynthesis, absorb waste products, grow, and divide within their coral host. The coral animals provide a safe and stable environment and a source of carbon dioxide and nutrients; the zooxanthellae reciprocate by providing oxygen, carbohydrates, and the alkaline pH necessary to enhance the rate of calcium carbonate deposition. The coral occasionally absorbs a cell, "harvesting" the organic compounds for its own use. The zooxanthellae are captive within the coral; so none of their nutrients are lost, as they would be if the zooxanthellae were planktonic plants that could drift away from the reef. Instead nutrients are used directly by the coral for its own needs. The cycling of materials is short, direct, quick, and very efficient.

Because of the needs of its zooxanthellae, hermatypic corals depend on light and warmth. Reef corals grow best in brightly lighted water about 5 to 10 meters (16 to 33 feet) deep. Coral reefs can form to depths of 90 meters (300 feet), but growth rates decline rapidly past the optimum 5- to 10-meter depth. In ideal conditions coral animals grow at a rate of about 1 centimeter (½ inch) per year. They prefer clear water because turbidity prevents light penetration, and suspended particles interfere with feeding. The animals are protected from the harmful effects of bright sunlight by a mucous coating that contains an ultraviolet-blocking "suntan lotion."

Hermatypic corals also prefer water of normal or slightly elevated salinity. Coral animals are highly susceptible to osmotic shock, and exposure to fresh water is rapidly fatal. (Reefs growing in shallow water have a flat upper surface because rain is lethal.) Fresh water and suspended sediments prevent reefs from forming near the mouths of rivers or in areas adjacent to islands or continents where rainfall is abundant. Reef corals are also susceptible to potentially devastating diseases, such as bleaching, about which scientists as yet know very little (see **Box 14.1**).

Nearly all of the reef-building corals are found within the 21°C (70°F) isotherm indicated in **Figure 14.12**, a zone

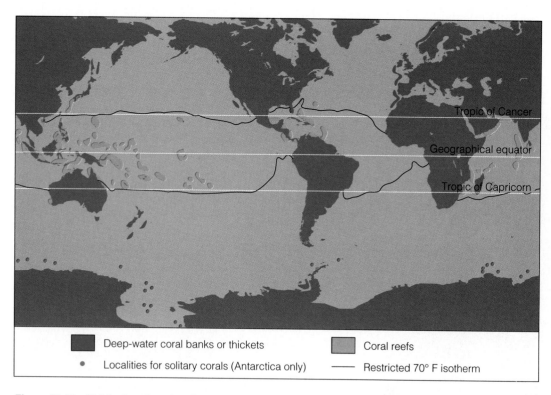

Deep-water coral banks or thickets   Coral reefs

· Localities for solitary corals (Antarctica only)   —— Restricted 70° F isotherm

Figure 14.12 Distribution of coral reefs and their relation to sea surface temperature. The 70°F isotherm connects ocean areas with a surface temperature of 70°F. The ocean surface toward the equator becomes progressively warmer.

BOX 14.1 ● *Coral Bleaching*

In 1983, marine scientists first encountered a baffling new disease of coral reefs. Because it causes coral animals to expel their brownish zooxanthellae and turn an uncharacteristic creamy color, the phenomenon is known as *coral bleaching*. Without zooxanthellae, the coral cannot secrete calcium carbonate and generate a skeleton or reef. After one bout of bleaching, corals usually recover, but while they are bleached, they stop growing, leaving the reef vulnerable to erosion.

Some 60 species of coral are known to be affected. After the first outbreak in eastern Pacific reefs, the mysterious disease reached the Caribbean in 1987, attacking reefs from the Florida Keys to Jamaica and as far east as the Virgin Islands.

The causes for bleaching are unknown. Initial suspicions centered on global warming. If seawater were becoming warmer, the supersensitive coral might respond by changing in ways incompatible with needs of their zooxanthellae. The 1987 episode was widespread, occur-

ring at distant sites at about the same time, leading scientists to believe local conditions of pollution may not be to blame. Other researchers believe that a combination of temperature increase, sediments, and chemical pollutants may be the cause. Statistical analysis of water temperature in places where bleaching has been observed suggest that not enough time has passed to discern a temperature trend. We'll have to wait and see.

Corals are subject to other unusual diseases as well. Elkhorn coral, the principal reef-building coral in the Caribbean, is affected by lethal white-band disease, and Pacific species are developing other peculiar maladies. Whether these problems are natural cyclical events or have been triggered by human activity remains to be discovered.

Most corals recover from these diseases, but by 1990 the recovery rate had slowed markedly. This trend raises fears for the future of the reefs themselves, and it lends urgency to the quest for a cause.

that corresponds roughly with the 25° latitude lines in both hemispheres. Poleward of this area, water is too cold for the zooxanthellae to survive; temperature below 18°C (65°F) causes their death. But as Figure 14.12 also shows, individual coral organisms are found in some cold, high-latitude waters as well. These corals lack zooxanthellae; so they deposit calcium carbonate much more slowly, and the structures they build do not resemble those found in the tropics. Instead, these deep-water corals, known as **ahermatypic** (*a* = without + *hermatos* = mound-building) corals, build smooth banks on the cold, dark outer edges of temperate continental shelves from Norway to the Cape Verde Islands, and off New Zealand and Japan. The rarest corals of all are large solitary organisms living on the abyssal floors and outer continental shelves of the Antarctic.

Other Denizens of the Reef

Tropical coral reefs typically form in areas of high wave energy; indeed, reef organisms preferentially grow into high-energy environments in an attempt to be first in ob-

taining dissolved and suspended material in the water. In most reefs there is an approximate balance between construction and destruction. The reef consists of actively growing coral colonies and fragments of material of different sizes, from worn coral boulders down to fine sand.

Corals are by no means the only participants in reef life, however; they may account for only about half of the biomass in these areas. Other reef residents include calcareous algae, whose secretions help "cement" the reef together, as well as a bewildering array of encrusting, burrowing, producing, and consuming creatures ranging upward in size from the microscopic. Some tunnel into the coral or shatter it in search of food, contributing to the erosion of the reef. Fierce competition exists among reef organisms for food, living space, protection from predators, and mates. The bright colors, protective camouflage, spines, and various toxins and venoms common to tropical organisms are probably related to the intense struggle for existence that goes on in these beautiful but deceptively calm-looking places. A typical reef scene is depicted in **Figure 14.13**.

a

Figure 14.13   (a) The coral reef habitat. (b) Key. (c) Life on a coral reef.

1 black-capped petrel
2 sea nettle
3 angelfish
4 lobed corals
5 sea whips and soft corals
6 triggerfish
7 sea fans
8 tube anemone
9 orange stone coral
10 bryozoans
11 brain coral
12 butterfly fish
13 moray eel
14 cleaner fish
15 tube corals

16 muricid snail
17 nudibranch
18 sponges
19 colonial tunicate
20 giant clam
21 purple pseudochromid fish
22 cobalt sea star
23 soft corals
24 barber pole shrimp
25 sea anemones
26 clown fish
27 worm tubes
28 cowry
29 sea fan

b

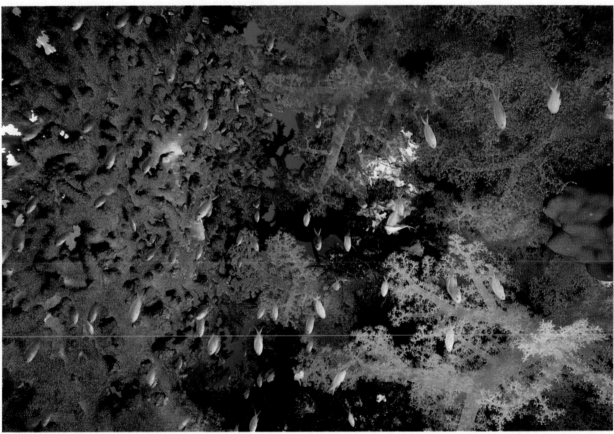

c

Reef Types

In 1842, Charles Darwin classified tropical reef structures into three types: fringing reefs, barrier reefs, and atolls (see **Figure 14.14**). We still use this classification today.

Fringing Reefs As their name implies, **fringing reefs** cling to the margin of land. As can be seen in **Figure 14.14a**, a fringing reef connects to shore near the water surface. Fringing reefs form in areas of low rainfall runoff, primarily on the lee (downwind side) of tropical islands. The greatest concentration of living material will be at the reef's seaward edge, where plankton and clear water of normal salinity are dependably available. Most new islands anywhere in the tropics have fringing reefs as their first reef form. Permanent fringing reefs are common in the Hawaiian Islands and in similar areas near the boundaries of the tropics.

Barrier Reefs **Barrier reefs** are separated from land by a lagoon (**Figure 14.14b**). They tend to occur at lower latitudes than fringing reefs, and they can form around islands or in lines parallel to continental shores. The outer edge—the barrier—is raised because the seaward part of the reef is supplied with more food and is able to grow more rapidly than the shore side. The lagoon may be from a few meters to 60 meters (200 feet) deep, and it may separate the barrier from shore by only tens of meters or by 300 kilometers (190 miles), in the case of northeastern Australia's Great Barrier Reef. Coral grows slowly within the lagoon because fewer nutrients are available and because sediments and fresh water run off from shore. As you would expect, conditions and species within the lagoon are much different from those of the wave-swept barrier. The calm lagoon is often littered with eroded coral debris moved from the barrier by storms.

The Great Barrier Reef is the largest biological construction on the planet. It extends along the northeast coast of Queensland for 2,000 kilometers (1,250 miles) and is up to 150 kilometers (95 miles) wide. It isn't a single reef, but a conglomeration of thousands of interlinked segments. The segments present a steep outer wall to the prevailing currents and trade winds. At a growth rate of 1 centimeter (½ inch) per year, the structure is obviously of great age. The variety of organisms within the Australian Barrier Reef staggers the imagination. About 500 species of hermatypic coral live there, nearly ten times the number found in the western Atlantic. Over 1,000 species of bivalve mollusks, 3,000 species of shore fishes, and 40 species of sea snakes also live in the area.

Atolls An **atoll** (**Figure 14.14c**) is a ring-shaped island of coral reefs and coral debris enclosing, or almost enclosing, a shallow lagoon above which no land protrudes.

Coral debris may be driven onto the reef by waves and wind to form an emergent arc on which coconut palms and other land plants take root. These plants stabilize the sand and lead to colonization by birds and other species. This is the tropical island of the travel posters.

Though an atoll's central lagoon connects to the deep water outside through a series of channels or grooves, coral does not usually thrive there because the water may become too fresh during rains or too hot, and because feeding opportunities for the coral in the enclosed lagoon are limited. Despite these drawbacks, some lagoons have enough resources within to sustain small *patch reefs*, miniatures of the larger reef.

Some atolls are isolated, but most occur in loose groups (**Figure 14.15**) in shallow continental shelf areas or in the deep open ocean. More than 300 atolls exist—most in the Pacific. They range in size from a few kilometers in diameter to Kwajalein in the Marshall Islands, whose slender 280-kilometer (176-mile) ring of coral encloses a lagoon of 2,850 square kilometers (1,100 square miles).

Atoll Formation

How do atolls form? Scientists began speculating on the cause of their ring shape soon after the first scientific voyages published their reports. Charles Darwin imagined a volcanic island growing from the sea, accumulating a skirt of coral around its shore, and then slowly subsiding at a rate equal to the growth rate of the coral. The central volcanic island would eventually sink from view, but the coral could grow continuously atop skeletons of past generations to maintain a living presence near the surface. **Figure 14.16** shows the progression. Note that the island begins with a fringing reef, passes through a barrier reef stage as it sinks, and eventually becomes an atoll as the peak disappears beneath the ocean surface. Should the island subside faster than about 1 centimeter (½ inch) per year (the growth rate of coral), all trace of the island and the reefs will disappear. (The submerged island may become a guyot.)

This theory seemed reasonable, but Darwin couldn't explain what would cause volcanic islands to subside because he didn't know about plate tectonics. Now we know that volcanoes can form near spreading centers, ride outward and downward from their birthplaces, cease to be active as they leave their source of mantle heat, and sink as they are carried into deeper water—just slowly enough to permit coral growth to continue around their edges as they go. An alternate hypothesis suggested that a slow but continuous rise in sea level could cause the growth of atolls. But this explanation could be correct only if the thickness of the accumulated coral was less than about 200 meters (660 feet), the minimum sea level of comparatively recent geologic time.

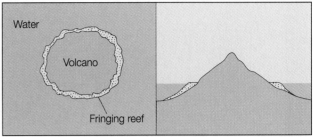

Figure 14.14 The three types of coral reefs. (a) The structure of a fringing reef like that on Moorea, in the Society Islands. (b) The structure of a barrier reef, an example of which is Bora Bora, also in the Society Islands. (c) An atoll, such as Aratika, in the Tuamotu Archipelago.

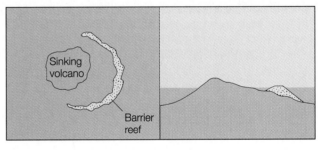

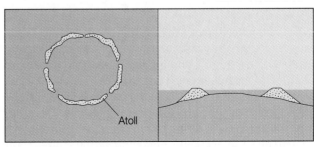

Figure 14.15 A group of atolls in the Tuamotu Archipelago in the South Pacific. The photo was taken looking southeast from *Apollo 7* from an altitude of 153 kilometers (96 miles).

Drilling at Eniwetok Atoll in 1952, in preparation for thermonuclear bomb testing, settled the issue. The strength and composition of the island was probed in order to calibrate the yield of the bombs. The drills hit volcanic rock beneath the coral at depths of 1,267 meters (4,156 feet) and 1,458 meters (4,782 feet), much deeper than the lowest sea level of any known ice age. Darwin's hypothesis appears to be correct.

THE DEEP-SEA FLOOR

Most of the deep-ocean floor is an area of endless sameness. It is eternally dark, almost always very cold, slightly hypersaline (to 36‰), and highly pressurized. Life there is more plentiful and obvious than in the bathypelagic water above; but compared to other bottom areas, the density of life at great depths is extremely low. For example, 5,000 grams of living organisms can be found on a square meter of nearshore seafloor, and 200 grams might be recovered from the same area of continental shelf, but less than *1 milligram per square meter* is typical for deep-ocean benthic communities! Currents are usually so weak that they impose no strain on residents, but they do transport the few drifting organisms and waft the smell of food. Most bottom dwellers are blind.

The feeding strategies of animals living on the deep-ocean floor are often bizarre. Tripod fish (**Figure 14.17**) use sensitive extensions of their fins and gill coverings to detect the movement of prey many meters away. Some organisms whose mouths blend with the natural contours of the ooze act as living caves into which small creatures crawl for protection. The predator need not even swallow to get the prey into its gut; back-pointing spines direct the victim along a one-way path to the stomach! Other species are capable of smelling sunken dead organisms for miles downcurrent, then spending weeks or months slowly following the scent to its source. The metabolic rate of organisms in cold water tends to be low; so most deep animals require relatively little food,

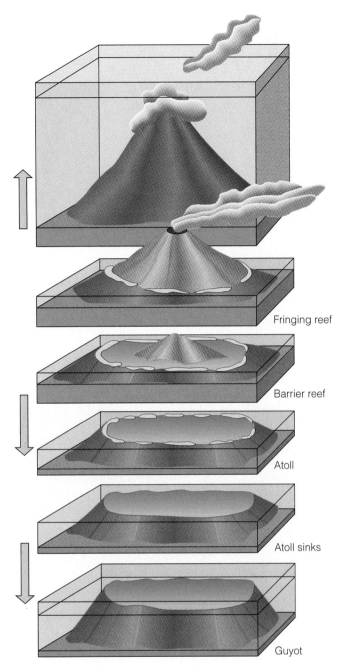

Figure 14.16 The history of an atoll. Volcanic activity at a spreading center builds the island, which acquires a fringing reef. As the island moves away from the spreading center, the volcano becomes inactive, the island slowly subsides, and the coral animals continue to build, forming first a barrier reef and then an atoll. If the subsidence rate increases above about 1 centimeter (½ inch) a year, the coral dies and the atoll becomes a guyot.

Fringing reef

Barrier reef

Atoll

Atoll sinks

Guyot

Figure 14.17 A blind tripod fish, an abyssal benthic species. The long, curved projections are thought to aid in sensing the distant vibrations of prospective prey.

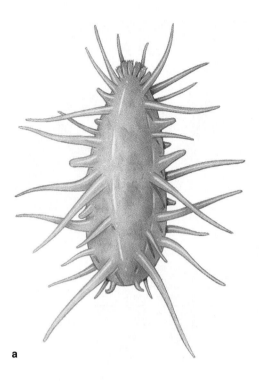

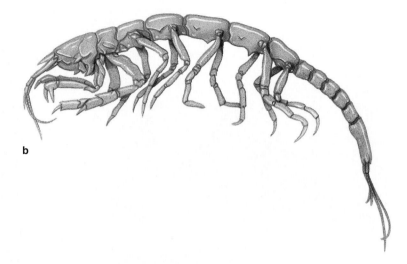

b

Figure 14.18 Abyssal benthic animals. (a) *Oneirophanta*, a 10-centimeter (4-inch) holothuran (sea cucumber) found on the abyssal plains of the North Atlantic. (b) *Apseudes galatheae*, a blind, thumb-sized crustacean found in the Kermadec Trench, north of New Zealand.

a

move slowly, and live very long lives. Some may feed less than once in a year and may live to be hundreds of years old. Two deep benthic representatives are seen in **Figure 14.18**.

This deep uniform environment, though rigorous, holds benefits for those few species adapted to it. Perhaps the best adapted organisms are the ubiquitous ophiuroids (brittle stars, **Figure 14.19**) that inhabit sedimentary bottoms at nearly all latitudes, sometimes at arm-tip to arm-tip densities. They feed by collecting tiny nutritive particles—some of which have fallen through the water for many months—along grooves beneath their arms and then transporting them to their mouths. Brittle stars are among the most widely distributed of Earth's animals.

The organisms within deep pelagic and benthic communities share some curious adaptations. Gigantism is a common characteristic. Individuals of representative families in deep water often tend to be much larger than related individuals in the shallow ocean. Fragility is also common in the depths. Not only are heavy support structures unnecessary in the calm deep environment, but the low water pH and deficiency of dissolved calcium discourage skeletal development. Some animals have slender legs or stalks to raise them above the sediment, and some come apart like warm gelatin at the slightest touch. Except for its influence on enzyme activity, hydrostatic pressure is not a problem for these animals. Because they lack gas-filled internal spaces, their internal pressure is precisely the same as that outside their bodies.

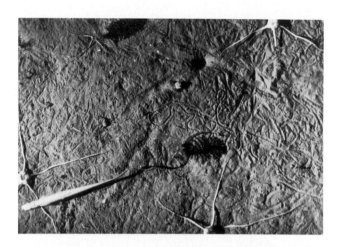

Figure 14.19 Brittle stars and their tracks on the continental slope off New England. The depth here is 1,476 meters (4,842 feet).

VENT COMMUNITIES

The oceanographic world was excited in 1977 when scientists in Woods Hole Oceanographic Institution's submersible *Alvin* (Figure c, Box 4.1) discovered an entirely new type of marine community over 3,000 meters (10,000 feet) below the surface.[1] They were searching the near-

[1]Ironically, there wasn't a single marine biologist on this research cruise—only geologists, geophysicists, chemists, physicists, and a science writer!

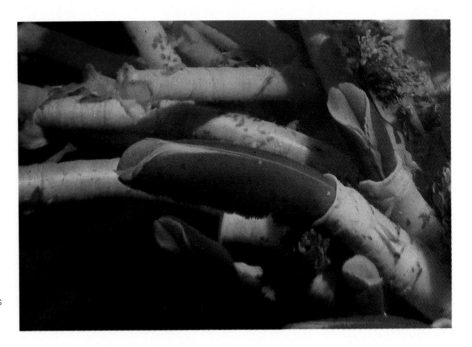

Figure 14.20 *Riftia*, the genus of pogonophorans found around hydrothermal vents on oceanic ridges.

freezing bottom for the source of some unusually warm water, which had been detected by remote probes. What they found were jets of superheated water (to 350°C, 650°F) blasting from rift vents in the young oceanic ridge 350 kilometers (220 miles) north and east of the Galápagos Islands. Clustered around the vents were dense aggregations of large, previously unknown animals. Bottom water in the area was laden with hydrogen sulfide, carbon dioxide, and oxygen, upon which specialized bacteria were found to live. These bacteria evidently form the base of a food chain that extends to the unique animals. Large crabs, clams, sea anemones, shrimp, and unusual worms were found in this warm oasis.

Some of the tube worms, contained in their own long parchmentlike tubes, measured 3 to 4 meters (10 to 13 feet) in length and were the diameter of a human arm. These strange animals have been tentatively identified as pogonophorans, members of a small phylum of invertebrates also found in fairly shallow water. Three species of this newly and appropriately named genus *Riftia* (**Figure 14.20**) have been identified so far. The tubes of these pogonophorans are flexible and capable of housing the length of the animal when it retracts. The animals extend tufts of tentacles from the openings of their tubes. Feeding was something of a puzzle because these animals have no mouth, digestive tract, or anus. The trunks of the worms were found to contain large "feeding bodies" tightly packed with bacteria similar to those seen in the water and on the bottom near the geothermal vents. The worms' tentacles absorb hydrogen sulfide from the water and transport it to the bacteria, which then use the hydrogen sulfide as an energy source to convert carbon dioxide to organic molecules. The ultimate source of the worms' energy (and the energy of most other residents in this community) is this energy-binding process, called chemosynthesis, which replaces photosynthesis in the world of darkness.

The clams and shrimp of the vent communities are equally unusual. For example, the large white clam *Calyptogena* grows among uneven basaltic mounds (**Figure 14.21**). Each the size of a shoe, the clams shelter the same kinds of bacteria as *Riftia*. Though the clam retains its filter-feeding structures, it too derives nutrition from the bacteria. Small shrimp discovered at the vents in 1985 have been found to possess special organs that may allow them to sense heat from the vents. Such an adaptation would permit them to range away from the vents for food, yet return to the warmth and richness of the home community.

Similar vent communities have now been found off Florida, California, and Oregon and in several other locations along oceanic ridges. Could vent communities occupy the active central rift valleys of a significant percentage of the 65,000 kilometers (41,000 miles) of Earth's oceanic ridges? Perhaps the deep-vent communities will prove to be more important in marine biology than has been previously supposed. Marine biologists are eager to continue their explorations.

Figure 14.21 A vent clam field dominated by the giant white clam *Calyptogena magnifica*.

CHAPTER SUMMARY

Benthic organisms live on or in the ocean bottom. They may be distributed through their habitats randomly, uniformly, or (most commonly) in clumped distributions. Nearshore benthic habitats in the temperate zones often contain seaweeds, nonvascular plants known as multicellular algae. Carbohydrates produced by these algae (and other plantlike organisms) provide energy to the animals of the benthic communities.

Rocky intertidal communities are among the ocean's richest and most diverse. Although the problems of rocky-shore living are formidable, hundreds of organisms have adapted to its rigors because of the wealth of food available there. Sand and cobble beaches may seem more benign, but the difficulty of maintaining a dependable foothold and separating food from inedible particles limits the number of organisms able to live there. Salt marshes and estuaries are among the ocean's most productive habitats, and estuaries shelter a remarkable variety of juvenile forms, some of which are refugees from the forbidding and competitive open-ocean environment. Except for vent communities associated with mid-ocean ridges, the deep-sea bed is the most sparsely populated benthic habitat, due largely to a limited food supply. It stands in remarkable contrast to the world of tropical coral reefs, places of overwhelming productivity, diversity, and beauty.

Terms and Concepts to Remember

accessory pigments	desiccation	polyp
ahermatypic	estuary	random
algae	fringing reefs	distribution
atoll	gas bladders	Rhodophyta
barrier reefs	hermatypic	sessile
benthic	holdfast	stipes
blades	intertidal zone	thallus
Chlorophyta	kelps	unicellular algae
clumped	mangroves	uniform
distribution	motile	distribution
coral reef	multicellular algae	wave shock
corals	Phaeophyta	zooxanthellae

Study Questions

1. What factors influence the distribution of organisms within a benthic community? How are these distributions described? Why is random distribution so rare?

2. What are algae? Are all algae seaweeds? How are seaweeds classified? Which seaweeds live at the greatest depths? Why?

3. What problems confront the inhabitants of the intertidal zone? How do you explain the richness of the intertidal zone in spite of these rigors? Which intertidal area has larger numbers of species and individuals: sand beach or rocky? Why?

4. Which benthic marine habitat is the most sparsely populated? Why?

5. If tropical oceans generally support very little life, why do coral reefs contain such astonishing biological diversity and density?

6. Explain Charles Darwin's classification scheme for coral reefs. Is the classification still in use?

For Further Study

Ballard, R. D. 1977. "Notes on a Major Oceanographic Find." *Oceanus* 20 (no. 3): 35–44. News of the startling discovery of the rift community.

Gage, J. D., and P. A. Tyler. 1991. *A Natural History of Organisms at the Deep-Sea Floor*. New York: Cambridge University Press. A scholarly celebration of the growth of our knowledge about biological systems on the seafloor.

Levinton, J. S. 1982. *Marine Ecology*. New York: Prentice-Hall. An excellent technical introduction to marine ecological theory. Chapter 5 summarizes marine community diversity and evolution.

MacGinitie, G., and N. MacGinitie. 1968. *The Natural History of Marine Animals*. Chapters 4, 5, and 8 cover animal groupings, relationships, and habitats.

Miller, G. T. 1994. *Living in the Environment*. 8th ed. Belmont, CA: Wadsworth. Part 2, on basic ecological concepts, is especially recommended.

Nybakken, J. W. 1992. *Marine Biology: An Ecological Approach*. 3d ed. New York: Harper & Row. Excellent general marine biology text with an ecological emphasis.

Ricketts, E. F., J. Calvin, J. Hedgpeth, and D. W. Phillips. 1985. *Between Pacific Tides*. 5th ed. Stanford, CA: Stanford University Press. A gracefully written classic applicable especially to the western American coast.

USES AND ABUSES OF THE OCEAN

"This is <u>Exxon Valdez</u>. We are hard aground."

The supertanker *Exxon Valdez* ran aground in Alaska's Prince William Sound on 24 March 1989. More than 40 million liters (almost 10.8 million gallons; 29,000 metric tons) of Alaskan crude oil—about 22% of her cargo—escaped from the crippled hull. Only about 17% of this oil was recovered by a work crew of more than 10,000 people using containment booms, skimmer ships, bottom scrapers, and absorbent sheets. About 35% of the oil evaporated, 8% was burned, 5% was dispersed by strong detergents, and 5% biodegraded in the first five months. The rest of the oil, some 30% of the spill, formed oil slicks on Prince William Sound and fouled more than 450 kilometers (300 miles) of coastline.

Recent analysis of the affected parts of Prince William Sound shows the cleaned areas to be in generally worse shape than areas left alone. Most of the small animals that make up the base of the food chain in these areas were cooked by the 65°C (150°F) water used to blast oil from between the rocks. Others were smothered when the high-pressure jets rearranged sand and mud. It appears that an overambitious cleanup program can be counterproductive. Sylvia Earle, NOAA's chief on-site scientist, has said: "Sometimes the best, and ironically the most difficult, thing to do in the face of an ecological disaster is to do nothing." The biological cost of the spill will not be known for a decade. The financial cost exceeded $2.5 billion by September 1993.

Cleaning up the March 1989 *Exxon Valdez* oil spill, Prince William Sound, Alaska.

There are two sides to humanity's use of the ocean. On the one hand, we find the ocean's resources useful, convenient, and essential. On the other hand, we find we cannot exploit those resources without damaging their source. Through the last few generations our increasingly anxious efforts at resource extraction and utilization have affected the ocean and atmosphere on a planetary scale. World economies are now dependent on oceanic materials, and we are unwilling to abandon or diminish their use until we see clear signs of severe environmental damage. We have thus embarked on an unintentional global experiment in marine resource exploitation and waste management. We only hesitatingly adjust course as we rush into the unknown.

MARINE RESOURCES

We will discuss three groups of marine resources in this chapter. **Physical resources** result from the deposition, precipitation, or accumulation of useful substances in the ocean or seabed. Most physical resources are mineral deposits, but petroleum and natural gas, mostly remnants of once-living organisms, are included in this category. Fresh water obtained from the ocean is also a physical resource. **Biological resources** are living animals and plants collected for human use. As the term suggests, **nonextractive resources** are uses of the ocean in place; recreation, waste disposal, and transportation of people and commodities by sea are examples.

Marine resources can be classified as either renewable or nonrenewable. **Renewable resources** are naturally replaced on a seasonal basis by the growth of marine organisms or by other natural processes. **Nonrenewable resources** such as oil, gas, and solid mineral deposits are present in the ocean in fixed amounts and cannot be replenished over time spans as short as human lifetimes. Exploitation of any resource—renewable or nonrenewable—may result in pollution.

MARINE POLLUTION

The utilization of ocean resources can result in the accidental (or intentional) release of harmful substances. We define **marine pollution** as the introduction into the ocean by humans of substances or energy that change the quality of the water or affect the physical and biological environment. It is not always easy to identify a pollutant; some materials labeled as pollutants are produced in large quantities by natural processes. For example, a volcanic eruption can produce immense quantities of carbon dioxide, methane, sulfur compounds, and oxides of nitrogen. Excess amounts of these substances produced

by human activity may cause global warming and acid rain. For this reason we need to distinguish between "natural pollutants" and human-generated ones.

We have long used the sea as a dump for our wastes. The ocean's great volume and relentless motion dissipate and distribute natural and synthetic substances, but its ability to absorb is not inexhaustible. No one knows to what extent we have contaminated the ocean. By the time the first oceanographers began widespread testing, the Industrial Revolution was well under way and changes had already occurred. Traces of synthetic compounds have now found their way into every oceanic corner. We will never know what the natural ocean was like or what remarkable plants and animals may have vanished as a result of human activity. Our limited knowledge of pristine conditions is gleaned from small seawater samples recovered from deep within the polar ice pack and from tiny air bubbles trapped in glaciers. There are few undisturbed habitats left to study, and few marine organisms are completely free of the effects of ocean pollutants.

Characteristics of a Pollutant

A **pollutant** causes damage by interfering directly or indirectly with the biochemical processes of an organism. Some pollution-induced changes may be instantly lethal; other changes may weaken an organism over weeks or months, or alter the dynamics of the population of which it is a part, or gradually unbalance the entire community.

In most cases, an organism's response to a particular pollutant will depend on its sensitivity to the *combination* of quantity and toxicity of that pollutant. Some pollutants are toxic to organisms in tiny concentrations. For example, the photosynthetic ability of some species of diatoms is diminished when chlorinated hydrocarbon compounds are present in parts-per-trillion quantities. Other pollutants seem harmless, as when fertilizers flowing from agricultural land stimulate plant growth in estuaries. Still other pollutants may be hazardous to some organisms but not to others. For example, crude oil interferes with the delicate feeding structures of zooplankton and coats the feathers of birds, but it simultaneously serves as a feast for certain bacteria.

Pollutants also vary in their persistence; some reside in the environment for thousands of years, while others last only a few minutes. Some pollutants break down into harmless substances spontaneously or through physical processes (like the shattering of large molecules by sunlight). Sometimes pollutants are removed from the environment through biological activity. For example, some marine organisms escape permanent damage by metabolizing hazardous substances to harmless ones. Indeed, many pollutants are ultimately **biodegradable**, that is, able to be broken down by natural processes into simpler

Figure 15.1 Bits of plastic and other debris litter the approach to a salt marsh.

compounds. Most pollutants, however, resist attack by water, air, sunlight, or living organisms because the synthetic compounds of which they are composed resemble nothing in nature.

The ways in which pollutants are changing the ocean and the atmosphere are often difficult for researchers to determine. Environmental impact cannot always be predicted or explained. As a result, marine scientists vary widely in their opinions about what pollutants are doing to the ocean and atmosphere, and what to do about it. Environmental issues are frequently emotional, and media reports tend to sensationalize short-term incidents (like oil spills) rather than more serious, long-term problems (like atmospheric changes or the effects of long-lived chlorinated hydrocarbon compounds).

The Costs of Pollution

In 1985 government and industry in the United States spent about $70 billion to control atmospheric, terrestrial,

and marine pollution—an average of $289 for each American. This figure is equivalent to about 1.6% of the gross national product, or 2.7% of capital expenditures by U.S. business. That same year the U.S. lost 4% of its gross national product through environmental damage. Clearly, the financial costs of pollution will continue to increase.

But there are other costs. Failure to control pollution will eventually threaten our food supply (marine and terrestrial), destroy whole industries, produce a greater disparity between have and have-not nations, and cause a decline in the health of all the planet's citizens. To these must be added the aesthetic costs of an ocean despoiled by pollution; few of us look forward to sharing the beach with oiled birds, jettisoned diapers, or clumps of medical waste (**Figure 15.1**).

PHYSICAL RESOURCES

Table 15.1 shows the value of the most important marine mineral resources to the world economy. Petroleum and natural gas are the most valuable by far; mineral resources from all other offshore sources amount to less than 2% of the annual value of offshore oil and gas production.

Petroleum and Natural Gas

The ocean makes a significant contribution to present world needs for petroleum and natural gas; 28% of the crude oil and 19% of the natural gas produced in 1987 came from the seabed. About one-third of known world reserves of oil and natural gas lie along the continental margins. Major U.S. marine reserves are located on the continental shelf of central and southern California, off the Texas and Louisiana Gulf Coast, and along the north slope of Alaska. The deep-sea floors probably contain little or no oil or natural gas.

Oil is one of the few naturally occurring liquids, a complex chemical soup containing perhaps a thousand compounds, mostly hydrocarbons. Petroleum is almost always associated with marine sediments, suggesting that the organic substances from which it was formed were once marine. Planktonic organisms or soft-bodied benthic marine animals are the most likely candidates. Their bodies apparently accumulated in quiet basins where the supply of oxygen was low and there were few bottom scavengers. The action of anaerobic bacteria converted the original tissues into simpler, relatively insoluble organic compounds that were probably buried, possibly first by turbidity currents, then later by the continuous fall of sediments from the ocean above. Slow cooking under this thick sedimentary blanket for millions of years completed the chemical changes that produce oil.

If the organic material cooked too long, or at too high a temperature, the mixture turned to methane, the domi-

Table 15.1   Value of Materials from Marine Sources Worldwide, 1987

Material	Revenue from Marine Sources ($ millions)	Percentage of Total Revenue Provided from Marine Sources	Estimated Percentage of Total Resources Located in the Ocean
Crude oil	55,218	28%	34%
Natural gas	23,434	19%	26%
Sand and gravel	334	1%	25%
Magnesium (metal)	310	50%	very large
Table salt (NaCl)	210	33%	very large
Tin	185	14%	7%
Magnesium (compounds)	95	25%	very large
Manganese	0	<1%	very large
Sulfur	40	<1%	<1%
Shell (CaCO$_3$)	100	1%	<5%

Source: Broadus, 1987; Craig, Vaughan, and Skinner, 1988; Mangone, 1986; U.S. Dept. of the Interior, 1989.

nant component of natural gas. Deep sedimentary layers are older and hotter than shallow ones and have higher proportions of natural gas to oil. The maximum depth at which oil deposits have been located is about 7 kilometers (4.4 miles) beneath the bottom. Below that, only natural gas is found.

Oil is less dense than the surrounding sediments; so it can migrate from its source rock through porous overlying formations. It collects in the pore spaces of reservoir rocks when an impermeable overlying layer prevents further upward migration of the oil (**Figure 15.2**). When searching for oil, geologists use sound reflected off subsurface structures to look for the signature combination of layered sediments, depth, and reservoir structure before they drill.

Drilling for oil offshore is far more costly than drilling on land because special drilling equipment and transport systems are required. Most offshore platforms rest in water less than 100 meters (330 feet) deep, but as oil demand (and therefore price) continues to rise deeper deposits further offshore will be exploited from larger platforms. The tallest drilling platform presently in position is deployed off the Louisiana coast 341 kilometers (214 miles) southwest of New Orleans. The Shell Oil Company's tension leg platform *Auger* is held in position by steel cables anchored into the seabed and pulling against the semisubmerged platform's buoyancy (see **Figure 15.3**). It is anchored in 872 meters (2,860 feet) of water and—with its cables—is nearly 1,000 meters (3,275 feet) tall! By the year 2001, some 20 wells are scheduled to produce 46,000 barrels of oil and 125 million cubic feet of natural gas daily. Total cost of the platform: $1.3 billion.

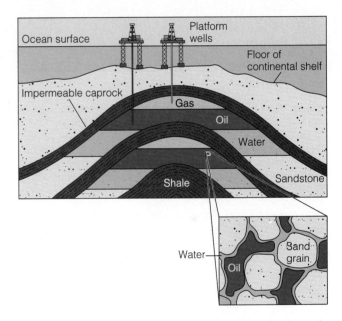

Figure 15.2   Oil and natural gas are often found together beneath a dome of impermeable caprock. Oil and gas are not found in great hollow reservoirs, but within pore spaces in rock.

a

Figure 15.3 (a) Shell Oil Company's tension leg platform *Auger*, deployed in 872 meters (2,860 feet) of water on the Garden Banks of the Gulf of Mexico. (b) Tension leg platforms are held in position by steel cables anchored into the seabed and pulling against the semisubmerged platform's buoyancy. Platform *Mars*, due for deployment off Louisiana in 912 meters (2,933 feet) of water in 1996, is pictured in this imaginary view in relation to the Houston skyline. Both platforms are designed to withstand hurricane-force waves of 22 meters (71 feet) and winds of 225 kilometers (140 miles) per hour.

b

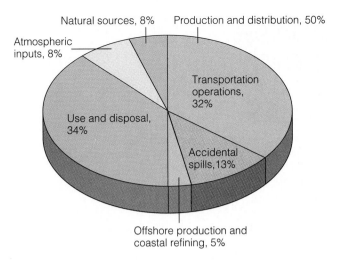

Figure 15.4 Sources of ocean oil pollution.

Figure 15.5 A bird coated with crude oil. Such birds die unless the oil is removed with a mild detergent solution. Many will die of cold or toxic effects even when the oil is removed.

Oil Pollution

Natural seeps have been leaking large quantities of oil into the sea for millions of years. The amount of oil entering the ocean has increased greatly in recent years, however, because of our growing dependence on marine transportation for petroleum products, offshore drilling, nearshore refining, and street runoff carrying waste oil from automobiles (see **Figure 15.4**).

In the late 1980s people were using nearly 5 billion metric tons (5.5 billion tons) of crude oil each year, about half of which was transported to market in large tankers. In 1985 about 3.2 million metric tons (3.5 million tons) of oil entered the world ocean. Natural seeps accounted for only about 8% of this annual input, about a quarter of a million metric tons. Some was released intentionally, quietly, and routinely, during the loading, discharging, and flushing of tanker ships. This oil is particularly harmful to seabirds; between 150,000 and 450,000 marine birds are killed in the North Sea and North Atlantic regions each year by the routine release of oil from tankers (see **Figure 15.5**). It's no wonder that a sea surface completely free of an oil film is quite rare.

It is difficult to generalize about the effects that a concentrated release of oil—an oil spill from a tanker, coastal storage facility, or well—will have in the marine environment. The consequences of a spill vary with its location and proximity to shore; with the quantity and composition of the oil; with the season of the year, currents, and weather conditions at the time of release; and with the composition and diversity of the affected communities. Intertidal and shallow-water subtidal communities are most sensitive to the effects of an oil spill.

Spills of *crude* oil are generally larger in volume and more frequent than spills of refined oil. Most components of crude oil do not dissolve easily in water, but those that do can harm the delicate juvenile forms of marine organisms even in minute concentrations. The remaining insoluble components form sticky layers on the surface that prevent free diffusion of gases, clog adult organisms' feeding structures, and decrease the sunlight available for photosynthesis. Even so, crude oil is not highly toxic, and it is biodegradable. Though crude oil spills look terrible and generate great media attention, most forms of marine life in an area recover from the effects of a moderate spill within about five years. For example, the 126 million gallons of light crude oil released into the Persian Gulf during the 1991 Gulf War (see **Table 15.2**) have dissipated relatively quickly and will probably cause little long-term biological damage.

Spills of *refined* oil, especially near shore where marine life is abundant, can be more disruptive for longer periods of time. The refining process removes and breaks the heavier components of crude oil and concentrates the remaining lighter, more biologically active ones. Components added to oil during the refining process also make it more deadly. A 1969 spill of refined oil at West Falmouth, Massachusetts, killed countless marine organisms and caused grave environmental damage that is still apparent in the form of reduced species diversity and fertility. In a similar accident in Mexico, the tanker *Tampico* went aground and dumped her load of refined gasoline into an unspoiled Baja California bay. The area still shows biological scars after nearly three decades.

The volatile components of any oil spill eventually evaporate into the air, leaving the heavier tars behind.

Table 15.2 The 21 Worst Oil Spills in History

The *Exxon Valdez* disaster of March 1989 was the worst in U.S. history. But when ranked with other massive oil spills involving both shipping and well disasters, it falls to the bottom of the top 21.

Date	Incident and Location	Size of Spill (millions of gallons)
3 June 1979	Well spill in Bay of Campeche off coast of Mexico	140
19–28 Jan. 1991	Discharged into Persian Gulf during Gulf War	126
4 Feb. 1983	Well spill at Nowruz, Iran, that dumped oil into Persian Gulf	80
6 Aug. 1978	Fire aboard *Castillo de Vellver*, off Cape Town, South Africa	78.5
16 March 1978	Tanker *Amoco Cadiz* ran aground off coast of France	68.7
19 July 1979	Collision of two ships off Trinidad and Tobago, *Atlantic Empress* and *Aegean Captain*	48.8
Aug. 1980	Leak at well No. D103, Libya	42
2 Aug. 1979	Wreck of *Atlantic Empress*, Barbados	41.5
23 Feb. 1980	Wreck of *Irenes Serenade*, Greece	36.6
20 Aug. 1981	Leak at Kuwait National Petrol Tank, Kuwait	31.2
15 Nov. 1979	Wreck of *Independence*, Istanbul, Turkey	28.9
25 May 1978	Leak at No. 126 well/pipe, Iran	28
Jan. 1993	Wreck, *Braer*, Northern Scotland	25
6 July 1979	Leak at British Petroleum storage tank, Nigeria	23.9
15 Aug. 1985	Wreck of *Nova*, Kharg Island, Persian Gulf	21.4
11 Dec. 1978	Leak at BP-Shell fuel depot, Zimbabwe	20
7 Jan. 1983	Wreck of *Assimi*, off Oman	15.8
12 June 1978	Tohoku Oil Co., Japan	15
31 Dec. 1978	Wreck of *Andros Patria*, Spain	14.6
10 Dec. 1983	Wreck of *Peracles*, Qatar	14
March 1989	Wreck of *Exxon Valdez*, Alaska	12.7

Source: Oil Spill Intelligence Report; National Public Radio; *Los Angeles Times*, 28 Jan. 1991.

Wave action causes the tar to form into balls of varying sizes. Some of the tar balls fall to the bottom, where they may be assimilated by bottom organisms or incorporated into sediments. Bacteria will eventually decompose these spheres, but the process may take years to complete, especially in cold polar waters. This oil residue—especially if derived from refined oil—can have long-lasting effects on seafloor communities. The fate of spilled oil is summarized in **Figure 15.6**.

The methods used to contain and clean up an oil spill sometimes cause more damage than the oil itself. Detergents used to disperse oil are especially harmful to living things. Cleanup of the 1969 *Torrey Canyon* accident off the southern coast of England, one of the first large tanker accidents, did much more environmental damage than the 100,000 metric tons (110,000 tons) of crude oil released. Some resort beaches in the south of England were closed for two seasons, not because of oil residue but because of the stench of decaying marine life killed by the chemicals used to make the shore look clean. Even the more sophisticated methods used in dealing with the *Exxon Valdez* disaster, the worst oil spill in U.S. history (and the twentieth worst spill ever; see Table 15.2), seem to have done more harm than the oil itself (see the vignette that opens this chapter).

Of course the best way to deal with oil pollution is to prevent it from happening in the first place. Tanker design is being modified to limit the amount of oil intentionally released in transport. Legislation is being considered that will limit new tanker construction to stronger double-hull designs (although shipping experts are uncertain that even a double-hull design could have prevented the *Exxon Valdez* spill). Perhaps most important, crew testing and training are being upgraded.

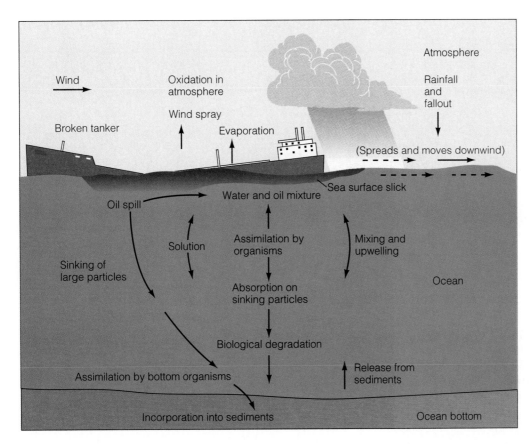

Figure 15.6 Some of the possible pathways that an oil spill may travel.

Labels in figure:
Wind
Oxidation in atmosphere
Wind spray
Atmosphere
Rainfall and fallout
Broken tanker
Evaporation
(Spreads and moves downwind)
Sea surface slick
Oil spill
Water and oil mixture
Solution
Assimilation by organisms
Mixing and upwelling
Sinking of large particles
Absorption on sinking particles
Ocean
Biological degradation
Assimilation by bottom organisms
Release from sediments
Incorporation into sediments
Ocean bottom

Sand and Gravel

Sand and gravel are not very glamorous marine resources, but they are second in dollar value only to oil and natural gas. More than 112 million metric tons (123 million tons) of sand and gravel, valued at about a third of a billion dollars, were mined offshore in 1987. Only about 1% of the world's total sand and gravel production is scraped and dredged from continental shelves each year, but the seafloor supplies about 20% of the sand and gravel used in the island nations of Japan and the United Kingdom.

Most of the exploitable U.S. deposits of marine sand and gravel are found off the coasts of Alaska, California, Washington, and the East Coast states from Virginia to Maine. Nearshore deposits are widespread, easily accessible, and used extensively in the building industry to manufacture concrete and to surface highways. Offshore oil wells in Alaska are built on huge human-made gravel platforms, and the large quantities of gravel available at those locations make offshore drilling practical there.

Mining gravel and sand is not without environmental consequence. The direct removal of sediments destroys nearshore marine communities, and the clouds of particles churned up by the extraction process can overwhelm habitats and clog the feeding and respiratory structures of organisms kilometers downcurrent. Even when gravel and sand are mined on land, dislodged materials running off the land after heavy rains can shower estuaries with silt.

Fresh Water

Only 0.017% of Earth's water is liquid, fresh, and available at the surface for easy use by humans. Another 0.6% is available as groundwater within half a mile of the surface. Unfortunately much of this is polluted or otherwise unfit for human consumption. More than any other factor in nature, the availability of fresh, pure drinking water determines the number of people who can inhabit any geographic area, their use of other natural resources, and their life-style.

Fresh water is becoming an important marine resource. Exploitation of that resource by **desalination**, the separation of pure water from seawater, is already under way, mainly in the Middle East, West Africa, Peru, Florida, Texas, and California. More than 1,500 desalination plants are currently operating worldwide, producing a total of about 13.3 billion liters (3.5 billion gallons) of fresh water per day. The largest desalination plant, in Saudi Arabia, produces about 114 million liters (30 million gallons) daily! Desalination, water conservation, and perhaps iceberg harvesting will become more common as water becomes more polluted, scarcer, and more valuable.

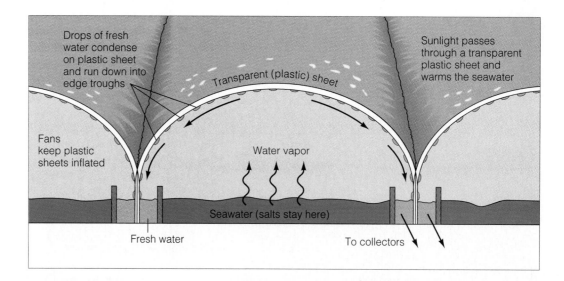

Figure 15.7 A low-tech scheme for desalination.

As always, there are trade-offs. Desalination is a costly, energy-intensive process, and more often than not the source of that energy will be oil and gas. One of the most elegant ideas for inexpensive desalination is illustrated in **Figure 15.7**. Other than the sunlight needed to make water vapor, this passive system requires no energy except that required by pumps to move water through plastic-covered troughs and small fans to keep the plastic sheets inflated. Also, the salty brine that remains after freshwater extraction must be disposed of, and highly saline water can be lethal to marine life.

BIOLOGICAL RESOURCES

Ancient kitchen middens (garbage dumps of bones and shells) found in many coastal regions demonstrate that humans have used the sea for thousands of years as a source of food and medicines. Today fish, crustaceans, and mollusks contribute about 14.5% of the total animal protein consumed by humans; fish meal and by-products included in the diets of animals raised for food account for another 3.5%. About 85% of the annual catch of fish, crustaceans, and mollusks comes from the ocean, the rest from fresh water.

The sea will probably not be able to provide substantially more food to help alleviate future problems of malnutrition and starvation caused by human overpopulation; indeed, population growth would likely absorb any increase in harvesting quantity or efficiency. Nevertheless, these resources currently sustain a great many people.

Fishes, Crustaceans, and Mollusks

Fishes, crustaceans, and mollusks are the most valuable living marine resources. Commercial fishermen took

Table 15.3 World Commercial Catch of Marine Fish, Crustaceans, and Mollusks, by Species Group, 1989

Species Group	Millions of Metric Tons, Live Weight
Herring, sardines, anchovies	24.6
Cods, hakes, haddocks	12.8
Jacks, mullets, sauries	9.2
Mollusks	7.9
Redfish, basses, conger eels	5.9
Crustaceans	4.5
Tunas, bonitos, billfishes	4.0
Mackerel, snooks, cutlass fishes	3.8
Flounders, halibuts, soles	2.0
Miscellaneous marine fishes	11.0
Total for all marine sources, excluding marine mammals and aquatic plants	85.7

Source: Food and Agriculture Organization of the United Nations, *Yearbook of Fisheries Statistics,* 1990, vol. 67; Rome. U.S. Department of Commerce, *Fisheries of the United States, 1990.*

nearly 83 million metric tons (91 million tons) of these animals in 1987. The largest share of the global catch of marine fish, crustaceans, and mollusks—over 66%—comes from the temperate waters over the continental shelves in the northwest, southeast, and west central Pacific, and the northeast Atlantic (see **Figure 15.8**).

Of the thousands of species of marine fishes, crustaceans, and mollusks, fewer than 500 species are regularly caught and processed. The ten major groups listed in **Table 15.3** supply more than 95% of the commercial marine catch. Note that the largest commercial harvest is

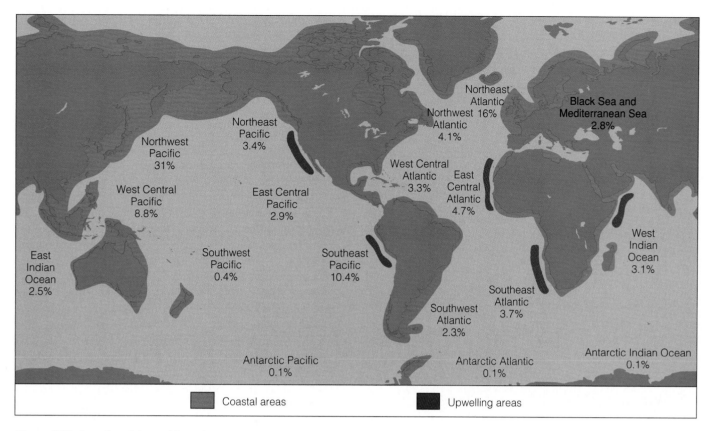

Figure 15.8 Location of the world's major commercial marine fisheries and distribution of the catch by fishery for 1984. Fishing is best in coastal areas over the continental shelves and in areas of upwelling.

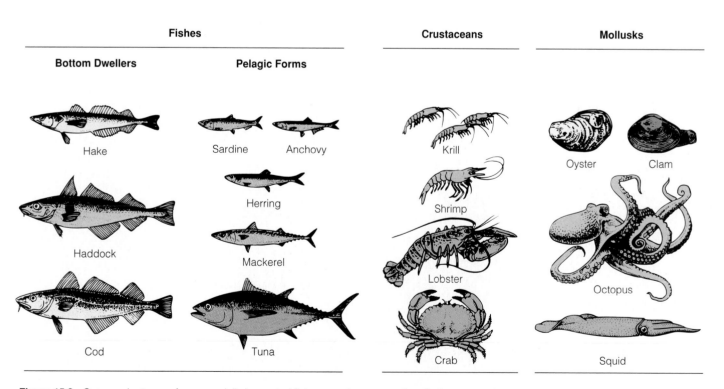

Figure 15.9 Some major types of commercially harvested fishes, crustaceans, and mollusks.

of the herring and its relatives, which accounts for more than a quarter of the live weight of all living marine resources caught each year. **Figure 15.9** shows some of the most important commercial species.

The fastest growing fishery is that for the small euphausiid crustaceans called *krill* (see again Figure 13.15). The ocean around Antarctica supports tremendous numbers of krill, which eat the abundant diatoms of the surface waters and are in turn eaten by baleen whales. Krill have proliferated as the whale population has declined. The Soviets and Japanese pioneered krill harvesting and processing in the late 1960s. Today the countries of the CIS (former Soviet Union) field about 100 processing ships; they produce krill "butter" and krill "cheese" spread, as well as large quantities of feed for poultry, cattle, and fur-bearing animals. Japan has about 14 processing ships; the Japanese are experimenting with krill cakes and a type of krill sausage thickened with algin. Krill is also sold raw and fresh-frozen. About 500,000 metric tons (550,000 tons) of krill are now being taken each year, making the krill fishery the fifteenth largest in landed catch. Some experts feel that the krill harvest might ultimately match the total production of the rest of the global marine fishery!

Fishing is a big business, employing more than 15 million people worldwide. Though estimates vary widely, the value in 1987 for the worldwide marine catch was thought to be around $45 billion! Nearly half of the world's commercial marine catch is taken by only five countries: the far-ranging ships of Japan lead the world with 16% of the catch, followed by the CIS with 13%, China (7%), the United States (6%), and Chile (6%). About 75% of the annual harvest is taken by commercial fishermen who operate vast fleets equipped with large trawl nets (**Figure 15.10**), drift nets (**Figure 15.11**), purse seines, or gill nets. They work year-round, using satellite sensors, aerial photography, scouting vessels, and sonar to pinpoint the location of fish schools. Huge factory ships

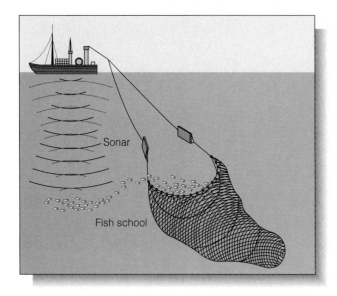

Figure 15.10 Stern trawler fishing. After sonar on the trawler finds the fish, they're captured by a trawl net towed along the bottom. Boards angled to the water flow keep the mouth of the net open.

Figure 15.11 Drift net fishing. Typically, a fleet of huge factory ships uses sophisticated electronic detection devices and aircraft to find large schools of fish. Each ship then launches 20 to 50 fast, small catcher boats to set hundreds to thousands of miles of drift nets, weighted to stay at the wanted depth. After drifting overnight, the nets are hauled in by the factory ships, and the catch is processed on board by canning or freezing.

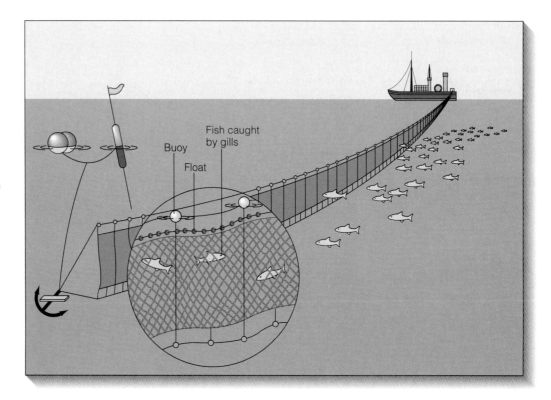

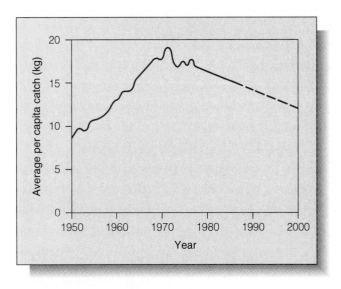

Figure 15.12 Average per capita world fish catch has declined in most years since 1970 and is projected to decline further by the end of the century.

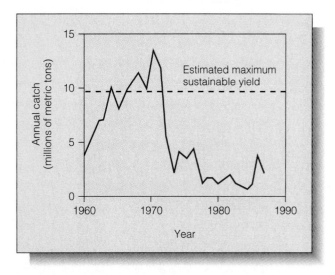

Figure 15.13 The Peruvian anchovy catch, showing the combined effects of overfishing and El Niño.

often follow the largest fleets to process, can, or freeze the animals on the run. The living resources of the ocean are under furious assault.

The cost per unit of seafood has risen dramatically in spite of all this high-tech assistance. The increasing expense of fuel for the fishing fleets and processing plants, the cost of wages for the crews, and the greater distances that boats must cover to catch each ton of fish—all have helped drive up the cost of seafood. In spite of greater efforts, the total marine catch leveled off in about 1970 and remained surprisingly stable until 1980, when greater demand and increasing prices began to drive the tonnage upward again. But since 1970 the world population has grown and the average *per capita* world fish catch has declined alarmingly (see **Figure 15.12**).

Overfishing

Can we continue to take this much food from the ocean year after year? The **maximum sustainable yield**, the maximum amount of each type of fish, crustacean, and mollusk that can be caught without impairing future populations, probably lies between 100 and 170 million metric tons annually. As can be seen in Table 15.3, current yield is nudging the lower figure. We may be perilously close to the catastrophic collapse of some fisheries. Fleets are currently obtaining fewer tons per-unit-effort and are ranging farther afield in their urgent search for food.

The experience of the Peruvian anchovy fishery illustrates the effects of **overfishing**, harvesting so many fish that there is not enough breeding stock left to replenish the species. In 1960 the anchoveta catch in Peru was about 4 million metric tons (4.4 million tons); only ten

years later it had reached nearly 13 million metric tons a year (see **Figure 15.13**). The combination of overfishing and an El Niño that began in 1971 dropped the 1972 catch to about 4.8 million metric tons. By 1980 less than 1 million metric tons were recovered, and the 1985 catch, further decimated by the 1983–84 El Niño, was less than 100,000 metric tons. Peru no longer has a commercial anchovy fishery, and in terms of total tonnage of fish caught, the country slipped from first place to fourth.

By the early 1980s overfishing had depleted the stocks of 24 other valuable fisheries. Cod and herring had been overfished in the North Atlantic, and salmon and Alaskan king crab were becoming scarce in the Pacific Northwest. The Pacific sardine had been reduced to **commercial extinction**, depletion of a resource species to a point where it is no longer profitable to harvest.

Even when faced with evidence that it is depleting a stock and disrupting the equilibrium of a fragile ecosystem, the fishing industry's response was to increase the number of boats and develop more efficient techniques for capturing animals. A particularly disruptive fishing technique employs **drift nets** (Figure 15.11)—fine, vertically suspended nylon nets as much as 7 meters (25 feet) high and 80 kilometers (50 miles) long. More than 800 Taiwanese, Korean, and Japanese vessels now deploy some 48,000 kilometers (30,000 miles) of these "walls of death" each night—more than enough to encircle the Earth! Drift nets catch the fish and squid for which they were designed, but they also entangle everything else that touches them, including turtles, birds, and marine mammals (**Figure 15.14**). The process has been compared to stripmining: The ocean is literally sieved of its contents. An estimated 30 kilometers (18 miles) of these nearly invisible

Figure 15.14 An unintentional but unavoidable consequence of drift net fishing.

Figure 15.15 The whaling industry has pushed most of the dozen or so species of great whales to the brink of extinction through overharvesting. This photograph shows a blue whale being butchered at the Coal Harbor Whaling Station, British Columbia, Canada, in 1963. Commercial whaling of great whales was banned in 1985 but began again in early 1993.

nets are lost each night. These remnants, made of non-biodegradable plastic, become "ghost nets," fishing for decades. In November 1991, Japan agreed to cease drift net fishing by the end of 1992. Pressure is being applied to Taiwan and South Korea. Pirate netting continues in the Pacific and has begun in the Atlantic.

Whaling

Since the 1880s whales have been hunted to provide meat for human and animal consumption; oil for lubrication, illumination, industrial products, cosmetics, and margarine; bone for fertilizers and food supplements; and baleen for corset stays. **Figure 15.15** shows whales being butchered for food and oil. An estimated 4.4 million large whales existed in 1900; eight of the 11 species of large whales once hunted by the whaling industry are commercially extinct, and fewer than a million large whales of all species are thought to remain in the ocean. **Table 15.4** indicates the extent of their slaughter.

Substitutes exist for all whale products, but the harvest of most commercial species did not stop until whaling became uneconomical. In 1985 the International Whaling Commission, an organization of whaling countries established to manage whale stocks, placed a moratorium on the slaughter of large whales. Except for a suspiciously large harvest of whales taken by the Japanese for "scientific purposes," commercial whaling ceased in 1987. Fewer than 700 large whales were taken in 1988 (see **Table 15.5**). For many species—including the great blue whale, largest animal ever to have existed on Earth—protection may have come too late to save them from extinction.

Table 15.4 Worldwide Whale Catches, 1868–1984

Year	Number of Whales	Year	Number of Whales
1868[a]	30	1963–64	63,001
1873	36	1964–65	64,680
1878	116	1965–66	57,891
1883	569	1966–67	52,238
1888	709	1967–68	46,645
1893	1,607	1968–69	42,126
1898	1,993	1969–70	42,480
1903	3,867	1970–71	38,771
1908	5,509	1971–72	32,133
1913	25,673	1972–73[d]	32,605
1918	9,468	1973–74	31,629
1923	18,120	1974–75	29,961
1928	23,593	1975–76	22,049
1933	28,907	1976–77	16,309
1938[b]	54,835	1977–78	13,638
1943	8,372	1978–79	10,668
1948	44,002	1979–80	3,542
1953	53,642	1980–81	2,928
1958	64,075	1981–82	2,050
1960–61	65,641	1982–83	1,683
1961–62[c]	66,090	1983–84	1,683
1962–63	63,579		

Source: Committee for Whaling Statistics, Oslo.
[a] The year the harpoon gun was invented.
[b] World catch record, in tonnage.
[c] Season catch record in individuals taken.
[d] U.S. ceases to import whale products.

Table 15.5 Total Value of Whaling in 1988

Iceland agreed to stop all whaling in 1990, but, along with Norway, resumed commercial whaling in 1993. Under the guise of "scientific whaling," Japan increased its 1991 take to 875 whales. Subsistence whaling is practiced by aboriginal peoples to provide whale products for their own use.

Number of Whales	Species	Nation	Estimated Value per Whale ($ millions)	Total Value ($ millions)
276	Minke	Japan	0.2	55.2
130	Minke	Greenland	0.2	26
68	Finback	Iceland	1.0	68
10	Sei	Iceland	0.5	5
168	Gray	USSR		Subsistence
3	Humpback	Bequia (Eastern Caribbean)		Subsistence
29	Bowhead	USA		Subsistence
684				**$154.2**

Source: Kraus, 1989.

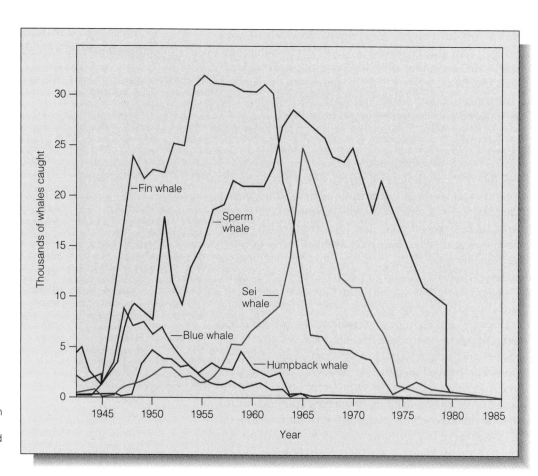

Figure 15.16 All five species included in this graph are commercially extinct, and the blue whale is still in danger of total extinction even though commercial blue whale hunting ended in 1964. At that time, only about 1,000 blue whales were left, and experts fear that the population may still be too small to recover.

Many more small whales have been killed than large ones, but not for food or raw materials. For reasons that are not well understood, porpoises (which are small whales) gather above schools of yellowfin tuna in the open sea. Fishermen have learned to find the tuna by spotting the porpoises. Nets cast to catch the tuna also entangle the air-breathing porpoises, and the mammals drown. Nearly all the dead porpoises are simply pitched over the side as waste. More than 6 million small whales have been killed in association with tuna fishing since 1971!

Passage of the U.S. Marine Mammals Protection Act in 1972 brought a drop in the number of porpoise deaths, but from 1977 to 1987 the number of tuna boats in the U.S. fleet declined from more than 100 boats to 34 boats, while the foreign fleet rose from fewer than 10 to more than 70 boats. Foreign fishermen do not always abide by the porpoise-saving provisions of the act, but the U.S. Congress voted in October 1988 to ban importation of tuna not caught in accordance with new methods designed to reduce the kill. In April 1990, American tuna-processing companies agreed to buy tuna only from fish-ermen whose methods do not result in the deaths of porpoises. American commercial fishermen have agreed to comply, and conservationists hope the foreign fleets will follow their lead.

Whaling continues in spite of the moratorium. Scarcity has driven up the cost of whale meat, a delicacy that makes up about 10% of the meat eaten by Japanese, and conservationists have feared that unrestricted whaling would resume when the moratorium expired. Norway resumed whaling operations in 1993. Again, a short-sighted industry is willing to take immediate profits despite obvious signs that the fishery is collapsing (see **Figure 15.16**). And even in the unlikely event that whaling truly ends, will the rapidly expanding krill fishery leave the recovering baleen whale populations anything to eat?

Drugs

Modern medical researchers estimate that perhaps 10% of all marine organisms are likely to yield clinically useful compounds. One derived from a Caribbean sponge is already in use: Acyclovir, the first antiviral compound

Figure 15.17 The tanker *Universe Ireland*, one of the world's largest, has a capacity of 300,000 tons. Most tankers now being built are smaller because the giant tankers have maneuverability problems and are too large to enter some ports.

approved for humans, has been fighting herpes infections of the skin and nervous system since 1982.

Newly discovered compounds are also being tested. Extracts from 30% of all tunicate species investigated show antiviral and antitumor activity; one of these extracts, Didemnin-B, shows promise as a treatment for malignant melanoma, the deadliest form of skin cancer. A South Pacific sponge makes a chemical that can kill *Candida*, a fungus that causes vaginitis and the throat infection known as *thrush*. A species of coral produces a powerful anti-inflammatory painkiller, a promising and nonaddictive drug for arthritis. Another compound, derived from cyanobacteria, stimulated the immune system of test animals by 225% and cells in culture by 2,000%; the drug may be useful in treating AIDS. Vidabarine, another antiviral drug developed from sponges, may attack the AIDS virus directly. Human trials of some of these drugs are already underway. In the not-too-distant future, drugs from marine organisms may be among our most valuable marine resources.

NONEXTRACTIVE RESOURCES

Transportation and recreation are the main nonextractive resources the oceans provide. People have been using the ocean for transportation for thousands of years. Through most of this time the transport of cargo has been responsible for far more revenue than the movement of passengers. At present, oil tankers ship the greatest gross tonnage of any type of cargo (over 260 million metric tons annually). Oil accounts for about 51% of the total value of world trade transported by sea; iron, coal, and grain make up 25% of the rest.

Nearly half of the world's crude oil production is transported to market by ships. Tankers are needed because very few of the major oil-drilling sites are close to areas where the demand for refined oil products is highest. The largest tankers (**Figure 15.17**) are more than 430 meters (1,300 feet) long and 66 meters (206 feet) wide; they carry more than 500,000 metric tons (3.5 million barrels) of oil. The complexity of international trade routes (and volumes transported) is shown in **Figure 15.18**.

Harbors are an essential adjunct to transportation. Cargoes are no longer loaded and offloaded piece-by-piece by teams of longshoremen. Today's harbors bristle with automated bulk terminals, high-volume tanker terminals (both offshore and dockside), containership facilities, roll-on-roll-off ports for automobiles and trucks, and passenger facilities required by the growing popularity of cruising. Most of this specialized construction has occurred since 1960. New Orleans is now the greatest U.S. port; nearly 168 millions tons of cargo passed through its docks in 1987.

Transportation is sometimes combined with recreation. In the last decade, the cruise industry has experienced spectacular growth. Passengers on luxurious liners can enjoy a few relaxing days on the ocean crossing the North Atlantic, visiting tropical islands, or touring places accessible to the public only by ship.

Other ocean-related leisure pursuits, including sport fishing, surfing, diving, day cruising, sunbathing, dining in seaside restaurants, and just plain relaxing also contribute to the economy. As an example, consider the recreational boating industry in California (the world's largest). As of 1 January 1988 that state had 711,193 registered recreational boats, most used in the ocean. Recreational boating supports 106,000 jobs, pays $188 million in annual state and local taxes and fees, and has a yearly economic impact estimated at $5.6 billion!

Public interest and curiosity about whales are a small but important source of recreational revenue. Whale-watching trips generate an annual revenue of $56 million; and dolphin displays, whale artwork and books, and conservation group donations amount to another $46 million.

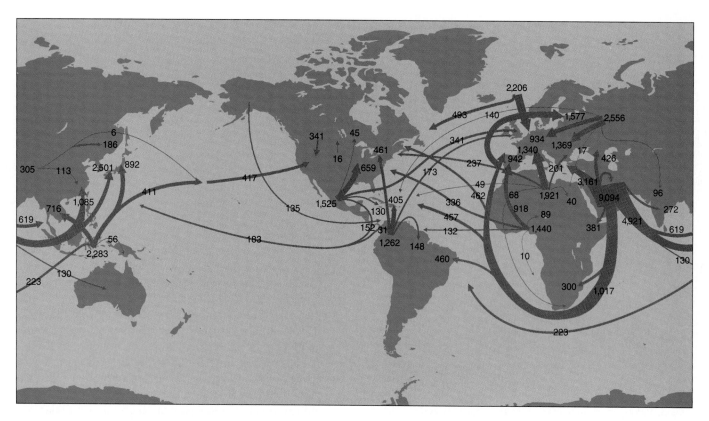

Figure 15.18 The complexity of international crude oil trade routes in 1984; all numbers are in thousands of barrels per day. Arrows indicate origins and destinations of the major shipments, but not necessarily specific routes.

THE OCEAN AS WASTE DUMP

Oil is not the only substance polluting the ocean. Heavy metals and synthetic organic chemicals, both capable of causing serious damage to organisms, are present in growing quantities.

Heavy Metals and Synthetic Organic Chemicals

Among the dangerous heavy metals being introduced into the ocean are mercury and lead. Human activity releases about 5 times as much mercury and 17 times as much lead as is derived from natural sources, and incidents of mercury and lead poisoning (major causes of brain damage and behavioral disturbances in children) have increased dramatically over the last two decades. Lead particles from industrial wastes, landfills, and gasoline residue reach the the ocean through runoff from land during rains, and the lead concentration in some shallow-water bottom-feeding species is increasing at an alarming rate. Consumers should be wary of seafoods taken near shore in industrialized regions.

Copper, another heavy metal, is so effective in killing marine organisms that it has long been used in marine antifouling paints. The ship *Pac Baroness*, a freighter carrying 21,000 metric tons (23,000 tons) of finely powdered copper, sank in 448 meters (1,480 feet) of water off the coast of central California after a collision in 1987. Her toxic cargo was scattered over the seabed, and a plume of copper-tainted water has been detected 40 kilometers (24 miles) downcurrent from the wreck. Marine life has been significantly disrupted in the area, which is a major fishing zone for Dover sole and rock cod. A similar incident off Holland in 1965 killed more than 100,000 fish and destroyed commercial mussel beds.

Canadian and Norwegian researchers have found that coal combustion, electric utilities, steel and iron manufacturing, fuel oils, fuel additives, and incineration of urban refuse are the major sources of oceanic and atmospheric contamination by heavy metals. They now believe that the toxicity of heavy metals is a greater health problem than the toxicity of all organic and radioactive wastes. In the wellness-conscious 1990s, fish are seen as a safe and healthful food. But with the ocean still receiving heavy-metal-contaminated runoff from the land, a rain of pollutants from the air, and fallout from shipwrecks, we can only wonder how much longer seafood will be safe.

BOX 15.1

Minamata's Tragedy

For the people of Minamata, Japan, who were poisoned by mercury released into the ocean from a nearby factory, heavy metal pollution is a continuing horror story. Between 1953 and 1960 more than 100 people who ate shellfish taken from Minamata Bay were afflicted by a form of mercury poisoning now called Minamata disease (**Figure a**). Their symptoms included kidney damage, neuromuscular deterioration, birth defects, insanity, and, eventually, death. The source of the mercury was a plastics plant that was dumping waste mercuric chloride into the bay.

a A victim of Minamata disease.

Bacterial action changed this chemical into a form that could be taken up and concentrated by benthic organisms. Villagers who ate clams and oysters, especially pregnant women and young children, were seriously affected. The discharge of mercury has been stopped, but the bay, which had been a source of food for the poor, is still completely unusable for fishing and clamming, even after more than 30 years. This situation will not change for generations; mercury is not easily removed from sediments and has a long residence time.

Many synthetic organic chemicals also enter the ocean and become incorporated into its organisms. Some of these synthetic organic substances are listed in **Table 15.6**. Ingestion of even small amounts of these compounds can cause illness or even death.

Halogenated hydrocarbons—a class of synthetic hydrocarbon compounds that contain chlorine, bromine, or iodine—are used in pesticides, flame retardants, industrial solvents, and cleaning fluids. The concentration of **chlorinated hydrocarbons**—the most abundant and dangerous halogenated hydrocarbons—is so high in the water off New York State that officials have warned women of childbearing age and children under 15 not to consume more than half a pound of local bluefish a week. (They are told *never* to eat striped bass caught in the area.) One administrator for the U.S. Environmental Protection Agency has written, "Anyone who eats the liver from a lobster taken from an urban area is living dangerously."

The level of synthetic organic chemicals in seawater is usually very low, but some organisms at higher levels in the food chain can concentrate these toxic substances in their flesh. This **biological amplification** is especially hazardous to top carnivores in a food web. The damage caused by biological amplification of DDT, a chlorinated hydrocarbon pesticide, is particularly instructive. In the early 1960s California pelicans began producing eggs with thin shells containing less than normal amounts of calcium carbonate. The eggs broke easily, no chicks were hatched, and the nests were eventually abandoned. The pelicans were disappearing. The trail led investigators to DDT (which also affected pelicans, as depicted in **Figure 15.19**). Plankton absorbed DDT from the water; fishes that fed on these microscopic organisms accumulated DDT in their tissues; and the birds that fed on the fishes ingested it, too. The whole food chain was contaminated, but because of biological amplification the top carnivores were most strongly affected. A chemical interaction between DDT and the birds' calcium-depositing tissues

Table 15.6 Synthetic Organic Chemicals That Have Been Detected in the Ocean

Name	Major Health Effects
Aldicarb (Temik)	High toxicity to the nervous system
Benzene	Chromosomal damage, anemia, blood disorders, and leukemia
Carbon tetrachloride	Cancer; liver, kidney, lung, and central nervous system damage
Chloroform	Liver and kidney damage; suspected cancer
Dioxin	Skin disorders, cancer, genetic mutations
Ethylene dibromide (EDB)	Cancer and male sterility
Polychlorinated biphenyls (PCBs)	Liver, kidney, and lung damage
Trichloroethylene (TCE)	In high concentrations, liver and kidney damage, central nervous system depression, skin problems, and suspected cancer and mutations
Vinyl chloride	Liver, kidney, and lung damage; lung, cardiovascular, and gastrointestinal problems; cancer and suspected mutations

Source: Miller, *Living in the Environment*, 5th ed. (Belmont, CA: Wadsworth, 1988).

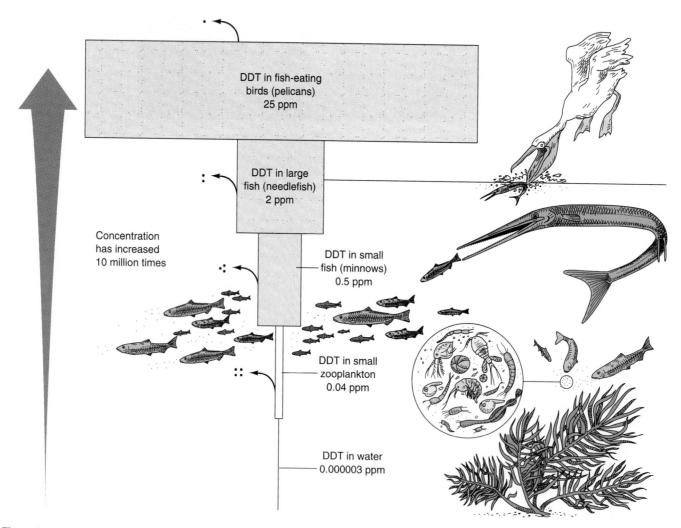

DDT in fish-eating
birds (pelicans)
25 ppm

DDT in large
fish (needlefish)
2 ppm

Concentration
has increased
10 million times

DDT in small
fish (minnows)
0.5 ppm

DDT in small
zooplankton
0.04 ppm

DDT in water
0.000003 ppm

Figure 15.19 The concentration of DDT in the fatty tissues of organisms was biologically amplified approximately 10 million times in this food chain of an estuary adjacent to Long Island Sound, near New York City. Dots represent DDT, and arrows show small losses of DDT through respiration and excretion.

Figure 15.20 Chemical assault on major U.S. coastal areas.

Map legend:
- High levels of toxic chemicals in fish livers
- Contamination by PCBs or pesticides
- Oxygen depletion
- Areas in which more than one-third of acreage is closed to commercial shellfish harvest

prevented the formation of proper eggshells. DDT was eventually banned in the United States, and the pelican and osprey populations are recovering.

Biological amplification of other chlorinated hydrocarbons has also affected other species. **Polychlorinated biphenyls (PCBs)**, fluids once widely used to cool and insulate electrical devices and to strengthen wood or concrete, may be responsible for the behavior changes and declining fertility of some populations of seals and sea lions on islands off the California coast. PCBs have also been implicated in a deadly viral epidemic among dolphins in the western Mediterranean. Ingestion of the chemical may have severely weakened the dolphins' immune system and made it impossible for them to fight the infection—up to 10,000 dolphins died during the summer of 1990. Even more alarming to biologists is the recent discovery that the nearshore dolphins off U.S. coasts are intensely contaminated. The concentration of chlorinated hydrocarbons in these animals exceeds 6,900 parts-per-million (ppm), concentrations high enough to disrupt the dolphins' immune systems, hormone production, reproductive success, neural function, and ability to stave off cancers.[1] These levels vastly exceed the 50 ppm

limit the U.S. government considers hazardous, and the 5 ppm considered the maximum acceptable level for humans! Investigations are continuing, as are probes into the effects of dioxin and other synthetic organic poisons accumulating in the oceanic sink.

Figure 15.20 indicates the locations of U.S. coastal areas degraded by synthetic organic chemicals. Production of such chemicals subject to biological amplification in food chains currently exceeds 91 million metric tons (100 million tons) each year. Even vaster volumes of ocean will be affected if the predicted 1% of that production reaches the sea.

Solid Waste

Not all pollutants enter the ocean in a dissolved state; much of the burden arrives in solid form. Some solid waste is ultimately biodegradable, but plastic, which now makes up almost 6% of the solid waste stream, is not. Scientists estimate that some kinds of synthetic materials—plastic six-pack holders, for example—will not decompose for about 400 years!

Americans generate 120 million metric tons of plastic waste, about 500 kilograms (1,100 pounds) per person, each year. Much of this waste plastic finds its way to the sea. In September 1987, volunteers collected 307 tons of litter along the Texas Gulf Coast, most of it plastic. The

---

[1] If you drag a beached porpoise into the ocean, you could theoretically receive a $10,000 fine for improper disposal of polluted materials!

Figure 15.21 Plastic mounds on an isolated shore of Niihau, one of the Hawaiian Islands.

Figure 15.22 Sea lions (seen here) and seals die by the thousands each year after becoming entangled in plastic debris, especially discarded and broken fishing nets.

haul included 31,733 bags, 30,295 bottles, and 15,631 six-pack yokes. In isolated areas the dunes of plastic debris can build to surreal proportions (see **Figure 15.21**).

A 1987 survey by the Woods Hole Oceanographic Institution found that each square mile of ocean surface off the northeast coast of the United States has more than 46,000 pieces of plastic floating on the surface. This material included ropes, fishing line and nets, plastic sheeting and bags, and granules of broken plastic cups. A staggering 100,000 marine mammals and 2 million seabirds die *each year* after ingesting or being caught in plastic debris! Sea turtles mistake plastic bags for their jellyfish prey and die from intestinal blockages. Seals and sea lions starve after becoming entangled in nets (see **Figure 15.22**) or muzzled by six-pack rings. The same kinds of rings strangle fish and seabirds. Adding ingredients to plastics that would hasten their decomposition would add only 5% to 7% to their cost, but this price increase is currently unacceptable to industry.

What should we do with plastic and other solid wastes such as glass and paper, disposable diapers, scrap metal, building debris, and all the rest? Dumping it into the ocean is clearly unacceptable, yet places to deposit this material are becoming scarce. In 1988 the average New Yorker threw away 1.03 metric tons (1.13 tons) of waste, up 14% in only two years. California's Los Angeles and Orange counties generate enough solid waste to fill Dodger Stadium every nine days. Transportation of waste to sanitary landfills becomes more expensive as nearby landfills reach capacity.

Is incineration the answer? We currently burn about 7% of our trash, a figure projected to reach 30% by the year 2000. By contrast, Sweden burns half its solid wastes.

Unfortunately, even with air-pollution control devices, incinerators still emit great quantities of tiny particles, heavy metal residue, and immense amounts of carbon dioxide. Dumping of the residual ash is also a problem; ash from Philadelphia's garbage incinerators has been turned away by states as far away as South Carolina, and one shipload of the stuff was refused entry at an African port!

Is recycling the answer? The Japanese currently recycle about 50% of their solid waste and are importing even more; scrap metal and wastepaper headed for Japan are the two biggest exports from the Port of New York. Americans are buying back their own refuse in the form of appliances, automobiles, and the cardboard boxes that hold their televisions and compact disk players. Massachusetts has set a goal of recycling 25% of its waste. The direct savings to consumers, as well as the environmental rewards to ocean and air, will be significant.

The *best* solution is a combination of recycling and reducing of the amount of debris we generate by our daily activities. We will soon have no other choice.

Sewage

About 98% of sewage is water. Sewage treatment separates the fluid component from the solids, treats the water to kill disease organisms and reduce the levels of nutrients, and releases it into a river or the ocean. The remaining **sewage sludge**—a semisolid mixture of organic matter containing bacteria and viruses, toxic metal compounds, synthetic organic chemicals, and other debris—is digested, thickened, dried, and then shipped to landfills, burned to generate electricity, or dumped into the ocean. The liquid effluent wanders from its outlet and circulates with currents, but sludge and other insoluble residues may stay near the outfall or dump site for years. The amount of wastewater and sewage sludge has increased by 60% in the last decade.

Almost 2 billion liters (half a billion gallons) of partially treated sewage pours into Boston Harbor every day; a new, effective sewage treatment plant was opened in 1992, but this was 22 years after the deadline mandated by the U.S. Clean Water Act. About the same amount of treated sewage pours from outfall pipes off Los Angeles. Treatment plants in southern California are sometimes overwhelmed after rare heavy rainstorms when raw sewage enters the ocean in large quantity. Rain or shine, areas around the entrance to San Diego harbor are often so contaminated with sewage that anyone who enters the water runs the risk of bacterial or viral infection.

Sludge may be an even greater problem than raw sewage. The water just south of Long Island has been one of the most intensely used ocean dumping sites in the world. The area is covered with sludge, which creates an oxygen-poor environment in which few animals can survive. During storms some of this material washes up on local beaches, contaminates shellfish beds, and routinely causes disease outbreaks among people consuming raw oysters and clams from the area. After a public outcry, a new dumping site was selected, beyond the edge of the continental shelf. The 10 million tons of wet sludge produced by treatment plants in New York and New Jersey since March 1986 have been dropped there by huge barges. The plume from this sludge now contaminates the Gulf Stream.

Liquid effluent and sludge also contribute to eutrophication (*eu* = good + *trophos* = feeding). As we've seen, some forms of algae thrive in nutrient-rich wastewater, where they multiply to prodigious numbers, secrete toxins, or deplete oxygen in the surrounding water and cause the death of fish and crustaceans. A huge anoxic zone in the Gulf of Mexico appears to have been triggered by wastewater.

ATMOSPHERIC CHANGES

The ocean and the atmosphere are extensions of each other, and human activity has changed the atmosphere as it has changed the ocean. Pollutants injected into the air can have global consequences for the ocean and for all the Earth's inhabitants. Potentially two of the most destructive atmospheric problems are depletion of the ozone layer and global warming.

Ozone Layer Depletion

Ozone is a molecule formed of three atoms of oxygen. Ozone forms naturally when lightning strikes through air; larger quantities are generated spontaneously in the stratosphere (the uppermost layer of atmosphere). A diffuse layer of ozone mixed with other gases—the **ozone layer**—surrounds the world at a height of about 20 to 40 kilometers (12 to 25 miles).

"Harmless" synthetic chemicals released into the atmosphere—primarily **chlorofluorocarbons** (**CFCs**) used as cleaning agents, refrigerants, fire-extinguishing fluids, spray-can propellants, and insulating foams—are converted by the energy of sunlight into compounds that attack and partially deplete the Earth's atmospheric ozone. Ozone levels in the stratosphere have decreased by at least 3% over most of the United States since 1969 (see **Figure 15.23**). A 4% drop has been noted over Australia and New Zealand, and a 50% decrease has been observed near the North and South poles. The amount of depletion varies with latitude (and with the seasons) because of variations in the intensity of sunlight.

This decline in ozone alarms scientists because stratospheric ozone intercepts some of the high-energy ultraviolet

BOX 15.2 ● *The Law of the Sea*

Prehistoric peoples living near the shore were the earliest users of marine resources. With the rise of nation states and military establishments, the fight for control of marine resources began. Some nations assumed the ocean belonged to all and endeavored to guarantee free right to passage and resources. Others decided it belonged to none and tried to control access to ports and resources by force. In 1604 Hugo Grotius, a learned Dutch jurist, wrote *De Jure Praedae* ("On the Law of Prize and Booty"), a treatise justifying the action of a Dutch admiral who had successfully defended Dutch trading rights in a dispute with Portugal. One chapter of this work, in which Grotius defended free ocean access for all nations, was reprinted in 1609 under the title *Mare Liberum* ("A Free Ocean"). *Mare Liberum* formed the base for all modern international laws of the sea.

About a century later, in 1703, the concept of territorial seas adjacent to land was recognized. A country's seaward boundary was set at about 5 kilometers (3 miles)—the distance a cannonball could be fired from shore. This 5-kilometer (3-mile) limit stood until 1945.

The United Nations and the International Law of the Sea

After World War II the technology became available to search for oil and natural gas on continental shelves. After U.S. oil companies found rich deposits beyond the 5-kilometer limit off Louisiana, President Harry Truman issued a proclamation annexing the physical and biological resources of the continental shelf contiguous to the United States. Other nations rushed to make similar claims.

The United Nations then became involved. A committee of the General Assembly began to formulate policy that was later presented at the First United Nations Conference on Law of the Sea in 1958 in New York. Twenty-four years of effort by delegates from many interested nations resulted in the 1982 Draft Convention on the Law of the Sea. In April 1982 the United Nations adopted the Convention by a vote of 130 to 4, with 17 abstentions. (The United States, Turkey, Venezuela, and Israel voted against the Convention.) By 1988 more than 140 countries had signed all or most parts of the treaty. It is now legally binding, but signatories have selectively chosen to respect or ignore its individual provisions.

Some important features of the 1982 Draft Convention are as follows:

- **Territorial waters** are defined as extending 12 miles from shore. A nation has the right to jurisdiction within its territorial waters. Straits used for international navigation are excluded from a nation's territorial waters in that any vessel has the right to innocent passage.

- The 200 nautical miles (370 kilometers) from a nation's shoreline is its **Exclusive Economic Zone (EEZ)**. Nations hold sovereignty over resources, economic activity, and environmental protection within their EEZ.

- All ocean areas outside the EEZ are considered the **high seas**. In tradition, the high seas are common property to be shared by the citizens of the world. An International Seabed Authority was established to oversee the extraction of mineral resources from the deep sea.

- The values of protecting the ocean and preventing marine pollution were endorsed.

- Subject to some conditions, the freedom of scientific research in the ocean was encouraged.

The treaty places about 40% of the world ocean under the control of the coastal countries, within the EEZs. The resources of the remaining 60%, the high seas, are to be shared by the citizens of the world.

The United States Exclusive Economic Zone

The United States did not sign the 1982 Draft Convention for a variety of reasons. Among these was concern that private enterprise would be deprived of profits if it were made to share high seas resources with other countries. Instead, the United States unilaterally claimed sovereign rights and jurisdiction over all marine resources within its own 200 nautical-mile region, which it called the **United States Exclusive Economic Zone**. The proclamation—similar in most ways to the 1982 United Nations treaty but lacking the provision of shared high seas resources—was signed by President Ronald Reagan on 10 March 1983. The United States EEZ brings within national domain over 10.3 million square kilometers (3 million square nautical miles) of continental margins, an area 30% larger than the land area of the United States and a region of diverse geological and oceanographic settings (see **Figure a**).

The first step in exploring this new region has been to map the surface of the seafloor, a project that is still going on. The first bottom surveys were conducted off the West Coast of the United States because of the diverse energy and mineral resources known to exist there. Massive deposits of various metallic sulfides are being explored at the hydrothermal vents on the Gorda and Juan de Fuca ridges. Over 100 previously unknown volcanoes have been mapped within the United States EEZ of our West Coast. Huge faults, submarine landslides, seamounts, and details of the spreading crest of

the oceanic ridges are a few of the features that have been discovered so far. Knowledge of the modern tectonic setting for ore formation has been valuable in locating new ore deposits on land.

Manganese nodules and crusts have been discovered within the United States EEZ off the East and West coasts, Hawaii, and the Pacific Island territories. Cobalt, present in the manganese crusts, is being studied to determine how the deposit is formed and how it can be retrieved economically.

The EEZ project also includes research into meteorology, accurate weather forecasting, environmental studies, effects of plate movement on the ocean floor, effects of mining on seafloor organisms, and geohazards such as submarine landslides and earthquakes. It is hoped that this new era of marine exploration and research will lead to informed decisions about offshore activities to benefit not only the United States but other maritime nations.

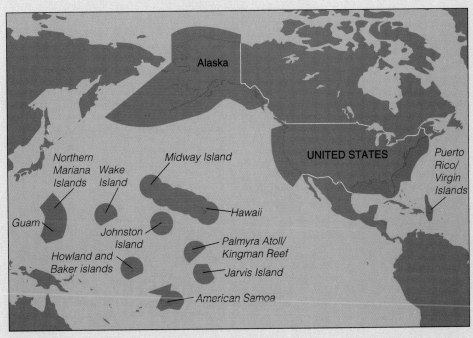

a The U.S. Exclusive Economic Zone (EEZ), shown here in dark blue, covers a vast area of the continental margin. The edge of the continent as it extends out under the ocean contains a vast wealth of resources.

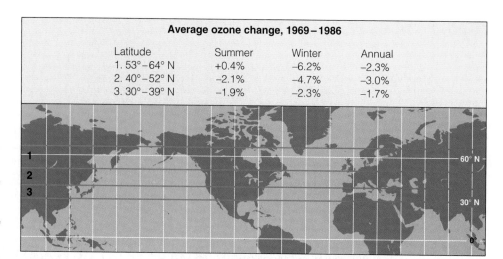

Average ozone change, 1969–1986			
Latitude	Summer	Winter	Annual
1. 53°–64° N	+0.4%	–6.2%	–2.3%
2. 40°–52° N	–2.1%	–4.7%	–3.0%
3. 30°–39° N	–1.9%	–2.3%	–1.7%

Figure 15.23 Ozone depletion in the stratosphere above various latitudes of the Northern Hemisphere between 1969 and 1986. A 1991 NASA study showed that between 1978 and mid-1990, the decrease in ozone was about twice the percentage shown here.

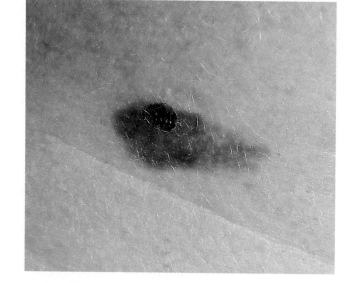

Figure 15.24 A malignant melanoma in an early stage. Left untreated, this most dangerous form of skin cancer could spread and cause the death of the patient. Skin cancer appears to be associated with exposure to the sun. If you spend much time in the sun you increase your risk for developing skin cancer. Be alert for changes in moles, or for any pigmented lesion that is asymmetrical, has an irregular border, is an unusual color, grows in size, or begins to itch. If you notice any of these characteristics, see a physician.

radiation coming from the sun. Ultraviolet radiation injures living things by breaking strands of DNA and unfolding protein molecules. Species normally exposed to sunlight have evolved defenses against average amounts of ultraviolet radiation, but increased amounts could overwhelm those defenses. Land plants such as soybeans and rice would be subjected to sunburn that decreases their yields. Even plankton in the uppermost 2 meters of ocean would be affected; recent research indicates an alarming decrease in phytoplankton primary productivity of between 6% and 12% in the coastal waters around Antarctica.[2]

Our own species would not escape: A 1% decrease in atmospheric ozone would probably be accompanied by a

[2]See the 1992 article by Smith et al., listed in this chapter's bibliography.

" I MISS THE OZONE LAYER...."

Figure 15.25

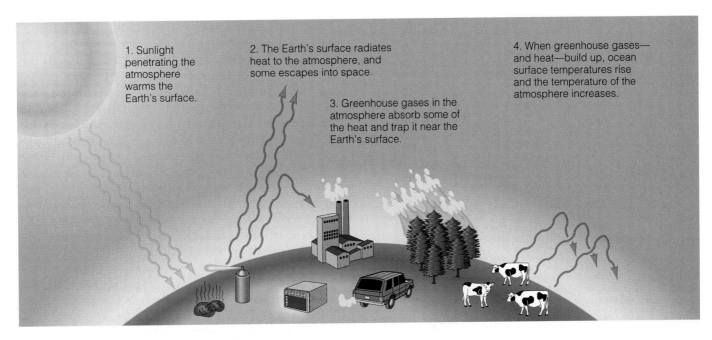

1. Sunlight penetrating the atmosphere warms the Earth's surface.

2. The Earth's surface radiates heat to the atmosphere, and some escapes into space.

3. Greenhouse gases in the atmosphere absorb some of the heat and trap it near the Earth's surface.

4. When greenhouse gases—and heat—build up, ocean surface temperatures rise and the temperature of the atmosphere increases.

Figure 15.26 How the greenhouse effect works.

5% to 7% increase in human skin cancer (see **Figure 15.24**). According to medical researchers' estimates, the 4% ozone depletion over Australia and New Zealand will cause at least a 20% increase in human skin cancers over the next two decades. Strong ultraviolet light can also suppress the immune system and cause eye cataracts. The quantity of photochemical smog shrouding our cities will also increase as ozone levels fall.

In June 1990, representatives of 53 nations agreed to ban major production and use of ozone-destroying chemicals by the year 2000. Legislation is being proposed to phase out CFC production in the United States, and a sense of great urgency surrounds ongoing research to find safe substitutes for these substances. Yet if their production were completely halted tomorrow, chlorine levels in the stratosphere would continue to rise for another 20 to 30 years, according to NASA sources. Moreover, CFCs stay in the atmosphere for up to 110 years; so their damaging effect will continue. It may already be too late.

Global Warming

The surface temperature of the Earth fluctuates slowly over time. The global temperature trend has been generally upward in the 18,000 years since the last ice age, but the *rate* of increase has recently accelerated. This rapid warming may be the result of an enhanced **greenhouse effect**, the trapping of heat by the atmosphere. Glass in a greenhouse is transparent to light but not to heat. The temperature inside a greenhouse rises because the heat is unable to escape. On Earth **greenhouse gases**—carbon

dioxide, methane, CFCs, and others—take the place of glass. Heat that would otherwise radiate away from the planet is absorbed and trapped by these gases, causing a rise in surface temperature. **Figure 15.26** shows this mechanism.

A certain amount of greenhouse effect is necessary for life; without it, Earth's average atmospheric temperature would be about $-18°C$ (0°F). Earth has been kept warm by natural greenhouse gases. The sources of these gases are volcanic and geothermal processes, the decay and burning of organic matter, and respiration and other biological sources. The removal of these gases by photosynthesis and absorption by seawater appears to prevent the planet from overheating. But the human demand for quick energy to fuel industrial growth, especially since the beginning of the Industrial Revolution, has injected unnatural amounts of new carbon dioxide into the atmosphere from the combustion of fossil fuels. Reckless burning of forests and jungles exacerbates the problem. Carbon dioxide is now being produced at a greater rate than it can be absorbed. **Figure 15.27** shows how much carbon dioxide has increased in the atmosphere since about 1750. The atmosphere's carbon dioxide content now rises at the rate of 0.4% each year. CFCs and methane are increasing even more rapidly, and one molecule of CFC is equivalent in greenhouse heating effect to 10,000 carbon dioxide molecules.

There has been a 4°C (7°F) rise in global temperature from the end of the last ice age until today. Carbon dioxide and other human-generated greenhouse gases produced since 1880 are thought to be responsible for about 1°C

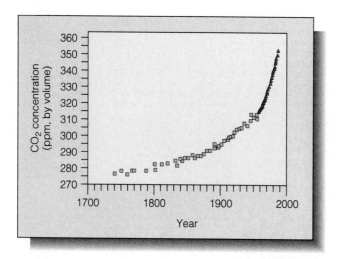

Figure 15.27 The rise in atmospheric carbon dioxide since 1750. Measurements up to about 1960 (■) came from samples of air trapped in the Antarctic ice; later measurements (▲) were made directly at Mauna Loa, Hawaii. (After Watson et al., 1990)

(1.8°F) of that increase (see **Figure 15.28**). If current models of greenhouse warming are correct, we can expect global temperature to rise another 2.5°C (4.5°F) by the year 2030. A temperature increase of this magnitude would cause water in the ocean to expand; average sea level would rise between 8 and 29 centimeters (3 and 11 inches). (One European model predicts a 4°C (7°F) rise by 2070, with an ocean rise of 0.5 to 1.5 meters, or 1.6 to 5 feet.) If the West Antarctic ice sheet were to melt during the next century, as happened during a warm period 120,000 years ago, sea level would climb as much as 6 meters (20 feet) over several hundred years. Imagine the effect on the harbors, coastal cities, river deltas, and wetlands where one-third of the world's people now live. The costs to society would be incalculable.

But scientists are still uncertain about the magnitude of human-induced global warming. Unpredictable natural processes, such as fluctuations in the energy output of the sun or the eruption of Mt. Pinatubo in the Philippines in 1991, can greatly influence temperature predic-

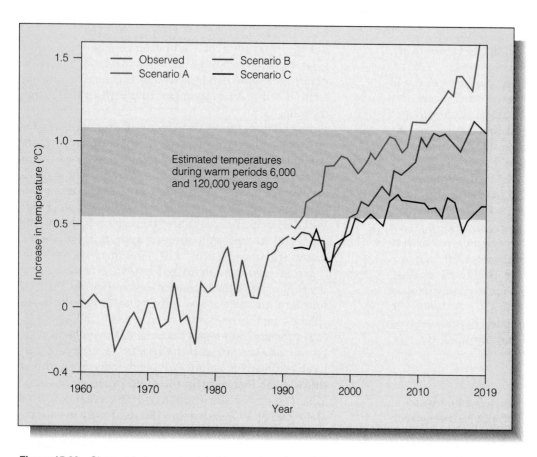

Figure 15.28 Observed changes in global temperature since 1960 and projected changes through the year 2019 as a result of the greenhouse effect. Scenario A assumes continued growth in the rate of release of carbon dioxide and similar gases. Scenario B assumes the implementation of some controls on the release of greenhouse gases. Scenario C assumes a complete halt in their release. Even under Scenario C, temperatures would still climb into the shaded area, which represents the average temperatures during extremely warm periods about 6,000 and 120,000 years ago.

tions. Researchers have estimated that the huge volume of dust and gas injected by Mt. Pinatubo into the stratosphere will more than compensate for greenhouse heating and will bring Earth's atmospheric temperature down by about 1°C (1.8°F) through the next five years. Most researchers are counseling caution—and a reduction in the generation of greenhouse gases—until more data can be accumulated.

Unfortunately, it will be exceedingly difficult to curtail our production of carbon dioxide. In the last hundred years, industrial production has increased by 50 times; we have burned 1,000 million barrels of oil, 1,000 million metric tons of coal, and 10,000 million cubic meters of natural gas. Carbon dioxide is a major product of combustion for these hydrocarbon compounds. The world's energy demand is projected to increase 3.5 times between now and the year 2025, with carbon dioxide emissions 65% higher than today.

Alternatives to fossil fuel must be found if we are to maintain world economies and prevent an increase in global temperature with all its uncertainties. The only alternate source of energy that currently produces significant amounts of power is nuclear energy, which now generates about 17% of U.S. electricity. Despite much publicity to the contrary, these pressurized water reactors have good records of dependable power production and safety.[3] The problem with nuclear power lies not so much in the everyday operation of the reactors, but in disposing of the nuclear wastes they produce. By 1989 about 51,000 highly radioactive spent fuel assemblies were in temporary storage in deep pools of cooled water; they must be stored for 10,000 years before their levels of radioactivity will be low enough to pose no environmental hazard. Radioactive substances emit **ionizing radiation**, a form of energy able to penetrate and permanently damage cells. It is essential that artificial sources of ionizing radiation be isolated from the environment.

Of course the ocean has been investigated as a potential dumping site. Some low-level medical and industrial nuclear wastes have already been jettisoned at sea, and the results have not been promising. A few barrels of radioactive waste dropped to the seabed near the Farallon Islands off San Francisco have broken open. Drums of radioactive waste from another source have washed ashore in Oregon. Small amounts of radioactive waste now enter the ocean at the mouth of the Columbia River after having leaked from storage containers near Hanford, Washington. Fishermen far from the North Atlantic coast have snagged containers of radioactive residue in trawl nets.

Will citizens of industrial countries (and countries wishing to become industrialized) agree to lessen the

[3]The Soviet reactor at Chernobyl that exploded in 1986 was of a much different design.

danger of increased global warming by slowing their economic growth, decreasing their dependence on fossil fuel combustion, and developing safe alternate sources of energy? Some insight may be gained from the behavior of ranchers and industrialists in the rain forests of New Guinea, the Philippines, and Brazil. The Amazon rain forest of Brazil is being burned at a rate of about 12 square kilometers *per hour* (almost 5 square miles *each hour*), acreage equivalent to the area of West Virginia every year. Huge stands of trees that should be nurtured to absorb excess carbon dioxide are being destroyed. The cleared land is used for farms, cattle ranches, roads, and cities. The priorities are obvious.

WHAT CAN BE DONE?

In a pivotal paper published in 1968, Garrett Hardin examined what he termed "The Tragedy of the Commons." Hardin's title was suggested by his study of societies in which some agricultural areas were held "in common"; that is, jointly owned by all residents. Citizens of these societies owned small homes, plots of land, and perhaps a cow that was put to pasture on the commons. Each farmer *kept* the milk and cheese given by his cow but *distributed* the costs of cow ownership—overgrazing of the commons, cow excrement, fouled drinking water, and so on—among all the citizens. This arrangement worked well for centuries because wars, diseases, and poaching kept the numbers of people and cows well below the carrying capacity of the land. But eventually political stability and relative freedom from disease allowed the human (and cow) population to increase. Farmers pastured more cows on the commons and gained more benefits. Soon the overstressed commons could no longer sustain the growing numbers of cows, and the area held in common was ruined. Eventually no cows could survive there.

The lesson applies to our present situation. Hardin noted that in our social system each individual tends to act in ways that maximize his or her material gain. Each of us gladly keeps the *positive* benefit of work but willingly distributes the *costs* among all. For example, this morning I drove to my college office. The benefit to me was one trip to my office. A cost of this short drive was the air pollution generated by the fuel combustion in my car's engine. Did I route the exhaust fumes through a hose to a mask held tightly over my nose and mouth? That is, did I reserve the environmental *costs* of my actions for my own use just as I had reserved for myself the *benefit* of my ride to work? No. I shared those fumes with my fellow Californians, just as you shared your morning's sewage with your fellow citizens, or the factory down the road shared its carbon dioxide with all the world. Indeed, the world itself is our commons. The

BOX 15.3 ● *Some Things You Can Do for the Environment*

Each of us can make daily decisions that collectively make a difference. Some things we can do are listed here.

A. Energy

1. Drive fewer miles. Ride your bike or walk. Use public transportation. Carpool.

2. Replace lamps with smaller wattage lightbulbs. Replace incandescent bulbs with compact 18-watt fluorescent ones. Turn off lights in rooms not being used. Switch to low-voltage exterior lighting.

3. Purchase energy-efficient appliances and cars. Government-mandated stickers on cars and appliances allow you to compare the amounts of energy each will use. Spend a bit more now to save a lot later.

4. Check insulation and weatherstripping around your house, apartment, or dorm. Keep thermostats relatively high in summer, low in winter.

5. Check into alternative energy sources such as solar and wind power units.

6. Buy and use rechargeable batteries. The disposable ones contain heavy metals that can pollute.

B. Food

7. Try to avoid foods from areas where pesticides have been used. Never eat fish caught in developed harbors or bays.

8. Consider how your food choices affect the environment. Eat foods low on the food chain and avoid foods raised in places where endangered environments (such as rain forests) were destroyed.

9. Raise a garden.

10. Buy bulk or unpackaged goods, or buy food in reusable containers.

C. Toxic Materials and Pollutants

11. Check for toxic materials (old cans of paint, pesticides, engine oil, pool chemicals, and so on) you may have around the house. Take them to an approved toxic waste receiving station.

12. Read labels. Buy the least toxic materials available.

13. Swat flies—don't spray them.

14. Buy clothing that does not require dry cleaning. Dry clean only when necessary.

---

modern tragedy of the commons rests on these kinds of actions.

The carrying capacity of the whole "Earth-commons" may already have been exceeded. Births now exceed deaths by about three people per second. *Each year* there are 90 million more of us, a total equal to the combined populations of Mexico City, Tokyo and Yokohama, São Paulo, New York, Shanghai, and Calcutta. The number of people has tripled in this century and is expected to double again before reaching a plateau sometime in the next century. Another billion humans will join the world population in the next ten years, 92% of them in third-world countries (see **Figure 15.29**).[4]

[4]In the United States alone, the population is growing by the equivalent of four Washington D.C.s every year, another New Jersey every three years, another California every twelve. By the year 2000 another 40 million people will reside within U.S. boundaries.

This exploding population is not content with using the same proportion of resources used today. Citizens of the world's least-developed countries are influenced by education and advertising to demand a developed-world standard of living. They look with justifiable envy on the United States, a country with 4.5% of the planet's population that consumes 55% of its raw material resources and 25% of its energy, while generating 30% of industry-related carbon dioxide. Can the world support a population whose expectations are rising as rapidly as their numbers? In Garrett Hardin's words, "We can maximize the number of human beings living at the lowest possible level of comfort, or we can try to optimize the quality of life for a much smaller population." The burgeoning human population is the greatest environmental problem of all.

We cannot expect science to solve the problem for us. Most of the decisions and necessary actions fall outside

D. Water

15. Use water-saving showerheads, sink faucet aerators, and toilets.

16. Sweep walks and driveways; don't hose them down.

17. Water lawns infrequently but thoroughly. Better yet, replace them with drought-tolerant ground covers.

18. Take shorter showers. Shower with a friend.

19. Run only full dishwasher and clothes washer loads.

E. Recycling and Waste Reduction

20. Recycle newspapers, cans, glass, and plastics.

21. Buy products that are refillable and recyclable.

22. Volunteer your time in a beach cleanup campaign.

23. Use cloth rags and diapers instead of paper diapers and paper towels.

24. Take your grocery bags back to the market for another trip home.

F. Preservation of Life and the Environment

25. Plant trees.

26. Preserve open space. Become involved in parkland and open space movements.

27. Avoid CFCs. Check air conditioners to see if they leak.

28. Boycott organizations that violate sound environmental practices.

29. Don't litter. Pack your trash out of campsites.

G. Other Things

30. Write your legislators. Better yet, visit them!

31. Get involved in an environmental group such as the Sierra Club or Greenpeace.

32. Talk to your friends about environmental matters.

33. Buy quality. Purchase a well-built item, maintain it, repair it, keep it forever.

34. Sit quietly on a beach at sunset or sunrise.

35. Read. Listen.

36. Be positive about results. Be encouraged.

Source: Thanks to Jack Mertz.

pure science in the areas of values, ethics, morality, and philosophy. The solution to environmental problems, if one exists, *lies in education and action.* Each of us is obliged to become informed on issues that affect the Earth, its ocean, and its air—to learn the arguments and weigh the evidence. Once informed, we must act in rational ways. Chaining yourself to an oil tanker is not rational, but selecting well-designed, long-lasting, and recyclable products made by responsible companies with minimal impact on the environment (and encouraging others to do so) certainly is. (A few other things you can do are listed in **Box 15.3**).

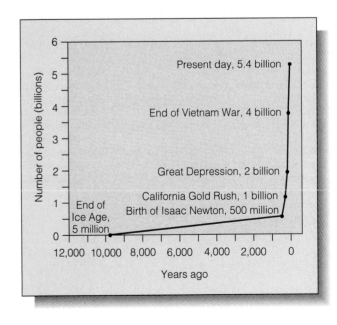

Figure 15.29 Growth of the human population. It took 4 million years for humanity to reach the 2 billion mark, but only about 50 years to double that number. With our present growth rate of about 1.8% per year, we will double our 1995 size of 5.7 billion people in less than 36 years.

Obvious answers and quick solutions are often misleading; a great deal of research and work are needed to give reliable insight into the many difficult questions that confront us. The present trade-off between financial and ecological considerations is often strongly tilted in the direction of immediate gain, of short-term profit, of immediate convenience. Education may be the only way to modify these destructive behaviors. *Hardin suggests that absolute freedom in a commons brings ruin to all.*

Humanity is part of the natural world, not its master. We may be able to learn to live in harmony with this small, beautiful, blue world. We need not relive the "tragedy of the commons" on a planetary scale. True convenience and true progress depend on the preservation of open space, serious and sustained attention to population control, conversion to a steady-state economy instead of one that must grow to stay alive, business incentives for preservation, the use of renewable resources, and, above all, *public education in environmental issues.* We must ask ourselves difficult questions: "What is the optimal quality of life?" "How can I achieve balance between my material needs and the needs of the Earth?" "What do I want to leave for my children?" "How can I reserve the quiet, renewing ocean for myself, for all species, and for the future?" Our cities are crowded and our tempers are short. Times of turbulent change lie before us. The trials ahead will be severe.

Each of us, individually, needs to take a stand. We must preserve the sunsets and fog, the waves to ride, the cold clean spray on our faces. Margaret Mead summarized our summed potential for making a difference: "Never doubt that a small group of thoughtful, committed citizens can change the world. Indeed it is the only thing that ever has." We need to start now.

CHAPTER SUMMARY

Marine resources include physical resources such as oil, natural gas, building materials, and chemicals; biological resources such as seafood and kelp; and nonextractive resources like transportation and recreation. The contribution of marine resources to the world economy has become so large that international laws now govern their allocation.

In spite of their abundance, marine resources provide only a fraction of the worldwide demand for raw materials, human food, and energy. Similar resources on land can usually be obtained more safely and at lower cost.

Our species has always exercised its capacity to consume resources and pollute its surroundings, but only in the last few generations have our efforts affected the ocean and atmosphere on a planetary scale. The introduction into the biosphere of unnatural compounds (or natural compounds in unnatural quantities) has had, and will continue to have, unexpected detrimental effects. The destruction of marine habitats and the uncontrolled harvesting of the ocean's living resources have also disturbed delicate ecological balances. We find ourselves in difficult situations for which solutions do not come easily.

Terms and Concepts to Remember

biodegradable
biological
    amplification
biological resources
chlorinated
    hydrocarbons
chlorofluorocarbons
    (CFCs)
commercial
    extinction
desalination
drift nets
Exclusive Economic
    Zone (EEZ)

greenhouse effect
greenhouse gases
high seas
ionizing radiation
marine pollution
maximum
    sustainable yield
nonextractive
    resources
nonrenewable
    resources
overfishing
ozone
ozone layer

physical resources
pollutant
polychlorinated
    biphenyls (PCBs)
renewable
    resources
sewage sludge
territorial waters
United States
    Exclusive
    Economic Zone

Study Questions

1. Distinguish between physical and biological resources, and between renewable and nonrenewable resources.

2. What are the three most valuable physical resources of the ocean? How does the contribution of each to the world economy compare to the contribution of that resource derived from land?

3. Does the ocean provide a substantial percentage of all protein needed in human nutrition? Of all animal protein? What is the most valuable biological resource? The fastest growing fishery?

4. What is a nonextractive resource? Give some examples.

5. What is pollution? What factors determine how dangerous a pollutant is?

6. Why is refined oil more hazardous to the marine environment than crude oil? Which is spilled more often?

7. What heavy metals are most toxic? How do these substances enter the ocean? How do they move from the ocean to marine organisms and people?

8. Few synthetic organic chemicals are dangerous in the very low concentrations in which they enter the ocean. How are these concentrations increased? What can be the outcome when these substances are ingested by organisms in a marine food chain?

9. What are the signs of overfishing? How does the fishing industry often respond to these signs? What is the result?

10. What synthetic chemicals appear to be causing depletion of the Earth's protective ozone layer? What is the likely result?

11. What is the greenhouse effect? Is it always detrimental? What gases contribute to the greenhouse effect? Why do most scientists believe the Earth's average surface temperature will increase over the next few decades? What may result?

12. What is the "tragedy of the commons"? Do you think Garrett Hardin was right in applying the old idea to modern times? What will you do to minimize your negative impact on the ocean and atmosphere?

For Further Study

Broadus, J. M. 1987. "Seabed Materials." *Science* 235 (no. 4791). Thorough summary of material resources.

Broadus, J. M. 1991. "The Sea Environment: Good News, Bad News." *United States Naval Institute Proceedings* 117 (no. 10): 50–55. Presentation of a "dirty dozen" environmental problems requiring attention.

Cherfas, J. 1990. "The Fringe of the Ocean Under Siege from Land." *Science* 248 (no. 4952): 163–65. Excellent summary of the present assault on the ecology of ocean margins.

Hardin, G. 1968. "The Tragedy of the Commons." *Science* 162 (13 Dec.): 1243–48. Hardin makes the case that freedom in a commons brings ruin to all.

Holmes, B. 1994. "Biologists Sort the Lessons of Fisheries Collapse." *Science* 264 (no. 5163): 1252–53. News reports of the recent collapse of many North American fisheries should come as no surprise.

Jones, P. D., and T. M. W. Wigley. 1990. "Global Warming Trends." *Scientific American*, August, 84–91. Part of a special issue entitled "Energy for Planet Earth."

Kerr, R. A. 1991. "A Lesson Learned, Again, at Valdez." *Science* 252 (no. 5004): 371. The cleanup at Prince William Sound was more damaging than the effects of the crude oil spilled.

Kraus, S. D. 1989. "Whales for Profit." *Whalewatcher* 23 (no. 2). Source of Table 19.5, Total Value of Whaling in 1988.

McKibben, Bill. 1989. *The End of Nature*. New York: Random House. A beautifully written plea for rationality and change.

Miller, G. T. 1994. *Living in the Environment*. 8th ed. Belmont, CA: Wadsworth. Excellent reference and general text on environmental science.

O'Bannon, B. K. 1988. *Fisheries of the United States, 1987*. Washington, DC: U.S. Department of Commerce.

Ross, D. A. 1980. *Opportunities and Uses of the Ocean*. New York: Springer-Verlag. Excellent and concise coverage of marine resources.

Smith, R. C., et al. 1992. "Ozone Depletion: Ultraviolet Radiation and Phytoplankton Biology in Antarctic Waters." *Science* 255 (no. 5047): 952–59. Reduced concentration of atmospheric ozone, and consequently higher ultraviolet irradiance, has reduced phytoplanktonic productivity in the far south by as much as 6% to 12%.

Toufexis, A. 1988. "The Dirty Seas." *Time*, 1 Aug., 44–50. A fine summary of the situation.

Watson, R. T., et al. 1990. "Greenhouse Gases and Aerosols." In *Climate Change: The IPCC Scientific Assessment*, ed. J. T. Houghton et al., 1–40. New York: Cambridge University Press.

Weisskopf, M. 1988. "Plastic Reaps a Grim Harvest in the Oceans of the World." *Smithsonian*, March, 59–66. Excellent review of the growing crisis of plastic in the ocean.

AFTERWORD

The marine sciences are at the threshold of a new age. The recent revolutions in biology and geology are being assimilated and the road ahead seems clearer. A renaissance in the design of sampling devices, robot submersible vehicles, and data processing has brought new vigor to oceanography. Satellite-borne sensors can provide data in an instant that would have taken years to collect using surface ships. Shipboard technology has become so sophisticated that Wyville Thomson or Fridtjof Nansen would hardly recognize our sensors or sampling devices.

The tools may be different, but the spirits of those who use them remain the same. Today's marine scientists are like all the men and women who have gone before: *We want to know about the ocean.* We haunt our mailboxes for journals bearing the latest research news, search television listings for any new ocean shows, inspect new samples with the enthusiasm of little kids, and share our insights with anyone at the drop of a hat. I am personally delighted that you have traveled with me this far. Those of us who enjoy an oceanographic background (and this now includes you) look at the Earth with greater understanding than we did before we began. The whole concept of an ocean world appeals to us, gives us profound pleasure, and sobers us with a deep sense of responsibility. In no other field of science do so many ideas interweave to form so rich a tapestry.

Our journey together is over, but before we go our separate ways, I have three last ideas to share:

- Change has been a recurrent theme of this book. The Earth's climate has changed with time, as has its atmospheric composition, its ocean chemistry, the size and positions of its continents, and its life forms. The Earth may seem a calm and stable home, but it is really a violent place for inhabitation by such seemingly delicate objects as living things. Even so, life and the ocean have grown old together. The story of the Earth is the story of change and chance; its history is written in the rocks, the water, and the genes of the millions of organisms that have evolved here. We are survivors.

But that survival may now be in question. Change is now progressing at an unnatural rate, and these human-induced changes are imposing stress on natural systems. What we do *with* and *to* the ocean is literally of planetary consequence. In the last century we have developed the physical, chemical, and biological machinery to destroy or rejuvenate the world ocean and all of its life. A painful time of inadvertent global experimentation lies just ahead.

All of us who love the colors and textures of this small wet world need to act to moderate the negative effects of the looming environmental crisis. In Chinese, the written character for the word *crisis* has two components: danger and opportunity. Informed citizens will express their concern, discuss this concern with others, and act whenever possible to minimize the threats and take advantage of new opportunities. Intelligence and beauty must triumph; we have no other rational alternative.

- Appreciation of the ocean doesn't come exclusively from the realm of science. Philosophers, artists, composers and poets have had much to say about the sea. Read Homer's description of the ocean in *The Odyssey* (try books iv, x, and xi). See how Lord Byron's feeling for the ocean colors his poetry (see, for instance, *Childe Harold's Pilgrimage*, stanzas 183 and 184). Read modern poet Robinson Jeffers's powerful *Continent's End*. Share Prospero's marine magic in Shakespeare's *The Tempest*. Find some of the evocative woodcuts of Rockwell Kent and the impressionistic ocean paintings of English artist J. M. W. Turner. Listen to Benjamin Britten's *Four Sea Interludes* from *Peter Grimes*, and Ralph Vaughan-Williams's *Sea Symphony* and *Sinfonia Antartica* (but take care not to blow out your sound system). Read the ocean novels of Herman Melville and Jack London, and try reading the journals and accounts of the famous explorers and scientists you have met in this book. Sit on a quiet beach at night with the stars of the Milky Way shining softly overhead. The pervasive inspiration of the wave-breathing ocean is never far away.

- Don't let your involvement stop here. Lifelong learning is the truest joy, a pleasure that does not diminish with age, a source of wisdom and calm. We can learn much about patience, hope, and optimism from the ocean. We can learn much about the world—and about ourselves—by looking for the oceanic connections between things. I hope your interest in learning about the ocean has just been kindled. There is much good in the world. Go and add to it.

APPENDIX I

Measurements and Conversions

Other than the United States, only two countries in the world—Liberia and South Yemen—do not use metric measurements. The metric system, a contribution of the French Revolution, conquered Europe along with Napoleon. It is based on a decimal system, a system familiar to Americans because of our decimal money system: 10 cents to a dime, 10 dimes to the dollar.

The first move toward a rational system of measurement was made in 1670 by Gabriel Mouton, the vicar of St. Paul's Church in Lyon, France. Instead of the then-prevalent measurement system based on the width of the king's hand, or the length of his outstretched right arm, or the weight of a particular basket of stones kept in the palace, Mouton suggested a length measure based on the arc of 1 minute of longitude, to be subdivided decimally. Other measurements would follow from this unit of length. His proposal contained the three major characteristics of the metric system: using the Earth itself as a basis for measurement, subdividing decimally (by 10s), and using standard prefixes (*kilo, centi, milli,* and so on). These ideas were debated for 125 years before being implemented by a commission appointed by Louis XVI in one of his last official acts before being imprisoned during the French Revolution. One ten-millionth of the distance from the North Pole to the equator (on the line of longitude passing through Paris) was selected as the standard unit of length, the meter. A new unit of weight was derived from the weight of 1 cubic meter of pure water. Temperature was to be based on pure water's boiling and freezing points. A list of prefixes for decimal multiples and submultiples was proposed. In 1795, a firm decision was made to establish the system throughout France, and in 1799, the metric system was implemented "for all people, for all time."

At first, people objected to the changes, but the government insisted that old measurements be included side by side with the equivalent new (metric) ones. In everyday competition, the advantages of the metric system proved decisive; in 1840, it was declared a legal monopoly in France. The French public had been won over to the new, simple, rational system of measurement. All of Europe followed, and eventually, virtually all other countries.

Not the United States, however. Though Ben Franklin proposed that the country convert in the eighteenth century, the people of the United States have continued to insist that the metric system—now known as the Système International (SI)—is too difficult to learn and work with. The federal government has urged conversion to metric units to increase opportunities for international trade. In August 1988, President Ronald Reagan signed the Omnibus Trade and Competitiveness Act. This act amended the 1975 Metric Conversion Act, stating that by 1992, all federal agencies must, wherever feasible, use the metric (SI) system in their purchases, grants, and other business.

The government may be making the change, but the public clings tenaciously to inches, pints, and pounds. Why? Is it really simpler to add $\frac{1}{16}$ of an inch, $\frac{1}{32}$ of an inch, and $\frac{3}{8}$ of an inch to cut a bookshelf to length? Can you remember how many cups to a quart? How many pints to a gallon? How many miles to a league?[1] The reason we continue to use the English system (which, of course, the English have long since abandoned) is because it is familiar to us. We know how long 5 inches is, and how much a quart is, and what 72° Fahrenheit represents. Perhaps by following the French example—by having measurements expressed everywhere in English *and* metric measurements—we may be able to make a complete conversion within a generation or two. That's why English and metric measurement are used together throughout this book. The process has already begun, of course: You use 35mm film, 2-liter soft drink containers, 750-milliliter wine bottles, 100-watt light bulbs—and you might run a 10-K (10-kilometer) race on Saturday.

The conversion factors listed here will give you an idea of how English and metric (SI) units are equivalent. Don't panic—the system is as rational and logical as it has always been. Note that 1 meter equals 100 centimeters and that 1 centimeter equals 10 millimeters. (See the table showing multiples and submultiples for an explanation of the relationship between prefixes like *centi* and *milli.*) Note that 2.54 centimeters equals 1 inch. (See the table of conversion factors if you wish to convert from one system to another.) Some numerical oceanographic data are included in supplemental tables.

---

[1] A mile is 5,280 feet, and a league is 5,280 yards. Captain Nemo, in his fictional submarine *Nautilus,* traveled a distance greater than twice the circumference of the Earth in Jules Verne's fantasy *20,000 Leagues Under the Sea,* a task made even more formidable by his total dependence on English units of measurement!

Scientific Notation

Multiples and Submultiples

		Name	Common Prefixes
10^{12}	= 1,000,000,000,000	trillion	tera
10^9	= 1,000,000,000	billion	giga
10^6	= 1,000,000	million	mega
10^3	= 1,000	thousand	kilo
10^2	= 100	hundred	hecto
10^1	= 10	ten	deka
10^{-1}	= 0.1	tenth	deci
10^{-2}	= 0.01	hundredth	centi
10^{-3}	= 0.001	thousandth	milli
10^{-6}	= 0.000001	millionth	micro
10^{-9}	= 0.000000001	billionth	nano
10^{-12}	= 0.000000000001	trillionth	pico

Conversion Factors

Area

1 square inch (in.^{2})	6.45 square centimeters
1 square foot (ft^2)	144 square inches
1 square centimeter (cm^2)	0.155 square inch
	100 square millimeters
1 square meter (m^2)	10^4 square centimeters
	10.8 square feet
1 square kilometer (km^2)	247.1 acres
	0.386 square mile
	0.292 square nautical mile

Mass

1 kilogram (kg)	2.2 pounds
	1,000 grams
1 metric ton	2,205 pounds
	1,000 kilograms
	1.1 tons
1 pound	16 ounces
	454 grams
	0.45 kilogram
1 ton	2,000 pounds
	907.2 kilograms
	0.91 metric ton

Length

1 micrometer (μm)	0.001 millimeter
1 millimeter (mm)	1,000 micrometers
	0.1 centimeter
	0.001 meter
1 centimeter (cm)	10 millimeters
	0.394 inch
	10,000 micrometers
1 meter (m)	100 centimeters
	39.4 inches
	3.28 feet
	1.09 yards
1 kilometer (km)	1,000 meters
	1,093 yards
	3,280 feet
	0.62 statute mile
	0.54 nautical mile
1 inch (in.)	25.4 millimeters
	2.54 centimeters
1 foot (ft)	30.5 centimeters
	0.305 meter
1 yard	3 feet
	0.91 meter
1 fathom	6 feet
	2 yards
	1.83 meters
1 statute mile	5,280 feet
	1,760 yards
	1,609 meters
	1.609 kilometers
	0.87 nautical mile
1 nautical mile	6,076 feet
	2,025 yards
	1,852 meters
	1.15 statute miles
1 league	15,840 feet
	5,280 yards
	4,804.8 meters
	3 statute miles
	2.61 nautical miles

Pressure

1 atmosphere (sea level)	760 millimeters of mercury at 0°C
	14.7 pounds per square inch
	33.9 feet of water (fresh)
	29.9 inches of mercury
	33 feet of seawater

Temperature

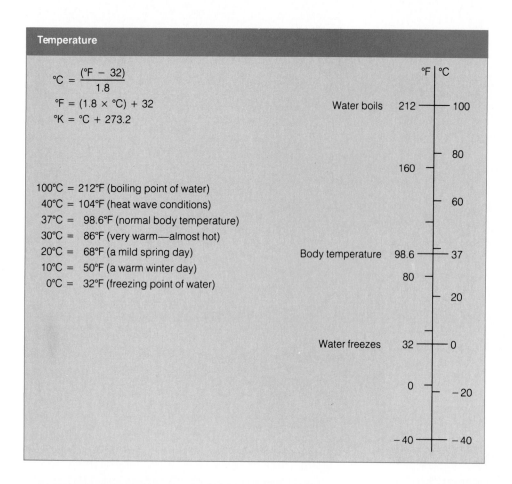

$$°C = \frac{(°F - 32)}{1.8}$$

$$°F = (1.8 \times °C) + 32$$

$$°K = °C + 273.2$$

100°C = 212°F (boiling point of water)
40°C = 104°F (heat wave conditions)
37°C = 98.6°F (normal body temperature)
30°C = 86°F (very warm—almost hot)
20°C = 68°F (a mild spring day)
10°C = 50°F (a warm winter day)
0°C = 32°F (freezing point of water)

	°F	°C
Water boils	212	100
		80
	160	
		60
Body temperature	98.6	37
	80	
		20
Water freezes	32	0
	0	−20
	−40	−40

Volume

1 cubic inch (in.³)	16.4 cubic centimeters
1 cubic foot (ft³)	1,728 cubic inches
	28.32 liters
	7.48 gallons
1 cubic centimeter (cc; cm³)	1 milliliter
	0.061 cubic inch
1 liter	1,000 cubic centimeters
	61 cubic inches
	1.06 quarts
	0.264 gallon
1 cubic meter (m³)	10^6 cubic centimeters
	264.2 gallons
	1,000 liters
1 cubic kilometer (km³)	10^9 cubic meters
	10^{15} cubic centimeters
	0.24 cubic mile

Time

1 hour	3,600 seconds
1 day	24 hours
	1,440 minutes
	86,400 seconds
1 calendar year	31,536,000 seconds
	525,600 minutes
	8,760 hours
	365 days

Speed

1 statute mile per hour	1.61 kilometers per hour
	0.87 knot
1 knot (nautical mile per hour)	51.5 centimeters per second
	1.15 miles per hour
	1.85 kilometers per hour
1 kilometer per hour	27.8 centimeters per second
	0.62 mile per hour
	0.54 knot

Some Familiar Metric Approximations

Measurement	Metric Unit	Approximate Size of Unit
Length	millimeter	diameter of a paper clip wire
	centimeter	a little more than the width of a paper clip (about 0.4 inch)
	meter	a little longer than a yard (about 1.1 yards)
	kilometer	somewhat further than $\frac{1}{2}$ mile (about 0.6 mile)
Mass (Weight)	gram	a little more than the mass (weight) of a paper clip
	kilogram	a little more than 2 pounds (about 2.2 pounds)
	metric ton	a little more than a short ton (about 2,200 pounds)
Volume	milliliter	five of them make a teaspoon
	liter	a little larger than a quart (about 1.06 quarts)
Pressure	kilopascal	atmospheric pressure is about 100 kilopascals

Source: U.S. Metric Board Report.

Numerical Oceanographic Data

Equivalences in Concentration of Seawater

Seawater with 35 grams of salt per kilogram of seawater	3.5 percent
	35 parts per thousand (‰)
	35,000 parts per million (ppm)

Speed of Sound

Velocity of sound in seawater at 34.85 parts per thousand (‰)	4,945 feet per second
	1,507 meters per second
	824 fathoms per second

Area, Volume, and Depth of the World Ocean

Body of Water	Area (10^6 km^2)	Volume (10^6 km^3)	Mean Depth (m)
Atlantic Ocean	82.4	323.6	3,926
Pacific Ocean	165.2	707.6	4,282
Indian Ocean	73.4	291.0	3,963
All oceans and seas	361	1,370	3,796

Geological Time

As we saw in Chapter 2, astronomers and geologists have determined that the Earth originated about 4.6 billion years ago. They have divided the Earth's age into eras, roughly corresponding to major geological and evolutionary changes that have taken place, as shown in the chart in **Figure 1**. Note that the time spans of the different eras are not shown to scale; if they were, the chart would run off the page.

Recall that life began fairly soon after the Earth's crust, atmosphere, and ocean formed. One way to conceive of the time span over which life evolved is to imagine it as a 24-hour clock, with life originating at midnight (**Figure 2**). In this scheme, invertebrates with hard parts (which make good fossils) became abundant about 4:30 P.M., and animals began to leave the ocean for land about 8 P.M. Our closest human ancestors (*Homo sapiens*) appeared about 2 seconds before midnight, agriculture began only ¼ second before midnight, and the industrial revolution has been around for ⁷⁄₁₀₀₀ of a second.

Era	Period	Epoch	Millions of Years Ago (mya)	
CENOZOIC	Quaternary	Recent		
			0.01	
		Pleistocene		
			1.65	
	Tertiary	Pliocene		
			5	
		Miocene		
			25	
		Oligocene		
			38	
		Eocene		
			54	
		Paleocene		
			65	
MESOZOIC	Cretaceous	Late		
			100	
		Early		
			138	
	Jurassic			
			205	
	Triassic			
			240	
PALEOZOIC	Permian			
			290	
	Carboniferous			
			360	
	Devonian			
			410	
	Silurian			
			435	
	Ordovician			
			505	
	Cambrian			
			550	
PROTEROZOIC				
			2,500	
ARCHEAN				

Figure 1   A Geological Time Scale

Times of Major Geological and Biological Events

1.65 mya to present. Major glaciations. Modern humans emerge and begin what may be greatest mass extinction of all time on land, starting with Ice Age hunters.

65–1.65 mya. Unprecedented mountain building as continents rupture, drift, collide. Major climatic shifts; vast grasslands emerge. Major radiations of flowering plants, insects, birds, mammals. Origin of earliest human forms.

65 mya. Asteroid impact? Mass extinction of all dinosaurs and many marine organisms.

135–65 mya. Pangea breakup continues, broad inland seas form. Major radiations of marine invertebrates, fishes, insects, dinosaurs. Origin of flowering plants.

181–135 mya. Pangea breakup begins. Rich marine communities. Major radiations of dinosaurs.

205 mya. Asteroid impact? Mass extinction of many organisms in seas, some on land; dinosaurs, mammals survive.

240–205 mya. Recovery, radiations of marine invertebrates, fishes, dinosaurs. Gymnosperms the dominant land plants. Origin of mammals.

240 mya. Mass extinction. Nearly all species in seas and on land perish.

280–240 mya. Pangea, worldwide ocean forms; shallow seas squeezed out. Major radiations of reptiles, gymnosperms.

360–280 mya. Tethys Sea forms. Recurring glaciations. Major radiations of insects, amphibians. Spore-bearing plants dominate. Gymnosperms present. Origin of reptiles.

370 mya. Mass extinction of many marine invertebrates, most fishes.

435–360 mya. Laurasia forms, Gondwana moves north. Vast swamplands, early vascular plants. Radiations of fishes continue. Origin of amphibians.

435 mya. Glaciations as Gondwana crosses South Pole. Mass extinction of many marine organisms.

500–435 mya. Gondwana moves south. Major radiations of marine invertebrates, early fishes.

550–500 mya. Landmasses dispersed near equator. Simple marine communities. Origin of animals with hard parts.

700–550 mya. Supercontinent Laurentia breaks up; widespread glaciations.

2,500–570 mya. Oxygen present in atmosphere. Origin of aerobic metabolism. Origin of protistans, algae, fungi, animals.

3,800–2,500 mya. Origin of photosynthetic bacteria.

4,600–3,800 mya. Formation of Earth's crust, early atmosphere, oceans. Chemical evolution leading to origin of life (anaerobic bacteria).

4,600 mya. Origin of Earth.

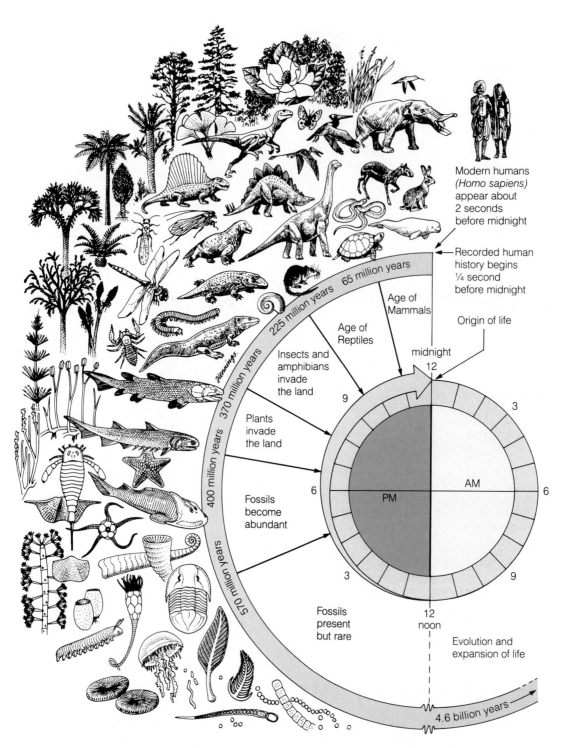

Modern humans
(*Homo sapiens*)
appear about
2 seconds
before midnight

Recorded human
history begins
¼ second
before midnight

Origin of life

65 million years

225 million years

Age of
Mammals

Age of
Reptiles

Insects and
amphibians
invade
the land

midnight
12

370 million years

Plants
invade
the land

400 million years

Fossils
become
abundant

AM

PM

6

6

3

3

9

9

570 million years

Fossils
present
but rare

12
noon

Evolution and
expansion of life

4.6 billion years

Figure 2 Greatly simplified history of the development of different forms of life on Earth through biological evolution, compressed to a 24-hour time scale.

Latitude and Longitude, Time, and Navigation

The ocean is large and easy to get lost in. A backyard, like that shown in **Figure 1**, is smaller, but we can still be lost in it if we don't have a frame of reference. Note that the yard is framed by a fence. We can refer to this frame to establish our position—in this case, at the intersection of perpendicular lines drawn from fence posts 2 and C. Many towns are arranged in this way: Fourth and D streets intersect at a precise spot based on the municipal frame of reference.

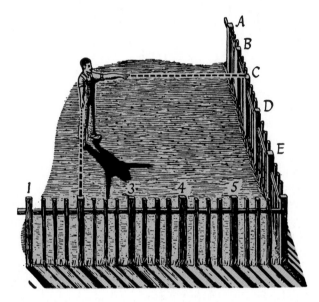

Figure 1

But the World Is Round: Spherical Coordinates

If the world were flat, a simple scheme of rectangular coordinates would serve all mapping purposes—a rectangle, like the yard in Figure 1, has four sides from which to measure. A sphere has no edges, no beginnings or ends, so what shall we use as a frame of reference for the Earth? Since the Earth turns, the poles, the axis of rotation, are the only absolute points of reference. We can draw an imaginary line equidistant from the North and South poles, a line that *equates* the globe into north-

ern and southern halves: the equator. Other lines, drawn parallel to the equator, further divide the sphere north and south of the equator. These lines, or parallels, are lines of **latitude** (**Figure 2**).

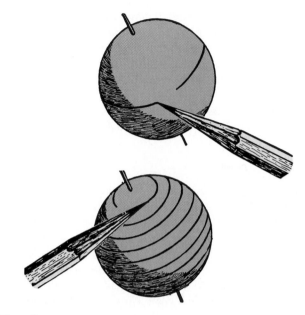

Figure 2

We can further subdivide the Earth by drawing lines at regular intervals through both poles. Note that unlike the parallels, these lines, called meridians, are all equally long. Meridians are lines of **longitude** (**Figure 3**).

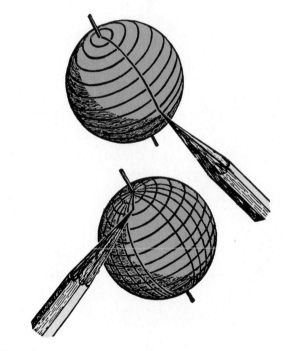

Figure 3

If you travel north from the equator, you can count the parallels (lines of latitude) that cross your path to find out how far you have gone. Likewise, if you travel east from a reference meridian, you can count the meridians (lines of longitude) that cross your path to find out how far you have gone. Just as a football player on the field knows his distance from the goal by the yard lines that cross his run, so you know how far north or east you have gone by the lines that have crossed your path.

Since there are no continuous lines of fence posts on the spherical Earth, our reference frame for latitude and longitude must be marked from the equator and poles by some other means. This is done by degrees.

Why Degrees?

Degrees measure fractions of a circle. We need to know what fraction of the Earth's circumference separates us from the equator and from the reference meridian to have a definite idea of our location.

Babylonian astronomers first divided the circle into 360 degrees (°). Why 360? The moon cycles around the Earth every 30 days. It takes about 12 months ("moonths") to make a year. Thus, 30 × 12 = 360, the number of days they supposed was in a year. Circles were divided the same way. As we saw in Chapter 2, the Greek librarian Hipparchus applied this division to the surface of the Earth.

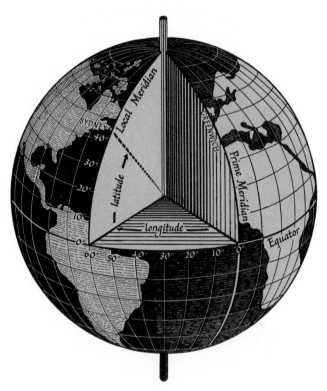

Figure 4 The latitude and longitude of Sydney, Canada.

In **Figure 4**, we have marked the position of Sydney, Canada. A line drawn from Sydney to the center of the Earth intersects the plane of the equator at an angle of 46° to the north. That is its latitude.

The reference meridian, the meridian from which all others are marked, is known as the prime meridian. Unlike the equator, there is no earthly reason why the prime meridian should pass through any particular place. It passes through Greenwich, England, because an international agreement signed in 1884 decreed it so. The meridian on which Sydney, Canada, lies intersects the plane of the prime meridian at an angle of 60°. The angular distance of Sydney from the prime meridian is 60° to the west. That is its longitude. So its position is 46°N 60°W.

We can do this for each hemisphere. A line drawn to the center of the Earth from Sydney, *Australia*, intersects the plane of the equator at an angle of 34° south latitude. Sydney, Australia, lies 151° east of the prime meridian. Thus, its position is 34°S 151°E. (Note that the greatest possible longitude is 180°; once you pass 180°, the line opposite the prime meridian, you begin to come around the other side of the Earth, and the angle to Greenwich decreases.)

What Does Time Have to Do with This?

Meridians are often numbered from the prime meridian in 15s. The Earth takes 24 hours to complete a 360° rotation. Divide 360° by 24 hours and you get 15, the number of degrees the sun moves across the sky in 1 hour. Meridians on a globe are often spaced to represent 1 hour's turning of the Earth toward or away from the sun, toward or away from noon.

You can use this fact to find your east-west position, your longitude. Imagine that you have a radio that can tell you the precise time of noon at Greenwich.[1] If your local noon comes *before* Greenwich noon, you are east of Greenwich. For instance, if the sun is highest in your sky at 10 A.M. Greenwich time, you are 2 hours before—30° east—of Greenwich. The Earth must turn 2 more hours before the sun will shine directly above the Greenwich meridian. If your local noon is *after* Greenwich noon, you are west of Greenwich. Suppose that the sun is at high noon and your chronometer, set at Greenwich time, says 6 P.M. That means that the Earth has been turning 6 hours since noon at Greenwich, and 6 hours times 15° per hour is 90°. That's your longitude relative to Greenwich: 90°W.

[1]Any shortwave radio will do. Tune it to 2.5, 5, 10, 15, or 20 mHz for radio stations WWV (Colorado) or WWVH (Hawaii). These stations broadcast time signals giving a measure of coordinated universal time, an international time standard based on the time at Greenwich. For a telephone report of coordinated universal time, call WWV at (303) 499-7111.

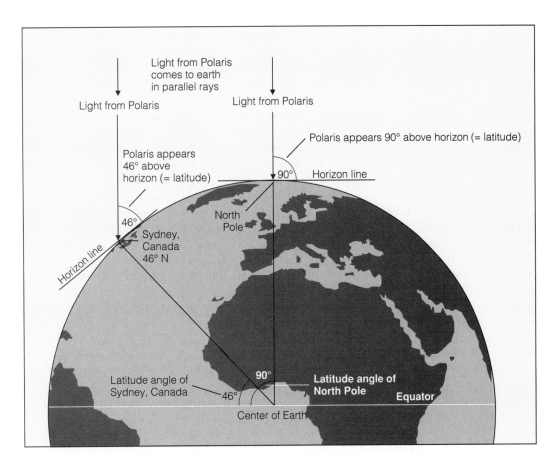

Navigation

Longitude is half the problem. To find latitude and obtain a position, we need to measure the angle north or south of the equator. But we can't use the time difference between local noon and Greenwich noon to determine longitude because the sun moves from east to west and we want to measure north-south position. Instead, we use the angle of the North Star above the horizon. Polaris, the current North Star, lies almost exactly above the North Pole. If we were standing at the North Pole, the North Star would appear almost directly overhead; ideally, the angle from the horizon to the star would be 90°, the same as the latitude of the North Pole (**Figure 5**). At Sydney, Canada, the angle from the horizon to the star would be about 46°—again, the same as the latitude. What would the angle of Polaris be at the equator, 0° latitude? (If you enjoyed your high school geometry course, you might try to prove that the angle from the horizon to Polaris is equal to the latitude at any position in the Northern Hemisphere.)

Polaris is not visible in the Southern Hemisphere, so how can we find south latitude? By finding the angle above the horizon of other stars. In practice, navigators in both hemispheres use a sextant to measure angles from the horizon to selected stars, planets, the moon, and the sun. The time of the observation is carefully noted. The navigator takes these readings to his or her stateroom, consults a series of mathematical tables, does some relatively simple calculations to compensate for observational errors, and comes up to the pilothouse with the vessel's latitude and longitude, accurate (in the best of circumstances) to within $\frac{1}{2}$ mile, marked on a small slip of paper. The daily results are always entered into the ship's log.

New Tricks

Discovering position by measuring the angular positions of heavenly bodies—celestial navigation—is a dying art. Global positioning satellites, loran-C, inertial platforms, radar, and other electronic wonders have largely replaced the romance of a navigator standing on the bridge squinting through a sextant. The slip of paper has been supplanted by the glow of back-lit liquid-crystal readouts or a chart with an X marking the ship's position, accurate to within 50 feet, feeding out of a slot. Still, when the power fails, the human navigator becomes the most popular person on board.

APPENDIX IV

Maps and Charts

It is easier to draw a diagram to show someone how to get to a place than to describe the process in words. For centuries, travelers have made special diagrams—maps and charts—to jog their own memories and to show others how to reach distant destinations. A **map** is a representation of some part of the Earth's surface, showing political boundaries, physical features, cities and towns, and other geographical information. A **chart** is also a representation of the Earth's surface, but it has been specially designed for convenient use in navigation. It is intended to be worked on, not merely looked at. A **nautical chart** is primarily concerned with navigable water areas. It includes information such as coastlines and harbors, channels, obstructions, currents, depths of water, and the positions of aids to navigation.

Any flat map or chart is necessarily a distortion of the spherical Earth. If we roll a flat sheet of paper around a globe to form a cylinder, the paper will contact the globe only along one curve. Let's assume that it's the equator. If

the lines of latitude and longitude on the globe are covered with ink, only the equator will contact the paper and print an exact replica of itself. Unroll the cylinder, and that part of the new map will be a perfect representation of the Earth. To include areas north and south of the equator, we will have to "throw them forward" onto the paper; we need to *project* them in some way.

Now imagine our globe to be a translucent sphere. If we place a bright light at its center, we can project the lines of latitude and longitude onto the rolled paper cylinder (**Figure 1**). Careful tracing of these lines will result in a map, but the areas away from the equator will be distorted: The farther from the equator, the greater the distortion. A useful modification of this projection—one that does not distort high latitudes as dramatically—was devised by Gerhardus Mercator, a Flemish cartographer who published a map of the world in 1569. Though landmasses and ocean areas are not depicted as accurately in a Mercator projection as they would be on a globe, such a map is still useful because it enables mariners to steer a course over long distances by plotting straight lines.

The distortion in Mercator projections has led generations of school children to believe that Greenland is the same size as South America (**Figure 2**). Mercator charts can distort our perceptions of the ocean as well: The area of the continental shelves at high latitudes, the amount of primary productivity in the polar regions, and the importance of ocean currents at the northerly or southerly extremes of an ocean basin may be exaggerated if presented in Mercator projection. The projection used in this

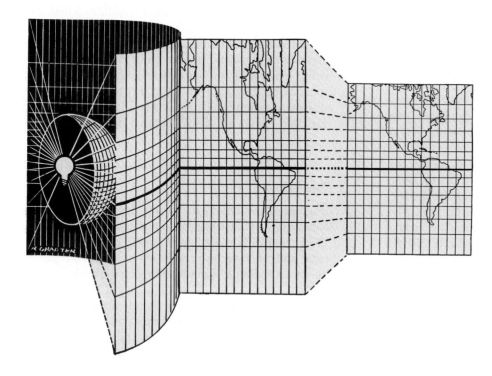

Figure 1 Central projection of a globe upon a cylinder, and a modified map structure, the Mercator, made to the same scale along the equator.

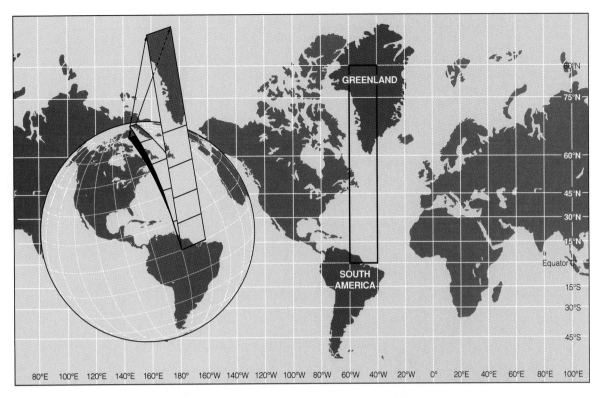

Figure 2 A gore of the globe peeled and projected according to the scheme devised by Gerhardus Mercator. This is the projection used on modern sailing charts. Note that this projection's distortion at high latitudes makes Greenland and South America appear about the same size. Next time you're near a globe, check their real sizes.

book—a further modification of the mercator projection known as the Miller projection—was chosen for its more accurate representation of surface area at high latitudes.

Mapmakers have invented other projections, each with advantages and disadvantages for particular uses. Some are conical projections: a flat sheet of paper wrapped into a cone with its edge touching the globe at a line of latitude north (or south) of the equator and the point of the cone above the North (or South) Pole. Conical projections do not distort high-latitude areas in the same way a Mercator projection does, and if drawn for the ocean area in which a mariner is sailing, can be used to draw great circle routes as straight lines. However, the distortions inherent in a conical projection prevent it from being used to represent more than about one-third of the globe on a single sheet of paper. Other projections, like the point-contact projection shown in **Figure 3**, try to minimize distortion around a specific location. All map and chart projections are distorted in some way; a sphere cannot be flattened onto a plane without deformation. Marine scientists necessarily become familiar with various chart projections and are careful to use the proper chart for its intended purpose.

Figure 4 is a Mercator projection of the world. On it are indicated areas of interest discussed in this book.

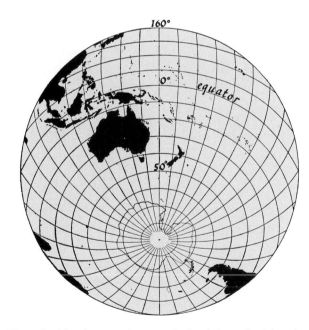

Figure 3 A Lambert equal-area projection (a type of point-contact projection), centered at 50°S, 160°E.

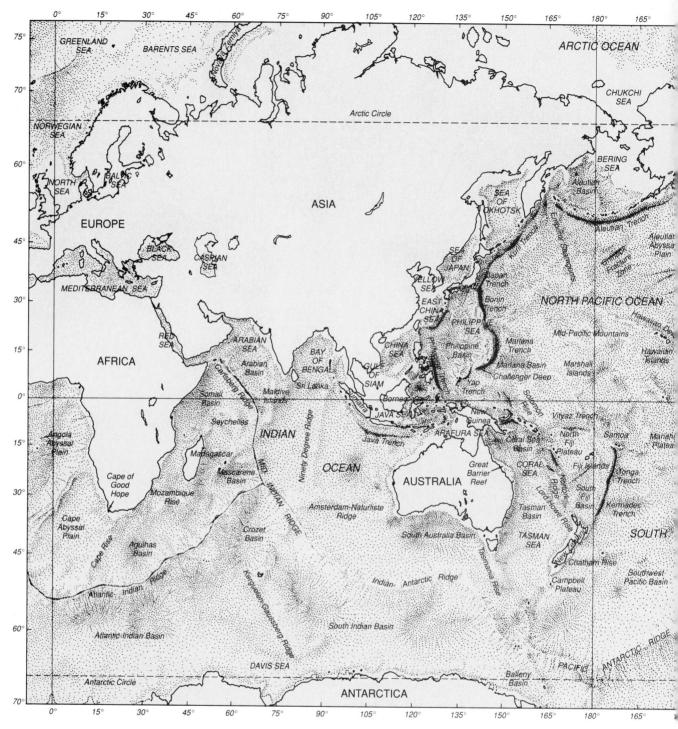

Figure 4 A map of the world, with many oceanic features labeled.

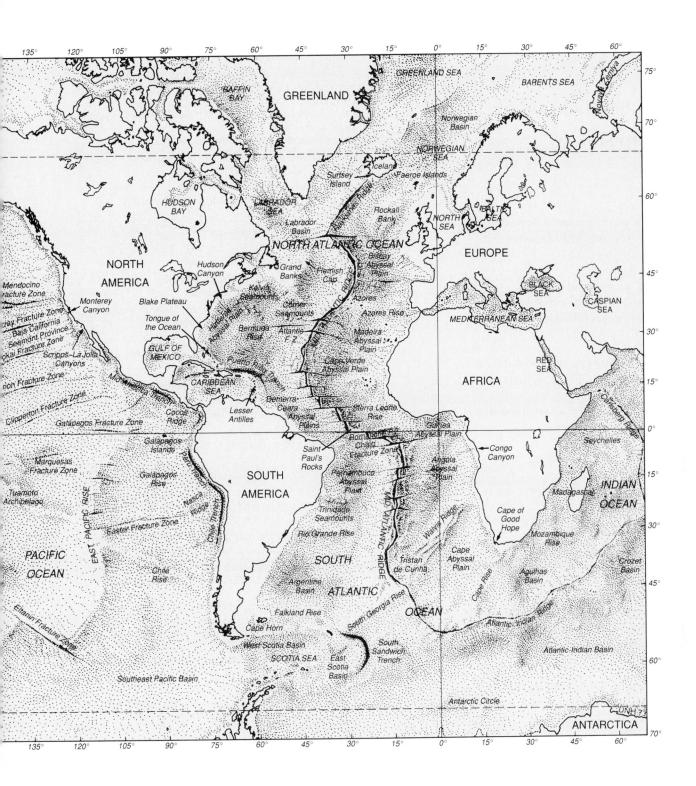

PACIFIC OCEAN

135° 120° 105° 90° 75° 60° 45° 30° 15° 0° 15° 30° 45° 60°

GREENLAND SEA
BARENTS SEA
BAFFIN BAY
GREENLAND
Novaya Zemlya
Norwegian Basin
NORWEGIAN SEA
Surtsey Island
Iceland
Faeroe Islands
HUDSON BAY
LABRADOR SEA
Labrador Basin
Rockall Bank
NORTH SEA
BALTIC SEA
NORTH ATLANTIC OCEAN
Biscay Abyssal Plain
EUROPE
Hudson Canyon
Grand Banks
Flemish Cap
Azores
NORTH AMERICA
Mendocino Fracture Zone
Monterey Canyon
Blake Plateau
Kelvin Seamounts
Corner Seamounts
Azores Rise
BLACK SEA
CASPIAN SEA
Murray Fracture Zone
Baja California
Seamont Province
Molokai Fracture Zone
Tongue of the Ocean
Hatteras Abyssal Plain
Bermuda Rise
Atlantis F.Z.
Madeira Abyssal Plain
MEDITERRANEAN SEA
RED SEA
Scripps–La Jolla Canyons
GULF OF MEXICO
Puerto Rico Trench
Cape Verde Abyssal Plain
AFRICA
Clarion Fracture Zone
Mid-America Trench
CARIBBEAN SEA
Lesser Antilles
Demerara–Ceara Abyssal Plains
Sierra Leone Rise
Clipperton Fracture Zone
Cocos Ridge
Galapagos Fracture Zone
Carlsberg Ridge
Galapagos Islands
Saint Paul's Rocks
Romanche Fracture Zone
Guinea Abyssal Plain
Congo Canyon
Seychelles
Marquesas Fracture Zone
Galapagos Rise
SOUTH AMERICA
Pernambuco Abyssal Plain
Angola Abyssal Plain
INDIAN OCEAN
Tuamoto Archipelago
Nazca Ridge
Trinidade Seamounts
MID-ATLANTIC RIDGE
Madagascar
Mozambique Rise
Easter Fracture Zone
Chile Trench
Rio Grande Rise
Walvis Ridge
Cape of Good Hope
Crozet Basin
PACIFIC OCEAN
Chile Rise
SOUTH ATLANTIC OCEAN
Tristan de Cunha
Cape Abyssal Plain
Agulhas Basin
Argentine Basin
Cape Rise
Atlantic-Indian Ridge
Atlantic-Indian Basin
Eltanin Fracture Zone
Falkland Rise
South Georgia Rise
Cape Horn
West Scotia Basin
SCOTIA SEA
East Scotia Basin
South Sandwich Trench
Southeast Pacific Basin
Antarctic Circle
UNH 73
ANTARCTICA

135° 120° 105° 90° 75° 60° 45° 30° 15° 0° 15° 30° 45° 60°

75° 70° 60° 45° 30° 15° 0° 15° 30° 45° 60° 70°

APPENDIX V

Taxonomic Classification of Marine Organisms

Exclusively nonmarine phyla generally have been omitted, along with most extinct phyla and classes.

KINGDOM MONERA. Bacteria and cyanobacteria, either single cells or simple associations of cells; autotrophic and heterotrophic forms.

KINGDOM PROTISTA. Eukaryotic single-celled forms.

PHYLUM CHRYSOPHYTA. Diatoms, coccolithophores, silicoflagellates.

PHYLUM PYRROPHYTA. Dinoflagellates, zooxanthellae.

PHYLUM CRYPTOPHYTA. Some "microflagellates"; cryptomonads.

PHYLUM EUGLENOPHYTA. A few "microflagellates"; mostly freshwater.

PHYLUM ZOOMASTIGINA. Nonphotosynthesizing flagellated protozoa.

PHYLUM SARCODINA. Amoebas and their relatives.
Class Rhizopodea. Foraminiferans.
Class Actinopodea. Radiolarians.

PHYLUM CILIOPHORA. Ciliated protozoa.

KINGDOM FUNGI. Fungi, mushrooms, molds, lichens; mostly land, freshwater, or highest supratidal organisms; heterotrophic.

KINGDOM PLANTAE. Photosynthetic autotrophs.

DIVISION CHLOROPHYTA. Multicellular green algae.

DIVISION PHAEOPHYTA. Brown algae, kelps.

DIVISION RHODOPHYTA. Red algae, encrusting and coralline forms.

DIVISION ANTHOPHYTA. Flowering plants (angiosperms). Most species are freshwater or terrestrial. Marine eelgrass, manatee grass, surfgrass, turtle grass, salt marsh grasses, mangroves.

KINGDOM ANIMALIA. Multicellular heterotrophs.

PHYLUM PLACOZOA. Amoeba-like multicellular animals.

PHYLUM MESOZOA. Wormlike parasites of cephalopods.

PHYLUM PORIFERA. Sponges.

PHYLUM CNIDARIA. Jellyfish and their kin; all are equipped with stinging cells.
Class Hydrozoa. Polyplike animals that often have a medusalike stage in their life cycle, such as Portuguese man-of-war.
Class Scyphozoa. Jellyfish with no (or reduced) polyp stage in life cycle.
Class Anthozoa. Sea anemones, coral.

PHYLUM CTENOPHORA. "Sea gooseberries," comb jellies; round, gelatinous, predatory, common.

PHYLUM PLATYHELMINTHES. Flatworms, tapeworms, flukes; many free-living predatory forms, many parasites.

PHYLUM NEMERTEA. Ribbon worms.

PHYLUM GNATHOSTOMULIDA. Microscopic, wormlike; live between grains in marine sediments.

PHYLUM GASTROTRICHA. Microscopic, ciliated; live between grains in marine sediments.

PHYLUM ROTIFERA. Ciliated; common in fresh water, plankton, and attached to benthic objects.

PHYLUM KINORYNCHA. Small, spiny, segmented, wormlike; live between grains in marine sediments; all marine.

PHYLUM ACANTHOCEPHALA. Spiny-headed worms; all parasitic in vertebrate intestines.

PHYLUM ENTOPROCTA. Polyplike, small, benthic suspension feeders.

PHYLUM NEMATODA. Roundworms. Common, free-living, parasitic.

PHYLUM BRYOZOA. Common, small, encrusting colonial marine forms.

PHYLUM PHORONIDA. Shallow-water tube worms; suspension feeders; a few centimeters long; all marine.

PHYLUM BRACHIOPODA. Lampshells; bivalved animals, superficially like clams; scarce, mainly in deep water.

PHYLUM MOLLUSCA. Mollusks.
Class Monoplacophora. Rare deep-water forms with limpetlike shells.
Class Polyplacophora. Chitons.
Class Aplacophora. Tusk shells, sand burrowing.
Class Gastropoda. Snails, limpets, abalones, sea slugs, pteropods.
Class Bivalvia. Clams, oysters, scallops, mussels.
Class Cephalopoda. Squid, octopuses.

PHYLUM ARTHROPODA.
Subphylum Crustacea. Copepods, barnacles, krill, isopods, amphipods, shrimp, lobsters, crabs.
Subphylum Chelicerata. Horseshoe crabs, sea spiders.
Subphylum Uniramia. Insects, centipedes, millipedes; one genus and five species in the ocean.

PHYLUM PRIAPULIDA. Small, rare, wormlike, subtidal.

PHYLUM SIPUNCULA. Peanut worms; all marine.

PHYLUM ECHIURA. Spoon worms.

PHYLUM ANNELIDA. Segmented worms; includes polychaetes such as feather duster worms.

PHYLUM TARDIGRADA. "Water bears"; tiny, eight-legged animals with the ability to survive long periods of hibernation.

PHYLUM PENTASTOMA. Tongue worms; parasites of vertebrates.

PHYLUM POGONOPHORA. Beard worms; no digestive system; deep-water tube worms; all marine.

PHYLUM ECHINODERMATA. Spiny-skinned, benthic, radially symmetrical, most with a water-vascular system.

Class Asteroidea. Sea stars.
Class Ophiuroidea. Brittle stars, basket stars.
Class Echinoidea. Sea urchins, sand dollars, sea biscuits.
Class Holothuroidea. Sea cucumbers.
Class Crinoidea. Sea lilies.

PHYLUM CHAETOGNATHA. Arrowworms; stiff-bodied, planktonic, predaceous, common.

PHYLUM HEMICHORDATA. Acorn worms; unsegmented burrowers.

PHYLUM CHORDATA.

Subphylum Urochordata. Sea squirts, tunicates, salps.
Subphylum Cephalochordata. Lancelets, *Amphioxus*.
Subphylum Vertebrata.

Class Agnatha. Jawless fishes: lampreys, hagfishes; cartilaginous skeleton.

Class Chondrichthyes. Sharks, skates, rays, sawfish, chimeras; cartilaginous skeleton.
Class Osteichthyes. Bony fishes.
Class Amphibia. Frogs, toads, salamanders; no marine species.
Class Reptilia. Sea snakes, turtles, one species of crocodile.
Class Aves. The birds.

Order Sphenisciformes. Penguins.
Order Procellariiformes. Albatrosses, petrels.
Order Charadriiformes. The gulls.
Order Pelecaniformes. The pelicans.

Class Mammalia. Warm-blooded, with hair and mammary glands.

Order Cetacea. Whales, porpoises, dolphins.
Order Sirenia. Manatees.
Order Carnivora. Two marine families.

Suborder Pinnipedia. Seals, sea lions, walruses.
Suborder Fissipedia. Sea Otters.

Order Primates. One family that regularly enters the ocean.

Family Hominidae. Humans.

Working in Marine Science

Working in the marine sciences is wonderfully appealing to many people. They sometimes envision a life of diving in warm, clear water surrounded by tropical fish, or descending to the seabed in an exotic submersible outfitted like Captain Nemo's fictional submarine in *20,000 Leagues Under the Sea*, or living with intelligent dolphins in a marine life park. Then reality sets in. There are rewards from working in the marine sciences, but they tend to be less spectacular than the first dreams of students looking to the ocean for a life's work.

A marine science worker is paid to bring a specific skill to a problem. If that problem lies in warm, tropical water or in a marine park, fine. But more likely, the problem will yield only to prolonged study in an uncomfortable, cold, or dangerous environment. The intangible rewards can be great; the physical rewards are often slim. Having said that, let me add that no endeavor is more interesting or exciting, and few are more intellectually stimulating. Doing marine science is its own reward.

Training for a Job in Marine Science

Marine science is, of course, science. And science requires mathematics—you need math to do the chemistry, physics, measurements, and statistics that lie at the heart of science. Your first step in college should be to take a math placement test, enroll in an appropriate math class, and spend time doing math. *Math is the key to further progress in any area of marine science.*

With your math skills polished, start classes in chemistry, physics, and basic biology. Surprisingly, except for one or two introductory marine science classes, you probably won't take many marine science courses until your junior year. These introductory classes will be especially valuable because a balanced survey of the marine sciences can aid you in selecting an appealing specialty. Then, with a good foundation in basic science, you can begin to concentrate in that specialty.

Other skills are important too. The ability to write and speak well is crucial in any science job. Also critical is computer literacy—preferably DOS, not Macintosh, by the way. Expertise in photography or foreign languages

or the ability to field-strip and rebuild a diesel engine or hydraulic winch will put you a step above the competition at hiring time. Certification as a scuba diver is almost mandatory; you can never have too much diving experience. (Remember, though, diving is only a tool, a way to deliver an informed set of eyes and an educated brain to a work site.) You should be in good health. Indeed, good aerobic fitness is essential in most marine science jobs; stamina is often a crucial factor in long experiments under difficult conditions at sea. It is also desirable to be physically strong—marine equipment is heavy and often bunglesome. And it helps greatly if you are not prone to seasickness.

Deciding what school to attend will depend on your skills. Readers of this book will probably be enrolled in a general oceanography course in a college or university. The first step would be to discuss your interests with your professor (or his or her teaching assistants). You'll need to attend a four-year college or university to complete the first phase of your training. If you're attending a two-year institution, picking a specific transfer institution can come later, but keep a few things in mind: No matter where you take your first two years of training, you need thorough preparation in basic science. You should attend an institution with strengths in the area of your specialty (such as geology, biology, and marine chemistry). And you should be reasonable in your expectations of acceptance if you're a transfer student (that is, don't try for Stanford or Yale with a B average).

Another thing: Most marine scientists have completed a graduate degree (a master's degree or doctorate). Most graduate students hold teaching or research assistantships (that is, they get paid for being grad students). In all, progress to a final degree is a long road, but the journey is itself a pleasure.

If the thought of four or more years of higher education doesn't appeal, does that mean there's no hope? Not at all. Many students begin a program with the goal of becoming a marine technician, animal trainer at a marine life park, marina or boatyard employee or manager, or crew member on a private yacht. Those jobs don't always require a bachelor's degree. Jobs at Sea World and other marine theme parks do require athletic ability, extreme patience, public speaking skills, a love of animals, and, usually, diving experience. Few positions are available, but there is some turnover in the ranks of junior trainers, and being hired is certainly possible.

Becoming a marine technician is an especially attractive alternative to the all-out chemistry-physics-math academic route. For every highly trained marine scientist, there are perhaps five technical assistants who actually do the experiments, maintain the equipment, work daily with organisms, and build special apparatus. Marine technicians tend to spend more time at hands-on tasks than marine scientists. Most of these folks (including the author of the letter that ends this appendix) have the equivalent of a two-year technical degree, usually from a community college.

Don't quit your job, burn your bridges, leave your family, sell your possessions, and dedicate yourself monklike to marine science. Do some investigation. Nothing is as valuable as *actually going out and talking to people who do things that you'd like to do*. Ask them if they enjoy their work. Is the pay okay? Would they start down the same road if they had it to do all over again? You may decide to expand your involvement in marine science in a more informal way, by becoming a volunteer; joining the Sierra Club, Audubon Society, Greenpeace, or other environmental group; working for your state's fish and game office as a seasonal aide; or attending lectures at local colleges and universities.

If you decide to continue your education, don't be discouraged by the time it will take. Have a general view of the big picture, but proceed one semester at a time. Again, remember that the educational journey is itself a great pleasure. Don Quixote reminds us of the joys of the road, not the inn.

The Job Market

Marine science is very attractive to the general public. People are naturally drawn to thoughts of working in the field. Unfortunately, there aren't a great many jobs in the marine sciences. But there will always be some jobs, and people will fill them. Those people will be the best prepared, most versatile, and most highly motivated of those who apply. Perhaps not surprisingly, marine biology is the most popular marine science specialty. Unfortunately, it is also the area with the smallest number of nonacademic jobs. Museums, aquariums, and marine theme parks employ biologists to care for animals and oversee interpretive programs for the public. A few marine biologists are employed as monitoring specialists by water management agencies like sanitation districts, which discharge waste into the ocean. Electrical utilities that use seawater to cool the condensers in power generating plants almost always have a handful of marine biologists on staff to watch the effects of discharged heat on local marine life and to write the reports required by watchdog agencies. State and federal agencies employ marine biologists to read and interpret those documents and to set standards. Relatively small businesses, like private shipyards, agricultural concerns, and chemical plants, can't afford their own staff biologists, so private consulting firms staffed by marine biologists and other specialists have arisen to assist in the preparation of the environmental impact reports required of businesses under various legislation.

There are more jobs in physical oceanography: marine geology, ocean engineering, and marine chemistry and physics. Thousands of marine geologists work for oil and mineral companies; indeed, with the increasing emphasis on offshore resources, the market for these people may be increasing. Marine engineers are needed to design, construct, and maintain offshore oil rigs, ships, and harbor structures. Marine chemists are hard at work figuring ways to stop corrosion and to extract chemicals from seawater. Physicists are vitally interested in the transmission of underwater sound and light, in the movement of the ocean, and in the role the ocean plays in global weather and climate. Economists, lawyers, writers, and mathematicians also work in the marine science field.

Many biological and physical oceanographers are teachers and professors. Indeed, there are nearly as many marine scientists employed in the academic world as there are in private industry and government. If you like the idea of teaching, you might consider this avenue. The demand for science teachers at all educational levels is already great and is expected to increase.

Four factors will be significant in influencing your employability:

1. *Experience*. Employers are favorably impressed by experience, especially work experience related to the duties of the position for which you are applying. Volunteer work counts.

2. *Grades*. Good grades are important, especially for positions in government agencies. A grade point average of 3.0 or higher in all college work increases your chances of employment and should give you a higher starting salary.

3. *Geographical availability*. Don't restrict yourself geographically. Not everyone can work in Hawaii or California, but four out of ten marine scientists work in just three states: California, Maryland, and Virginia.

4. *Diversity*. Again, mastery of more than one specialty gives you an employment edge. Being a plankton connoisseur *and* being able to repair a balky computer while ordering in-port supplies over a radiotelephone in Spanish makes a lasting impression.

Report from a Student

Students in marine science programs graduate, get jobs, and move on. One of the pleasures of being a professor is hearing from them. One of our former students, an employee of the Marine Science Institute at the University of California, Santa Barbara, recently reported his activities as part of a team using the submersible *Alvin* to

Figure 1 Dan Dion attaching hydraulic actuators to water-sampling bottles, in preparation for a dive by *Alvin*. The bottles were part of a sampling program that included measurements of water conductivity, transmissivity, temperature, and iron and manganese ion content near a hydrothermal vent. This information was later merged with data from transponder navigation to obtain a three-dimensional map of plume structure and chemistry.

investigate plumes of warm water issuing from hydrothermal vents along the southern Juan de Fuca Ridge. The nature of his work—and his enthusiasm for it—is clearly evident in this excerpt. Dan Dion writes:

The buoyant plume experiment wasn't going very well. The chemistry dives were pushed back because of technical difficulties and poor weather (rough seas cut two dives). The first two buoyant plume dives ended in failure. The first one because of mechanical/electrical problems, the second because of a computer crash. Everyone worked around the clock to get things in order for dive 2440. I was scheduled to go down with John Trefrey, from the Florida Institute of Technology. Cindy Van Dover was our pilot. We launched *Alvin* right on schedule at 0800, and descended from the glacier blue water into the bioluminescent snowstorm of the euphotic zone. During the hour and a half descent we listened to the music of Enya in the soft light of the sub as we busily prepared ourselves for the experiment: booting up the computers, loading film in the cameras, tapes in the recorders, etc. We had three laptop computers to deal with in the cramped spaces of the sub. I was in charge of two of them, one that plotted our in-sub navigation (from transponders), and the other that controlled

and recorded data from the [continuous temperature-depth-conductivity probe]. The third laptop was connected to the chemical analyzer and John was in control of that. All of the instruments were operating perfectly. I periodically saved the computer file in the event of another crash. We reached the bottom right on target; Monolith Vent was in sight, 2261 meters below the surface. We did a video survey of the vent, especially a chimney that was rapidly growing back after the geologists had decapitated it just a few days earlier. We ascended to 55 meters-off-bottom and began our drive-throughs. To me, the navigator, it was the ultimate video game. From the computer screen I would guide *Alvin* through a dark abyss, calling out headings that would maneuver us into a "lawn mower" pattern crisscrossing the plume. It was quite visible, and even beautiful; wispy, intricate patterns of "smoke" which seemed to dance like graceful ghosts. We completed passes at 35, 20, 10, and 5 meters above the bottom, then one last one at 45 meters. Eight hours of sub time went by so quickly! Our dive was a huge success; in addition to all the samples we obtained, we generated over 25 megabytes of data. I used everything I learned . . . from computer skills to navigation and marlinspike seamanship (and, of course, chemistry!).

Marine science is equipment training sessions, long cruises, seminars and lectures, visiting experts, hot sand volleyball games, and chilly labs with classical music. Marine science is a long and demanding road; but it is, quite honestly, great fun. Captain Nemo in his sub never had it this good!

For More Information

1. Consult the catalog of any college or university offering a marine science curriculum.

2. Send for a copy of *Ocean Opportunities, a Guide to What the Oceans Have to Offer*. The pamphlet is available from The Marine Technology Society, 1730 M Street, NW, Washington, DC 20036.

3. See: Wunsch, Carl. 1993. "Marine Science in the Coming Decades." *Science*, January 15, vol. 259, no. 5093:296–297. An expert oceanographer's glimpse into the future.

GLOSSARY

absorption Conversion of sound or light energy into heat.

abyssal hill Small sediment-covered inactive volcano or intrusion of molten rock less than 200 meters (650 feet) high, thought to be associated with seafloor spreading. Abyssal hills punctuate the otherwise flat abyssal plain.

abyssal plain Flat, cold, sediment-covered ocean floor between the continental rise and the oceanic ridge at a depth of 3,700 to 5,500 meters (12,000 to 18,000 feet). Abyssal plains are more extensive in the Atlantic and Indian oceans than in the Pacific.

abyssal zone The ocean between about 4,000 and 5,000 meters (13,000 and 16,500 feet) deep.

accessory pigment One of a class of pigments (such as fucoxanthin, phycobilin, and xanthophyll) present in various photosynthetic plants and which assist in the absorption of light and the transfer of its energy to chlorophyll. Also called masking pigment.

accretion An increase in the mass of a body by accumulation or clumping of smaller particles.

acid A substance that releases a hydrogen ion (H^+) in solution.

acid rain Rain containing acids and acid-forming compounds such as sulfur dioxide and oxides of nitrogen.

acoustical tomography A technique for studying ocean structure that depends on pulses of low-frequency sound to sense differences in water temperature, salinity, and movement beneath the surface.

active margin Continental margin near an area of lithospheric plate convergence. Also called Pacific-type margin.

active sonar A device that generates underwater sound from special transducers and analyzes the returning echoes to gain information of geological, biological, or military importance.

active transport The movement of molecules from a region of low concentration to a region of high concentration through a semi-permeable membrane at the expense of energy.

adaptation An inheritable structural or behavioral modification. A favorable adaptation gives a species an advantage in survival and reproduction. An unfavorable adaptation lessens a species' ability to survive and reproduce.

adhesion Attachment of water molecules to other substances by hydrogen bonds. Wetting.

Agnatha The class of jawless fishes: hagfishes and lampreys.

ahermatypic Describing coral species lacking symbiotic zooxanthellae and incapable of secreting calcium carbonate at a rate suitable for reef production.

air mass A large mass of air with nearly uniform temperature, humidity, and density throughout.

algae Collective term for nonvascular plants possessing chlorophyll and capable of photosynthesis. (Singular, alga.)

algin A mucilaginous commercial product of multicellular marine algae. Widely used as a thickening and emulsifying agent.

alkaline Basic. *See* **base**.

alternation of generations A reproductive cycle in which a plant alternates between sexual and asexual stages.

amphidromic point A "no-tide" point in an ocean caused by basin resonances, friction, and other factors around which tide crests rotate. About a dozen amphidromic points exist in the world ocean. Sometimes called a node.

angiosperm A flowering vascular plant that reproduces by means of a seed-bearing fruit. Examples are sea grasses and mangroves.

angle of incidence In meteorology, the angle of the sun above the horizon.

animal A multicellular organism unable to synthesize its own food and often capable of movement.

Animalia The kingdom to which multicellular heterotrophs belong.

Annelida The phylum of animals to which segmented worms belong.

Antarctic Bottom Water The densest ocean water (1.0279 g/cm^3), formed primarily in Antarctica's Weddell Sea during Southern Hemisphere winters.

Antarctic Circle The imaginary line around the Earth, parallel to the equator at 66°33'S, marking the southernmost limit of sunlight at the June solstice. The Antarctic Circle marks the northern limit of the area within which, for one day or more each year, the sun does not set (around 21 December) or rise (around 21 June).

Antarctic Convergence Convergence zone encircling Antarctica between about 50° and 60°S, marking the boundary between Antarctic Circumpolar Water and Subantarctic Surface Water.

Antarctic Ocean An ocean in the Southern Hemisphere bounded to the north by the Antarctic Convergence and to the south by Antarctica.

aphotic zone The dark ocean below the depth to which light can penetrate.

aquaculture The growing or farming of plants and animals in a water environment under controlled conditions. *Compare* **mariculture**.

Arctic Circle The imaginary line around the Earth, parallel to the equator at 66°33'N, marking the northernmost limit of sunlight at the December solstice. The Arctic Circle marks the southern limit of the area within which, for one day or more each year, the sun does not set (around 21 June) or rise (around 21 December).

Arctic Convergence Convergence zone between Arctic Water and Subarctic Surface Water.

Arctic Ocean An ice-covered ocean north of the continents of North America and Eurasia.

Arthropoda The phylum of animals that includes shrimp, lobsters, krill, barnacles, and insects. The phylum Arthropoda is the world's most successful.

artificial system of classification A method of classifying an object based on attributes other than its reason for existence, its ancestry, or its origin. *Compare* **natural system of classification**.

Asteroidea The class of the phylum Echinodermata to which sea stars belong.

asthenosphere The hot plastic layer of the upper mantle below the lithosphere, extending some 700 kilometers (430 miles) below the surface. Convection currents within the asthenosphere power plate tectonics.

atmospheric circulation cell Large circuit of air driven by uneven solar heating and the Coriolis effect. Three circulation cells form in each hemisphere. *See also* **Hadley cell**; **Ferrel cell**; **polar cell**.

atoll A ring-shaped island of coral reefs and coral debris enclosing, or almost enclosing, a shallow lagoon from which no land protrudes. Atolls often form over sinking, inactive volcanoes.

atom The smallest particle of an element that exhibits the characteristics of that element.

ATP Adenosine triphosphate, the compound that acts as the immediate source of energy for all life on Earth. The energy stored in ATP is provided directly by photosynthesis or by respiration of glucose.

authigenic sediment Sediment formed directly by precipitation from seawater. Also called *hydrogenous sediment*.

autotroph An organism that makes its own food by photosynthesis or chemosynthesis.

auxospore A naked diatom cell without valves. Often a dormant stage in the life cycle following sexual reproduction.

Aves The class of birds.

backshore Sand on the shoreward side of the berm crest, sloping away from the ocean.

backwash Water returning to the ocean from waves washing onto a beach.

baleen The interleaved, hard, fibrous, hornlike filters within the mouth of baleen whales.

barrier island A long, narrow, wave-built island lying parallel to the mainland and separated from it by a lagoon or bay. *Compare* **sea island**.

barrier reef A coral reef surrounding an island or lying parallel to the shore of a continent, separated from land by a deep lagoon. Coral debris islands may form along the reef.

basalt The relatively heavy crustal rock that forms the seabeds, composed mostly of oxygen, silicon, magnesium, and iron. Its density is about 2.9 g/cm^3.

base A substance that combines with a hydrogen ion (H$^+$) in solution.

bathyal zone The ocean between about 200 and 4,000 meters (700 and 13,000 feet) deep.

bathyscaphe Deep-diving submersible designed like a blimp, which uses gasoline for buoyancy and can reach the bottom of the deepest ocean trenches. From the Greek *batheos* ("depth,") and *skaphidion* ("a small ship").

bay mouth bar An exposed sand bar attached to a headland adjacent to a bay and extending across the mouth of the bay.

beach A zone of unconsolidated (loose) particles extending from below water level to the edge of the coastal zone.

beach scarp Vertical wall of variable height marking the landward limit of the most recent high tides. Corresponds with the berm at extreme high tides.

benthic zone The zone of the ocean bottom. *See also* **pelagic zone**.

berm A nearly horizontal accumulation of sediment parallel to shore. Marks the normal limit of sand deposition by wave action.

berm crest The top of the berm; the highest point on most beaches. Corresponds to the shoreward limit of wave action during most high tides.

big bang The hypothetical event that started the expansion of the universe from a geometric point. The beginning of time.

bilateral symmetry Body structure having left and right sides that are approximate mirror images of each other. Examples are crabs and humans. *Compare* **radial symmetry**.

biodegradable Able to be broken by natural processes into simpler compounds.

biogenous sediment Sediment of biological origin. Organisms can deposit calcareous (calcium-containing) or siliceous (silicon-containing) residue.

biological amplification Increase in concentration of certain fat-soluble chemicals such as DDT or heavy-metal compounds, in successively higher trophic levels within a food web.

biological factor A biologically generated aspect of the environment, such as predation or metabolic waste products, that affects living organisms. Biological factors usually operate in association with purely physical factors such as light and temperature.

biological resource A living animal or plant collected for human use. Also called a *living resource*.

bioluminescence Biologically produced light.

biomass The mass of living material in a given area or volume of habitat.

biosynthesis The initial formation of life on the Earth.

Bivalvia The class of the phylum Mollusca that includes clams, oysters, and mussels.

Bjerknes, Vilhelm (1862–1951) Pioneering Norwegian physicist and discoverer of the nature and formation of extratropical cyclones, which cause most mid-latitude weather.

blade Algal equivalent of a vascular plant's leaf. Also called a *frond*.

bond *See* **chemical bond**.

brackish Describing water intermediate in salinity between seawater and fresh water.

breakwater An artificial structure of durable material that interrupts the progress of waves to shore. Harbors are often shielded by a breakwater.

buffer A group of substances that tends to resist change in the pH of a solution by combining with free ions.

buoyancy The ability of an object to float in a fluid by displacement of a volume of fluid equal in mass to the mass of the floating object.

calcareous ooze Ooze composed mostly of the hard remains of calcium-carbonate-containing organisms.

calcium carbonate compensation depth The depth at which the rate of accumulation of calcareous sediments equals the rate of dissolution of those sediments. Below this depth, sediment contains little or no calcium carbonate.

calorie The amount of heat needed to raise the temperature of 1 gram (0.035 ounce) of pure water by 1°C (1.8°F).

capillary wave A tiny wave with a wavelength of less than 1.73 centimeters (0.68 inch), whose restoring force is surface tension; the first type of wave to form when the wind blows.

Carnivora The order of mammals that includes seals, sea lions, walruses, and sea otters.

carrying capacity The size at which a particular population in a particular environment will stabilize when its supply of resources—including nutrients, energy, and living space—remains constant.

cartilage A tough, elastic tissue that stiffens or supports.

cartographer A person who makes maps and charts.

catastrophism The theory that the Earth's surface features are formed by catastrophic forces such as the biblical flood. Catastrophists believe in a young Earth and a literal interpretation of the biblical account of Creation.

celestial navigation The technique of finding one's position on Earth by reference to the apparent positions of stars, planets, the moon, and the sun.

cell The basic organizational unit of life on this planet.

Cephalopoda The class of the phylum Mollusca that includes squid, octopuses, and nautiluses.

Cetacea The order of mammals that includes porpoises, dolphins, and whales.

CFCs *See* **chlorofluorocarbons**.

Challenger **Expedition** The first wholly scientific oceanographic expedition, 1872–76. Named for the steam corvette used in the voyage.

chart A map that depicts mostly water and the adjoining land areas.

chemical bond An energy relationship that holds two atoms together as a result of changes in their electron distribution.

chemical equilibrium In seawater, the condition in which the proportion and amounts of dissolved salts per unit volume of ocean are nearly constant.

chemosynthesis The synthesis of organic compounds from inorganic compounds using energy stored in inorganic substances such as sulfur, ammonia, and hydrogen. Energy is released when these substances are oxidized by certain organisms.

chitin A complex nitrogen-rich carbohydrate from which parts of arthropod exoskeletons are constructed.

chiton A polyplacophoran mollusk.

chlorinated hydrocarbons The most abundant and dangerous class of halogenated hydrocarbons, synthetic organic chemicals hazardous to the marine environment.

chlorinity A measure of the content of chloride, bromine, and iodide ions in seawater. We may derive salinity from chlorinity by multiplying by 1.80655.

chlorofluorocarbons (CFCs) A class of halogenated hydrocarbons thought to be depleting the Earth's atmospheric ozone. CFCs are used as cleaning agents, refrigerants, fire-extinguishing fluids, spray-can propellants, and insulating foams.

chlorophyll A pigment responsible for trapping sunlight and transferring its energy to electrons, thus initiating photosynthesis.

Chlorophyta Green algae.

Chondrichthyes The class of fishes with cartilaginous skeletons: the sharks, skates, rays, and chimaeras.

Chordata The phylum of animals to which tunicates, *Amphioxus*, fishes, amphibians, reptiles, birds, and mammals belong.

chromatophore A pigmented skin cell that expands or contracts to affect color change.

chronometer A very consistent clock. It doesn't need to tell accurate time, but its rate of gain or loss must be constant and known exactly so that accurate time may be calculated.

clamshell sampler Sampling device used to take shallow samples of the ocean bottom.

classification A way of grouping objects according to some stated criteria.

clay Sediment particle smaller than 0.004 millimeter in diameter; the smallest sediment size category.

climate The long-term average of weather in an area.

climax community A stable, long-established community of self-perpetuating organisms that tends not to change with time.

clockwise Rotation around a point in the direction that clock hands move.

clumped distribution Distribution of organisms within a community in small, patchy aggregations, or clumps; the most common distribution pattern.

Cnidaria The phylum of animals to which corals, jellyfish, and sea anemones belong.

cnidoblast Type of cell found in members of the phylum Cnidaria that contains a stinging capsule. The threads that evert from the capsules assist in capturing prey and repelling aggressors.

coast The zone extending from the ocean inland as far as the environment is immediately affected by marine processes.

coastal cell The natural sector of a coastline in which sand input and sand outflow are balanced.

coastal upwelling Upwelling adjacent to a coast, usually induced by wind.

coccolithophore A very small planktonic alga carrying discs of calcium carbonate, which contributes to biogenous sediments.

cohesion Attachment of water molecules to each other by hydrogen bonds.

colligative properties Those characteristics of a solution that differ from those of pure water because of material held in solution.

Columbus, Christopher (1451–1506) Italian explorer in the service of Spain who discovered islands in the Caribbean in 1492. Although traditionally credited as the discoverer of America, he never actually sighted the North American continent.

commensalism A symbiotic interaction between two species in which only one species benefits and neither is harmed.

commercial extinction Depletion of a resource species to a point where it is no longer profitable to harvest the species.

community The populations of all species that occupy a particular habitat and interact within that habitat.

compass An instrument for showing direction by means of a magnetic needle swinging freely on a pivot and pointing to magnetic north.

compensation depth The depth in the water column at which the production of carbohydrates and oxygen by photosynthesis exactly equals the consumption of carbohydrates and oxygen by respiration. The break-even point for autotrophs. Generally a function of light level.

compound A substance composed of two or more elements in a fixed proportion.

condensation theory Premise that stars and planets accumulate from contracting, accreting clouds of galactic gas, dust, and debris.

conduction The transfer of heat through matter by the collision of one atom with another.

conservative constituent An element that occurs in constant proportion in seawater. For example, chlorine, sodium, and magnesium.

constructive interference The addition of wave energy as waves interact, producing larger waves.

consumer A heterotrophic organism.

continental crust The solid masses of the continents, composed primarily of granite.

continental drift The theory that the continents move slowly across the surface of the Earth.

continental margin The submerged outer edge of a continent, made of granitic crust. Includes the continental shelf and continental slope. *Compare* **ocean basin**.

continental rise The wedge of sediment forming the gentle transition from the outer (lower) edge of the continental slope to the abyssal plain. Usually associated with passive margins.

continental shelf Gradually sloping submerged extension of a continent, composed of granitic rock overlain by sediments. Has features similar to the edge of the nearby continent.

continental slope The sloping transition between the granite of the continent and the basalt of the seabed. The true edge of a continent.

contour current A bottom current made up of dense water that flows around (rather than over) seabed projections.

convection Movement within a fluid resulting from differential heating and cooling of the fluid. Convection produces mass transport or mixing of the fluid.

convection current A single closed-flow circuit of rising warm material and falling cool material.

convergence zone The line along which waters of different density converge. Convergence zones form the boundaries of tropical, subtropical, temperate, and polar areas.

convergent evolution The evolution of similar characteristics in organisms of different ancestry; the body shape of a porpoise and a shark, for instance.

convergent plate boundary A region where plates are pushing together and where a mountain range, island arc, and/or trench will eventually form. Often a site of much seismic and volcanic activity.

Cook, James (1728–1779) Officer in the British Royal Navy who led the first European voyages of scientific discovery.

coral Any of over 6,000 species of small cnidarians, many of which are capable of generating hard calcareous (aragonite, $CaCO_3$) skeletons.

coral reef A linear mass of calcium carbonate (aragonite and calcite) assembled from coral organisms, algae, mollusks, worms, and so on. Coral may contribute less than half of the reef material.

core The innermost layer of the Earth, composed primarily of iron, with nickel and heavy elements. The inner core is thought to be a solid 6,000°C (11,000°F) sphere, the outer core a 5,000°C (9,000°F) liquid mass. The average density of the outer core is about 11.8 g/cm^3, and that of the inner core is about 16 g/cm^3.

Coriolis, Gaspard Gustave de (1792–1843) The French scientist who in 1835 worked out the mathematics of the motion of bodies on a rotating surface. *See* **Coriolis effect**.

Coriolis effect The apparent deflection of a moving object from its initial course when its speed and direction are measured in refer-
ence to the surface of the rotating Earth. The object is deflected to the right of its anticipated course in the Northern Hemisphere and to the left in the Southern Hemisphere. The deflection occurs for any horizontal movement of objects with mass and has no effect at the equator.

cosmogenous sediment Sediment of extraterrestrial origin.

counterclockwise Rotation around a point in the direction opposite to that in which clock hands move. Also called *anticlockwise*.

countercurrent A surface current flowing in the opposite direction from an adjacent surface current.

covalent bond A chemical bond formed between two atoms by electron sharing.

crest *See* **wave crest**.

crust The outermost solid layer of the Earth, composed mostly of granite and basalt. The top of the lithosphere. The crust has a density of 2.7–2.9 g/cm^3 and accounts for 0.4% of the Earth's mass.

Crustacea The class of phylum Arthropoda to which lobsters, shrimp, crabs, barnacles, and copepods belong.

cryptic coloration Camouflage. May be active (under control of the animal) or passive (an unalterable color or shape).

current Mass flow of water. (The term is usually reserved for horizontal movement.)

cyclone A weather system with a low-pressure area in the center around which winds blow counterclockwise in the Northern Hemisphere and clockwise in the Southern Hemisphere. Not to be confused with a tornado, a much smaller weather phenomenon associated with severe thunderstorms. *See also* **tropical cyclone**; **extratropical cyclone**.

deep scattering layer (DSL) A relatively dense aggregation of fishes, squid, and other mesopelagic organisms capable of reflecting a sonar pulse that resembles a false bottom in the ocean. Its position varies with the time of day.

deep-water wave A wave in water deeper than one-half its wavelength.

deep zone The zone of the ocean below the pycnocline, in which there is little additional change of density with increasing depth. Contains about 80% of the world's water.

degree An arbitrary measure of temperature. One degree Celsius (°C) = 1.8 degrees Fahrenheit (°F).

delta The deposit of sediments found at a river mouth, sometimes triangular in shape (hence the name after the Greek letter).

density The mass per unit volume of a substance, usually expressed in grams per cubic centimeter (g/cm^3).

density curve A graph showing the relationship between a fluid's temperature or salinity and its density.

density stratification The formation of layers in a material, with each deeper layer being denser (weighing more per unit of volume) than the layer above.

dependency A feeding relationship in which an organism is limited to feeding on one species or, in extreme cases, on one size phase of one species.

deposition Accumulation, usually of sediments.

depositional coast A coast in which processes that deposit sediment exceed erosive processes.

desalination The process of removing salt from seawater or brackish water.

desiccation Drying.

destructive interference The subtraction of wave energy as waves interact, producing smaller waves.

diatom The Earth's most abundant, successful, and efficient single-celled phytoplankton. Diatoms possess two interlocking valves made primarily of silica. The valves contribute to biogenous sediments.

diffusion The movement—driven by heat—of molecules from a region of high concentration to a region of low concentration.

dinoflagellate One of a class of microscopic single-celled flagellates, not all of which are autotrophic. The outer covering is often of stiff cellulose. Planktonic dinoflagellates are responsible for "red tides."

dissolution The dissolving by water of minerals in rocks.

disthermal zone The zone of stable temperature below the thermocline.

disturbing force The energy that causes a wave to form.

diurnal tide A tidal cycle of one high tide and one low tide per day.

divergent evolution Evolutionary radiation of different species from a common ancestor.

divergent plate boundary A region where plates are moving apart and where new ocean or rift valley will eventually form. A spreading center forms the junction.

doldrums The zone of rising air near the equator known for sultry air and variable breezes. Also known as the *intertropical convergence zone*. *See also* **intertropical convergence zone** (ITCZ).

downwelling   Circulation pattern in which surface water moves vertically downward.

drag   The resistance to movement of an organism induced by the fluid through which it swims.

drift net   Fine vertically suspended net that may be 7 meters (25 feet) high and 80 kilometers (50 miles) long.

DSL   *See* **deep scattering layer**.

earthquake   A sudden motion of the Earth's crust resulting from waves in the Earth caused by faulting of the rocks or by volcanic activity.

eastern boundary current   Weak, cold, diffuse, slow-moving current at the eastern boundary of an ocean (off the west coast of a continent). Examples include the Canary Current and the Humboldt Current.

ebb current   Water rushing out of an enclosed harbor or bay because of the fall in sea level as a tide trough approaches.

Echinodermata   The phylum of exclusively marine animals to which sea stars, brittle stars, sea urchins, and sea cucumbers belong.

Echinoidea   The class of the phylum Echinodermata to which sea urchins and sand dollars belong.

echolocation   The use of reflected sound to detect environmental objects. Cetaceans use echolocation to detect prey and avoid obstacles.

echo sounder   A device that reflects sound off the ocean bottom to sense water depth. Its accuracy is affected by the variability of the speed of sound through water.

ecology   Study of the interactions of organisms with one another and with their environment.

ectoderm   The outermost layer of cells in a developing embryo.

ectotherm   An organism incapable of generating and maintaining steady internal temperature from metabolic heat and therefore whose internal body temperature is approximately the same as that of the surrounding environment. A cold-blooded organism.

eddy   A circular movement of water usually formed where currents pass obstructions, or between two adjacent currents flowing in opposite directions, or along the edge of a permanent current.

EEZ   *See* **exclusive economic zone**.

Ekman spiral   A theoretical model of the effect on water of wind blowing over the ocean. Because of the Coriolis effect, the surface layer is expected to drift at an angle of 45° to the right of the wind in the Northern Hemisphere and 45° to the left in the Southern Hemisphere. Water at successively lower layers drifts progressively to the right (N), or left (S), though not as swiftly as the surface flow.

Ekman transport   Net water transport, the sum of layer movement due to the Ekman spiral. Theoretical Ekman transport in the Northern Hemisphere is 90° to the right of the wind direction.

electron   A tiny negatively charged particle in an atom responsible for chemical bonding.

element   A substance composed of identical atoms that cannot be broken into simpler substances by chemical means.

El Niño   A southward-flowing nutrient-poor current of warm water off the coast of western South America, caused by a breakdown of trade wind circulation.

endoderm   The innermost layer of cells in a developing embryo.

endotherm   An organism capable of generating and regulating metabolic heat to maintain a steady internal temperature. Birds and mammals are the only animals capable of true endothermy. A warm blooded organism.

energy   The capacity to do work.

ENSO   Acronym for the coupled phenomena of El Niño and the Southern Oscillation. *See also* **El Niño**; **Southern Oscillation**.

entropy   A measure of the disorder in a system.

environmental resistance   All the limiting factors that act together to regulate the maximum allowable size, or carrying capacity, of a population.

epicenter   The point on the Earth's surface directly above the focus of an earthquake.

epipelagic zone   The lighted, or photic, zone in the ocean.

equator   *See* **geographical equator**; **meteorological equator**.

equatorial upwelling   Upwelling in which water moving westward on either side of the geographical equator tends to be deflected slightly poleward and replaced by deep water often rich in nutrients. *See also* **upwelling**.

Eratosthenes of Cyrene (276–192 B.C.) Greek scholar and librarian at Alexandria who first calculated the circumference of the Earth about 230 B.C.

erosion   A process of being gradually worn away.

erosional coast   A coast in which erosive processes exceed depositional ones.

estuary   A body of water partially surrounded by land where fresh water from a river mixes with ocean water, creating an area of remarkable biological productivity.

euphotic zone   The upper layer of the photic zone in which net photosynthetic gain occurs. *Compare* **photic zone**.

euryhaline   Describing an organism able to tolerate a wide range in salinity.

eurythermal   Describing an organism able to tolerate wide variance in temperature.

eurythermal zone   The upper layer of water, where temperature changes with the seasons.

eustatic change   A worldwide change in sea level, as distinct from local changes.

eutrophication   A set of physical, chemical, and biological changes brought about when excessive nutrients are released into water.

evaporite   Deposit formed by the evaporation of ocean water.

evolution   Change. The maintenance of life under constantly changing conditions by continuous adaptation of successive generations of a species to its environment.

excess volatiles   A compound found in the ocean and atmosphere in quantities greater than can be accounted for by the weathering of surface rock. Such compounds probably entered the atmosphere and ocean from deep crustal and upper mantle sources through volcanism.

exclusive economic zone (EEZ)   The offshore zone claimed by signatories to the 1982 United Nations Draft Convention on the Law of the Sea. The EEZ extends 200 nautical miles (370 kilometers) from a contiguous shoreline. *See also* **United States exclusive economic zone**.

exoskeleton   A strong, lightweight, form-fitted external covering and support common to animals of the phylum Arthropoda. The exoskeleton is made partly of chitin and may be strengthened by calcium carbonate.

extratropical cyclone   A low-pressure midlatitude weather system characterized by converging winds and ascending air rotating counterclockwise in the Northern Hemisphere and clockwise in the Southern Hemisphere. An extratropical cyclone forms at the front between the polar and Ferrel cells.

fault   A fracture in a rock mass along which movement has occurred.

Ferrel, William (1817–1891) The American scientist who discovered the mid-latitude circulation cells of each hemisphere.

Ferrel cell The middle atmospheric circulation cell in each hemisphere. Air in these cells rises at 60° latitude and falls at 30° latitude. *See also* **westerlies**.

fetch The uninterrupted distance over which the wind blows without a significant change in direction, a factor in wind wave development.

Fissipedia The carnivoran suborder that includes sea otters.

fjord A deep, narrow estuary in a valley originally cut by a glacier.

flagellum A whiplike structure used by some small organisms and gametes to move through the environment. (Plural, *flagella*).

float method A method of current study that depends on the movement of a drift bottle or other free-floating object.

flood current Water rushing into an enclosed harbor or bay because of the rise in sea level as a tide crest approaches.

flow method A method of current study that measures the current as it flows past a fixed object.

food General term for organic molecules capable of providing energy to heterotrophs when combined with oxygen during biochemical respiration.

food web A group of organisms associated by a complex set of feeding relationships in which the flow of food energy can be followed from primary producers through consumers.

foraminiferan One of a group of planktonic amoeba-like animals with a calcareous shell, which contributes to biogenous sediments.

Forchhammer's principle *See* **principle of constant proportions**.

foreshore Sand on the seaward side of the berm, sloping toward the ocean, to the low tide mark.

fracture zone Area of irregular, seismically inactive topography marking the position of a once-active transform fault.

freezing point The temperature at which a solid can begin to form as a liquid is cooled.

fringing reef A reef attached to the shore of a continent or island.

front The boundary between two air masses of different density. The density difference can be caused by differences in temperature and/or humidity.

frontal storm Precipitation and wind caused by the meeting of two air masses, associated with an extratropical cyclone. Generally, one air mass will slide over or under the other, and the resulting expansion of air will cause cooling and, consequently, rain or snow.

frustule Siliceous external cell wall of a diatom consisting of two interlocking valves fitted together like the halves of a box.

fucoxanthin A brown or tan accessory pigment found in many species of brown algae and some species of diatoms.

fully developed sea The theoretical maximum height attainable by ocean waves given wind of a specific strength, duration, and fetch. Longer exposure to wind will not increase the size of the waves.

galaxy A large rotating aggregation of stars, dust, gas, and other debris held together by gravity. There are perhaps 50 billion galaxies in the universe and 50 billion stars in each galaxy.

gas bladder In multicellular algae, an air-filled structure that assists in flotation.

gas exchange Simultaneous passage, through a semipermeable membrane, of oxygen into an animal and carbon dioxide out of it.

Gastropoda The class of the phylum Mollusca that includes snails and sea slugs.

geographical equator 0° latitude, an imaginary line equidistant from the geographical poles.

geostrophic Describing a gyre or current in balance between the Coriolis effect and gravity; literally, "turned by the Earth."

glucose A carbohydrate ($C_6H_{12}O_6$) produced by autotrophs during photosynthesis. It can be used for energy or converted into other organic compounds.

granite The relatively light crustal rock—composed mainly of oxygen, silicon, and aluminum—that forms the continents. Its density is about 2.7 g/cm^3.

gravimeter A sensitive device that measures variations in the pull of gravity at different places on the Earth's surface.

gravity wave A wave with wavelength greater than 1.73 centimeters (0.68 inch), whose restoring forces are gravity and momentum.

greenhouse effect Trapping of heat in the atmosphere. Incoming short-wavelength solar radiation penetrates the atmosphere, but the longer-wavelength outgoing radiation is absorbed by greenhouse gases and reradiated to Earth, causing a rise in surface temperature.

greenhouse gases Gases in the Earth's atmosphere that cause the greenhouse effect; include carbon dioxide, methane, and CFCs.

groin A short, artificial projection of durable material placed at a right angle to shore in an attempt to slow longshore transport of sand from a beach. Usually deployed in repeating units.

group velocity Speed of advance of a wave train; for deep-water waves, half the speed of individual waves within the group.

Gulf Stream The strong western boundary current of the North Atlantic, off the east coast of the United States.

guyot A flat-topped, submerged inactive volcano.

gyre Circuit of mid-latitude currents around the periphery of an ocean basin. Most oceanographers recognize five gyres plus the West Wind Drift.

habitat The place where an individual or population of a given species lives. Its "mailing address."

hadal zone The deepest zone of the ocean, below a depth of 5,000 meters (16,500 feet).

Hadley, George (1685–1768) A London lawyer and philosopher who worked out the overall scheme of wind circulation in an effort to explain the trade winds.

Hadley cell The atmospheric circulation cell nearest the equator in each hemisphere. Air in these cells rises near the equator because of strong solar heating there and falls because of cooling at about 30° latitude. *See also* **trade winds**.

halocline The zone of the ocean in which salinity increases rapidly with depth. *See also* **pycnocline**.

Harrison, John (1693—1776) British clockmaker who invented the modern chronometer in 1760.

heat A form of energy produced by the random vibration of atoms or molecules.

heat budget An expression of the total solar energy received on the Earth during some period of time and the total heat lost from the Earth by reflection and radiation into space through the same period.

heat capacity The amount of heat required to raise the temperature of 1 gram of a substance by 1°C (1.8°F).

Henry the Navigator (1394–1460) Prince of Portugal who established a school for the study of geography, seamanship, shipbuilding, and navigation.

hermatypic Describing coral species possessing symbiotic zooxanthellae within their tissues and capable of secreting calcium carbonate at a rate suitable for reef production.

heterotroph An organism that derives nourishment from other organisms because it is unable to synthesize its own food molecules.

hierarchy Grouping of objects by degrees of complexity, grade, or class. A hierarchical system of nomenclature is based on distinctions within groups and between groups.

high-energy coast A coast exposed to large waves.

high seas That part of the ocean past the exclusive economic zone, which is considered common property to be shared by the citizens of the world. About 60% of the ocean area.

high tide The high-water position corresponding to a tidal crest.

holdfast A complex branching structure that anchors many kinds of multicellular algae to the substrate.

holoplankton Permanent members of the plankton community. Examples are diatoms and copepods. *Compare* **meroplankton**.

Holothuroidea The class of the phylum Echinodermata to which sea cucumbers belong.

horse latitudes Zones of erratic horizontal surface air circulation near 30°N and 30°S latitudes. Over land, dry air falling from high altitudes produces deserts at these latitudes (for example, the Sahara).

hot spot A small stationary heat source in the Earth's mantle. Hot spots are not always located at a plate boundary.

hurricane A large tropical cyclone in the North Atlantic or eastern Pacific, whose winds exceed 118 kilometers (74 miles) per hour.

hydrogen bond Relatively weak bond formed between a partially positive hydrogen atom and a partially negative oxygen, fluorine, or nitrogen atom of an adjacent molecule.

hydrogenous sediment Sediment formed directly by precipitation from seawater. Also called *authigenic sediment*.

hydrostatic pressure The constant pressure of water around a submerged organism.

hydrothermal vent Spring of hot, mineral- and gas-rich seawater found on some oceanic ridges in zones of active seafloor spreading.

hypertonic Referring to a solution having a higher concentration of dissolved substances than the solution that surrounds it.

hypothesis A speculation about the natural world that may be verified or disproved by observation and experiment.

hypotonic Referring to a solution having a lower concentration of dissolved substances than the solution that surrounds it.

hypsographic curve A graph of the area of the Earth's surface above any given elevation or depth above or below sea level.

ice age One of several periods (lasting several thousand years each) of low temperature during the last million years. Glaciers and polar ice were derived from ocean water, lowering sea level at least 100 meters (328 feet). (See Appendix 2, Geological time.)

iceberg Large mass of ice floating in the ocean that was formed on or adjacent to land. Tabular icebergs are tablelike or flat; pinnacled icebergs are castellated, or jagged. Southern icebergs are often tabular; northern icebergs are often pinnacled.

ice cap Permanent cover of ice. Formally limited to ice atop land, but informally applied also to floating ice in the Arctic Ocean.

ice floe A mass of firm sea ice floating as a unit.

ice pack A large, floating expanse of broken ice masses pressed and frozen together.

inlet A passage giving the ocean access to an enclosed lagoon, harbor, or bay.

insolation rate The amount of solar energy reaching the Earth's surface per unit time.

interference Addition or subtraction of wave energy as waves interact. Also called *resonance*. *See also* **constructive interference**; **destructive interference**.

intermediate-depth water wave A wave moving through water deeper than 1/20 but shallower than ½ its wavelength. Also called a *transitional wave*.

internal wave A progressive wave occurring at the boundary between liquids of different densities.

intertidal zone The marine zone between the highest high tide point on a shoreline and the lowest low tide point. The intertidal zone is sometimes subdivided into four separate habitats by height above tidal datum, typically numbered 1 to 4, land to sea.

intertropical convergence zone (ITCZ) The equatorial area at which the trade winds converge. The ITCZ usually lies at or near the meteorological equator. Also called the *doldrums*.

invertebrate Animal lacking a backbone.

ion An atom (or small group of atoms) that becomes electrically charged by gaining or losing one or more electrons.

ionic bond A chemical bond resulting from attraction between oppositely charged ions. These forces are said to be "electrostatic" in nature.

ionizing radiation Fast-moving particles or high-energy electromagnetic radiation emitted as unstable atomic nuclei disintegrate. The radiation has enough energy to dislodge one or more electrons from atoms it hits to form charged ions, which can react with and damage living tissue.

island arc Curving chain of volcanic islands and seamounts almost always found paralleling the concave edge of a trench.

isostatic equilibrium Balanced support of lighter material in a heavier, displaced supporting matrix. Analogous to buoyancy in a liquid.

isotonic Referring to a solution having the same concentration of dissolved substances as the solution that surrounds it.

ITCZ *See* **intertropical convergence zone**.

kelp Informal name for any species of large phaeophyte.

kingdom The largest category of biological classification. Five kingdoms are presently recognized.

knot A speed of 1 nautical mile per hour. *See* **nautical mile**.

krill *Euphausia superba*, a thumb-sized crustacean common in Antarctic waters.

lagoon A shallow body of seawater generally isolated from the ocean by a barrier island. Also the body of water enclosed within an atoll, or the water within a reverse estuary.

land breeze Movement of air offshore as marine air heats and rises.

latent heat of evaporation Heat added to a liquid during evaporation (or released from a gas during condensation) that produces a change in state but not a change in temperature. For pure water, 585 calories per gram at 20°C (68°F).

latent heat of fusion Heat removed from a liquid during freezing (or added to a solid during thawing) that produces a change in state but not a change in temperature. For pure water, 80 calories per gram at 0°C (32°F).

lateral-line system A system of sensors and nerves in the head and midbody of fishes and some amphibians that functions to detect low-frequency vibrations in water.

latitude Regularly spaced imaginary lines on the Earth's surface running parallel to the equator.

law A large construct explaining events in nature that have been observed to occur with unvarying uniformity under the same conditions.

Library of Alexandria The greatest collection of writings in the ancient world, founded in the third century B.C. by Alexander the Great. Could be considered the first university.

light Electromagnetic radiation propagated as small, nearly massless particles that behave like both a wave and a stream of particles.

limiting factor A physical or biological environmental factor whose absence or presence in an inappropriate amount limits the normal actions of an organism.

Linnaeus, Carolus Carl von Linné (1707–1778). Swedish "father" of modern taxonomy.

lithification Conversion of sediment into sedimentary rock by pressure or by the introduction of a mineral cement.

lithosphere The brittle, relatively cool outer layer of the Earth, consisting of the oceanic and continental crust and the outermost, rigid layer of mantle.

littoral zone The band of coast alternately covered and uncovered by tidal action; the intertidal zone.

longitude Regularly spaced imaginary lines on the Earth's surface running north and south and converging at the poles.

longshore bar A submerged or exposed line of sand lying parallel to shore and accumulated by wave action.

longshore current A current running parallel to shore in the surf zone, caused by the incomplete refraction of waves approaching the beach at an angle.

longshore drift Movement of sediments parallel to shore, driven by wave energy.

longshore trough Submerged excavation parallel to shore adjacent to an exposed sandy beach. Caused by the turbulence of water returning to the ocean after each wave.

low-energy coast A coast only rarely exposed to large waves.

low tide The low-water position corresponding to a tidal trough.

low tide terrace The smooth, hard-packed beach seaward of the beach scarp on which waves expend most of their energy. Site of the most vigorous onshore and offshore movement of sand.

lunar tide Tide caused by gravitational and inertial interaction of moon and Earth.

macroplankton Animal planters larger than 1 to 2 centimeters (½ to 1 inch). An example is the jellyfish.

Magellan, Ferdinand (c. 1480–1521) Portuguese navigator in the service of Spain who led the first expedition to circumnavigate the Earth, 1519–22. He was killed in the Philippines.

magma Molten rock capable of fluid flow. Called lava aboveground.

magnetometer A device that measures the amount and direction of residual magnetism in a rock sample.

Mammalia The class of mammals.

mangrove Large flowering shrub or tree that grows in dense thickets or forests along muddy or silty tropical coasts.

mantle The layer of the Earth between the crust and the core, composed of silicates of iron and magnesium. The mantle has an average density of about 4.5 g/cm^3 and accounts for about 68% of the Earth's mass.

map A representation of the Earth's surface, usually depicting mostly land areas. *See also* **chart**.

mariculture The farming of marine organisms, usually in estuaries, bays, or nearshore environments or in specially designed structures using circulating seawater. *Compare* **aquaculture**.

marine energy resource Any resource resulting from the direct extraction of energy from the heat or movement of ocean water.

marine pollution The introduction by humans of substances or energy into the ocean that change the quality of the water or affect the physical and biological environment.

marine science The process (or result) of applying the scientific method to the ocean, its surroundings, and the life forms within it. Also called oceanography or oceanology.

masking pigment *See* **accessory pigment**.

mass A measure of the quantity of matter.

Maury, Matthew (1806–1873) "Father" of physical oceanography. Probably the first person to undertake the systematic study of the ocean as a full-time occupation, and probably the first to understand the global interlocking of currents, wind flow, and weather.

maximum sustainable yield The maximum amount of fish, crustaceans, and mollusks that can be caught without impairing future populations.

mean sea level The height of the ocean surface averaged over a few years' time.

medusa Free-swimming body form of many members of the phylum Cnidaria.

membrane A complex structure of proteins and lipids that forms boundaries around and within the cell. It is usually semipermeable, allowing some kinds of molecules to pass through but not others.

meroplankton Temporary members of the plankton community. Examples are very young fishes and barnacle larvae. *Compare* **holoplankton**.

mesoderm The middle layer of cells in a developing embryo.

mesosphere The rigid inner mantle, similar in chemical composition to the asthenosphere.

metabolic rate The rate at which energy-releasing reactions proceed within an organism.

metamerism Segmentation; repeating body parts.

Meteor **Expedition** German Atlantic expedition begun in 1925; the first to use an echo sounder and other modern optical and electronic instrumentation.

meteorological equator Also called the thermal equator. The irregular imaginary line of thermal equilibrium between hemispheres. It is situated about 5° north of the geographical equator, and its position changes with the seasons, moving slightly north in northern summer.

meteorological tide A tide influenced by the weather. Arrival of a storm surge will alter the estimate of a tide's height or arrival time, as will a strong, steady onshore or offshore wind.

Milky Way The name of our galaxy. Sometimes applied to the field of stars in our home spiral arm, which is correctly called the Orion arm.

mixed layer *See* **surface zone**.

mixed tide A complex tidal cycle, usually with two high tides and two low tides of unequal height per day.

mixing time The time necessary to mix a substance through the ocean, about 1,000 years.

mixture A close intermingling of different substances that still retain separate identities. The properties of a mixture are heterogeneous; they may vary within the mixture.

molecule A group of atoms held together by chemical bonds. The smallest unit of a compound that retains the characteristics of the compound.

Mollusca The phylum of animals that includes chitons, snails, clams, and octopuses.

molt To shed an external covering.

monsoon A pattern of wind circulation that changes with the season. Also, the rainy season in areas with monsoon wind patterns.

moon tide *See* **lunar tide**.

motile Able to move about.

multicellular Consisting of more than one cell.

multicellular algae Algae with bodies consisting of more than one cell. Examples are kelp and *Ulva*.

mutation A heritable change in an organism's genes.

mutualism A symbiotic interaction between two species that is beneficial to both.

Mysticeti The suborder of baleen whales.

nanoplankton Very small members of the plankton community. Examples are coccolithophores and silicoflagellates.

Nansen bottle A water-sampling instrument perfected early in this century by the Norwegian scientist and explorer Fridtjof Nansen.

natural selection A mechanism of evolution that results in the continuation of only those forms of life best adapted to survive and reproduce in their environment.

natural system of classification A method of classifying an organism based on its ancestry or origin.

nautical chart A chart used for marine navigation.

nautical mile The length of 1 minute of latitude, 6,076 feet, 1.15 statute miles, or 1.85 kilometers. (See Appendixes 1 and 4.)

neap tide The time of smallest variation between high and low tides occurring when Earth, moon, and sun align at right angles. Neap tides alternate with spring tides, occurring at two-week intervals.

nebula Diffuse cloud of dust and gas.

Nematoda The phylum of animals to which roundworms belong.

neritic Of the shore or coast. Refers to continental margins and the water covering them, or to nearshore organisms.

neritic zone The zone of open water near shore, over the continental shelf.

niche Description of an organism's functional role in a habitat. Its "job."

node The line or point of no wave action in a standing pattern. *See also* **amphidromic point**.

nodule Solid mass of hydrogenous sediment, most commonly manganese or ferro-manganese nodules and phosphorite nodules.

nonconservative constituent An element whose proportion in seawater varies with time and place, depending on biological demand or chemical reactivity. An element with a short residence time. For example, iron, aluminum, silicon, trace nutrients, dissolved oxygen, and carbon dioxide.

nonconservative nutrient A compound or ion needed by autotrophs for primary productivity and which changes in concentration with biological activity.

nonextractive resource Any use of the ocean in place, such as transportation of people and commodities by sea, recreation, or waste disposal.

nonrenewable resource Any resource that is present on Earth in fixed amounts and cannot be replenished.

nonvascular plant Plant having no obvious vessels for the transport of fluid and lacking leaves, stems, and roots. Examples are algae.

nor'easter (northeaster) Any energetic extratropical cyclone that sweeps the eastern seaboard of North America in winter.

notochord Stiffening structure found at some time in the life cycle of all members of the phylum Chordata.

nutrient Any needed substance that an organism obtains from its environment *except* oxygen, carbon dioxide, and water.

ocean (1) The great body of saline water that covers 70.78% of the surface of the Earth. (2) One of its primary subdivisions, bounded by continents, the equator, and other imaginary lines.

ocean basin Deep-ocean floor made of basaltic crust. *Compare* **continental margin**.

oceanic crust The outermost solid surface of the Earth beneath ocean floor sediments, composed primarily of basalt.

oceanic ridge Young seabed at the active spreading center of an ocean, often unmasked by sediment, bulging above the abyssal plain. The boundary between diverging plates. Often called a mid-ocean ridge, though less than 60% of the length exists at mid-ocean.

oceanic zone The zone of open water away from shore, past the continental shelf.

oceanography The science of the ocean. *See also* **marine science**.

oceanus Latin form of *okeanos*, the Greek name for the "ocean river" past Gibraltar.

Odontoceti The suborder of toothed whales.

oolite sand Hydrogenous sediment formed when calcium carbonate precipitates from warmed seawater as pH rises, forming rounded grains around a shell fragment or other particle.

ooze Sediment of at least 30% biological origin.

Ophiuroidea The class of the phylum Echinodermata to which brittle stars belong.

orbit In ocean waves, the circular pattern of water particle movement at the air-sea interface. Orbital motion contrasts with the side-to-side or back-and-forth motion of pure transverse or longitudinal waves.

orbital inclination The 23°27' "tilt" of the Earth's rotational axis relative to the plane of its orbit around the sun.

orbital wave A progressive wave in which particles of the medium move in closed circles.

osmoregulation The ability to adjust internal salt concentration.

osmosis The diffusion of water from a region of high water concentration to a region of lower water concentration through a semipermeable membrane.

Osteichthyes The class of fishes with bony skeletons.

outgassing The volcanic venting of volatile substances.

overfishing Harvesting so many fish that there is not enough breeding stock left to replenish the species.

oxygen minimum zone A zone in which oxygen is depleted by animals and not replaced by phytoplankton.

oxygen revolution The time span, from about 2 billion to 400 million years ago, during which photosynthetic autotrophs changed the composition of the Earth's atmosphere to its current oxygen-rich mixture.

ozone O_3, the triatomic form of oxygen. Ozone in the upper atmosphere protects living things from some of the harmful effects of the sun's ultraviolet radiation.

ozone layer A diffuse layer of ozone mixed with other gases surrounding the world at a height of about 20 to 40 kilometers (12 to 25 miles).

P wave Primary wave. A compressional wave associated with an earthquake and which can move through both liquid and rock.

Pacific Ring of Fire The zone of seismic and volcanic activity that encircles the Pacific Ocean.

paleomagnetism   The "fossil," or remanent, magnetic field of a rock.

Pangaea   Name given by Alfred Wegener to the original "protocontinent." The breakup of Pangaea gave rise to the Atlantic Ocean and to the continents we see today.

Panthalassa   Name given by Alfred Wegener to the ocean surrounding Pangaea.

parasitism   A symbiotic relationship in which one species spends part or all of its life cycle on or within another, using the host species (or food within the host) as a source of nutrients. The most common form of symbiosis.

partially mixed estuary   An estuary in which an influx of seawater occurs beneath a surface layer of fresh water flowing seaward. Mixing occurs along the junction.

passive margin   Continental margin near an area of lithospheric plate divergence. Also called *Atlantic-type margin.*

passive sonar   A device that detects the intensity and direction of underwater sounds.

PCBs   *See* **polychlorinated biphenyls**.

pelagic   Of the open ocean. Refers to the water above the deep-ocean basins, sediments of oceanic origin, or organisms of the open ocean.

pelagic zone   The realm of open water. *See also* **benthic zone**.

period   *See* **wave period**.

pH   A measure of the acidity or alkalinity of a solution. Numerically, the negative logarithm of the concentration of hydrogen ions in an aqueous solution. A pH of 7 is neutral; lower numbers indicate acidity, and higher numbers indicate alkalinity.

Phaeophyta   Brown multicellular algae, including kelps.

photic zone   The thin film of lighted water at the top of the world ocean. The photic zone rarely extends deeper than 200 meters (660 feet). *Compare* **euphotic zone**.

photon   The smallest unit of light energy.

photosynthesis   The process by which autotrophs bind light energy into the chemical bonds of food with the aid of chlorophyll and other substances. The process uses carbon dioxide and water as raw materials and yields glucose and oxygen.

phycobilin   A reddish accessory pigment found in red algae.

phylum   One of the major groups of the animal kingdom whose members share a similar body plan, level of complexity, and evolution-

ary history (see Appendix 5). (Plural, *phyla*) (The major groups of the plant kingdom are called divisions.)

physical factor   An aspect of the physical environment that affects living organisms, such as light, salinity, or temperature.

physical resource   Any resource that has resulted from the deposition, precipitation, or accumulation of a useful nonliving substance in the ocean or seabed. Also called a *nonliving resource.*

phytoplankton   Plantlike, usually single-celled members of the plankton community.

Pinnipedia   The carnivoran suborder that contains the seals, sea lions, and walruses.

piston corer   A seabed-sampling device capable of punching through up to 25 meters (80 feet) of sediment and returning an intact plug of material.

planet   A smaller, usually nonluminous body orbiting a star.

plankter   Informal name for a member of the plankton community.

plankton   Drifting or weakly swimming organisms suspended in water. Their horizontal position is to a large extent dependent on the mass flow of water rather than on their own swimming efforts.

plankton bloom   A sudden increase in the number of phytoplankton cells in a volume of water.

plankton net   Conical net of fine nylon or dacron fabric used to collect plankton.

Plantae   The kingdom to which multicellular autotrophs belong.

plate   One of about a dozen rigid segments of the Earth's lithosphere that move independently. The plate consists of continental or oceanic crust and the cool, rigid upper mantle directly below the crust.

plate tectonics   The theory that the Earth's lithosphere is fractured into plates, which move relative to each other and are driven by convection currents in the mantle. Most volcanic and seismic activity occurs at plate margins.

Platyhelminthes   The phylum of animals to which flatworms belong.

plunging wave   Breaking wave in which the upper section topples forward and away from the bottom, forming an air-filled tube.

polar cell   The atmospheric circulation cell centered over each pole.

polar front   Boundary between the polar cell and the Ferrel cell in each hemisphere.

polar molecule   A molecule with unbalanced charge. One end of the molecule has a slight negative charge, and the other end has a slight positive charge.

polar ocean areas   Zones to the south of the Antarctic Convergence and to the north of the Arctic Convergence.

polar regions   Earth's cold area poleward of either the Arctic or Antarctic Circle.

pollutant   A substance that causes damage by interfering directly or indirectly with an organism's biochemical processes.

Polychaeta   The largest and most diverse class of phylum Annelida. Nearly all polychaetes are marine.

polychlorinated biphenyls (PCBs)   Chlorinated hydrocarbons once widely used to cool and insulate electrical devices and to strengthen wood or concrete. PCBs may be responsible for the changes and declining fertility of some marine mammals.

Polynesia   A large group of Pacific islands lying east of Melanesia and Micronesia and extending from the Hawaiian Islands south to New Zealand and east to Easter Island.

polynya   A gap in polar pack ice at which liquid water contacts the atmosphere.

polyp   One of two body forms of Cnidaria. Polyps are cup-shaped and possess rings of tentacles. Coral animals are polyps.

Polyplacophora   The class of the phylum Mollusca that includes the chitons.

poorly sorted sediment   A sediment in which particles of many sizes are found.

population   A group of individuals of the same species occupying the same area.

population density   The number of individuals per unit area.

Porifera   The phylum of animals to which sponges belong.

potable water   Water suitable for drinking.

precipitate   (1) A solid substance formed in an aqueous reaction. (2) The process by which a solute forms in and falls from a solution. The falling of water or ice from the atmosphere.

precipitation   Liquid or solid water that falls from the air and reaches the surface as rain, hail, or snowfall.

pressure   Force per unit area.

prey   An organism consumed by a predator.

primary coast   Coasts on which terrestrial influences dominate. *See also* **secondary coast**.

primary consumer Initial consumer of primary producers. The consumers of autotrophs; the second level in food webs.

primary forces The forces that induce and maintain water flow in ocean current systems: thermal expansion, wind friction, and density differences.

primary producer An organism capable of using energy from light or energy-rich chemicals in the environment to produce energy-rich organic compounds. An autotroph.

primary productivity The synthesis of organic materials from inorganic substances by photosynthesis or chemosynthesis. Expressed in grams of carbon bound into carbohydrate per unit area per unit time (g C/m^2/yr).

principle of constant proportions The proportions of major conservative elements in seawater remain nearly constant, though total salinity may change with location. Also called *Forchhammer's principle*.

progressive wave A wave of moving energy in which the wave form moves in one direction along the surface (or junction) of the transmission medium (or media).

Protista The kingdom of single-celled nucleated organisms to which protozoa, diatoms, and dinoflagellates belong. Also called *Protoctista*.

proton A positively charged particle at the center of an atom.

protostar Tightly condensed knot of material that has not yet attained fusion temperature.

protozoa Multiphyletic animal group within the kingdom Protista. Protozoa include amoebas, paramecia, foraminiferans, and radiolarians.

pteropod Small planktonic mollusk with a calcareous shell, which contributes to biogenous sediments.

pycnocline The middle zone of the ocean in which density increases rapidly with depth. Temperature falls and salinity rises in this zone.

radial symmetry Body structure in which the body parts radiate from a central axis like spokes from a wheel. An example is a sea star. *Compare* **bilateral symmetry**.

radioactive decay The disintegration of unstable forms of elements, which releases subatomic particles and heat.

radiolarian One of a group of usually planktonic amoeba-like animals with a siliceous shell, which contributes to biogenous sediments.

radiometric dating A technique using the constant rate at which naturally radioactive elements decay to determine the age of a material containing those elements.

random distribution Distribution of organisms within a community whereby the position of one organism is in no way influenced by the positions of other organisms or by physical variations within that community. A very rare distribution pattern.

reef A hazard to navigation. A shoal, a shallow area, or a mass of fish or other marine life.

refraction Bending of light or sound waves as they move at an angle other than 90° between media of different optical or acoustical densities. *See also* **wave refraction**.

refractive index The degree of refraction from one medium to another expressed as a ratio. The higher the ratio (refractive index), the greater the bending of waves between media.

refractometer A compact optical device that determines the salinity of a water sample by comparing the refractive index of the sample to the refractive index of water of known salinity.

renewable resource Any resource that is naturally replaced on a seasonal basis by the growth of living organisms or by other natural processes.

Reptilia The class of reptiles, including turtles, crocodiles, iguanas, and snakes.

residence time The average length of time a dissolved substance spends in the ocean.

respiration Release of stored energy from chemical bonds in food; carbon dioxide and water are formed as by-products. (Respiration is a biochemical process and is not the same as the mechanical process of breathing.)

restoring force The dominant force trying to return water to flatness after formation of a wave.

reverse estuary An estuary along an arid coast in which salinity increases from the ocean to the estuary's upper reaches because of evaporation of seawater and a lack of freshwater input.

Rhodophyta Red multicellular algae.

Richter scale A logarithmic measure of earthquake magnitude. A great earthquake measures above 8 on the Richter scale.

rip current A strong, narrow surface current that flows seaward through the surf zone and is caused by the escape of excess water that has piled up in a longshore trough.

rogue wave A single wave crest much higher than usual, caused by constructive interference.

S wave Secondary wave. A transverse wave associated with an earthquake and which cannot move through liquid.

salinity A measure of the dissolved solids in seawater, usually expressed in grams per kilogram or parts per thousand by weight. Standard seawater has a salinity of 35 ‰ at 0°C (32°F).

salinometer An electronic device that determines salinity by measuring the electrical conductivity of a seawater sample.

salt gland Specialized tissue responsible for concentration and excretion of excess salt from blood and other body fluids.

salt wedge estuary An estuary in which rapid river flow and small tidal range cause an inclined wedge of seawater to form at the mouth.

sand Sediment particle between 0.062 and 2 millimeters in diameter.

sandbar A submerged or exposed line of sand accumulated by wave action.

sand spit An accumulation of sand and gravel deposited downcurrent from a headland. Sand spits often have a curl at the tips.

saturation State of a solution in which no more of the solute will dissolve in the solvent. The rate at which molecules of the solute are being dissolved equals the rate at which they are being precipitated from the solution.

scattering The dispersion (or "bounce") of sound or light waves when they strike particles suspended in water or air. The amount of scatter depends on the number, size, and composition of the particles.

schooling Tendency of small fish of a single species, size, and age to mass in groups. The school moves as a unit, which confuses predators and reduces the effort spent searching for mates.

science A systematic way of asking questions about the natural world and testing the answers to those questions.

scientific method The orderly process by which theories explaining the operation of the natural world are verified or rejected.

scientific name The genus and species name of an organism.

sea Simultaneous wind waves of many wavelengths forming a chaotic ocean surface. Sea is common in an area of wind wave origin.

sea breeze Onshore movement of air as inland air heats and rises.

sea cave A cave near sea level in a sea cliff cut by processes of marine erosion.

sea cliff Cliff marking the landward limit of marine erosion on an erosional coast.

seafloor spreading The theory that new ocean crust forms at spreading centers, most of which are on the ocean floor, and pushes the continents aside. Power is thought to be provided by convection currents in the Earth's upper mantle.

sea grass Any of several marine angiosperms. Examples are *Zostera* (eelgrass) and *Phyllospadix* (surfgrass). Sea grasses are not seaweeds.

sea ice Ice formed by the freezing of seawater.

sea island Island whose central core was connected to the mainland when sea level was lower. Rising ocean separates these high points from land, and sedimentary processes surround them with beaches. *Compare* **barrier island**.

sea level The height of the ocean surface. *See also* **mean sea level**.

seamount Circular or elliptical projection from the seafloor, more than 1 kilometer (0.6 mile) in height, with a relatively steep slope of 20° to 25°.

Seasat First satellite dedicated to oceanic research, launched in 1978.

seaweed Informal term for large marine multicellular algae.

secondary coast Coasts dominated by marine processes. *See also* **primary coast**.

secondary consumer Consumer of primary consumers.

secondary forces The forces that influence the direction and nature of water flow in ocean current systems: the Coriolis effect, gravity, friction, and the shape of ocean basins.

second law of thermodynamics Disorder (entropy) in a closed system must increase over time. If disorder decreases, it does so at the expense of energy. Since the universe as a whole may be considered a closed system, it follows that an increase in order in one part must result in a decrease in order in another.

sediment Loose particles of inorganic or organic origin that accumulate on the seabed. Usually sand- or dust-sized, but occasionally as large as gravel, pebbles, or even boulders.

seiche Pendulum-like rocking of water in an enclosed area; a form of standing wave that can be caused by meteorological or seismic forces, or that may result from normal resonances excited by tides.

seismic Referring to earthquakes and the shock of earthquakes.

seismic sea wave Tsunami caused by displacement of the Earth along a fault. (Earthquakes and seismic sea waves are caused by the same phenomenon.)

seismic wave A low-frequency wave generated by the forces that cause earthquakes. Some kinds of seismic waves can pass through the earth. *See also* **P wave; S wave**.

seismograph An instrument that detects and records Earth movement associated with earthquakes and other disturbances.

semidiurnal tide A tidal cycle of two high tides and two low tides each lunar day, with the high tides of nearly equal height.

sensible heat Heat whose gain or loss is detectable by a thermometer or other sensor.

sessile Attached. Nonmotile. Unable to move about.

sewage sludge Semisolid mixture of organic matter, microorganisms, toxic metals, and synthetic organic chemicals removed from wastewater at a sewage treatment plant.

shadow zone (1) The wide band at the Earth's surface 105°–143° away from an earthquake in which seismic waves are nearly absent. P waves are absent because they are refracted by the Earth's liquid outer core; S waves are absent from this band and the zone immediately opposite the earthquake site because they are absorbed by the outer core. (2) In sonar, the volume of ocean from which sound waves diverge and in which a submarine may hide.

shallow-water wave A wave in water shallower than $\frac{1}{20}$ its wavelength.

shelf break The abrupt increase in slope at the junction between continental shelf and continental slope.

shore The place where ocean meets land. On nautical charts, the limit of high tides.

side-scan sonar A high-resolution sound-imaging system used for geological investigations, archeological studies, and the location of sunken ships and airplanes.

siliceous ooze Ooze composed mostly of the hard remains of silica-containing organisms.

silicoflagellate A tiny single-celled phytoplankter with a siliceous skeleton.

silt Sediment particle between 0.004 and 0.062 millimeter in diameter.

Sirenia The order of mammals that includes manatees, dugongs, and the extinct sea cows.

sofar *Sound fixing and ranging*. An experimental U.S. Navy technique for locating survivors on life rafts, based on the fact that sound from explosive charges dropped into the layer of minimum sound velocity can be heard for great distances. *See* **sofar channel**.

sofar channel Layer of minimum sound velocity in which sound transmission is unusually efficient. Sounds leaving this depth tend to be refracted back into it. The sofar layer usually occurs at mid-latitude depths around 1,200 meters (4,000 feet).

solar nebula The diffuse cloud of dust and gas from which the solar system originated.

solar system The sun together with the planets and other bodies that revolve around it.

solar tide Tide caused by the gravitational and inertial interaction of the sun and Earth.

solstice One of two times of the year when the overhead position of the sun is farthest from the equator. The time of the solstice is midway between equinoxes.

solute A substance dissolved in a solvent. *See* **solution**.

solution A homogeneous substance made of two components, the solvent and the solute.

solvent A substance able to dissolve other substances. *See* **solution**.

sonar *Sound navigation and ranging*.

sound A form of energy transmitted by rapid pressure changes in an elastic medium.

sounding Measurement of the depth of a body of water.

Southern Oscillation A reversal of airflow between normally low atmospheric pressure over the western Pacific and normally high pressure over the eastern Pacific. The cause of El Niño. *See* **El Niño**.

speciation The formation of new species. Charles Darwin suggested that this is accomplished through isolation and natural selection.

species Any group of actually or potentially interbreeding organisms reproductively isolated from all other groups and capable of producing fertile offspring. (*Note*: The word *species* is both singular and plural.)

species diversity Number of different species in a given area.

species-specific relationship An exclusive relationship between two species. Parasites are usually species-specific; that is, they can usually parasitize only one species of host.

spilling wave A breaking wave whose crest slides down the face of the wave.

spreading center The junction between diverging plates at which new ocean floor is being made. Also called *spreading zone*.

spring tide The time of greatest variation between high and low tides occurring when Earth, moon, and sun form a straight line. Spring tides alternate with neap tides throughout the year, occurring at two-week intervals.

standing wave A wave in which water oscillates without causing progressive wave forward movement. There is no net transmission of energy in a standing wave.

star A massive sphere of incandescent gases powered by the conversion of hydrogen to helium and other heavier elements.

state An expression of the internal form of matter. Water exists in three states: solid, liquid, and gas. A solid has a fixed volume and fixed shape; a liquid has a fixed volume but no fixed shape; and a gas has neither fixed volume nor fixed shape.

stenohaline Describing an organism unable to tolerate a wide range in salinity.

stenothermal Describing an organism unable to tolerate wide variance in temperature.

stipe Multicellular algal equivalent of a vascular plant's stem.

storm Local or regional atmospheric disturbance characterized by strong winds often accompanied by precipitation.

storm surge An unusual rise in sea level as a result of the low atmospheric pressure and strong winds associated with a tropical cyclone. Onrushing seawater precedes landfall of the tropical cyclone and causes most of the damage to life and property.

stratigraphy The branch of geology that deals with the definition and description of natural divisions of rocks. Specifically, the analysis of relationships of rock strata.

subduction The downward movement into the asthenosphere of a lithospheric plate.

subduction zone An area at which a lithospheric plate is descending into the asthenosphere. The zone is characterized by linear folds (trenches) in the ocean floor and strong deep-focus earthquakes. Also called a *Wadati-Benioff zone*.

sublittoral zone The ocean floor near shore. The inner sublittoral extends from the littoral (intertidal) zone to the depth at which wind waves have no influence; the outer sublittoral extends to the edge of the continental shelf.

submarine canyon A deep, V-shaped valley running roughly perpendicular to the shoreline and cutting across the edge of the continental shelf and slope.

subsidence Sinking, often of tectonic origin.

Subtropical Convergence Convergence zone marking the boundary between Central Water and either Subarctic or Subantarctic Surface Water. The northern Subtropical Convergence lies at about 45°N in the Pacific and 60°N in the Atlantic; the southern Subtropical Convergence lies at 40°–50°S.

succession The changes in species composition that lead to a climax community.

sun tide *See* **solar tide**.

supernova The explosive collapse of a massive star.

supralittoral zone The splash zone above the highest high tide; not technically part of the ocean bottom.

surf The confused mass of agitated water rushing shoreward during and after a wind wave breaks.

surface current Horizontal flow of water at the ocean's surface.

surface zone The upper layer of ocean in which temperature and salinity are relatively constant with depth. Depending on local conditions, the surface zone may reach to 1,000 meters (3,300 feet) or be absent entirely. Also called the *mixed layer*.

surface-to-volume ratio A physical constraint on the size of cells. As a cell's linear dimensions grow, its surface area does not increase at the same rate as its volume. As surface-to-volume ratio falls, each square unit of outer membrane must serve an increasing interior volume.

surf beat The pattern of constructive and destructive interference that causes successive breaking waves to grow, shrink, and grow again over a few minutes' time.

surf zone The region between the breaking waves and the shore.

surging wave A wave that surges ashore without breaking.

suspension feeder An animal that feeds by straining or otherwise collecting plankton and tiny food particles from the surrounding water.

sverdrup (sv) A unit of volume transport named in honor of oceanographer Harald U. Sverdrup: 1 million cubic meters of water flowing past a fixed point each second.

swash Water from waves washing onto a beach.

swell Mature wind waves of one wavelength that form orderly undulations of the ocean surface.

swim bladder A gas-filled organ that assists in maintaining neutral buoyancy in some bony fishes.

symbiosis The co-occurrence of two species in which the life of one is closely interwoven with the life of the other; mutualism, commensalism, or parasitism.

synoptic sampling Simultaneous sampling at many locations.

taxonomy In biology, the laws and principles covering the classification of organisms.

tektite A small, rounded glassy component of cosmogenous sediments, usually less than 1.5 millimeters (1/16 inch) in length. Thought to have formed from the impact of an asteroid or meteor on the crust of the Earth or moon.

Teleostei The osteichthyan order that contains the cod, tuna, halibut, perch, and other species of bony fishes.

temperate zone The mid-latitude area between the Tropic of Cancer and the Arctic Circle and between the Tropic of Capricorn and the Antarctic Circle.

temperature The response of a solid, liquid, or gas to the input or removal of heat energy. A measure of the atomic and molecular vibration in a substance, indicated in degrees.

temperature-salinity (T-S) diagram A graph showing the relationship of temperature and salinity with depth.

terrane An isolated segment of seafloor, island arc, plateau, continental crust, or sediment transported by seafloor spreading to a position adjacent to a larger continental mass. Usually different in composition from the larger mass.

terrigenous sediment Sediment derived from the land and transported to the ocean by wind and flowing water.

territorial waters Waters extending 12 miles from shore and in which a nation has the right to jurisdiction.

thallus The body of an alga or other simple plant.

theory A general explanation of a characteristic of nature consistently supported by observation or experiment.

thermal equator *See* **meteorological equator**.

thermal equilibrium The condition in which the total heat coming into a system (such as a planet) is balanced by the total heat leaving the system.

thermal inertia Tendency of a substance to resist change in temperature with the gain or loss of heat energy.

thermocline The zone of the ocean in which temperature decreases rapidly with depth. *See also* **pycnocline**.

thermohaline circulation Water circulation produced by differences in temperature and/or salinity (and therefore density).

thermostatic property A property of water that acts to moderate changes in temperature.

tidal bore A high, often breaking wave generated by a tide crest that advances rapidly up an estuary or river.

tidal current Mass flow of water induced by the raising or lowering of sea level owing to passage of tidal crests or troughs. *See also* **ebb current**; **flood current**.

tidal datum The reference level (0.0) from which tidal height is measured.

tidal range The difference in height between consecutive high and low tides.

tidal wave The crest of the wave causing tides. Another name for a tidal bore. *Not* a tsunami or seismic sea wave.

tide Periodic short-term change in the height of the ocean surface at a particular place, generated by long-wavelength progressive waves which are caused by the interaction of gravitational force and inertia. Movement of the Earth beneath tide crests results in the rhythmic rising and falling of sea level.

tombolo Above-water bridge of sand connecting an offshore feature to the mainland.

top consumer An organism at the apex of a trophic pyramid, usually a carnivore.

tornado Localized, narrow, violent funnel of fast-spinning wind, usually generated when two air masses collide. Not to be confused with a cyclone. (The tornado's oceanic equivalent is a waterspout.)

trace element A minor constituent of seawater present in amounts less than 1 part per million.

trade winds Surface winds within the Hadley cells, centered at about 15° latitude, which approach from the northeast in the Northern Hemisphere and from the southeast in the Southern Hemisphere.

transform fault A plane along which rock masses slide horizontally past one another.

transform plate boundary Places where crustal plates shear laterally past one another. Crust is neither produced nor destroyed at this type of junction.

transitional wave *See* **intermediate-depth water wave**.

transverse current East-to-west or west-to-east current linking the eastern and western boundary currents. An example is the North Equatorial Current.

trench An arc-shaped depression in the deep-ocean floor with very steep sides and a flat sediment-filled bottom coinciding with a subduction zone. Most trenches occur in the Pacific.

trophic level A feeding step within a trophic pyramid.

trophic pyramid A model of feeding relationships between organisms. Primary producers form the base of the pyramid; consumers eating one another form the higher levels, with the top consumer at the apex.

tropical cyclone A weather system of low atmospheric pressure around which winds blow counterclockwise in the Northern Hemisphere and clockwise in the Southern Hemisphere. Originates in the tropics within a single air mass, but may move into temperate waters if water temperature is high enough to sustain it. Small tropical cyclones are called tropical depressions, larger ones tropical storms, and great ones hurricanes, typhoons, or willy-willies, depending on location.

tropical ocean area Warm central zone of the ocean equatorward of the subtropical convergences where surface temperature is nearly always above 20°C (69°F).

Tropic of Cancer The imaginary line around the Earth, parallel to the equator at 23°27'N, marking the point where the sun shines directly overhead at the June solstice.

Tropic of Capricorn The imaginary line around the Earth, parallel to the equator at 23°27'S, marking the point where the sun shines directly overhead at the December solstice.

tropics The area between the Tropic of Cancer and the Tropic of Capricorn.

trough *See* **wave trough**.

tsunami Long-wavelength shallow-water wave caused by rapid displacement of water. *See also* **seismic sea wave**.

tunicate A type of suspension-feeding invertebrate chordate.

turbidite A terrigenous sediment deposited by a turbidity current; typically, coarse-grained layers of nearshore origin interleaved with finer sediments.

turbidity current An underwater "avalanche" of abrasive sediments thought responsible for the deep sculpturing of submarine canyons and a means of transport for sediments accumulating on abyssal plains.

turbulence Chaotic fluid flow.

ultraplankton Extremely small plankton, smaller than nanoplankton.

undercurrent A current flowing beneath a surface current, usually in the opposite direction.

unicellular Consisting of a single cell.

unicellular algae Algae with bodies consisting of a single cell. Examples are diatoms and dinoflagellates.

uniform distribution Distribution of organisms within a community characterized by equal space between individuals (the arrangement of trees in an orchard). The rarest natural distribution pattern.

uniformitarianism The theory that all of Earth's geological features and history can be explained by processes occurring today and that these processes must have been at work for a very long time.

United States exclusive economic zone The region extending seaward from the coast of the United States for 200 nautical miles, within which the United States claims sovereign rights and jurisdiction over all marine resources.

United States Exploring Expedition The first U.S. oceanographic research voyage, launched in 1838.

upwelling Circulation pattern in which deep, cold, usually nutrient-laden water moves toward the surface. Upwelling can be caused by winds blowing parallel to shore or offshore.

$V = \sqrt{gd}$ Relation between velocity (V), the acceleration due to gravity (g), and water depth (d) for shallow-water waves.

$V = L/T$ Relation between velocity (V), wavelength (L), and period (T) for deep-water waves; velocity increases as wavelength increases. Typically measured in meters per second.

valve In diatoms, each half of the protective silica-rich outer portion of the cell. The complete outer covering is called the frustule.

vascular plant Plant having vessels for transport of fluid through leaves, stems, and roots. Examples are sea grasses, mangroves, and maple trees.

velocity Speed in a specified direction.

Vertebrata The subphylum of the phylum Chordata that includes animals with segmented backbones.

vertebrate A chordate with a segmented backbone.

Vikings Seafaring Scandinavian raiders who ravaged the coasts of Europe around A.D. 780–1070.

viscosity Resistance to fluid flow. A measure of the internal friction in fluids.

voyaging Traveling (usually by sea) with a specific purpose.

Wadati-Benioff zone *See* **subduction zone**.

water mass A body of water identifiable by its salinity and temperature (and therefore its density) by its gas content or another indicator.

water vapor The gaseous, invisible form of water.

water-vascular system System of water-filled tubes and canals found in some representatives of the phylum Echinodermata and used for movement, defense, and feeding.

wave Disturbance caused by the movement of energy through a medium.

wave crest Highest part of a progressive wave above average water level.

wave-cut platform The smooth, level terrace sometimes found on erosional coasts that marks the submerged limit of rapid marine erosion.

wave diffraction Bending of waves around obstacles.

wave frequency The number of waves passing a fixed point per second.

wave height Vertical distance between a wave crest and the adjacent wave troughs.

wavelength The horizontal distance between two successive wave crests (or troughs) in a progressive wave.

wave period Time for successive wave crests to pass a fixed point.

wave reflection The reflection of progressive waves by a vertical barrier. Reflection occurs with little loss of energy.

wave refraction Slowing and bending of progressive waves in shallow water.

wave shock Physical movement, often sudden, violent, and of great force, caused by the crash of a wave against an organism.

wave steepness Height-to-wavelength ratio of a wave. The theoretical maximum steepness of deep-water waves is 1:7.

wave train A group of waves of similar wavelength and period moving in the same direction across the ocean surface. The group velocity of a wave train is half the velocity of the individual waves.

wave trough The valley between wave crests below the average water level in a progressive wave.

weather The state of the atmosphere at a specific place and time.

Wegener, Alfred (1880–1930) German scientist who proposed the theory of continental drift in 1912.

well-mixed estuary An estuary in which slow river flow and tidal turbulence mix fresh and salt water in a regular pattern through most of its length.

well-sorted sediment A sediment in which particles are of uniform size.

westerlies Surface winds within the Ferrel cells, centered around 45° latitude, which approach from the southwest in the Northern Hemisphere and from the northwest in the Southern Hemisphere.

western boundary current Strong, warm, concentrated, fast-moving current at the western boundary of an ocean (off the east coast of a continent). Examples include the Gulf Stream and the Japan (Kuroshio) Current.

westward intensification The increase in speed of geostrophic currents as they pass along the western boundary of an ocean basin.

West Wind Drift The Antarctic Circumpolar Current, driven by powerful westerly winds north of Antarctica. The largest of all ocean currents, it continues permanently eastward without changing direction.

Wilson, John Tuzo (b. 1908) Canadian geophysicist who proposed the theory of plate tectonics in 1965.

wind The mass movement of air.

wind duration The length of time the wind blows over the ocean surface, a factor in wind wave development.

wind-induced vertical circulation Vertical movement in surface water (upwelling or downwelling) caused by wind.

wind strength Average speed of the wind, a factor in wind wave development.

wind wave Gravity wave formed by transfer of wind energy into water. Wavelengths from 60 to 150 meters (200 to 500 feet) are most common in the open ocean.

world ocean The great body of saline water that covers 70.78% of the Earth's surface.

xanthophyll A yellow or brown accessory pigment that gives some marine autotrophs a yellow or tan appearance.

zone Division or province of the ocean with homogeneous characteristics.

zooplankton Animal members of the plankton community.

zooxanthellae Unicellular dinoflagellates that are symbiotic with coral and that produce the relatively high pH and some of the enzymes essential for rapid calcium carbonate deposition in coral reefs.

INDEX

Galaxy, **26**, *27*
Galveston (Texas), 167, 200
Garrison, Jeanne Kristin, *217*
Gas bladders, ***254***
Gas exchange, **239**–240
Gases
  atmospheric, 99–100
  oceanic dissolved, *99*–100, 211
  solubility of, *211*
Geneva, Lake, *168*, 181
Geological time, 314–315
Geostrophic (adj.), **136**
Gill membranes, **239**–240
Glaciers, 200
Global Positioning System (GPS), 319
Global thermostatic effects, 105–106
Global warming, 187, 301–303
Globigerina, 82
Glomar Challenger, R/V, 20, *22*, 43, 76, 85, 89
Glossopteris, 38, 39, 52
Golden Gate (California), 182
GPS (Global Positioning System), 319
Granite, **33**, 59, 78
Grassle, J. F., 70
Gravel (as marine resource), 283
Gravity waves, *154*
Great Barrier Reef (Australia), 268
Great Salt Lake (Utah), 96
Greek ships for exploration, *5*
Greenhouse effect, 105, 113, **301**–303
Greenhouse gases, **301**
Greenland Current, *136*
Greenwich (England), 318
Groin, ***201***
Grotius, Hugo, 298
Groundwater, 3, 96
Group velocity, **158**
Gulf Stream, 65, 132, *134*, *136*, ***137***–138, *139*, 141, 144
Gulf War (Persian), 281, 282
Gulper eel, *241*
Guyots, *51*, 73
Gypsum, 84
Gyres, **133**–140

H

Habitat, **218**
Hadal zone, ***213***
Hadley, George, 121
Hadley cell, *121*
Haldane, J. B. S., 29
Halocline, **108**
Hardin, Garrett, 303–304
Hawaii
  European discovery of, *13*
  Hawaiians' discovery of, *9–10*
  Wilkes' exploration of, 14
Hawaiian Islands
  extent of, *50*
  formation of, *48*–51
Heard Island, 113
Heat, **101**, 209
  flow from within Earth, 34–35
  internal, 34, 52
  living things and, 209–210
  water and, 101–108
Heat budget, **105**–*106*
Heat capacity, **102**
Heating, solar, 117–119
Heavy metal pollutants, 292–295

Heezen, Bruce, 67
Heirarchy, **216**
Henry, Prince of Portugal, "the Navigator," ***10***–11
Hermatypic coral, **263**–265
Hermit crab, *252*
Herodotus, 190
Hess, Harry, 40–41, 51
Heterotroph, **206**
High seas, **298**
High tide, **176**–*177*
High-energy coasts, **188**
Hilo (Hawaii), 171–*172*
Himalayas, formation of, *47*, 48
Hipparchus, 6, 318
History of marine science, 4–23
HIV, 291
Holdfast, ***254***
Holoplankton, **231**
Homer, 308
Homestead, Florida, *115*
Homo sapiens, 314–*315*
Horse latitudes, *121*–**122**
"Horse mackerel," *222*
Hot spots, 32, **48**–51
Hudson Canyon, *67*
Hugo (hurricane), 130
Humboldt Current, 244
Humpback whales, *248*
Hurricanes, **124**, *127*–130, 188, 197, 200
Hydrogen bonds, **94**–95, *103*
Hydrogenous sediments, *78*–79, 83–*84*
Hydrostatic pressure, 212, 270
Hydrothermal vents, **70**–*71*, 272–*273*, 328–329
Hypatia, 6
Hypothesis, **26**
Hypsographic curve, *73*

I

Ice, color of, 94
Ice age, **65**, 67
Ice crystal, *103*
Iceberg, *93*
Iguanas, marine, 241
Indonesia, 7
Induction salinometer, ***98***
Industrial Revolution, 277, 301
Inertia, 176–178
  thermal, **105**
Inlet, *195*–*196*
Inman, Douglas, 202
Inner core, *33*–34, *37*
Interference (in waves), **161**–*162*, 165
Intermediate depth waves, 154–**155**, 156
Intermediate water, 145, 146–148
International Geophysical Year (1957), 40
Intertidal zone, 213, 253, **257**–260
Intertropical Convergence Zone (ITCZ), *121*–**122**
Ionizing radiation, **303**
Ions, 95–96, 98
Iron, descent of, 34
Island arcs, **73**
  distribution of, *47*
Isostatic equilibrium, ***35***–*36*, 73
ITCZ (*see* Intertropical Convergence Zone)

J

James Caird (boat), 151
Japan (tsunami in), 173
Jason, Jr., R/V, 61
Java Trench, *72*
Jeffers, Robinson, 308
Jobs, marine related, 326–329
JOIDES Resolution, R/V, 20, *85*, 88–89
Jurassic Period, 314

K

Kamchatka Current, *136*
Kelly, Dennis, *224*
Kelp, *204*, **255**–256
Kent, Rockwell, 24, 308
Kermadec-Tonga Trench, *72*, 73, 272
Kingdom, **216**
Krakatoa (volcano), 173
Krill, *224*, 234, 244, 248, 286
Kurashio Current, *136*, 137
Kuril Trench, *72*

L

La Pérouse, Jean-François, 170
Labrador Current, *136*
Lagoon, *196*
Lake Geneva, 168, 181
Lamont-Doherty Earth Observatory, 20, 86
Larus sp., *217*
Lasiognathus, *242*
Latent heat of evaporation, **104**
Latent heat of fusion, *104*
Latent heat of vaporization, **104**
Lateral-line system, **240**
Latitude, 6–*7*, 117, **317**–*319*
Lava, 44
Law of the Sea, the, 298–299
Laws, **26**
Lead (as pollutant), 292–295
Learning, 308
Lehmann, Inge, 38
Liberia, 310
Library of Alexandria, **6**
Life, origin of, *29*–30, 314
Light in the ocean, 108–111, *204*, 205–206, 207, 209
Limiting factor, **212**
Linnaeus, Carolus (Carl von Linné), *216*
Lisbon (Portugal), 173
Lithification, **81**
Lithophyllum, 256, *257*
Lithosphere, *33*–34, *35*, 39, 43–48, 51, 70
Lithothamnium, 256, *257*
Littoral zone, *212*–*213*
Lituya Bay (Alaska), *170*–171
Lofoten Islands (Norway), *175*
Loihi (new Hawaiian island), 48, *50*–51
London, Jack, 308
Long waves, *166*–173
Longitude, 6–*7*, **317**–*319*
Longshore bar, ***192***, 195
Longshore current, **189**–*190*
Longshore drift, **189**–*190*
Longshore transport, *189*–*190*
Longshore trough, *192*

Lopez, Barry, 93
Loran-C, 319
Lovell, James, 2
Low tide, **176**–*177*
Low tide terrace, *192*
Low-energy coasts, **188**
Lunar tides, *177*

M

M'Quahe, Peter, 214–215
MacDonald seamounts, 32
Macrocystis, 255
Macroplankton, **231**
Maelstrom, *175*
Magellan, Ferdinand, ***12***
Magma, **44**
Magnesium chloride, 95–96
Magnetometer, **52**
Maili Point (Oahu), *165*
Makarov, S. O., 18
Mammalia, **244**–250
Mammals, marine, 244–250
Manganese nodules, *83*–*84*
Mangroves, 256, *262*–263
Mantle, *33*–34, *37*
Maps, 5, 320–323
Mare Liberum, 298
Marggraff, Lt. Frederick, USN, 160
Margin
  active, **63**–65
  continental, *63*–68
  passive, **63**–65
Mariana Trench, *72*, 322
Marine Biological Laboratory (Woods Hole, MA.), 21
Marine biologists, activities of, *25*
Marine engineers, activities of, 25
Marine geologists, activities of, 25
Marine Mammals Protection Act (1972), 290
Marine pollution, *276*, **277**, 281–*283*, 292–297
Marine resources, 277–291
Marine science, **25**
  history of, 4–23
  origin of, 5, 26
Marine Technology Society, 329
Marquesas Islands, 7, *8*, *9*
Mars (drilling platform), *280*
Marshall Islands, 268
Masses, water, **145**–149
Matagorda Peninsula (Texas), *197*
Maury, Matthew, *14*–15
Maximum sustainable yield, **287**–289
Mead, Margaret, 306
Mean sea level, *179*, 186
Mediterranean Sea, 3, 5, 76, 107, 144, 146–*147*, 179, 227
Melanesia, *8*
Melanocoetus, *242*
Melanoma, malignant, 291, *300*
Melville, Herman, 213, 308
Melville, R/V, 32
Mercator, Gerhardus, 320–321
Mercator projection, *320*–323
Mercury (as pollutant), 293
Meridians, *317*–*319*
Meroplankton, **231**–232
Mesopelagic zone, *212*–*213*
Mesosaurus, 52
Mesosphere, *33*–34, *35*

CREDITS AND ACKNOWLEDGMENTS

Chapter 1

Chapter opening courtesy of NASA. **Fig. 1.1** is reprinted with the permission of Macmillan Publishing Company from *Exploring the Planets* by Hamblin & Christiansen. Copyright © 1990 by Macmillan Publishing Company. **Fig. 1.3** is reprinted with permission from Doubleday, a division of Bantam, Doubleday, Dell Publishing Group, Inc. from *Sailing Ships* by Bjorn Landström, 1978. **Fig. 1.8** painting © Herb Kawainui Kane, Collection of the Hawaii State Foundation on Culture and the Arts. Used with permission. **Fig. 1.10** is reprinted by permission of the James Ford Bell Library, University of Minnesota. **Fig. 1.13** painting © Herb Kawainui Kane, Collection of the Kauai Westin Resort. Used with permission. **Fig. 1.14** is reprinted by permission of The United States Naval Academy Museum. **Figs. 1.15, 1.16** courtesy of The Library of Congress. **Fig. 1.17** is reprinted by permission of The Royal Geographical Society, London, Archives from the journal of Lt. Pelham Aldrich kept aboard *H.M.S. Challenger*, 1872–1875. **Fig. 1.20a** is reprinted by permission of Hulton Deutsch Collection Limited. **Fig. 1.20b** from *The Amundsen Photographs* by Roland Huntford, Hodder and Stoughton Ltd., 1987. Copyright © 1987 by Anne-Christine Jacobsen. **Fig. 1.23** is reprinted by permission of The Scripps Institution of Oceanography. **Box 1.1** photo courtesy of the Ocean Drilling Program.

Chapter 2

Fig. 2.1a photo courtesy of Dan Dion. **Fig. 2.1b** photo courtesy of Dennis Kelly. **Fig. 2.2** is reprinted with permission of the Lick Observatory. **Fig. 2.3** is reprinted with permission of Mike Seeds. **Figs. 2.4, 2.6** are reprinted with permission of William K. Hartmann.

Chapter 3

Chapter opening illustration by W.K. Hartmann and R. Miller from *History of the Earth* by W.K. Hartmann and R. Miller, Workman Publishing Company, 1992. Reprinted by permission of the authors. **Fig. 3.6** is reprinted by permission of the USGS. **Fig. 3.7** is reprinted with permission of Merrill, an imprint of Macmillan Publishing Company, from *The Earth: An Introduction to Physical Geology*, Fourth Edition, by Edward J. Tarbuck and Frederick K. Lutgens. © 1993 by Merrill Publishing Company. **Fig. 3.17** is reprinted by permission of The Woods Hole Oceanographic Institution. **Fig. 3.21** from *Ocean Science* by Keith Stowe. © 1983 John Wiley & Sons, Inc. **Fig. 3.25** is reprinted by permission of the NOAA/National Geophysical Data Center. **Fig. 3.26** courtesy of Johnson Space Center. **Fig. 3.30** is reprinted by permission of W.C. Pittman III.

Chapter 4

Chapter opening illustration courtesy of R. Detrick/*Oceanus*. **Fig. 4.1** from *Oceanography* by Steve Neshyba. Reprinted by permission of John Wiley & Sons, Inc. **Box 4.1b** photo is used by permission of Dr. Andreas Rechnitzer. **Box 4.1c** photo courtesy of The Woods Hole Oceanographic Institution. **Box 4.1d** photo courtesy of JAMSTEC. **Box 4.1e** photo courtesy of Ted Delaca, University of Alaska. **Fig. 4.7** is reprinted by permission of Taylor & Francis International Scientific and Educational Publishers. **Fig. 4.8** is reprinted by permission of the Society for Sedimentary Geology. **Fig. 4.9** is reprinted by permission of the U.S. Department of the Navy. **Fig. 4.10** is from Alyn C. Duxbury and Alison B. Duxbury, *An Introduction to the World's Oceans*, Third Edition, © 1991 Wm. C. Brown Communications, Inc., Dubuque, IA. All rights reserved. Reprinted by permission. **Fig. 4.11** courtesy of Aluminum Company of America. **Fig. 4.13** is reprinted by permission of The Woods Hole Oceanographic Institution. **Fig. 4.15** is reprinted by permission of Chet Raymo.

Chapter 5

Chapter opening illustration by R. Miller from *History of the Earth* by W. K. Hartmann and R. Miller, Workman Publishing Company, 1992. Reprinted by permission of the authors. **Figs. 5.1, 5.2** courtesy of Charles D. Hollister. **Fig. 5.3a** courtesy of the USGS. **Fig. 5.3b** reprinted by permission of The Bettmann Archive. **Fig. 5.4** courtesy of the Department of Geology, University of Delaware. **Fig. 5.5a** courtesy of Dr. Howard Spero. **Fig. 5.5b** courtesy of Jay Yett. **Figs. 5.5c** and **5.6a & b** courtesy of Roger Witmer. **Figs. 5.7a & b** courtesy of Unocal Research. **Figs. 5.9b–d** are reprinted with permission of Macmillan Publishing Company from *Explorations of the Oceans* by John G. Weihaupt. © 1979 by John G. Weihaupt. **Figs. 5.10b–e** courtesy of Charles D. Hollister. **Figs. 5.11, 5.12 & Box 5.1a & b** reprinted by permission of the Deep Sea Drilling Project, Texas A & M University. **Fig. 5.15** from A.G. Fischer, *Geologic History of the Western North Pacific*, published June 5, 1970, Vol. 168, p. 1210. © 1970 AAAS. Reprinted by permission.

Chapter 6

Chapter opening photo courtesy of the U.S. Coast Guard. **Fig. 6.13** from G.P. Kuiper, ed., *The Earth as a Planet*, © 1954 The University of Chicago Press. Reprinted by permission of the publisher. **Fig. 6.19** reprinted by permission of John Wiley & Sons, Inc. from *Oceanography: Perspectives on a Fluid Earth* by Steve Neshyba, © 1987.

Chapter 7

Chapter opening photos a) courtesy of Hank Brandli, b) courtesy of U.S. Army Corps of Engineers. **Fig. 7.10** from Frederick K. Lutgens and Edward J. Tarbuck, *The Atmosphere*, Fifth Edition, © 1992, p. 192. Reprinted by permission of Prentice-Hall, Englewood Cliffs, New Jersey. **Fig. 7.12** courtesy of Tom Loebl, Imaging Publications. **Box 7.1** photo courtesy of Alastair Black. **Fig. 7.14** photo courtesy of NASA.

Chapter 8

Chapter opening photo is reprinted by permission of The Bettmann Archive. **Fig. 8.5a** is from *Laboratory Exercises in Oceanography*, Second Edition, by Pipkin, Gorsline, Casey and Hammond. © 1987 by W. H. Freeman and Company. Reprinted by permission. **Fig. 8.11** is reprinted by permission of O. Brown, R. Evans, and M. Carle, University of Miami Rosenstiel School of Marine and Atmospheric Science. **Box 8.1c** photo courtesy of Dr. Philip Richardson, The Woods Hole Oceanographic Institution. **Fig. 8.19** courtesy of Prentice-Hall.

Chapter 9

Fig. 9.9 courtesy of NOAA. **Fig. 9.12** courtesy of the United States Navy. **Figs. 9.15a–b** courtesy of Ivan Narragon. **Box 9.1** courtesy of Ron Romanosky. **Figs. 9.16b, 9.23a** from G.A. MacDonald, et al, *Volcanoes in the Sea*, University of Hawaii Press. Reprinted by permission. **Figs. 9.19a–b** courtesy of Thames Barrier Visitor's Centre. **Box 9.2** courtesy of the USGS. **Fig. 9.22a** courtesy of the United States Coast Guard. **Fig. 9.22b** is reprinted by permission of The Bettmann Archive. **Fig. 9.23b** courtesy of Bishop Museum.

Chapter 10

Chapter opening photo (b) reprinted by permission of BODØ Arrangement. **Fig. 10.7** is reprinted by permission of F. W. Rowbotham. **Fig. 10.11** courtesy of Photothèque EDF.

Chapter 11

Chapter opening photo © Steve Rose/Rainbow. **Fig. 11.2** is used by permission of W. Woodworth/Superstock, Inc. **Fig. 11.8** courtesy of Jim Reese. **Figs. 11.9, 11.22a** from *Waves and Beaches* by Willard Bascom. © 1980 by Willard Bascom. Used by permission of Doubleday, a division of Bantam, Doubleday, Dell Publishing Group, Inc. **Figs. 11.10, 11.12, 11.16a & b, 11.22a** reprinted by permission of Gerald G. Kuhn. **Fig. 11.11a–c** courtesy of Dr. William Wallace. **Fig. 11.14** courtesy of Jack Mertz. **Fig. 11.18** from *Oceanography* by Steve Neshyba. © 1987. Reprinted by permission of John Wiley & Sons, Inc. and The American Fisheries Society. **Fig. 11.19** is reprinted by permission of John S. Shelton. **Fig. 11.20** courtesy of USGS.

Chapter 12

Chapter opening photo is used by permission of Bruce Hall. **Box 12.1b** painting by Charles R. Knight, Field Museum of Natural History, Neg. No. CK34.1T, Chicago. **Fig. 12.7** art by Lloyd K. Townsend. **Fig. 12.11** is from Sylvia S. Maden, *Biology: Evolution, Diversity, and the Environment*, Second Edition © 1987 Wm. C. Brown Communications, Inc., Dubuque, Iowa. All rights reserved. Reprinted by permission.

Chapter 13

Chapter opening photo from James Joseph, et al., *Tuna and Billfish —Fish Without A Country*, Fourth

Edition, painting by George Mattson, 1988. Used by permission of Inter-American Tropical Tuna Commission. **Fig. 13.3a** courtesy of Greta Fryxell. **Fig. 13.3b** is from C. Shih and R.G. Kessel, *Living Images*, © 1982 Jones and Bartlett Publishers, Inc. Used by permission. **Fig. 13.4** courtesy of The Scripps Institution of Oceanography. **Fig. 13.5** courtesy of the Marine Life Research Group. **Box 13.1** photos courtesy of Dr. William P. Cochlan, Hancock Institute for Marine Sciences, University of Southern California. **Fig. 13.7** from R.J. Livingston (Ed.) in *Ecological Processes in Coastal and Marine Systems*, 1979. Reprinted by permission of Plenum Publishing Corporation. **Fig. 13.11** Manfred Kage/Bruce Coleman Limited. **Fig. 13.12a** John Clegg/Ardea London. **Fig. 13.12b** courtesy of Dr. Howard Spero. **Figs. 13.13, 13.28** © 1992 Sea World of California, Inc. All rights reserved. **Fig. 13.14** © Bob Cranston. Used by permission. **Fig. 13.16a** courtesy of Carl Roessler. **Fig. 13.16b** courtesy of Al Giddings/Images Unlimited, Inc. **Fig. 13.17** courtesy of James D. Watt/Planet Earth Pictures. **Fig. 13.19** is reprinted with permission of Macmillan Publishing Company from *Vertebrate Life* by F.H. Pough, et al., © 1989 by Macmillan Publishing Company. **Figs. 13.23 a–b** are used by permission of Bruce Hall. **Figs. 13.24, 13.25a** are from J.C. Briggs, *Marine Zoogeography*, © 1974 by McGraw-Hill, Inc. Reprinted by permission of the publisher. **Figs. 13.26, 13.33c** courtesy of David B. Fleetham/Visuals Unlimited. **Fig. 13.27** courtesy of Don Stanford. **Fig. 13.31b** courtesy of Norman Cole. **Fig. 13.32** courtesy of Scott Mercer, New England Whale Watch. **Fig. 13.33a** courtesy of Friends of the Sea Lion. **Fig. 13.33b** is used by permission of Bruce Hall. **Fig. 13.34** courtesy of Friends of the Sea Otter, Laguna, California/Richard Bucich. **Fig. 13.35** courtesy of the Department of Natural Resources, Florida.

Chapter 14

Chapter opening photo is used by permission of Linda E. Tway, Ph.D. **Fig. 14.3b** is used by permission of Bruce Hall. **Fig. 14.7b** is reprinted from *Between Pacific Tides*, Fifth Edition, by Edward F. Ricketts, et al.; Revised by David W. Phillips, with permission of the publishers, Stanford University Press, © 1985 by the Board of Trustees of the Leland Stanford Junior University. **Fig. 14.9** courtesy of the National Park Service. **Fig. 14.10** courtesy of Pat Mason. **Fig. 14.13c** Jacana, Sophia de Wilde/Photo Researchers, Inc. **Fig. 14.14a** courtesy of C.B. & D.W. Frith/Bruce Coleman, Ltd. **Fig. 14.14b** © Douglas Faulkner/The NAS Collection/Photo Researchers, Inc. **Fig. 14.14c** courtesy of Douglas Faulkner/Sally Faulkner Collection. **Fig. 14.15** courtesy of Johnson Space Center. **Fig. 14.16** courtesy of Charles D. Hollister, The Woods Hole Oceanographic Institution. **Fig. 14.17** is reprinted by permission of The Woods Hole Oceanographic Institution. **Figs. 14.18 a–b** are from J.C. Briggs, *Marine Zoogeography*, © 1974 by McGraw-Hill, Inc. Reprinted by permission of the publisher. **Fig. 14.19** from *Face of the Deep* by Bruce C. Heezen and Charles D. Hollister, 1971, Oxford University Press. Reprinted by permission of The Woods Hole Oceanographic Institution. **Figs. 14.20, 14.21** courtesy of Dudley Foster, The Woods Hole Oceanographic Institution.

Chapter 15

Chapter opening photo © 1989 Tony Dawson. Reprinted by permission. **Figs. 15.3 a–b** courtesy of Shell Oil Company. **Figs. 15.5** courtesy of Matthew Perry, U.S. Fish and Wildlife Service. **Fig. 15.6** is from David A. Ross, *Opportunities and the Uses of the Ocean*, Springer-Verlag, 1980. Copyright David A. Ross. Used by permission. **Figs. 15.14, 15.22** courtesy of Tom Campbell's Photographic. **Fig. 15.15** courtesy of Don Stanford. **Box 15.1** photo courtesy of W. Eugene Smith/Black Star. **Fig. 15.21** courtesy of Ken Sakamoto/Black Star. **Fig. 15.24** courtesy of Dr. Ronald Barr, UC-Irvine Medical Center. **Fig. 15.25** cartoon by Bill Schorr reprinted by permission of UFS, Inc.

Appendixes

Appendix III, Figs. 1, 2, 3, 4 and **Appendix IV, Figs. 1, 3** from David Greenhood, *Down to Earth: Mapping for Everybody*. Reprinted by permission of The University of Chicago Press.